New Industry 4.0 Advances in Industrial IoT and Visual Computing for Manufacturing Processes

New Industry 4.0 Advances in Industrial IoT and Visual Computing for Manufacturing Processes

Special Issue Editors

Luis Norberto López de Lacalle
Jorge Posada

MDPI • Basel • Beijing • Wuhan • Barcelona • Belgrade • Manchester • Tokyo • Cluj • Tianjin

Special Issue Editors
Luis Norberto López de Lacalle
University of the Basque Country
Spain

Jorge Posada
Vicomtech Foundation, Basque Research and
Technology Alliance (BRTA)
Spain

Editorial Office
MDPI
St. Alban-Anlage 66
4052 Basel, Switzerland

This is a reprint of articles from the Special Issue published online in the open access journal *Applied Sciences* (ISSN 2076-3417) (available at: https://www.mdpi.com/journal/applsci/special_issues/IoT_Manufacturing_Processes).

For citation purposes, cite each article independently as indicated on the article page online and as indicated below:

LastName, A.A.; LastName, B.B.; LastName, C.C. Article Title. *Journal Name* **Year**, *Article Number*, Page Range.

ISBN 978-3-03928-290-6 (Pbk)
ISBN 978-3-03928-291-3 (PDF)

Contents

About the Special Issue Editors

Luis Norberto López de Lacalle is a full professor at the University of Basque Country, teaching Machine Dynamics, Manufacturing Systems, and Machine-Tools. He is the founder and manager of the CFAA (Advanced Manufacturing Research Centre) and responsible for leading international and national projects. His topics of interest include the following, among others: traditional and non-traditional machining processes, process reliability, fault diagnosis, sustainability and efficiency in manufacturing, vibrations, etc. He is also the leader of the High-Performance Manufacturing Group at the University of the Basque country and works for different national agencies as a project manager and reviewer. He is the author of a number of publications in high impact journals (more than 160) and international conferences.

Jorge Posada is the Scientific and Associate Director of Vicomtech Foundation since 2001, an internationally recognized applied research center in visual computing and artificial intelligence for industry and society (+170 people, +60 Ph.D.), and a member of the BRTA (Basque Research and Technology Alliance). He holds a Ph.D. in Engineering and Computer Science from the Technische Universität Darmstadt (Germany), and an Executive MBA from IE Business School. He is also president of the Board of GraphicsVision.ai, a network of applied research centers with similar scientific interests. His research lines include visual computing, digital manufacturing, industry 4.0, knowledge engineering, and other related fields. He is the author of over 90 scientific publications, and is a scientific advisor and keynote speaker of industry 4.0 subjects for several industrial and government organizations. Jorge Posada is a member of IEEE, ACM (Recognition of Service Award), and Eurographics. He has been the Chair of prestigious conferences such as the ACM Web3D and Knowledge Engineering Society KES. He is a member of the Editorial Board of the *International Journal on Interactive Design and Manufacturing* (Springer), *Multimedia Tool and Applications* (Springer), *Sensors* (MDPI), where he also serves as a topic editor, and the *Applied Sciences Journal* (MDPI) in the Applied Industrial Technologies Section.

Editorial

Special Issue on New Industry 4.0 Advances in Industrial IoT and Visual Computing for Manufacturing Processes

Luis Norberto López de Lacalle [1,]* and Jorge Posada [2]

[1] Department of Mechanical Engineering (High-Performance Manufacturing Group), University of the Basque Country (UPV/EHU), Parque Tecnológico de Zamudio 202, 48170 Bilbao, Spain
[2] Vicomtech Technological Center, Paseo Mikeletegi 57, E-20009 Donostia/San Sebastián, Spain; jposada@vicomtech.org
* Correspondence: norberto.lzlacalle@ehu.eus

Received: 8 October 2019; Accepted: 8 October 2019; Published: 14 October 2019

The new advances of IIOT (Industrial Internet of Things), together with the progress in visual computing technologies, are being addressed by the research community with interesting approaches and results in the Industry 4.0 domain.

IIoT, industry 4.0, smart factories, and many other related concepts are nowadays a hot topic in industry, far beyond the initial demonstrations and initiatives that started years ago in policy-making, exhibition fairs, and journals. The applied science community is now very active in the context of helping companies and industries, which realize that the connectivity, transmission, curation, storage, analysis and use of data, together with an advanced visual computing technologies, such as visual analytics, intelligent computer vision, and graphics, can empower day-to-day production, processes, final product quality, and post-sale services. The discoveries of new possibilities in the horizontal value chain between different actors factors, the vertical dimension of improving efficiency and productivity in the smart factory, and the end-to-end dimension of considering the full lifecycle (including service) in the re-design of products, are the most relevant Industry 4.0 aspects addressed.

The present special issue involves research groups with interesting contributions in fields such as artificial vision, data analytics, smart factories and case studies, technology surveillance, and other topics closely related to the new industrial revolution. Some authors such as Švarcová et al. [1], focus on macroeconomic indicators. The role of public-private collaboration is also tackled in [2], because new research and development approaches can be applied in a regional agenda, like in the case of the German Industrie 4.0 program, the Industria 4.0 Italian program, the French Alliance Industrie du Futur, the Basque Industry 4.0 strategy, and other regional and international initiatives. All levels of current factories from layout, production scheduling, and even marketing can be affected [3]. Lim et al. [4] analyses the South Korea scenario.

One enabling technology in Industry 4.0 is cyber-physical systems (CPS) and cyber-physical production systems (CPPS). In [5] an interesting approach is presented on low-cost solutions that may cover several needs in machine monitoring without complex hardware. More complex and complete hardware and software solutions are studied in [6,7]. The criteria for selection of maintenance operators are presented in [8]. The capacity adjustment of job shop manufacturing systems is addressed using the advanced control strategy of Operator Theory in [9]. Predictive analytic models are addressed by [10] with a good survey on feature set reduction. In [11], an optimization strategy is presented for a cutting insert using ANNs and a Genetic Algorithm‖

It is interesting to note that several contributions are related to the emergence of new types of *services* directly related to Industry 4.0 concepts. In [7], authors propose a PLC as a smart *service* in Industry 4.0 for non-critical processes. Roesch et al. [12] proposes an end-to-end connection between industrial machines and their actual market demand using IT platforms. Schimanski [13] proposes a bridge between the BIM (Bulding Information Modeling) specifications in the construction industry to the related services in the design of configure-to-order services for construction equipment. A marketing perspective is also given by [3] to address the impact on current enterprises of the new Industry 4.0 technologies.

Regarding visual computing solutions, it is interesting to note that eight papers addressed how computer vision techniques, with the support of new artificial intelligence algorithms, can have direct and straightforward benefits in specific industrial application scenarios. Indeed, Industry 4.0 solutions also focus on bringing a higher degree of intelligence for production problems.

In this sense, there are papers about defect detection in fabrics using L0 gradient minimization and fuzzy C-Means [14]. Surface defect detection in generic cases using bilinear models [15] is introduced, with good classification and localization results: Fibre contour detection for food industry cases (pickles) using dilated convolution [16] is a concrete application case with interesting algorithm improvements. Detection of defects in micro-armatures for mobiles using deep convolution neural networks (CNNs) [17], blister defect detection using CNNs for lithium-ion batteries [18], and object detection using neural networks for identification of cracks [19], are also very good examples of practical problems tackled by the new generation computer vision and machine learning (incl. deep learning) techniques

A special mention should be given to the algorithmic contributions on inline inspection of warm-die forged revolution workpieces using 3D reconstruction (car component case), since it approaches some novel concepts with industrial impact in computational geometry [20], and to the new self-calibration approach of elliptic paraboloid arrays frequently used in precision measurement [21]. A contribution on how to build knowledge graphs for industrial terminology in the automotive sector is presented in [22].

The success of this special issue has motivated us to propose a new edition—New Industry 4.0 Advances in Industrial IoT and Visual Computing for Manufacturing Processes: Volume II.

We invite the research community to submit novel contributions covering both IIOT and/or visual computing aspects in Industry 4.0, with clear preference to articles that address both aspects. Examples of expected papers that extend the current areas covered in this first volume include the semantic-based, digital media oriented Visual Analytics Solutions on IIoT data [23,24] and especially the participation of the Operator as a key area in Industry 4.0 implementations (*Operator 4.0*) as described in [25,26]. The impact of these applied research lines is more and more relevant in the industrial production of today and tomorrow.

Funding: This research received no external funding.

Conflicts of Interest: The authors declare no conflict of interest.

References

1. Švarcová, J.; Urbánek, T.; Povolná, L.; Sobotková, E. Implementation of R&D Results and Industry 4.0 Influenced by Selected Macroeconomic Indicators. *Appl. Sci.* **2019**, *9*, 1846. [CrossRef]

2. Gerrikagoitia, J.; Unamuno, G.; Urkia, E.; Serna, A. Digital Manufacturing Platforms in the Industry 4.0 from Private and Public Perspectives. *Appl. Sci.* **2019**, *9*, 2934. [CrossRef]

3. Ungerman, O.; Dědková, J. Marketing Innovations in Industry 4.0 and Their Impacts on Current Enterprises. *Appl. Sci.* **2019**, *9*, 3685. [CrossRef]

4. Lim, S.; Kim, J. Technology Portfolio and Role of Public Research Institutions in Industry 4.0: A Case of South Korea. *Appl. Sci.* **2019**, *9*, 2632. [CrossRef]

5. Chen, Y.; Ting, K.; Chen, Y.; Yang, D.; Chen, H.; Ying, J. A Low-Cost Add-On Sensor and Algorithm to Help Small- and Medium-Sized Enterprises Monitor Machinery and Schedule Processes. *Appl. Sci.* **2019**, *9*, 1549. [CrossRef]

6. Iglesias, A.; Sagardui, G.; Arellano, C. Industrial Cyber-Physical System Evolution Detection and Alert Generation. *Appl. Sci.* **2019**, *9*, 1586. [CrossRef]

7. Langmann, R.; Stiller, M. The PLC as a Smart Service in Industry 4.0 Production Systems. *Appl. Sci.* **2019**, *9*, 3815. [CrossRef]

8. Patalas-Maliszewska, J.; Kłos, S. An Approach to Supporting the Selection of Maintenance Experts in the Context of Industry 4.0. *Appl. Sci.* **2019**, *9*, 1848. [CrossRef]

9. Liu, P.; Zhang, Q.; Pannek, J. Development of Operator Theory in the Capacity Adjustment of Job Shop Manufacturing Systems. *Appl. Sci.* **2019**, *9*, 2249. [CrossRef]

10. LaCasse, P.; Otieno, W.; Maturana, F. A Survey of Feature Set Reduction Approaches for Predictive Analytics Models in the Connected Manufacturing Enterprise. *Appl. Sci.* **2019**, *9*, 843. [CrossRef]

11. Solarte-Pardo, B.; Hidalgo, D.; Yeh, S. Cutting Insert and Parameter Optimization for Turning Based on Artificial Neural Networks and a Genetic Algorithm. *Appl. Sci.* **2019**, *9*, 479. [CrossRef]

12. Roesch, M.; Bauer, D.; Haupt, L.; Keller, R.; Bauernhansl, T.; Fridgen, G.; Reinhart, G.; Sauer, A. Harnessing the Full Potential of Industrial Demand-Side Flexibility: An End-to-End Approach Connecting Machines with Markets through Service-Oriented IT Platforms. *Appl. Sci.* **2019**, *9*, 3796. [CrossRef]

13. Schimanski, C.; Pasetti Monizza, G.; Marcher, C.; Matt, D. Pushing Digital Automation of Configure-to-Order Services in Small and Medium Enterprises of the Construction Equipment Industry: A Design Science Research Approach. *Appl. Sci.* **2019**, *9*, 3780. [CrossRef]

14. Zhang, H.; Ma, J.; Jing, J.; Li, P. Fabric Defect Detection Using L0 Gradient Minimization and Fuzzy C-Means. *Appl. Sci.* **2019**, *9*, 3506. [CrossRef]

15. Zhou, F.; Liu, G.; Xu, F.; Deng, H. A Generic Automated Surface Defect Detection Based on a Bilinear Model. *Appl. Sci.* **2019**, *9*, 3159. [CrossRef]

16. Li, H.; Liu, L.; Han, Z.; Zhao, D. Contour Detection for Fibre of Preserved Szechuan Pickle Based on Dilated Convolution. *Appl. Sci.* **2019**, *9*, 2684. [CrossRef]

17. Liu, J.; Feng, T.; Fang, X.; Huang, S.; Wang, J. An Intelligent Vision System for Detecting Defects in Micro-Armatures for Smartphones. *Appl. Sci.* **2019**, *9*, 2185. [CrossRef]

18. Ma, L.; Xie, W.; Zhang, Y. Blister Defect Detection Based on Convolutional Neural Network for Polymer Lithium-Ion Battery. *Appl. Sci.* **2019**, *9*, 1085. [CrossRef]

19. Li, Y.; Han, Z.; Xu, H.; Liu, L.; Li, X.; Zhang, K. YOLOv3-Lite: A Lightweight Crack Detection Network for Aircraft Structure Based on Depthwise Separable Convolutions. *Appl. Sci.* **2019**, *9*, 3781. [CrossRef]

20. Mejia-Parra, D.; Sánchez, J.; Ruiz-Salguero, O.; Alonso, M.; Izaguirre, A.; Gil, E.; Palomar, J.; Posada, J. In-Line Dimensional Inspection of Warm-Die Forged Revolution Workpieces Using 3D Mesh Reconstruction. *Appl. Sci.* **2019**, *9*, 1069. [CrossRef]

21. Lv, Z.; Su, Z.; Zhang, D.; Gao, L.; Yang, Z.; Fang, F.; Zhang, H.; Li, X. The Self-Calibration Method for the Vertex Distance of the Elliptical Paraboloid Array. *Appl. Sci.* **2019**, *9*, 3485. [CrossRef]

22. Zhao, M.; Wang, H.; Guo, J.; Liu, D.; Xie, C.; Liu, Q.; Cheng, Z. Construction of an Industrial Knowledge Graph for Unstructured Chinese Text Learning. *Appl. Sci.* **2019**, *9*, 2720. [CrossRef]

23. Smithers, T.; Posada, J.; Stork, A.; Pianciamore, M.; Ferreira, N.; Grimm, S.; Jimenez, I.; Di Marca, S.; Marcos, G.; Mauri, M.; et al. Information Management and Knowledge Sharing in WIDE. In Proceedings of the European Workshop for the Integration of Knowledge, Semantics and Digital Media Technology, London, UK, 25–26 November 2004; pp. 351–358, ISBN 0-902-23810-8.

24. Graña, M.; Toro, C.; Posada, J.; Howlett, R.; Jain, L.C. (Eds.) *Advances in Knowledge-Based and Intelligent Information and Engineering Systems*; Frontiers in Artificial Intelligence and Applications; IOS Press: Amsterdam, The Netherlands, 2012; Volume 243.

25. Posada, J.; Zorrilla, M.; Dominguez, A.; Simoes, B.; Eisert, P.; Stricker, D.; Rambach, J.; Döllner, J.; Guevara, M. Graphics and Media Technologies for Operators in Industry 4.0. *IEEE Comput. Graph. Appl.* **2018**, *38*, 119–132. [CrossRef] [PubMed]

26. Segura, A.; Diez, H.; Barandiaran, I.; Arbelaiz, A.; Alvarez, H.; Simoes, B.; Posada, J.; García-Alonso, A.; Ugarte, R. Visual Computing Technologies to support the Operator 4.0. *Comput. Ind. Eng.* **2018**. [CrossRef]

Review

A Survey of Feature Set Reduction Approaches for Predictive Analytics Models in the Connected Manufacturing Enterprise

Phillip M. LaCasse [1,*], Wilkistar Otieno [1] and Francisco P. Maturana [2]

[1] Department of Industrial and Manufacturing Engineering, University of Wisconsin–Milwaukee, Milwaukee, WI 53211, USA; otieno@uwm.edu
[2] Rockwell Automation, Inc. Milwaukee, WI 53204, USA; fpmaturana@ra.rockwell.com
* Correspondence: placasse@uwm.edu

Received: 9 January 2019; Accepted: 22 February 2019; Published: 27 February 2019

Abstract: The broad context of this literature review is the connected manufacturing enterprise, characterized by a data environment such that the size, structure and variety of information strain the capability of traditional software and database tools to effectively capture, store, manage and analyze it. This paper surveys and discusses representative examples of existing research into approaches for feature set reduction in the big data environment, focusing on three contexts: general industrial applications; specific industrial applications such as fault detection or fault prediction; and data reduction. The conclusion from this review is that there is room for research into frameworks or approaches to feature filtration and prioritization, specifically with respect to providing quantitative or qualitative information about the individual features in the dataset that can be used to rank features against each other. A byproduct of this gap is a tendency for analysts not to holistically generalize results beyond the specific problem of interest, and, related, for manufacturers to possess only limited knowledge of the relative value of smart manufacturing data collected.

Keywords: connected enterprise; smart manufacturing; big data; machine learning; data reduction; predictive analytics

---

1. Introduction

In exploring recent advancements in manufacturing and industry, interested practitioners and researchers might find themselves deciphering a series of seemingly related, sometimes interchangeable, but actually distinct terms that mean different things to different parties in different contexts. In some cases, specific terminology might be used in one part of the world whereas another term is employed elsewhere. Consider the following examples, representative but not exhaustive, that are commonly seen today.

One familiar idiom, "smart manufacturing", is a general term for the use of sensors and wireless technologies to capture data in all stages of production or product lifecycle. Examples include vehicle engines collecting and transmitting diagnostic information or optical scanners detecting defects in printed circuits [1,2].

Another common term, "Industrial Internet of Things (IIOT)", initially coined by General Electric (GE) in 2012, refers to a network of industry devices connected by communications technologies for the purposes of monitoring, collection, exchange, analysis and delivery of insights to drive smarter, faster business decisions [3,4]. Examples of consortia targeting the IIOT include the Industrial Internet Consortium (IIC) [5] and the OpenFog Consortium [6].

Industry 4.0 [7,8], China Manufacturing 2025 [9], and Connected Industries [10,11] are specific paradigms of the IIOT applied in the manufacturing context with common fundamental concepts such as smart manufacturing, cyber-physical systems, self-organization, adaptability and corporate social responsibility.

Finally, the Connected Enterprise (CE), a Rockwell Automation term [12] to describe its vision for the future of industrial automation, is an enterprise-wide extension of Industry 4.0 and China Manufacturing 2025. The integration of information technology (IT) and operational technology (OT) enables the interaction between manufacturing process data, people, and the business enterprise to optimize key performance indicators (KPI) at all levels of the organization: factory level, enterprise level and global supply chain level. Figure 1 is a pictorial representation of the CE strategy that enables the seamless convergence of the information and operation functions of an enterprise.

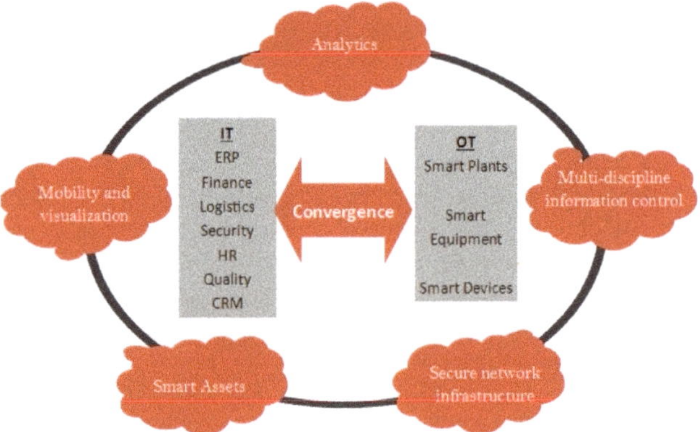

Figure 1. The Connected Enterprise strategy—enabling information technology (IT)—operational technology (OT) convergence.

The Connected Enterprise is one of: [13]:

- enterprise-wide visibility and collaboration;
- interconnected people, equipment, and processes;
- real-time learning of enterprise status;
- organizational agility by means of increased information to make informed, adaptive, proactive decisions.

The purpose of this literature review is to survey and discuss representative examples along the spectrum of existing research into a specific key enabler to the Connected Enterprise in manufacturing: big data. A good working definition of a "big data" environment is one such that the size, structure, or variety of information strains the capability of traditional software or database tools to capture, store, manage, and analyze it [14,15]. Big data is clearly both an enabler to the CE and an obstacle. It is an obstacle in that, by definition, it requires innovation to truly harness; it is an enabler in that it is precisely the availability of vast untapped data that undergirds the enormous potential of the CE. Bollier (2010) explores this duality in big data for The Aspen Institute in [16].

Three related factors provide the motivation for this research. First, the rapid advent of technology to capture and store manufacturing data without the parallel development of corresponding analytical capabilities has resulted in the circumstance by which vast quantities of data are collected but not effectively analyzed or interpreted [1]. Second, the dual nature of big data as both an enabler and an obstacle to the CE perpetuates the state of affairs by which manufacturers do not have a clear picture as to what manufacturing data, of the vast volumes collected, is truly valuable versus what can be discarded. This lack of clarity is due to disjoint or "siloed" data analytics capabilities within the organization [17] and by "where-to-start" paralysis brought on by the sheer volume of data and underdeveloped capability to visualize it [18]. Finally, it has been well established that limitations exist in how and to what extent the human analyst can process complex information [19–21]. The astounding development of the capabilities of automated analytical and artificial intelligence tools prompts interest in steps that can be taken to make them more palatable and digestible to human analysts and decision makers.

The objective of this literature survey is to determine whether, among the extensive body of knowledge on big data in the manufacturing environment, there is room for research into mechanisms or frameworks for feature filtration and prioritization when building applied machine learning models for predictive analytics. The interest is not so much in technology or architecture, which has been explored elsewhere [22–25], but rather in the human factor and how enablers to individual competencies can address the enterprise-level motivations for this research. It is true that certain machine learning algorithms can accommodate high dimensionality in input data, at the risk of potential issues such as overfitting. However, simply incorporating every potential feature into the model because it can algorithmically handle the calculations can shortchange the organization out of potentially useful information about the data at its disposal. The optimal subset problem is NP-Hard, which makes it impractical to iterate through all possible subsets of features to find the best subset for model training. For this reason, there is practical benefit in identifying how analysts and data scientists in manufacturing organizations decide how to select features for model inclusion and if that process is algorithmic and generalizable or if it is ad hoc, tailored to the specific problem of interest.

The remainder of the paper breaks down as follows. Section 2 describes the methodology employed in identifying which articles to review and how to group them. Section 3 contains the literature survey divided into three subsections. Each article receives short commentary in isolation regarding its applicability to the research motivations or objective. At the end of each subsection, a short discussion provides consolidated observations. Section 4 contains discussion with observations spanning the three subsections, and Section 5 provides brief concluding remarks.

2. Methodology

2.1. Data Collection

Articles in this review can be broadly categorized into two groups. The first group consists of featured articles that receive analysis and discussion as pertaining to the motivations and/or objective of this research. The second group consists of background or supporting work that provides context to the introduction, justification to the motivations, or theoretical foundations to techniques or algorithms referenced in the first group.

The process of identifying articles for the first group began with broad queries into databases of scholarly literature using a series of topically relevant keywords. The following keywords were used, typically in pairs but sometimes in groups of three or more: ["big data"], ["smart manufacturing"], [manufacturing], ["machine learning"], [deep learning], ["deep learning"], ["fault detection"], ["fault prediction"], ["fault diagnosis"], ["data reduction"], ["feature selection"], ["feature reduction"], ["instance selection"], and ["instance reduction"]. Quotation marks indicate that the phrase was searched in its entirety. Thus, the keyword ["deep learning"] would not return the phrase "deep neural network learning" but the keyword [deep learning] would.

Academic or scholarly databases searched include ScienceDirect, Institute of Electrical and Electronics Engineers (IEEE), Taylor & Francis, SpringerLink, Google Scholar, and the University of Wisconsin—Milwaukee library system. Time parameters were set for 2008 through 2018.

The keywords employed in database searches were selected to initially catch a wide scope of articles and then converge towards articles focusing more directly on the motivations and objective of the review.

A second means of identifying articles was to survey citations in articles identified in the database searches. For example, if a database search identified a survey paper on the use of machine learning for smart manufacturing, it would be possible that the articles cited therein might pose some relevance.

The intent is not to duplicate work but rather to complement it. Returning to the previous example, a list of articles analyzed from an algorithmic perspective on which machine learning technique was employed could be relevant to this review by seeing how those same papers approach the human dimension of the project.

To identify citations in the second category of articles, the process was ad hoc and tailored to the specific algorithm or technique that warranted additional background. This category made no restrictions to time window because many techniques employed today have their theoretical foundations in decades past. For example, much of the initial, exploratory research into human limitations in processing information took place decades ago.

2.2. Data Analysis

The first layer of analysis consisted of broadly categorizing or organizing the reviewed articles. In keeping with the general search methodology, this resulted in three general groups, selected for their intuitive sense in logically flowing from a broad, high-level search and then converging on the motivations and objective for the review.

- The first category explores big-data models for general industrial applications, specifically those featuring machine learning or deep learning.
- The second category focuses specifically on big data analyses and frameworks as applied to scenarios specific to smart manufacturing. Two subtopics emerged in the search results: fault detection and fault prediction.
- The third category addresses data reduction tools and techniques.

The three categories listed above came about partly by design and partly post hoc. From the beginning, the question of interest was data reduction, specifically feature filtration and prioritization. Upon conducting a high-level analysis of articles captured by queries described in Section 2.1, it became clear that it would be appropriate to organize by papers explicitly focused on data reduction and those not. Clearly, a paper that is explicitly on the topic of data reduction will cover the subject. However, this review is also interested in how articles approach the topic of data reduction as a step contained within some problem of interest, when the paper is not explicitly about data reduction. This would have resulted in two categories. It subsequently became clear upon examination that, of the papers not explicitly focused on data reduction, they could be subdivided into those focused on a specific manufacturing application and those focused on general applications independent of a specific problem type. This yielded the three categories that ultimately form the organization of Section 3.

The second layer of analysis consisted of identifying which articles merit discussion and how to organize that discussion.

The predominant theme for analyzing articles in the first category, general industrial applications, was the degree to which the article focused on enterprise capabilities that enable organizational competencies versus approaches or methodologies that relate to human competencies. The first two motivations for this research are predominantly organizational competencies that are developed by a combination of high-level, enterprise capabilities and low-level, individual competencies. Of interest to this review was whether the reviewed articles gave treatment to the research motivations and, if so, whether that treatment focused on the organizational or the individual competencies.

The focus for analyzing articles in the second category, specific manufacturing applications, was the extent to which data reduction was explicitly performed and, if so, the extent to which that reduction step received treatment in terms of analysis or generalizability. The working hypothesis was that most research would be focused on a specific application or problem of interest, with the input data treated in secondary fashion, being a means to some end and not as potentially an end unto itself. The reasoning behind the working hypothesis is that practitioners and researchers alike have priorities of work; solving the problem of interest is typically Priority #1. Time-constrained efforts to complete the task at hand can sometimes cause both researchers and practitioners alike to miss valuable nuggets of insight that could provide useful in subsequent future work.

The focus for the third category was the context for the data reduction and the type of data reduction performed. If the context was outside of the manufacturing realm, the question was if it would be possible to extend the technique to manufacturing contexts. If already contextualized within manufacturing, the question was how generalizable it might be to other contexts or if the technique was unique to the specific scenario or case study.

Within each section, individual articles receive commentary in isolation. Each section concludes with observations and discussion on themes contained in more than one paper therein. Finally, Section 4 provides consolidated observations and discussion for the entire set of reviewed literature.

3. Literature Survey

3.1. Big Data Approaches for General Industrial Applications

Existing research into big data utilization for general industrial applications may be broadly generalized to contain valuable work and insight into the state of technology, current challenges, and methodologies or high-level frameworks for big-data analytical projects. The following section contains examples that, while not intended to be exhaustive, are representative of the body of literature on the subject. These examples are reviewed with specific interest in how they treat the research motivations from the human versus the architectural or technological dimension.

Wuest et al. (2016) present an overview of machine learning in manufacturing, focusing specifically on advantages, challenges, and applications [26]. Of particular interest is a summary of several recent studies ([27–30]) on the key challenges currently faced by the larger global manufacturing industry, with agreement on the following key challenges:

- Adoption of advanced manufacturing technologies
- Growing importance of manufacturing of high value-added products
- Utilizing advanced knowledge, information management, and AI systems
- Sustainable manufacturing (processes) and products
- Agile and flexible enterprise capabilities and supply chains
- Innovation in products, services, and processes
- Close collaboration between industry and research to adopt new technologies
- New manufacturing paradigms.

It is interesting to observe, in addition to what is listed, what is not listed. Specifically, these recent studies did not identify data dimensionality as a key challenge. In other words, while there is recognition that voluminous manufacturing data is collected, there is not universal agreement that this is a problem that needs to be addressed on the front end [26]; rather, employment of various machine learning techniques is proposed as a means to deal with it [31], with methods towards this end dating as far back as the 1970s [32].

However, employment of machine learning algorithms to deal with the problem of high dimensionality can lead the analyst directly into one of the main challenges associated with machine learning that the paper identifies, which is that interpretation of results can be difficult. Especially when the model is intended to support real-time monitoring of parameters with respect to proximity to some threshold, the practical usefulness of the model is diminished when large numbers of irrelevant or redundant features are input into the model simply because the machine learning algorithm can accommodate them.

Alpaydin (2014) provides a comprehensive overview of machine learning, with specific techniques that apply to each of the needs described above [33]. It is pointed out, however, that existing applications of machine learning tend to narrowly focus on the problem at hand or on a specific process [34] and not holistically on the manufacturing enterprise or on generalizing the results to other processes. This observation is noteworthy, as it relates tangentially to the motivation for this literature review. One reason for the willingness to select machine learning algorithms that can handle high dimensionality may be a 'prisoner-of-the-moment' mentality. Analysts and data scientists perform real-world analyses to solve real-world problems, usually on a deadline imposed beyond their control. That deadline may be imposed by supervisors or it may be a function of outside constraints. Circumstances may not afford the luxury to step back, after completing the initial project, and thoroughly comb through the data to draw secondary conclusions about the nature of the input data. Rather, it is on to the next problem.

Wang et al. (2018) unpack the benefits and applications of deep learning for smart manufacturing, identifying benefits that include new visibility into operations for decision-makers and the availability of real-time performance measures and costs [35]. The authors provide, in addition to this practical information, a useful discussion on deep learning as a big-data analytical tool. In particular, they compare deep learning with traditional machine learning and offer three key distinctions between the two. Those distinctions are summarized in Table 1 [35].

Table 1. Distinction between traditional machine learning and deep learning.

Technique	Feature Learning	Model Construction	Model Training
Traditional Machine Learning	Features are identified, engineered and extracted manually through domain expert knowledge.	Models typically have shallow structures (few hidden layers) and are data-driven using selected features.	Modules are trained step by step.
Deep Learning	Features are learned by transforming the data into abstract representations.	Models are end-to-end, high hierarchies with nonlinear combinations of numerous hidden layers.	Model parameters are trained jointly.

Note the distinction in feature learning. Deep learning models do not explicitly engineer and extract features. Rather, they are learned abstractly. This is both an advantage and a tradeoff. The blessing is that model performance is typically superior. The tradeoff is in the transparency, traceability, and front-end verifiability of results.

The authors make an interesting observation, in that deep learning has shown itself to be most effective when it is applied to limited types of data and well-defined tasks [35]. This is notable in that conventional wisdom sometimes holds more data is better. Reducing the large data set to the most relevant subset of predictors may actually improve performance. This speaks directly to the motivation for this review and demonstrates the importance of the question. Not only does the capability to reduce a feature set to only the most relevant features enable an organization to build and increase institutional knowledge about the data at its disposal, but it also may lead to superior model performance.

Closely related, Tao et al. (2018) provide a comprehensive look at data-driven smart manufacturing, providing a historical perspective on the evolution of manufacturing data, a development perspective on the lifecycle of big manufacturing data, and a framework envisioning the future of data in manufacturing [2].

An observation is that Tao et al. also identify a gap and promising future research direction that aligns indirectly with the focus of this literature review: edge computing. Edge computing is, architecturally, an option for whittling down the volumes of production data into the core pieces that are truly meaningful and align with the key performance indicators (KPIs) of interest. Edge computing allows data to be analyzed at the "edge" of a network before being sent to a data center or cloud [36]. A related term, fog computing, was introduced by Cisco systems in 2014 and extends the cloud to be closer to devices that produce and act on IIOT data [37]. The distinction between the two concepts, as well as other emerging paradigms such as mobile edge computing (MEC) and mobile cloud computing (MCC) are not fully mature and are subject to overlap [38]. The commonality is that they represent means for an organization to operationalize the individual competencies that are the focus of this review.

A final framework for general industrial application of big data is presented by Flath and Stein (2017), specifically in the form of a data science "toolbox" for manufacturing prediction tasks. The objective is to bridge the gap between machine learning research and practical needs [39]. Feature engineering is identified as an important step that must take place prior to deriving useful patterns from the input data, and a case study employs Kullback–Leibler divergence to reduce 968 numeric features to 150 and 2140 categorical features to 27.

The preceding literature, summarized in Table 2 below, shows high-level analysis of trends and challenges. It also provides examples of methodologies and frameworks for applied big data analytics in manufacturing.

Table 2. Summary—big data for general industrial applications.

Author(s)	Focus
Wuest et al. [26]	Key challenges for global manufacturing industry
Alpaydin [33]	Machine learning overview
Wang et al. [35]	Deep learning for smart manufacturing
Tao et al. [2]	Data-driven smart manufacturing
Flath and Stein [39]	Data science "toolbox" for industrial analytics

A first observation is that there is not uniform agreement with regard to the question of dimensionality. At one extreme, the question is treated as a non-issue, to be handled by the machine learning algorithm selected. Other articles addressed the question at a high-level as important but always within the context of the larger problem-solving approach and not to the level of detail that would be useful to the data scientist.

A second observation is that the approaches for predictive analytics in this section are geared less towards the detailed steps that an analyst might perform and more towards the infrastructure, architecture, and general data landscape that an organization should possess in order to have the capability to perform applied predictive analytics projects. This is not entirely unexpected, as the articles in this section are selected specifically for their high-level, broad outlook. The expectation is that articles in Sections 3.2 and 3.3 will provide greater detail on the subject because articles reviewed in those sections focus more precisely on contexts that align better to activities at the level of the analyst or data scientist.

A third observation is gap identified by more than one researcher, which is the lack of holistic generalization of results beyond the specific, local problem under examination. This is related to manufacturers' limited knowledge of the relative utility or value contained among the different elements of the vast volumes of data that they collect in a somewhat mutually-reinforcing way. A lack of knowledge regarding the data landscape makes it difficult to generalize a dataset's utility from one application to the next. On the same token, not taking incremental steps to analyze projects after the fact for relevance and generalizability to other contexts perpetuates the deficiency in institutional knowledge.

3.2. Big Data Approaches for Specific Manufacturing Applications

This section moves from the higher level of general industrial or manufacturing applications to approaches geared towards specific smart manufacturing applications. The following literature instances fall into one of two subcategories: fault detection and fault prediction. Fault detection and fault prediction are important areas of interest, and it is not surprising that predictive analytics projects gravitate to those topics. Predictive analytics in any the context will naturally gravitate to the dominant interests or challenges facing decision makers in that context, and, for manufacturers, key performance indicators (KPIs) associated with cost, quality, and time are negatively influenced by faults in machinery or output. Most manufacturing processes involve some form of creation or assembly at a given stage followed by some manner of inspection or validation before moving on to the next stage. Components are assembled into some final product, which itself undergoes functional testing prior to distribution to the customer. Machine downtime for unscheduled maintenance will negatively impact cycle time and, by extension, cost. Undetected malfunctions or nonconformities in machinery can lead to defective products escaping from one stage of manufacture to the next. There is an ever-present need to reduce defective products, which creates a natural partnership between smart manufacturing and predictive analytics. It is therefore unsurprising that much of the literature in predictive analytics in the manufacturing context will be applied to case studies in either fault detection or fault prediction.

It will be observed that different publications employ different frameworks, techniques, models, or methodologies to address specific manufacturing applications, often addressing specialized subproblems or challenges. The focus in the ensuing section is how, from the human data scientist perspective, these analyses approach the challenge posed by big data. Is the big data challenge one of an excessive number of diverse features that may contain hidden predictive potential? Is the challenge one of data volume, with exceedingly large numbers of records produced? Neither? Both? Additionally, this review will analyze the ensuing articles with an eye towards knowledge management, or the extent to which there is opportunity to generalize beyond the specific problem of interest.

3.2.1. Fault Detection

In [40], a MapReduce framework is proposed and applied to the fault diagnosis problem in cloud-based manufacturing under the circumstance of a heavily unbalanced dataset. An unbalanced dataset is one in which a large number of examples but another class is represented by comparatively far fewer [41,42]. In terms of features for use in model training, each record of input data contains 27 independent variables and one fault type. There is no explicit discussion of reducing the 27 input variables to a smaller subset or what steps might be taken to do so for a scenario with higher dimensionality.

A hybridized CloudView framework is proposed in [43] for analyzing large quantities of machine maintenance data in a cloud computing environment. The hybridized framework contrasts with a global or offline approach [44] and a local or online approach [45], providing the advantage of being able to analyze sensor data in real-time while also predicting faults in machines using global information on previous faults from a large number of machines [43]. Feature selection is discussed at a high level, but the illustrative case study employs only three data inputs. The purpose of the case study is simply to illustrate the case-based reasoning applied and not apparently to address a specific situation.

In [46], Tamilselvan and Wang employ deep belief networks (DBN) for health state classification of manufacturing machines, with IIOT sensor data employed for model inputs. Specifically, signal data from seven different signals out of a possible 21 were selected for model training. Selection of which signals to include for model training was made based on literature and not on a specific methodological approach.

Deep belief networks are compared favorably to support vector machines (SVM), back-propagation neural networks (BNN), Mahalanobis distance (MD), and self-organizing maps (SOM) [46]. The deep belief network structure consists of a data layer, a network layer, and some number of hidden layers in between. This particular framework structures its hidden layers as a stacked network of restricted Boltzmann machines (RBMs) [47], with the hidden layer of the nth RBM as the data layer of the (n+1)th RBM.

A similar machine learning methodology is employed by Jia et al. (2016) for fault characterization of a rotating machinery in an environment characterized by massive data using deep neural networks (DNNs) [48]. A DNN is similar to the DBN, except that the layers are not constrained to be RBM. For an extensive overview of deep learning in neural networks, see [49]. In a case study in fault diagnosis of rolling element bearings, a total of 2400 features are extracted from 200 signals using fast Fourier transform (FFT); no explicit reduction step is performed or discussed. Rather, the full dataset is input into the DNN.

The DNN model achieves impressive results when compared with a back-propagation neural network (BPNN), with correct classification rates over 99% compared to 65–80% for the BPNN [48]. This indicates that the specific algorithm employed can have a non-trivial impact on the results, depending on the problem under study.

A framework for fault signal identification is proposed by Banerjee et al. (2010) in [50] using short term Fourier transforms (STFT) to separate the signal and SVM to classify it, and Banerjee and Das (2012) extend the approach in [51]. An explicit discussion on data preparation or feature filtration is absent due to the manageable feature set used for model training. However, the approach to extract features from signal data can lead to an excessive number of potential features, making such a step value-added.

Note also that this framework is a hybrid of several techniques, taking sensor data into the SVM after it has already been processed by signal processing and the time-based model. This is in contrast to frameworks relying exclusively on SVM [52,53] or exclusively on time series analysis [54].

Probabilistic frameworks for fault diagnosis grounded in Bayesian networks (BN) and the more generalized Dempster–Shafer theory (DST) are examined in [55] and [56], respectively. For background and additional information on DST, see [57]. The challenge explored by Xiong et al. (2016) in [56] is that of conflicting evidence, with the observation that, in practice, sensors are often disturbed by various factors. This can result in a conflict in the obtained evidence, specifically in a discrepancy between the observed results and the results obtained by fusion through Dempster's combination rule. This challenge reveals the need to reprocess the evidence using some framework or methodology prior to fusing it. Xiong et al. (2016) propose to do so with an information fusion fault diagnosis method based on the static discounting factor, and a combination of K-nearest neighbors (KNN) and dimensionless indicators [56].

Just as in Jia et al. (2016), Xiong et al. (2016)'s method is applied to fault diagnosis among rotating machinery in a large-scale petrochemical enterprise.

Khakifirooz et al. (2017) employ Bayesian inference to mine semiconductor manufacturing data for the purposes of detecting underperforming tool-chamber at a given production time. The authors use Cohen's kappa coefficient to eliminate the influence of extraneous variables [58].

The tool-chamber problem examined in [58] is relevant to this review in that it employs a large number of binary input variables in its model, one for each tool and each step, equal to 1 if the tool-chamber feature was used in a step and equal to 0 if not. The feature filtration approach employed is a two-fold application of Cohen's kappa coefficient, once for pairwise comparison of the features against each other and once for features against the target. Features exhibiting high agreement with each other are wrapped with peers into a group; feature exhibiting low agreement with the target are removed from the model, with 0.20 as the threshold for removal.

This method is appropriate when features and the target are both binary; a limitation is the method is not suitable for data in other forms. This required the target to be transformed from a continuous yield percentage to a categorical classification. A second possible limitation is that each variable is tested independently of the others, with no consideration for interaction. It is logically possible that a feature could have a poor Cohen's kappa coefficient but could interact with other features to produce an overall better model. An advantage of the approach, though not specifically discussed in the article, is that Cohen's kappa coefficient scores for each feature may be preserved from one analysis to the next and analyzed to see if they harbor latent relationships that might point to root causes of inadequate tool-chamber and not simply forecast it.

The final framework for fault detection that this literature review will explore is a cyber-physical system (CPS) architecture proposed by Lee (2017) for fault detection and classification (FDC) in manufacturing processes for vehicle high intensity discharge (HID) headlight and cable modules [59]. For additional background and exploration of CPS, see [60–63]. Although much of the article is devoted to material outside the scope of this review, such as network and database architecture, the manufacturing process explored is notable because it involves multiple subprocesses, some of which are performed in-house and some of which are outsourced to external parties. Furthermore, although there is a small set of main defects that may be observed (shorted cable, cable damage, insufficient soldering, and bad marking), those faults are not directly traceable to a single subprocess. Rather, any number of different subprocesses may result in any fault type. The impact, when performing fault detection and classification, is that the cause-effect relationships and the backwards tracing of faults to diagnoses must take place beforehand.

The input data for the case study consists of eight signals, three from torque sensors and five from proximity sensors, and three learning models are explored: support vector regression (SVR), radial bias function (RBF), and deep belief learning-based deep learning (DBL-DL). In the SVR and RBM models, no additional step in data filtration or feature extraction is performed; in the DBL-DL model, features are extracted in the form of two hidden layers. Unsurprisingly, the DBL-DL model outperforms the other two, with a classification error rate of 7% as compared to 8% for SVR and 9% for RBM [59].

3.2.2. Fault Prediction

In [64], Wan et al. (2017) present a manufacturing big data approach to the active preventive maintenance problem, which includes a proposed system architecture, analysis of data collection methods, and cloud-level data processing. The paper mainly focuses on data processing in the cloud, with pseudocode provided for a real-time processing algorithm. Two types of active maintenance are proposed as necessary: a real-time component to facilitate immediate responses to alarms and an offline component to analyze historic data to predict failures of equipment, workshops or factories.

Of interest to this review is to note that the aforementioned approach is in the context of an organization's ability to perform active preventive maintenance and not in the context of how a data scientist goes about performing his or her analysis. For example, 'data collection' in the context that Wan et al. describe refers to the required service-oriented architecture to integrate data from diverse sources. To the data scientist, 'data collection' is the employment of that architecture in identifying and obtaining specific data elements for model inclusion.

Munirathinam and Ramadoss (2014) apply big data predictive analytics to proactive semiconductor production equipment maintenance. Beginning with a review of maintenance strategies, the researchers present advantages and disadvantages for each of four different maintenance strategies: run to failure (R2F), preventive, predictive, and condition-based. Following this background, an approach for predictive maintenance is presented as follows [65]:

- Collect raw FDC, equipment tracking (ET), and metrology data
- Perform data reduction using a combination of principal component analysis (PCA) and subject matter expertise. This step, in the semiconductor case study, reduces the set of possible parameters from over 1000 to precisely 16
- Train model
- Display output to dashboard with a Maintenance/No Maintenance status

Two immediate observations are apparent when considering the data reduction step employed in this model. First, the use of PCA is effective but it carries with it the loss of interpretability after the fact. This limits the options associated with the dashboards created for visualization of model results. If there were an alternative to PCA that retains interpretability, it may be possible to identify specific thresholds in the input data that are triggers for required maintenance and then track proximity to those thresholds in a dashboard. A second observation is that PCA requires linearity among the parameters because it relies on Pearson correlation coefficients. It also assumes that a feature's contribution to variance relates directly to its predictive power [66]. It is not clear that this is always an appropriate assumption.

Ji and Wang (2017) present a big-data analytics-based fault prediction approach for shop floor scheduling. This application of the big data problem focuses less on the availability of machining resources and more on the problem of potential errors after scheduling [67]. Specifically, it is observed that task scheduling using traditional techniques considers currently available equipment, with time and cost saving as the main objectives. Missing from consideration is the condition prediction of the machines and their states. In other words, scheduling is made absent of any information on the expected condition of the machines during the production process. In the proposed framework, tasks are represented by a set of data attributes, which are then compared to fault patterns mined through big data analytics. This information is then used to assign a risk category to tasks based on generated probabilities. The model provides the opportunity for prediction of potential machine-related faults such as machine error, machine fault, or maintenance states based on scheduling patterns. This knowledge can lead to better machine utilization.

It should be noted that this particular framework, while creative, was not tested on actual data but rather on hypothetical datasets due to data proprietorship policy [67], hence providing clear opportunities for future research.

Neural networks are applied to recognize lubrication defects in a cold forging process, [68] predict ductile cast iron quality [69], optimize micro-milling parameters [70], predict flow behavior of aluminum alloys during hot compression [71], and predict dimensional error in precision machining [72]. Finally, a process approach is taken to improve reliability of high speed mould machining [73].

It was seen in the preceding models featuring NN that data reduction plays a role of minimal importance because the neural network accomplishes feature creation and selection in the hidden layers. In [68], a total of 20 features are selected for model input with no explicit data reduction step. Nor was any reduction step performed in [69], where the dataset was relatively small, consisting of only 700 instances of 14 independent variables in the training set. In [70] and [71], only three features are input into the artificial neural networks (ANN). In [72], an extension of a simulation and process planning approach in [73,74], the number of input variables is five.

Finally, quality and efficiency in freeform surface machining are driven by three primary issues: tool path, tool orientation, and tool geometry [75]. A feature-based approach to machining complex freeform surfaces in the cloud manufacturing environment yields the capability for adaptive, event-driven process adjustments to minimize surface finish errors [76].

An observation across the set of articles reviewed in Section 3.2 is that a specific data reduction step is rarely utilized, either because the feature set was small to begin with or because the machine learning technique could accommodate. The exceptions used either statistical measures (Cohen's kappa) or PCA to reduce the feature set. The article using the former technique did not report how many features the case study began with and how many were ultimately used for model training. It is, therefore, not clear the extent to which the technique is useful. In the case of PCA with subject matter expertise, a feature set of 1000 reduced to 16. Additional discussion and possible extension will be included in Section 4.

A second observation is that as in Section 3.1, variation exists in the frame of reference for which different articles approach the topic of predictive analytics. Some articles focus on the organizational capability to perform predictive analytics. These incorporate robust discussion on technology-centric elements such as architecture for data capture, storage, and extraction or at which levels different analyses may be performed (cloud, edge, real-time, offline, etc.). These typically featured commercially available technologies such as Hadoop or MapReduce and address some of the prerequisites for building organizational competencies in this area. Other articles, on the other hand, employed the term 'framework' to refer to a problem-solving approach or methodology, a sequence of actions to be performed by the analyst or data scientist. These articles more directly align with the objective of this literature review, but it is important to distinguish between the two perspectives as each are important. Indeed, the organizational capability for data capture, storage, and migration must necessarily precede any in-house capability to analyze smart manufacturing data or use it to train a machine learning model.

Table 3 provides a summary of the forgoing studies that approach big data analytics applied to specific manufacturing use cases. The table summarizes whether the paper focuses on organizational capabilities, methodological approaches for the analyst, case studies, or some combination. For case studies, the machine learning algorithm is listed.

Table 3. Big data approaches for specific industrial applications.

Authors(s)	Focus	Explicit Data Reduction Step
Kumar et al. [40]	Enterprise-level architecture; methodology to address class imbalance	No
Bahga and Madiseti [43]	Enterprise-level architecture	No
Tamilselvan and Wang [46]	Case study: Machine health states—DBN	No
Jia et al. [48]	Case study: Fault characterization—DNN	No
Banerjee et al. [50]	Case study: Fault signal identification—SVM	Discussed, not implemented
Xiong et al. [56]	Methodology: Information fusion to reconcile conflicting evidence in fault detection	No
Khakifirooz et al. [58]	Case study: Yield enhancement–Bayesian inference	Yes
Lee [59]	Enterprise-level architecture; Case study: Fault detection and classification—SVR, RBF, DBL-DL	No
Wan et al. [64]	Enterprise-level architecture; Methodology: Real-time and offline components; Case study: Fault prediction—Neural Network	No
Munirathinam & Ramadoss [65]	Enterprise-level architecture	Yes
Ji and Wang [67]	Enterprise-level architecture; Simulated proof of concept case study: Fault prediction for shop floor scheduling	No
Rolfe et al. [68]	Case study: Lubrication defects in cold forging process—NN	No
Perzyk and Kochanski [69]	Ductile cast iron quality—NN	No
Kilickap et al. [70]	Micro-milling parameter optimization—NN	No
Changqing et al. [71]	Alloy flow behavior-NN	No
Arnaiz-Gonzalez et al. [72]	Dimensional error in precision machining—NN	No
de Lacalle et al. [73]	High speed machining of moulds	No
Liu and Li [76]	Manufacturing freeform surfaces	No

3.3. Frameworks for Data Reduction

The third and final category of literature that this review will examine focuses on techniques or approaches specifically for data reduction, which includes feature reduction/selection and instance reduction/selection. There exists a substantial body of influential data preprocessing algorithms for missing values imputation, noise filtering, dimensionality reduction, instance reduction, and treatment of data for imbalanced processing [77]. Specific algorithms for feature selection include Las Vegas Filter/Wrapper [78], Mutual Information Feature Selection [79], Relief [80], and Minimum Redundancy Maximum Relevance (mRMR) [81]. Specific algorithms for instance reduction include condensed nearest neighbor (CNN) [82], edited nearest neighbor (ENN) [83], decremental reduction by ordered projections (DROP) [84], and iterative case filtering (ICF) [85].

The interest in the ensuing articles reviewed in this section is in their suitability for application to the CE. To this end, the domain in which the articles implement any applied case studies is also examined. It will be observed that the reviewed articles contain tasks that fall within Step 3 of the data source selection methodology outlined in [86] and broadly fall into one of three categories: sampling reduction, feature reduction, or instance reduction. Sampling reduction applies to contexts such as optical inspection or reengineering, where there is a need to obtain information for an entire component or surface. If that information may be obtained using fewer samples, then benefits in cost or efficiency follow. Instance reduction applies to contexts in which large numbers of data points are collected for a relatively smaller set of attributes or features. Feature selection is the process of reducing the number of attributes or columns to be input into a machine learning model for training.

Habib ur Rehman, et al. (2016) propose an enterprise-level data reduction framework for value creation in sustainable enterprises, which, while not contextualized to manufacturing, is easily extendable to this domain. The framework considers a traditional five-layer architecture for big data systems and adds three data reduction layers [87].

The first layer for local data reduction is intended for use in mobile devices to collect, preprocess, analyze, and store knowledge patterns. This physical layer can easily be conceptually translated to the CE. The second layer, for collaborative data reduction, is situated prior to the cloud level, with edge computing servers executing analytics to identify knowledge patterns. Note that "edge computing" may be referred to as "fog computing" in some cases [88]. This step will exist in varying degrees in the CE depending on the maturity of the process or organization. In the context of user Internet of Things (IoT) mobile data, as initially presented in the paper, there exists a body of data that must automatically be discarded in accordance with external constrains such as privacy laws. This brings a practical purpose to this initial filtration layer. In smart manufacturing, the physical layer represents IIOT machine or production data, all of which might theoretically harbor some purpose. It may not be prudent to automatically discard chunks of data until it has been definitively determined that there is little risk in doing so. Finally, a layer for remote data reduction is added to aggregate the knowledge patterns from edge servers that are then distributed to cloud data centers for data applications to access and further analyze [87].

It should be noted that this framework is at the institutional level and not at the level of the data scientist. The data reduction layers are presented as automated processes applied to the raw source data and not dependent on a specific project or problem of interest.

At the data scientist level, a second point-based data reduction scenario is presented in [89], in which Ma and Cripps (2011) develop a data reduction algorithm for 3D surface points for use in reverse engineering. In reverse engineering, data is captured from an existing surface on the order of millions of scanned points. There are challenges associated with volume of data, and there are challenges in the form of increased error associated with removing data. The data reduction algorithm is based on Hausdorff distance and works by first collecting a set of 3D point data from a surface using an optical device such as a laser scanner, iterating through the set of points, and determining if a point can be removed without causing the local estimation of surface characteristics to fall out of tolerance. This is done by comparing shape pre- and post-removal. The procedure is tested on an idealized aircraft wing but is extendable to any manner of reverse engineering that employs 3D measurement data. It is possible that this could also be extended to inspection-type applications, but the challenge is that the end-state number of required data points will be dependent on the nature of the surface. Additionally, it is not certain that Hausdorff distance would be the appropriate metric for other contexts such as automated optical inspection.

Considering data reduction with respect to the set of features to be used for model training, Jeong et al. (2016) propose a feature selection approach based on simulated annealing and apply it to a case study for detecting denial of service attacks [90]. This approach is similar to [91], which uses the same data set but a different local search algorithm.

The model starts with a randomly-generated set of features to include, trains a model on that set of features, and tests it by way of some pre-designated machine learning technique. The case study is a classification problem, and so examples used include SVM, multi-layer perceptron (MLP), and naïve Bayes classifier (NBC). After obtaining a solution and measure of performance using some cost function, neighborhood solutions are obtained and tested. Superior solutions are retained, and inferior solutions are either discarded or retained based on a probability calculation. This ability to retain an inferior solution allows the simulated annealing algorithm to "jump" out of a local extrema [92]. The intrusion detection case study employed 41 factors, which reduced to 14, 16, and 19 factors when using MLP, SVM, and NBC respectively. A limitation to this approach is that it requires model training at every iteration of the simulated annealing. This may limit the options for which machine learning technique to select; preference should be given to algorithms that quickly converge. Again, for only 41 factors, this is less of an issue. If there are hundreds or thousands, then this approach may be impractical.

Lalehpour, Berry, and Barari (2017) propose an approach for data reduction for coordinate measurement of planar surfaces for the purposes of reducing the number of samples required to adequately validate that a part has been build according to design specifications [93]. The larger context for this approach is manufacturing, but the applicability is narrowly scoped to an inspection station along an assembly line. Thus, this approach could be used in programming firmware for an optical inspection machine so that it can diagnose defective components as efficiently as possible. However, it would not be useful in performing root cause analysis to find the source of the defects or predict future occurrences.

Ul Haq, Wang, and Djurdjanovic (2016) develop a set of ten features that may be constructed from streaming signal data from semiconductor fabrication equipment. Technological developments allow the collection of inline data at ever increasing sampling rates. This has resulted in two effects, the first being an increase in the amount of data required to store, and the second being the ability to discern features that were previously not discernible [94]. Specifically, high sampling rates allow information to be gleaned from transient periods between steady signal states. This enables the extraction of features from the signal that could not be calculated with lower sampling rates.

The approach can be extended to any signal-style continuous data source from which samples are taken, although the implication is that the lower the sampling rate, the less likely that these new features will provide value. These constructed features are applied to case studies of tool and chamber matching and defect level prediction. A reasonable extension might be to apply the approach to machine diagnostic information for active preventive maintenance.

From a feature selection or dimensionality perspective, which is of most interest to this review, the ten features are calculated every time the signal transitions from one steady state to another. For relatively static signals, this will result in a manageable feature set; for more dynamic signals or for large time windows, the number of calculated features may become prohibitively large. This could be alleviated by adding an additional layer of features that employ various means to aggregate the values of the ten calculated features over the entire span of time.

Continuing on the topic of feature selection, Christ, Kempa-Liehr, and Feindt (2016) propose an algorithm for time series feature extraction, TSFRESH, that not only generates features but also employs a feature importance filter to screen out irrelevant features [95]. This framework, illustrated in Figure 2, begins by extracting up to 794 predefined features from time series data. Subsequently, the vector representing each individual feature is independently tested for significance against the target. This produces a vector of p-values with the same cardinality as the number of features. Finally, the vector of p-values is evaluated to decide which features to keep. The method for evaluating the vector of p-values is to control the false discovery rate (FDR) using the Benjamini-Hochberg procedure [96].

A case study using data from the UCI Machine Learning Repository [97] reduced an initial set of 4764 features to 623 [98].

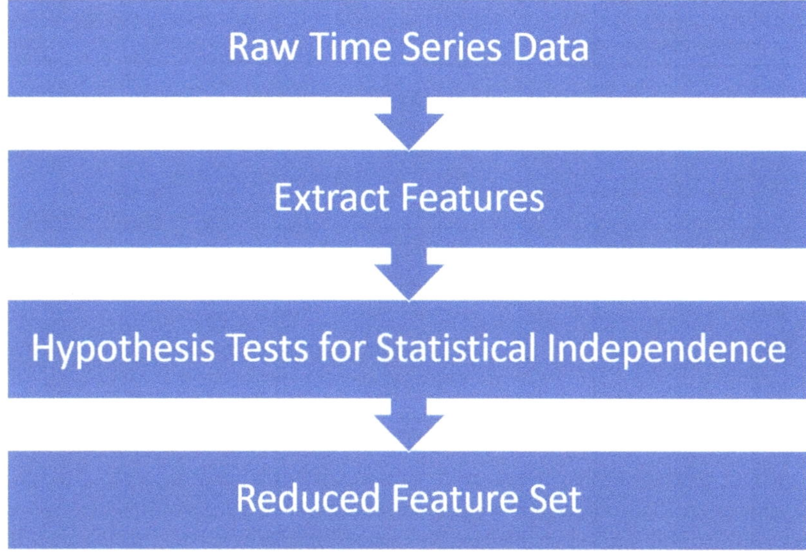

Figure 2. High-level TSFRESH process

For instance reduction, Wang et al. (2016) employ a framework based on two clustering algorithms, affinity propagation (AP) and k-nearest neighbor (k-NN), to extract "exemplars", or representations of some number of actual data points [99]. A clustering algorithm is employed to cluster the data instances into similar groups; an exemplar is then defined to represent the group. The context is in network security, specifically anomaly detection. The idea is that records for http traffic and network data under normal circumstances can be grouped or aggregated into representations of those conditions, which can produce cost savings in data storage. The technique is potentially extendable to other areas of manufacturing, although for mature processes there may not be the desire to perform aggregation of records because the "easy" relationships have already been discovered. Rather, a large number of records may be necessary to identify hidden structures or correlations in subgroups that might otherwise, in smaller sample sizes, be considered outliers [100,101].

A second instance reduction Nikolaidis, Goulermas, and Wu (2010) develop an instance reduction framework that draws a distinction between instances close to class boundaries and instances farther away from class boundaries [102]. The reasoning is that instances farther away from class boundaries are less critical to the classification process and are therefore more "expendable" from an instance reduction standpoint. The four-step framework first uses a filtering component such as ENN to smooth class boundaries and then classifies instances as "border" and "non-border". Following a pruning step for the border instances, the non-border instances are clustered using mean shift clustering (MSC).

As previously indicated, the reviewed articles from the Section 3.3, summarized in Table 4, cover reductions in the number of samples required to obtain a satisfactory result, techniques to reduce the number of instances or records, and techniques to reduce the number of features or attributes.

Of greatest interest to this review is the second category, feature selection, and two approaches seen in this section merit further discussion in relation to each other. The first approach, the TSFRESH approach, generates a list of up to 794 features from a single time series and, using statistical independence as the test, reduces the feature set by eliminating the features that do not exhibit a significant statistical dependence with the response. Using this approach, a model with N time series inputs would have 794 N features extracted by TSFRESH. Even if TSFRESH then filters out 50% of the features, there still could remain many hundreds of features in the model. This could be an excessive number of features that strains the capacity of the analyst to truly grasp what is going on or pinpoint the critical relationship(s) of interest. Extending the approach to include subsequent filter(s) could be a step in remedying this challenge.

Table 4. Frameworks for data reduction.

Author(s)	Focus
Habib ur Rehman, et al. [87]	High level/Institutional framework
Jeong et al. [90]	Feature selection meta-heuristic (simulated annealing)
Lalehpour, Berry, and Barari [93]	Sample reduction
Ma and Cripps [89]	Shape preservation with data reduction for 3D surface points
Ul Haq, Wang, and Djurdjanovic [94]	Feature extraction from streaming signal data
Christ, Kempa-Liehr, and Feindt [95]	Feature extraction and selection from time series data
Wang et al. [99]	Clustering algorithms to extract representative data instances
Nikolaidis, Goulermas, and Wu [102]	Instance reduction based on distance from class boundaries

The second approach of interest is the use of optimization heuristics to obtain a near-optimal subset of features for the problem at hand. It might be a reasonable extension to TSFRESH to incorporate a second filter that seeks to better optimize the feature set with respect to the objective function, possibly using a heuristic such as simulated annealing. This would also add the dimension of feature interaction, which is currently not present in the TSFRESH statistical independence filter.

A final observation from the third category of reviewed literature is that the set of literature on reducing or filtering the features that might go into a machine learning model is reasonably robust but is relatively less robust concerning the prioritization of the remaining features. This implies a gap in terms of approaches to quantitatively or qualitatively stack features against each other. An alternative explanation is that such approaches exist but were simply not employed in the reviewed literature. This seems unlikely, as, the benefit of such capability would be to see how a particular feature of interest fares in its utility from one problem to the next. In smart manufacturing, the same features of data are continually collected and used repeatedly in different analyses. It may be of interest to know which of those features tend to be valuable in harboring predictive power and which ones tend not to.

4. Discussion

This paper reviewed existing research into frameworks or approaches for big-data analytics as applied to three levels of projects, with increasing degrees of precision or detail. The first level reviewed was a high-level look at frameworks for general industrial applications. The second level focused specifically and local, lower-level smart manufacturing applications of fault detection and fault prediction. Finally, the third and most specialized level looked at approaches specifically oriented towards data reduction.

In each section, articles were discussed individually as they pertain to the motivation for this research and their applicability to the CE. At the end of each section, discussion followed to summarize any observations across the set of articles within the section and relate them to each other. The final level of discussion is to look at the full picture and identify any observations, trends, or commonalities that span the three levels.

The first observation is that there is a dichotomy in how the same verbiage can be applied to different contexts. Terms like 'framework', 'big data', and 'predictive analytics' in some cases are contextualized as architecture required to build organizational competencies and in other cases as approaches or methodologies to build individual competencies.

In the context of organizational competencies for big data analytics, much consideration is made as to the architecture for where the data exists, how it moves from one location to the next, and at which level or echelon the analysis takes place. In general, there is some layer or module at which the initial data is generated or collected but then options for what to do with it. Data may be migrated to a cloud-based data center and analyzed in a consolidated location, or it may be analyzed at local nodes. Whether to analyze at the edge or in the cloud will typically be a function of resources and of the time window available to perform corrective action. Actions that require real-time processing for quick action might not be performed at the cloud level because, by the time data is captured, cleaned, pre-processed, and run through a model, the window to correct an identified fault may have already passed. On the other hand, if there is sufficient time between the data collection point and the decision point, such as a manufacturing process in which there might be a gap of hours or days between assembly line procedures and testing, then analysis at the cloud level might be suitable.

It should be noted that, from an organizational competency perspective, the infrastructure is a prerequisite to the development of individual competencies in the form of data scientist best practices. However, it is those data scientist best practices that become contributing factors to other organizational competencies such as knowledge management and decisions on long term data retention. There is an iterative and cyclic relationship such that organizational competencies produce individual competencies which then build and reinforce other organizational competencies.

A second observation across the three sections of reviewed literature is that there was a conspicuous absence of any discussion of the generalization of results beyond the specific problem of interest. This is true on both the 'front' end and the 'back' end of the articles reviewed. In other words, upon conclusion of the experiment or analysis, there was no discussion in any reviewed article of knowledge management or steps to generalize results from an input data perspective. There was certainly discussion about future research opportunities in generalizing an overall approach or algorithm, but in no cases did that discussion manifest itself the form of practical reflection on a feature set's utility for the problem of interest and prospects for utility in other scenarios. Similarly, during model formulation, there was no discussion of institutional knowledge that might play a role in feature selection. Only one reviewed paper referenced a data screening decision that was made based on prior work. The context in that situation was 21 possible signals to use as model inputs, of which seven were selected based on reviewed literature.

This observation is not intended as a negative criticism of any past work. It is quite natural to expect that this might be the case because finite resources drive priorities, and in a fast-paced world there is often little time to breathe between the completion of one project and the start of another. Given this reality, there appears to be value in anything that can facilitate the creation and preservation of institutional knowledge in this domain.

A third observation is that feature selection approaches in most cases were performed using a single technique at a single point in the model building process. Feature filtration using Kullback–Leibler divergence reduced features sets of 1460 and 1460 to 198 and 175, respectively. Feature filtration using Cohen's kappa was stated as a step in one case study, but no results were provided as to how many features were filtered out. A combination of PCA and subject matter expertise reduced a feature set of 1000 to 16, although it was not clearly identified how many of those features were reduced from PCA and how many from subject matter expertise. In the case of TSFRESH, statistical hypothesis tests for independence filter out features that are statistically independent, reducing 4764 features to 623.

A natural next step for any of these techniques is to explore the possibility to layer one technique after another depending on how many features remain after a given filter. In the case of 1000 features reduced to 16, it is possible to successively iterate through all 65536 subsets of features to arrive at an optimal subset with minimal effort. In the case of 4764 features reduced to 623, however, this is computationally impractical. It is unlikely that the optimal subset of the 623 remaining features would be all 623 of those features; a layered approach to continue to weed out features would be a value-added step to analyses with large numbers of features remaining. This is especially true if there is the desire for the model and its results to be understandable and digestible on the human side of the enterprise. Furthermore, what is understandable and digestible to the data scientist may be neither understandable nor digestible to the decision maker. Communication and visualization are critical components to the human element, particularly for decision makers who may not have background in the technical aspects of data science.

5. Conclusions

The papers reviewed are not intended to represent an exhaustive list of all existing research on the subject. However, it is believed that the reviewed examples do provide a representative sample of the sort of research currently performed in this discipline.

A conclusion that may be drawn from the first general observation in Section 4 is that there is value in having a standard set of terminology when speaking about the big data environment in order to distinguish when one is referring to organizational capabilities or individual competencies. In the reviewed literature, terms like 'framework' or 'data collection' carried wide variance in their meaning depending on the context. It is likely that standard terms will be settled on over time, either bottom-up from common use or top-down from professional organizations in industry, academia, or government. At this point, it may suffice simply to be aware of the different contexts in which the topic may be broached. Attention to detail is always a good rule of thumb in any endeavor, and that may be a good temporary solution for now.

A second conclusion, following from the second general observation in Section 4, is that a generalized approach to provide clarity as to what input data is valuable and what input data is not valuable, perhaps with both a quantitative and qualitative dimension, can shape analysis decisions in the big-data environment. Those decisions might be localized to the problem of interest, as in deciding which features to include in the model. Those decisions might also extend to larger, resource-oriented decisions, such as start-up priorities for transitioning from a legacy manufacturing facility to a CE. From a knowledge management standpoint, there is value in building institutional knowledge regarding features that perform poorly as well as features that perform well. Knowing which features tend to habitually appear in good solutions and which features habitually appear in bad solutions, if such knowledge exists, would be tremendously helpful in long term data capture and storage decisions.

Finally, the third general observation in Section 4 lends itself to the conclusion that there is room for additional research into practical means for feature filtration and prioritization. On the surface, there appears to be no reason why the single-layer filtration techniques employed in the reviewed articles cannot be extended into a series of hierarchical filters. One possible limitation would be that several of the techniques employ similarly-themed filters that may produce only limited improvement when performed in sequence. For example, filtering once by Cohen's kappa and then by testing for statistical independence might not produce substantial improvement. However, following the initial filtration by way of Cohen's kappa with an optimization heuristic such as simulated annealing or genetic algorithm to find an optimal or near-optimal subset of features might be a promising avenue to explore.

It is also worth exploring, from a knowledge management standpoint, feature reduction and selection techniques that preserve as much interpretability as possible. It has already been discussed that PCA is a common approach, but the reduction in dimensions from M to K, where K < M, will necessarily take away the physical meaning from those K features. Techniques in feature reduction that preserve the nature of the original features are a value-added contribution to this question.

In closing, manufacturing in the 21st century is a highly competitive enterprise, and the business value in exploiting 21st century technologies in smart manufacturing, IIOT, Industry 4.0, and the CE cannot be overstated. At the core of this opportunity for the individual manufacturer is the untapped potential held in the volumes of smart manufacturing data collected and stored in its data repositories. For an organization to develop as a core competency a methodological approach or process to build and continually develop institutional knowledge about the data landscape at its disposal, it would move the body of input data for any given predictive analytics project from simply a means to an end to an end unto itself. This is something of a paradigm shift, but one that can produce meaningful advantage to the organization that harnesses it.

Future research will develop and explore the potential for frameworks of this nature in the smart manufacturing context.

Author Contributions: Conceptualization, P.M.L., W.O, and F.P.M.; methodology, P.M.L.; software, N/A; validation, P.M.L., W.O. and F.P.M.; formal analysis, P.M.L.; investigation, P.M.L.; resources, F.P.M.; data curation, N/A; writing—original draft preparation, P.M.L.; writing—review and editing, P.M.L., W.O., F.P.M.; visualization, N/A.; supervision, W.O.; project administration, W.O.; funding acquisition, N/A.

Funding: This research received no external funding.

Conflicts of Interest: The authors declare no conflicts of interest.

References

1. Kusiak, A. Smart manufacturing must embrace big data. *Nature* **2017**, *544*, 23–25. [CrossRef] [PubMed]
2. Tao, F.; Qi, Q.; Liu, A.; Kusiak, A. Data-driven smart manufacturing. *J. Manuf. Syst.* **2018**, *48*, 157–169. [CrossRef]
3. Everything You Need to Know about the Industrial Internet of Things. *G.E. Digital.* 2016. Available online: https://www.ge.com/digital/blog/everything-you-need-know-about-industrial-internet-things (accessed on 1 May 2018).
4. Schneider, S. The industrial internet of things (IIoT): Applications and taxonomy. In *Internet of Things and Data Analytics Handbook*; Wiley: Hoboken, NJ, USA, 2017; pp. 41–81.
5. Industrial Internet Consortium. Available online: https://www.iiconsortium.org/ (accessed on 2 May 2018).
6. OpenFog. Available online: https://www.openfogconsortium.org/ (accessed on 2 May 2018).
7. Lasi, H.; Fettke, P.; Kemper, H.G.; Feld, T.; Hoffmann, M. Industry 4.0. *Bus. Inf. Syst. Eng.* **2014**, *6*, 239–242. [CrossRef]
8. Gilchrist, A. *Industry 4.0: The Industrial Internet of Things*; Apress: New York, NY, USA, 2016.
9. Li, L. China's manufacturing locus in 2025: With a comparison of "Made-in-China 2025" and 'Industry 4.0'. *Technol. Forecast. Soc. Chang.* **2018**, *135*, 66–74. [CrossRef]
10. METI, Connected Industries. Ministry of Economy, Trade and Industry. 2017. Available online: http://www.meti.go.jp/english/policy/mono_info_service/connected_industries/index.html (accessed on 10 January 2019).
11. Granrath, L. Japan's Society 5.0: Going Beyond Industry 4.0. *Japan Industry News.* 2017. Available online: https://www.japanindustrynews.com/2017/08/japans-society-5-0-going-beyond-industry-4-0/ (accessed on 10 January 2019).
12. Rockwell Automation. *The Connected Enterprise eBook: Bringing People, Processes, and Technology Together*; Rockwell Automation: Milwaukee, WI, USA, 2015.
13. Otieno, W.; Cook, M.; Campbell-Kyureghyan, N. Novel approach to bridge the gaps of industrial and manufacturing engineering education: A case study of the connected enterprise concepts. In Proceedings of the 2017 IEEE Frontiers in Education Conference (FIE), Indianapolis, IN, USA, 18–21 October 2017; pp. 1–5.
14. Qin, S.J. Process data analytics in the era of big data. *AIChE J.* **2014**, *60*, 3092–3100. [CrossRef]
15. McKinsey & Company. *Big Data: The Next Frontier for Innovation, Competition, and Productivity*; McKinsey Global Institute: Washington, DC, USA, 2011; p. 156.
16. Bollier, D.; Firestone, C.M. *The Promise and Peril of Big Data*; The Aspen Institute: Washington, DC, USA, 2010.
17. Lenz, J.; Wuest, T.; Westkämper, E. Holistic approach to machine tool data analytics. *J. Manuf. Syst.* **2018**, *48*, 180–191. [CrossRef]
18. Thoben, K.; Wiesner, S.; Wuest, T. 'Industrie 4.0' and Smart Manufacturing—A Review of Research Issues and Application Examples. *Int. J. Autom. Technol.* **2017**, *11*, 4–19. [CrossRef]
19. Kaufman, E.L.; Lord, M.W.; Reese, T.W.; Volkmann, J. The Discrimination of Visual Number. *Am. J. Psychol.* **1949**, *62*, 498–525. [CrossRef] [PubMed]

20. Miller, G.A. The magical number seven, plus or minus two: Some limits on our capacity for processing information. *Psychol. Rev.* **1956**, *63*, 81–97. [CrossRef] [PubMed]

21. Simon, H.A. Designing organizations for an information-rich world. *Comput. Commun. Public Interes.* **1971**, *72*, 37.

22. Oussous, A.; Benjelloun, F.; Lahcen, A.A.; Belfkih, S. Big Data technologies: A survey. *J. King Saud Univ. Comput. Inf. Sci.* **2018**, *30*, 431–448. [CrossRef]

23. Honest, N. A Survey of Big Data Analytics. *Int. J. Inf. Sci. Tech.* **2016**, *6*, 35–43. [CrossRef]

24. Tsai, C.-W.; Lai, C.-F.; Chao, H.-C.; Vasilakos, A.V. Big data analytics: A survey. *J. Big Data* **2015**, *2*, 21. [CrossRef]

25. Spangenberg, N.; Roth, M.; Franczyk, B. Evaluating new approaches of big data analytics frameworks. In Proceedings of the International Conference on Business Information Systems, Poznań, Poland, 24–26 June 2015.

26. Wuest, T.; Weimer, D.; Irgens, C.; Thoben, K.-D. Machine learning in manufacturing: Advantages, challenges, and applications. *Prod. Manuf. Res.* **2016**, *4*, 23–45. [CrossRef]

27. Dingli, D.J. *The Manufacturing Industry—Coping with Challenges*; Working Paper No. 2012/05; 2012; p. 47. Available online: https://econpapers.repec.org/paper/msmwpaper/2012_2f05.htm (accessed on 27 February 2018).

28. Gordon, J.; Sohal, A.S. Assessing manufacturing plant competitiveness—An empirical field study. *Int. J. Oper. Prod. Manag.* **2001**, *21*, 233–253. [CrossRef]

29. Shiang, L.E.; Nagaraj, S. Impediments to innovation: Evidence from Malaysian manufacturing firms. *Asia Pac. Bus. Rev.* **2011**, *17*, 209–223. [CrossRef]

30. Thomas, A.J.; Byard, P.; Evans, R. Identifying the UK's manufacturing challenges as a benchmark for future growth. *J. Manuf. Technol. Manag.* **2012**, *23*, 142–156. [CrossRef]

31. Kotsiantis, S.B. Supervised Machine Learning: A Review of Classification Techniques. *Informatica* **2007**, *31*, 249–268.

32. Yang, K.; Trewn, J. *Multivariate Statistical Methods in Quality Management*; McGraw-Hill: New York, NY, USA, 2004.

33. Alpaydin, E. *Introduction to Machine Learning*, 3rd ed.; MIT Press: Cambridge, MA, USA, 2014.

34. Doltsinis, S.; Ferreira, P.; Lohse, N. Reinforcement learning for production ramp-up: A Q-batch learning approach. In Proceedings of the 11th International Conference on Machine Learning and Applications, Boca Raton, FL, USA, 12–15 December 2012; pp. 610–615.

35. Wang, J.; Ma, Y.; Zhang, L.; Gao, R.X.; Wu, D. Deep learning for smart manufacturing: Methods and applications. *J. Manuf. Syst.* **2018**, *48*, 144–156. [CrossRef]

36. Butler, B. What Is Edge Computing and How It's Changing the Network. *Network World.* 2017. Available online: https://www.networkworld.com/article/3224893/internet-of-things/what-is-edge-computing-and-how-it-s-changing-the-network.html (accessed on 7 March 2018).

37. Linthicum, D. Responsive Data Architecture for the Internet of Things. *Computer* **2016**, *49*, 72–75. [CrossRef]

38. Mahmud, R.; Kotagiri, R.; Buyya, R. Fog Computing: A Taxonomy, Survey and Future Directions. In *Internet of Everything*; Springer: Singapore, 2018; pp. 103–130.

39. Flath, C.M.; Stein, N. Towards a data science toolbox for industrial analytics applications. *Comput. Ind.* **2018**, *94*, 16–25. [CrossRef]

40. Kumar, A.; Shankar, R.; Choudhary, A.; Thakur, L.S. A big data MapReduce framework for fault diagnosis in cloud-based manufacturing. *Int. J. Prod. Res.* **2016**, *54*, 7060–7073. [CrossRef]

41. Japkowicz, N.; Stephen, S. The class imbalance problem: A systematic study. *Intell. Data Anal.* **2002**, *6*, 429–449. [CrossRef]

42. Longadge, R.; Dongre, S.S.; Malik, L. Class imbalance problem in data mining: Review. *Int. J. Comput. Sci. Netw.* **2013**, *2*, 83–87.

43. Bahga, A.; Madisetti, V.K. Analyzing massive machine maintenance data in a computing cloud. *IEEE Trans. Parallel Distrib. Syst.* **2012**, *23*, 1831–1843. [CrossRef]

44. Devaney, M.; Cheetham, B. Case-Based Reasoning for Gas Turbine Diagnostics. In Proceedings of the 18th International FLAIRS Conference (FLAIRS-05), Clearwater Beach, FL, USA, 16–18 May 2005.

45. Timmerman, H. SKF WindCon Condition Monitoring System for Wind Turbines. In Proceedings of the New Zealand Wind Energy Conference, Wellington, NZ, USA, 20–22 April 2009.

46. Tamilselvan, P.; Wang, P. Failure diagnosis using deep belief learning based health state classification. *Reliab. Eng. Syst. Saf.* **2013**, *115*, 124–135. [CrossRef]

47. Hinton, G.E. A Practical Guide to Training Restricted Boltzmann Machines. *Computer* **2012**, *9*, 599–619.

48. Jia, F.; Lei, Y.; Lin, J.; Zhou, X.; Lu, N. Deep neural networks: A promising tool for fault characteristic mining and intelligent diagnosis of rotating machinery with massive data. *Mech. Syst. Signal Process.* **2016**, *72–73*, 303–315. [CrossRef]

49. Schmidhuber, J. Deep Learning in neural networks: An overview. *Neural Netw.* **2015**, *61*, 85–117. [CrossRef] [PubMed]

50. Banerjee, T.; Das, S.; Roychoudhury, J.; Abraham, A. Implementation of a New Hybrid Methodology for Fault Signal Classification Using Short-Time Fourier Transform and Support Vector Machines. In Proceedings of the 5th International Workshop on Soft Computing Models in Industrial Environment Application (SOCO 2010), Guimarães, Portugal, 16–18 June 2010; Volume 73, pp. 219–225.

51. Banerjee, T.P.; Das, S. Multi-sensor data fusion using support vector machine for motor fault detection. *Inf. Sci.* **2012**, *217*, 96–107. [CrossRef]

52. Jack, L.B.; Nandi, A.K. Fault detection using support vector machines and artificial neural networks, augmented by genetic algorithms. *Mech. Syst. Signal Process.* **2002**, *16*, 373–390. [CrossRef]

53. Rychetsky, M.; Ortmann, S.; Glesner, M. Support vector approaches for engine knock detection. In Proceedings of the IJCNN'99. International Joint Conference on Neural Networks, Washington, DC, USA, 10–16 July 1999; Volume 2.

54. Altintas, Y. In-process detection of tool breakages using time series monitoring of cutting forces. *Int. J. Mach. Tools Manuf.* **1988**, *28*, 157–172. [CrossRef]

55. Wang, H.; Zhoui, J.; He, I.; Sha, J. An uncertain information fusion method for fault diagnosis of complex system. In Proceedings of the 2003 International Conference on Machine Learning and Cybernetics, Xi'an, China, 5 November 2003; pp. 1505–1510.

56. Xiong, J.; Zhang, Q.; Sun, G.; Zhu, X.; Liu, M.; Li, Z. An Information Fusion Fault Diagnosis Method Based on Dimensionless Indicators with Static Discounting Factor and KNN. *IEEE Sens. J.* **2016**, *16*, 2060–2069. [CrossRef]

57. Dempster, A.P. A Generalization of Bayesian Inference. *J. R. Stat. Soc.* **1968**, *30*, 205–247. [CrossRef]

58. Khakifirooz, M.; Chien, C.F.; Chen, Y.J. Bayesian inference for mining semiconductor manufacturing big data for yield enhancement and smart production to empower industry 4.0. *Appl. Soft Comput. J.* **2017**, *68*, 990–999. [CrossRef]

59. Lee, H. Framework and development of fault detection classification using IoT device and cloud environment. *J. Manuf. Syst.* **2017**, *43*, 257–270. [CrossRef]

60. Gunes, V.; Peter, S.; Givargis, T.; Vahid, F. A Survey on Concepts, Applications, and Challenges in Cyber-Physical Systems. *KSII Trans. Internet Inf. Syst.* **2014**, *8*, 120–132.

61. Rajkumar, R.; Lee, I.; Sha, L.; Stankovic, J. Cyber-physical systems. In Proceedings of the 47th Design Automation Conference on—DAC '10, Anaheim, CA, 13–18 June 2010; p. 731.

62. Saez, M.; Maturana, F.; Barton, K.; Tilbury, D. Modeling and Analysis of Cyber-Physical Manufacturing Systems for Anomaly Detection and Diagnosis. 2018. Available online: https://www.nist.gov/sites/default/files/documents/2018/05/22/univ_michigan_miguel_saez.pdf (accessed on 25 February 2019).

63. Saez, M.; Maturana, F.; Barton, K.; Tilbury, D. Anomaly detection and productivity analysis for cyber-physical systems in manufacturing. In Proceedings of the 2017 13th IEEE Conference on Automation Science and Engineering (CASE), Xi'an, China, 20–23 August 2017; pp. 23–29.

64. Wan, J.; Tang, S.; Li, D.; Wang, S.; Liu, C.; Abbas, H.; Vasilakos, A.V. A Manufacturing Big Data Solution for Active Preventive Maintenance. *IEEE Trans. Ind. Inform.* **2017**, *13*, 2039–2047. [CrossRef]

65. Munirathinam, S.; Ramadoss, B. Big data predictive analtyics for proactive semiconductor equipment maintenance. In Proceedings of the 2014 IEEE International Conference on Big Data (IEEE Big Data 2014), Washington, DC, USA, 27–30 October 2014; pp. 893–902.

66. Franklin, J. Signalling and anti-proliferative effects mediated by gonadotrophin-releasing hormone receptors after expression in prostate cancer cells using recombinant adenovirus. *J. Endocrinol.* **2003**, *176*, 275–284. [CrossRef] [PubMed]

67. Ji, W.; Wang, L. Big data analytics based fault prediction for shop floor scheduling. *J. Manuf. Syst.* **2017**, *43*, 187–194. [CrossRef]

68. Rolfe, B.F.; Frayman, Y.; Kelly, G.L.; Nahavandi, S. Recognition of Lubrication Defects in Cold Forging Process with a Neural Network. In *Artificial Neural Networks in Finance and Manufacturing*; IGI Global: Hershey, PA, USA, 2006; pp. 262–275.

69. Perzyk, M.; Kochański, A.W. Prediction of ductile cast iron quality by artificial neural networks. *J. Mater. Process. Technol.* **2001**, *109*, 305–307. [CrossRef]

70. Kilickap, E.; Yardimeden, A.; Çelik, Y.H. Mathematical Modelling and Optimization of Cutting Force, Tool Wear and Surface Roughness by Using Artificial Neural Network and Response Surface Methodology in Milling of Ti-6242S. *Appl. Sci.* **2017**, *7*, 1064. [CrossRef]

71. Huang, C.; Jia, X.; Zhang, Z. A modified back propagation artificial neural network model based on genetic algorithm to predict the flow behavior of 5754 aluminum alloy. *Materials* **2018**, *11*, 855. [CrossRef] [PubMed]

72. Arnaiz-González, Á.; Fernández-Valdivielso, A.; Bustillo, A.; de Lacalle, L.N.L. Using artificial neural networks for the prediction of dimensional error on inclined surfaces manufactured by ball-end milling. *Int. J. Adv. Manuf. Technol.* **2016**, *83*, 847–859. [CrossRef]

73. De Lacalle, L.N.L.; Lamikiz, A.; Salgado, M.A.; Herranz, S.; Rivero, A. Process planning for reliable high-speed machining of moulds. *Int. J. Prod. Res.* **2002**, *40*, 2789–2809. [CrossRef]

74. De Lacalle, L.N.L.; Lamikiz, A.; Sánchez, J.A.; Salgado, M.A. Effects of tool deflection in the high-speed milling of inclined surfaces. *Int. J. Adv. Manuf. Technol.* **2004**, *24*, 621–631. [CrossRef]

75. Lasemi, A.; Xue, D.; Gu, P. Recent development in CNC machining of freeform surfaces: A state-of-the-art review. *CAD Comput. Aided Des.* **2010**, *42*, 641–654. [CrossRef]

76. Liu, X.; Li, Y. Feature-based adaptive machining for complex freeform surfaces under cloud environment. *Robot. Comput. Integr. Manuf.* **2019**, *56*, 254–263. [CrossRef]

77. García, S.; Luengo, J.; Herrera, F. Tutorial on practical tips of the most influential data preprocessing algorithms in data mining. *Knowl.-Based Syst.* **2015**, *98*, 1–29. [CrossRef]

78. Liu, H.; Setiono, R. A Probabilistic Approach to Feature Selection—A Filter Solution. In Proceedings of the Thirteenth International Conference on Machine and Learning, Bari, Italy, 3–6 July 1996; pp. 319–327.

79. Battiti, R. Using Mutual Information for Selecting Features in Supervised Neural-Net Learning. *IEEE Trans. Neural Netw.* **1994**, *5*, 537–550. [CrossRef] [PubMed]

80. Kira, K.; Rendell, L. A practical approach to feature selection. In Proceedings of the Ninth International Conference on Machine Learning, Aberdeen, UK, 1–3 July 1992; pp. 249–256.

81. Peng, H.; Long, F.; Ding, C. Feature selection based on mutual information: Criteria of Max-Dependency, Max-Relevance, and Min-Redundancy. *IEEE Trans. Pattern Anal. Mach. Intell.* **2005**, *27*, 1226–1238. [CrossRef] [PubMed]

82. Hart, P. The condensed nearest neighbor rule (Corresp.). *IEEE Trans. Inf. Theory* **1968**, *14*, 515–516. [CrossRef]

83. Wilson, D.L. Asymptotic Properties of Nearest Neighbor Rules Using Edited Data. *IEEE Trans. Syst. Man Cybern.* **1972**, *2*, 408–421. [CrossRef]

84. Wilson, D.R.; Martinez, T.R. Reduction Techniques for Instance-Based Learning Algorithms. *Mach. Learn.* **2000**, *38*, 257–286. [CrossRef]

85. Brighton, H.; Mellish, C. Advances in Instance Selection for Instance-Based Learning Algorithms. *Data Min. Knowl. Discov.* **2002**, *6*, 153–172. [CrossRef]

86. Stanula, P.; Ziegenbein, A.; Metternich, J. Machine learning algorithms in production: A guideline for efficient data source selection. *Procedia CIRP* **2018**, *78*, 261–266. [CrossRef]

87. Rehman, M.H.U.; Chang, V.; Batool, A.; Wah, T.Y. Big data reduction framework for value creation in sustainable enterprises. *Int. J. Inf. Manag.* **2016**, *36*, 917–928. [CrossRef]

88. Luan, T.H.; Gao, L.; Li, Z.; Xiang, Y.; Wei, G.; Sun, L. Fog Computing: Focusing on Mobile Users at the Edge. *arXiv*, 2015; arXiv:1502.01815.

89. Ma, X.; Cripps, R.J. Shape preserving data reduction for 3D surface points. *CAD Comput. Aided Des.* **2011**, *43*, 902–909. [CrossRef]

90. Jeong, I.-S.; Kim, H.-K.; Kim, T.-H.; Lee, D.H.; Kim, K.J.; Kang, S.-H. A Feature Selection Approach Based on Simulated Annealing for Detecting Various Denial of Service Attacks. *Converg. Secur.* **2016**, *2016*, 1–18. [CrossRef]

91. Kang, S.-H.; Kim, K.J. A feature selection approach to find optimal feature subsets for the network intrusion detection system. *Cluster Comput.* **2016**, *19*, 325–333. [CrossRef]

92. Du, K.L.; Swamy, M.N.S. *Search and Optimization by Metaheuristics: Techniques and Algorithms Inspired by Nature*; Springer: Basel, Switzerland, 2016; pp. 1–434.
93. Lalehpour, A.; Berry, C.; Barari, A. Adaptive data reduction with neighbourhood search approach in coordinate measurement of planar surfaces. *J. Manuf. Syst.* **2017**, *45*, 28–47. [CrossRef]
94. Haq, A.A.U.; Wang, K.; Djurdjanovic, D. Feature Construction for Dense Inline Data in Semiconductor Manufacturing Processes. *IFAC-PapersOnLine* **2016**, *49*, 274–279. [CrossRef]
95. Christ, M.; Kempa-Liehr, A.W.; Feindt, M. Distributed and parallel time series feature extraction for industrial big data applications. *arXiv*, 2016; arXiv:1610.07717.
96. Benjamini, Y.; Yekutieli, D. The control of the false discovery rate in multiple testing under dependency. *Ann. Stat.* **2001**, *29*, 1165–1188.
97. Dheeru, D.; Taniskidou, E.K. *UCI Machine Learning Repository*; School of Information and Computer Sciences, University of California: Irvine, CA, USA, 2017.
98. Christ, M.; Braun, N.; Neuffer, J.; Kempa-Liehr, A.W. Time Series FeatuRe Extraction on basis of Scalable Hypothesis tests (tsfresh—A Python package). *Neurocomputing* **2018**, *307*, 72–77. [CrossRef]
99. Wang, W.; Liu, J.; Pitsilis, G.; Zhang, X. Abstracting massive data for lightweight intrusion detection in computer networks. *Inf. Sci.* **2018**, *433–434*, 1339–1351. [CrossRef]
100. Fan, J.; Han, F.; Liu, H. Challenges of Big Data analysis. *Natl. Sci. Rev.* **2014**, *1*, 293–314. [CrossRef] [PubMed]
101. Campos, J.; Sharma, P.; Gabiria, U.G.; Jantunen, E.; Baglee, D. A Big Data Analytical Architecture for the Asset Management. *Procedia CIRP* **2017**, *64*, 369–374. [CrossRef]
102. Nikolaidis, K.; Goulermas, J.Y.; Wu, Q.H. A class boundary preserving algorithm for data condensation. *Pattern Recognit.* **2011**, *44*, 704–715. [CrossRef]

Article

The PLC as a Smart Service in Industry 4.0 Production Systems [†]

Reinhard Langmann [1,*] and Michael Stiller [2]

[1] Faculty of Electrical Engineering & Information Technoloigy, Hochschule Duesseldorf University of Applied Sciences, 40476 Duesseldorf, Germany
[2] Fraunhofer Institute for Embedded Systems and Communication Technologies ESK, 80686 Munich, Germany; michael.stiller@esk.fraunhofer.de
* Correspondence: langmann@ccad.eu
† This paper is an extended version of paper published in the 14th International Conference on Remote Engineering and Virtual Instrumentation, NY, USA 15–17 March 2017.

Received: 17 July 2019; Accepted: 6 September 2019; Published: 11 September 2019

Abstract: Industrial controls, and in particular, Programmable Logic controllers (PLC) currently form an important technological basis for the automation of industrial processes. Even in the age of industry 4.0 and industrial internet, it can be assumed that these controllers will continue to be required to a considerable extent for the production of tomorrow. However, the controllers must fulfill a range of additional requirements, resulting from the new production conditions. Thereby, the introduction of the service paradigm plays an important role. This paper presents the concept of smart industrial control services (SICS) as a new type of a PLC. As a distributed service-oriented control system in an IP network, a SICS controller can replace the traditional PLC for applications with uncritical timing in terms of Industry 4.0. The SICS are programmed as usual in industry, according to the standard IEC 61131-3, and run in a SICS runtime on a server or in a cloud. The term Smart Service is introduced and the uses of SICS as a smart service, including a clearing system for the creation of new business models based on control as a service, are described. As a result, two different SICS prototype implementations are described and two application examples from manufacturing automation, as well as the evaluation of the real-time features and the engineering of a SICS controller, are discussed in the paper.

Keywords: control service; smart service; control as a service; cloud-based control system; automation system

---

1. Introduction

Industrial controls, and in particular, PLC controllers currently form an important technological basis for the automation of industrial processes. Even in the age of industry 4.0 (I40) and industrial internet, it can be assumed that these controllers will continue to be required to a considerable extent for the production of tomorrow. However, the controllers must fulfill a range of additional requirements, resulting from the new production conditions.

When applying Industry 4.0 principles [1], high-quality networked production systems result, based on cyber physical systems (CPS), also referred to as cyber physical production systems (CPPS). A series of I40 requirements are placed on the future controllers used in these systems. These include:

- Introduction of the service paradigm in production automation (production services);
- Autonomy, reconfigurability and agility (plug and work);
- Overcoming the strict information encapsulation of controllers;
- Networking in local and global networks;

- Interoperability between heterogeneous control systems;
- Dependencies are to be changeable dynamically at runtime;
- Use of models for the development of "higher-quality" control approaches;
- Orchestration of heterogeneous controllers.

Current PLC controllers cannot yet fulfill the majority of these requirements or can only do so on a rudimentary basis or at extremely high expense.

The paper describes the concept and two prototype implementations for a new type of a PLC controller in which the controller functions (control programs) will be implemented as smart control services in a cloud. The programming of this new PLC occurs as is usual in industry, pursuant to the standard IEC 61131-3.

2. State-of-the-Art

Resulting from the historical development of PLC controllers, they have been developed as proprietary device systems that are operated locally under real-time conditions. If a networking of these controllers is necessary from a user viewpoint, proprietary protocols that operate on top of TCP/IP or standardized protocols, such as Modbus TCP, Profinet, etc., are used for this. The standard technologies widespread from the Internet and Web have so far hardly played any role for PLC controllers.

For a number of years, however, a transformation has been underway, with PLC manufacturers increasingly integrating Information and Communication (IC) technologies from the web in their systems, such as web server and HTML pages for diagnosis and configuration, in order to adapt the controllers incrementally to the new requirements.

Four different approaches to make PLC controllers I40 compatible can essentially be revealed from state-of-the-art technologies. These include the introduction of basic web technologies, the global networking of process data, the introduction of service principles and the virtualization of PLCs.

2.1. Introduction of Basic Web Technologies

Most of the newer PLC controllers already contain a web server and special HTML pages on the device—enabling a browser-based configuration and diagnosis of the controller. Process data or program variables form the control program and can be read, sometimes also written, with restrictions. The solutions are proprietary and adapted to the relevant controller. Open and consistent web interfaces are not available. The above I40 requirements cannot, therefore, be fulfilled.

2.2. Global Networking of Process Data

For integration of the PLC controllers in supervising, management and coordination systems (e.g., SCADA or MES systems), which are partly based on web technologies, additional modules are integrated in the PLC controllers, enabling a bidirectional and event-based process data transmission between the controller, supervisor and management system. Those include solutions such as the web connector with the MQTT broker in WAGO controllers [2], or access to controllers that already contain an OPC UA server (e.g., Siemens PLC S7-1500).

These solutions also involve proprietary and closed control-integrated modules. Although the modules utilize web technologies, they cannot be transferred to other controllers. The global process data communication is used for HMIs (human machine interfaces) and/or supervisor and coordination functions in the higher level of the automation hierarchy (e.g., plant management level). Authoritative statements regarding the time response of the process data transmission are not available. However, different statements result from the reaction times (latencies) of 200 to 500 ms, or greater. Open and consistent web interfaces are not available. The above I40 requirements can only partly be fulfilled, sometimes with high adaptation expenses for integration into a CPPS.

2.3. Introduction of Service Principles

Based on the I40 requirement for the service capability of an I40 controller, some projects [3,4] are involved with the integration of service functions in PLC controllers. Thus, the Device Protocol for Web Services (DPWS) enables, as standardized protocol, service-based access to PLC controllers [5] also for reading/writing process data. The internal functional system of a PLC is to be equipped correspondingly for that, with the support of the controller's manufacturer.

However, the DPWS solutions have a principle disadvantage: Instead of reducing or removing the information encapsulation (I40 requirement), further functionalities (service functions) are encapsulated in the controller. Moreover, DPWS uses the very heavy-duty and complex Microsoft web service protocols. The attainable transmission time of process data via a global network, therefore, tend to be in the upper range. It is difficult to obtain any definite data.

2.4. Virtualisation of PLCs

Current R&D work deals with the virtualization of complete PLC controllers and their outsourcing into the cloud. A scalable control platform for cyber-physical systems in industrial productions is researched and realized in [6]. In [7], a cloud-based controller is presented, which also uses a virtual control system in an Infrastructure as a Service (IaaS) cloud. The work of [8] also uses virtualized PLC controls in the cloud and connects these to OPC UA-based automation devices using web technologies.

Problems with the virtualization of PLCs result especially from the fact that already available manufacturer-specific PLCs are virtualized. These controllers, however, are closed systems, which were originally not developed considering the aspects of web technologies. Adjustments, modifications or extensions of these controllers by third parties are hardly possible. Functionality cannot be resolved as services. The flexibility of virtualization is very limited.

Reference [9] proposed a methodology for converting an automation plant managed by PLCs onto an EFSM control module (EFSM—extended finite state machine) that is driven by single board computers or SoC (system-on-a-chip). The EFSM Control Module can use IoT devices, but in that solution, the functionality (control program) cannot be resolved as a service from the cloud.

In summary, it can be estimated that there are different solutions and efforts to equip PLC controllers with additional functions in order to be able to use the controllers in an Industry 4.0-type IP network. To this end, the known work already uses web technologies in part, in a manufacturer specific and/or limited way, and increasingly also tries to use the service principle and cloud structures as a new paradigm for the realization of control functions. However, there are still the following deficits that result in corresponding needs for research:

- Although web technologies are used, a flexible distribution of the structure and function of the control functionality is not used. The information encapsulation of industrial control programs in local or virtualized devices (PLCs) is not called into question.
- For smart control services using cloud technologies as an essential feature of a future networked industry, systematic investigations, architectures, interfaces and demonstration solutions are lacking.
- Available standard technologies from the world of IP networks for increasing flexibility and efficiency are not or insufficiently used in the control level.

3. Concept for a Smart Control Service

The following section describes the concept of a smart control service and introduces the basic model of a SICS.

3.1. Control Classification

To assess the I40 capabilities of a PLC controller, categories were introduced which divide an industrial controller according to its abilities: service ability (SA) and control locality (CL). The properties are divided according to class C (Controller) = <SA><CL>. With the proposed methodology, I40 control classes can be defined and structural configurations for a PLC as a smart industrial control service (SICS) can be indicated (Table 1).

Table 1. Industry 4.0 (I40) Capability of a PLC controller.

Class	Service Ability (SA)	Control Locality (CL)
0	No service	All control programs are encapsulated locally in the PLC hardware.
1	Services only for non-critical and overarching functionalities	Some control programs that include non-critical and overarching functionalities are not located on the local hardware but are instead distributed to other systems (for example, in the network).
2	Services for most functions available	Most control programs are distributed in the network. Control programs which are critical in terms of time and safety remain in the local PLC hardware.
3	All control functions as services.	All control programs are distributed in the network. Third instances can access all the control algorithms in real time.

Looking at a PLC as a CPS component, the traditional IEC 61131 control program (CP) can be divided into three parts:

- Basic functional program part (CP basic—CPb);
- A program part which performs superior, administrative and/or user interface functions (CP supervisory—CPs);
- Critical part of the program regarding real-time and security (CP critical—CPc).

In order to evaluate the I40 capabilities of a PLC or a control system, this 3-part structuring of the control program is used. Figure 1 shows the structure of such a PLC.

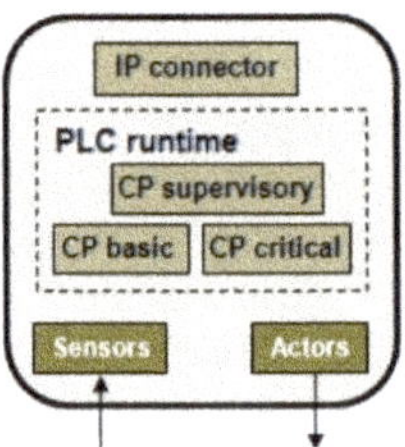

Figure 1. Structure of a PLC as a cyber physical systems (CPS) component.

If the control system, as shown in Figure 1, is used as the basis, and modified as a result of the increasing displacement of the control programs into a cloud as services, it leads to the evolution of a PLC as a CPS component Industrial Control, as shown in Figure 2.

Three types of CPS components (Figure 2) are produced according to the aforementioned disassembly of the PLC program into three parts:

(a) The controller hardware only implements the program components CPb and CPc. The traditional runtime environment of a PLC is still required.

(b) For safety reasons, only the CPc program parts are implemented in the controller hardware. The classic PLC runtime machine can be used but this is not mandatory. The implementation of the CPc could also be carried out with specific embedded program parts (e.g., in C).

(c) The controller hardware no longer contains a control part, but only sensors and actuators. All control programs are distributed as smart services in the network.

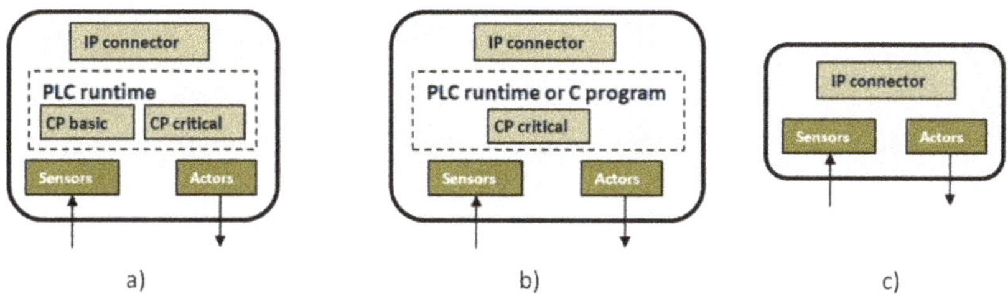

Figure 2. Evolution of a PLC as a CPS component Industrial Control. (**a**) PLC with program components CPb and CPc (**b**) PLC only with CPc component (**c**) The PLC hardware contains only sensors and actuators

The Service Ability considers the ability of a controller to utilize control functionalities (control programs) as services. According to Table 1, the program parts CPb, CPs and CPc can be distributed unequally. In a class C11 controller, for example, the uncritical and overlapping functionalities (CPs) are not located on the local PLC hardware, but distributed on other systems in the network (corresponds to a traditional, distributed control system). However, part of the control programs could also be used as a service from the cloud.

As the focus of the research is on cloud-based control services, work is focused on C3x controllers, in particular C33 (Figure 3). With this I40 control class, the complete control program, with all program parts (including the PLC runtime), is outsourced as services to a cloud or to any server in the network. The control hardware consists only of sensors and actuators connected to an IP network (see Figure 2, type c).

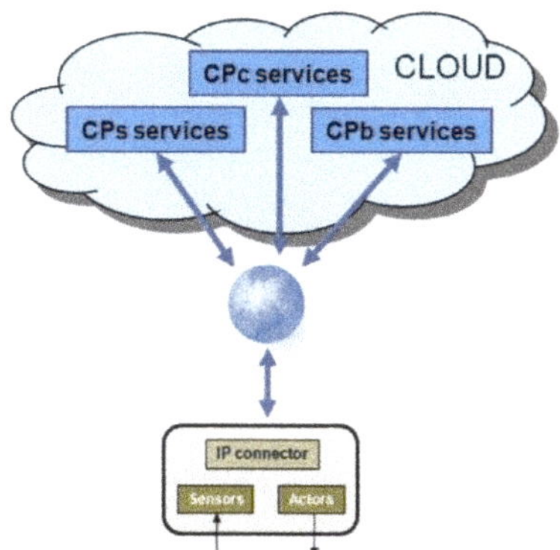

Figure 3. C33 controller.

If one realizes C3x controllers, such as the C33 controller, according to Figure 3 via smart services (see Section 4.3), such a control system could be referred to as a smart industrial control service controller, or in short, a SICS controller.

3.2. SICS Base Model

A SICS base model must take into account both the aspects of control engineering and the web technology features.

From a control engineering point of view, a SICS controller based on a traditional PLC consists of the following components (Figure 4):

- SISC program: IEC61131-3 control program in the PLCopen XML notation. It includes only the program and the variables, but not the I/O configuration.

- SICS runtime: Execution environment for the SICS program. It can be cycle controlled or event-based.
- SICS router: Device or I/O configuration for a SICS controller; i.e., it is determined which CPS components (which automation devices) are connected to the controller. Since the real devices are already virtualized in an Internet-based environment via the CPS component, a SICS configuration only includes the assignment between the absolute IN/OUT addresses in the SICS program and their assignment (routing) to the IN/OUTs of the CPS components. A SICS device configuration is, therefore, also referred to as a SICS router.

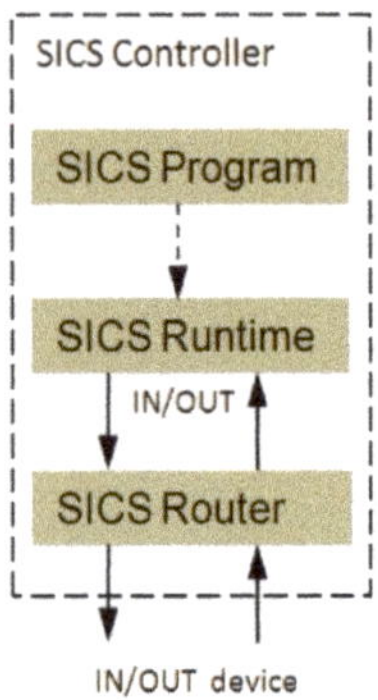

Figure 4. General structure of a smart industrial control service (SICS) controller.

By separating of SICS program, SICS runtime and SICS router in a SICS controller, and distributing the components across an IP network using cloud technologies, it is possible to change, in real time, the control program (control algorithm). This also applies to the device configuration (e.g., for replacements of modules or plug and work). A SICS program and SICS router can be exchanged on-the-fly during one program cycle.

For the identification of a viable SICS basic model, it is also necessary to show possible solution variants for a SISC controller, starting with the basic principles on the web, and then to reflect those in the available web technologies. As explained in detail in [10], if the control technology based SICS structure is adapted to the web-based functional systems, four general SICS base models are obtained:

(1) Server mode (SM): The SICS router is linked to fixed CPS components in a configuration process. After the SICS runtime has been started via the client, the SICS router automatically connects to the associated CPS component via the IP network and the SICS runtime executes the SICS program (Figure 5a).
(2) Server-based mixed mode (SMM): Before starting the SICS runtime, a SICS router is loaded from the server to the client. After the SICS runtime is started, this router dynamically connects the CPS component with the SICS runtime on the server. All process date from the automation device are now routed to the server via the client (Figure 5b).
(3) Client-based mixed mode (CMM): The control program runs in the SICS runtime on the client, but the communication to the CPS component runs over a configurable SICS router in the server (Figure 6a).
(4) Client Mode (CM): SICS runtime and SICS routers are executed as one instance on the client (web browser). The client is an inherent part of the SICS control system and is necessarily required for executing the control program. The server is no longer required at the runtime (Figure 6b).

Figures 5 and 6 illustrate the four basic models of a SICS control system.

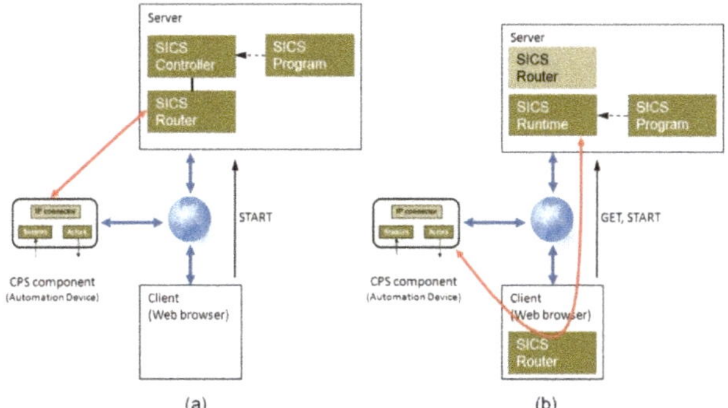

Figure 5. Component structure and communication paths for server-based SICS solutions. (**a**) Server mode (SM); (**b**) server-based mixed mode (SMM).

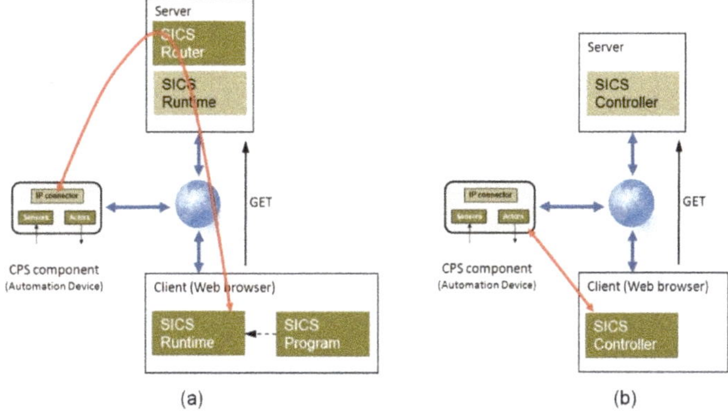

Figure 6. Component structure and communication paths for client-based SICS solutions. (**a**) Client-based mixed mode (CMM); (**b**) client mode (CM).

During the execution of the control program, the automation device, as a CPS component, has to be connected with the SISC runtime engine on the server/client by as IP network with as low latency and as much reliably as possible (see also the marked communication paths in the Figures 5 and 6).

3.3. Control Services

The control features of a SICS controller are no longer available as classic control functions, but rather as control services according to the service paradigm. A SICS (literally: smart industrial control service) can, thereby, use all the features of cloud computing, thus enabling the creation of new business models, such as the rental of control services.

In terms of information technology, SICS has to be produced, parameterized, distributed, stored and recalled as objects by means of software methods. Since the components of a SICS controller are no longer available as hardware, but only as software objects in the IC or Internet/Intranet and generally are also stored there in databases, it makes sense to use data models for the modeling of the SICS's service architecture. Figure 7 shows the SICS structure, for the controller depicted in Figure 4, as an entity relationship diagram (ERD).

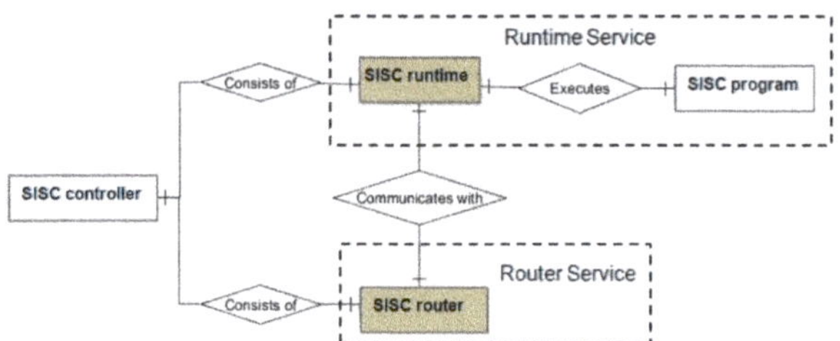

Figure 7. SICS services architecture according Figure 4 presented as an entity relationship diagram (ERD) diagram.

According Figure 7 a SICS controller is realized with two services:

- Runtime service (SICS runtime or SICS RT);
- Router service (SICS router or SICS-R).

A SICS controller usually requires both services. However, it can also be realized by only a runtime service if the connected CPS component (automation device) already has a SICS runtime interface. The SICS router service is comparable to the I/O configuration part of a classical PLC.

Both SICS services are built according to the principle of web-oriented automation services (WOAS) [11].

If required, the SICS service instances can be displayed in the client (web browser) for operation and visualization. Therefore, both services include a graphical user interface (HMI proxy). Via that parameter, the HMI proxy can be switched on or off.

3.3.1. SICS Runtime

Corresponding to the state machine in a traditional PLC, the SICS runtime also has a defined sequence behavior as the most important component in order to execute a control program. A SICS-RT implements the operating states, listed in Table 2, in accordance with a PLC.

Table 2. Operating modes of a SICS runtime.

Mode	Description
EXISTENT	The SICS-RT service is instantiated and is available for operation.
READY	The PLCopen XML program, whose URL is specified in the SICS-RT instance, is loaded. Depending on the type of execution of the control program (interpretative or as a compiled program), the XML program is loaded, translated or compiled in this state and stored in an internal format suitable for runtime operation.
AUTOMATIC	In the AUTOMATIC state, the control program is processed and the internal I/O image of the control is updated.
STEP	The STEP status allows step-by-step processing of the PLC program for debugging purposes. Each current step is displayed in the user interface of the SICS-RT instance (HMI proxy) and/or in the web console of the browser.
SUSPEND	The SICS-RT is stopped and all outputs are set to FALSE/0. The loaded control program is deleted. The status is automatically exited after approximately 2 s.

A SICS runtime can be operated in cyclic mode and event-based mode. In cycle mode, the I/O image is updated, equivalent to a traditional PLC. In event-based operation, the control program is executed only when the value of an input variable changes or an internal event occurs (for example, the execution of a timer).

3.3.2. SICS Router

The I/O configuration service of a SICS controller (SICS router) is separated from the SICS runtime for the following reasons:

- Securing a dynamic reconfiguration; i.e., in the case of an identical control program, the I/O configuration can be changed within a program cycle.
- Identical machines/systems can be operated with the same control program, despite different I/O modules.
- A distributed separate configuration service forms the basis for a future automatic IIoT-based device configuration (IIoT—industrial internet of things).

SICS routing works according to the following two principles:

- A SICS program (PLCopen XML program) works with absolute I/O addresses.
- The SICS router connects the absolute I/O addresses to the real I/O addresses of the devices (CPS components).

Figure 8 illustrates the functionality of a SICS router.

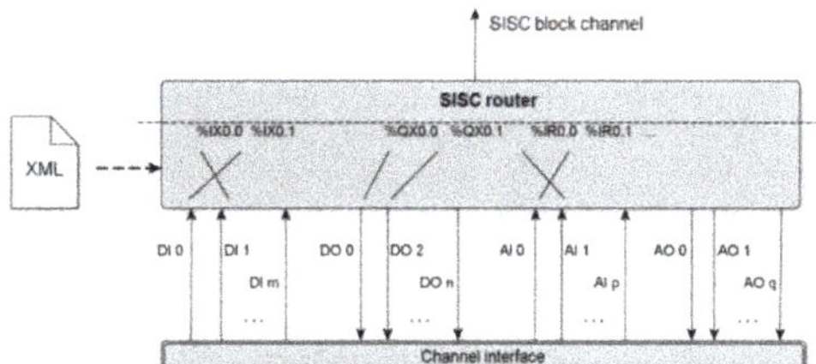

Figure 8. SICS router.

The digital and analogue inputs and outputs of a device, connected via the channel interface, are routed to absolute I/O program addresses and transferred to the SICS runtime via a SICS block channel. The routing rules (interconnection matrix) are defined via an XML file.

4. Results

To evaluate the SICS concept, three prototypical implementations were built: A SICS controller in the server mode (SM—Figure 5a), a SICS controller in the server-based mixed mode (SMM—Figure 5b) and a SICS controller in the client mode (CM—Figure 6b).

In the following, only the two SICS controllers in SMM and SM mode are considered in more detail, since these could be particularly suitable for future control applications.

4.1. SICS Controller in Server-Based Mixed Mode (SMM)

In the case of a SMM solution, the SICS runtime is executed in the server (cloud) as an instance and the SICS router in the client (browser) as an instance.

Here, as well, a direct process data communication takes place only between the client and devices. Between SICS router and SICS runtime, there is a special bidirectional block channel for the transmission of I/O images. This channel transmits the process data as strings over a secure WebSocket. The SICS runtime is operated via an HMI proxy on the client. Figure 9 illustrates the SMM solution.

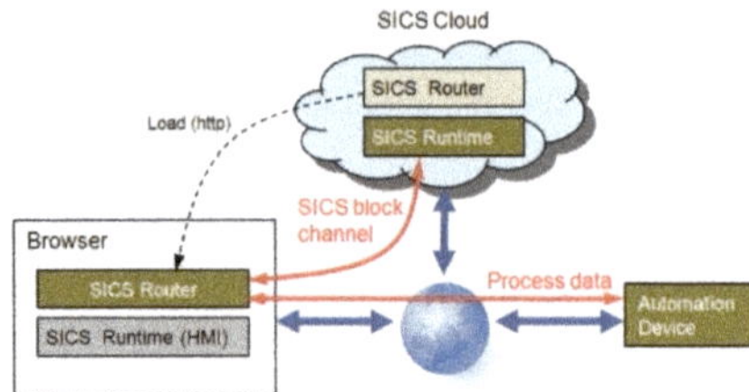

Figure 9. SICS controller in the server-based mixed mode (SMM).

In terms of technical implementation, the SICS-SMM controller is a distributed elaborate solution. However, process data connection to the devices can also be performed locally and the SICS runtime can use the full performance of the server anyway. A dynamic re-configuration is easy and possible.

Figure 10 shows the simplified implementation structure of a SICS-SMM controller at runtime.

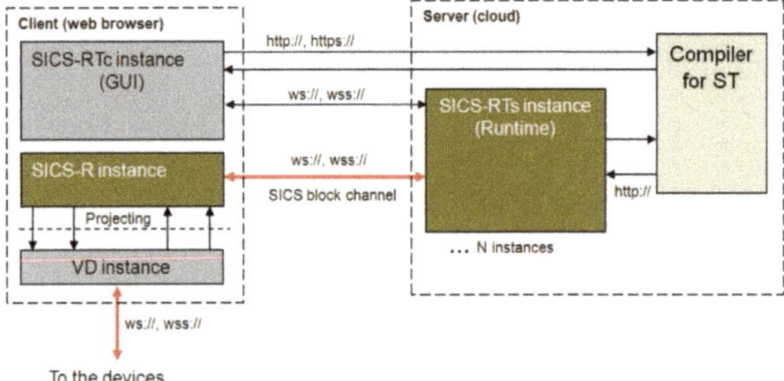

Figure 10. Implementation structure of a SICS SMM controller at runtime (s—server-based part; c—client-based part).

After opening the corresponding web page in the client, a SICS-RTs instance is generated dynamically as a runtime engine in the server, and simultaneously, the corresponding control program is compiled into an executable JavaScript (JS) program by the structured text (ST) compiler in the server. This JS program is then loaded into the CICS-RTs instance.

The following instances are created dynamically in the client:

- VD instance: This instance operates as a virtual device according to [12] and connects to the CPS components (devices).
- SICS-R instance: Routes the absolute I/O addresses to the physical device addresses.
- SICS-RTc instance: Serves as an HMI proxy for the visualization/operation of the CICS-RT instance in the server.

Generally, any number of SICS-RTs instances can run in the server, each of which can work with other clients. The ST compiler works together with all SICS-RTs instances and can also be moved to another server (cloud).

Some of the advantages of the SMM solution are:

- The configuration of a SICS SMM controller can be carried out completely in an IoT or IIoT platform because all required instances (in the server and in the client) can be created and deleted via the client.
- Service-like C3x controllers can be built, because all SICS instances are available and manageable as separate service objects. The cloud model SaaS (software as a service) can be implemented.

- Different SICS routers, which can also be generated dynamically, enable a dynamic reconfiguration of the control system online without time delay. This forms the basis for adaptive and self-adaptive systems.
- SICS runtime can use the full performance and stability of the server.

A detailed description of the implementation for the SICS-SMM is documented in [13].

4.2. SICS Controller in the Server-based Mode (SM)

The SICS SMM controller always requires an active web browser (client) during runtime, as this is where I/O routing to the devices takes place. This is eliminated with the SISC-SM controller: All required service entities of a SICS controller run exclusively in the server.

Based on the general structure in Figure 5a, a SICS controller in the server-based mode (SM) is shown in Figure 11.

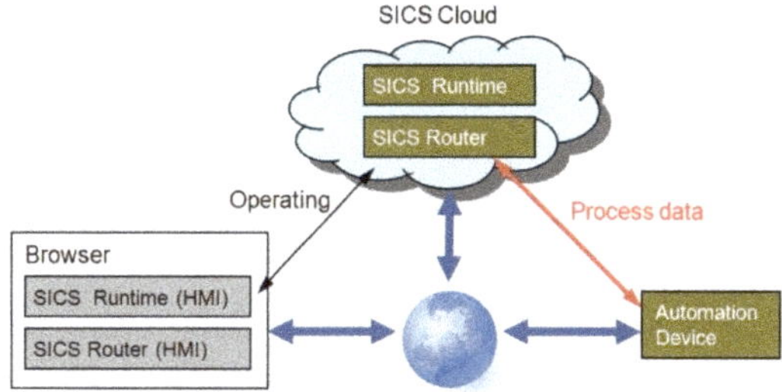

Figure 11. SICS controller in the server-based mode (SM).

Process data communication takes place only between server (cloud) and devices. The client (web browser) has no influence on this communication. In the client, which is temporarily required for operating purposes, there are only HMI proxy objects for operation and visualization of the SICS runtime and router instances.

The prototype implementation is based on the SMM structure of Figure 10, and it relocates the SISC router to the server. In addition, a SISC SM manager is required in the server, via which the SISC runtime instances must be generated explicitly by means of an administration tool. The simplified implementation structure of a SISC-SM controller is shown in Figure 12. A complete description of the SISC-SM can be found in [14].

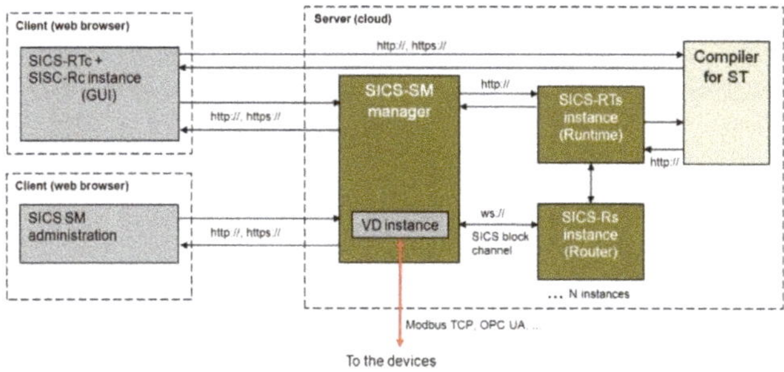

Figure 12. Implementation structure of a SICS SM controller at runtime (s—server-based part; c—client-based part).

A SISC-SM controller is operated and visualized via a web page. However, this is only temporary if necessary for monitoring the control. The SISC controller runs independently in the server or in a cloud after startup. As with the SICS-SMM solution, any number of SISC controller instances can be distributed on any servers, e.g., also running on the local network via Edge or Fog computing.

Some advantages of the SM solution are:

- Controllers can run as software instances independent of web browsers distributed on any server system.
- Communication to the devices may be via any TCP protocols (e.g., Modbus TCP or OPC UA), as far as is supported by the servers. The devices do not need a web-enabled communication.
- Different SICS router instances enable dynamic reconfiguration of the control system online without delay.
- The SICS runtime can use the full performance and stability of the server.

However, the prototype implementation still has a few disadvantages:

- The configuration of a SICS-SM controller cannot yet be realized completely via an IoT or IIoT platform. There is still a need for an additional administration tool.
- It is basically possible to set up C3x controllers as a smart service, since all SICS-SM instances are also available and manageable as separate service objects. However, the complete SaaS model (software as a service) is much more difficult to implement than in a SISC-SMM solution. Therefore, an implementation is not yet available.

4.3. PLC as a Smart Service

The SISC controllers described in the previous two sections as a new kind of PLC in the Industry 4.0 era make sense only if the controllers available as services can now also be handled, managed and billed as a smart service.

According to [15], smart services are characterized by the following properties:

- Connection to a technically suitable service infrastructure (e.g., IoT or IIoT platform or cloud system).
- Efficiency for the end customer directly or indirectly with the help of a service provider.
- Transparency, disclosure and constant discourse with the users.

Thus, according to those characteristics, a management and execution environment (middleware) is required with which to configure and operate a SICS control system. This could be, e.g., a web-based SCADA system, a cloud system or an IoT or IIoT system, which is provided by a third party as a provider.

The IoT platform FlexIOT (http://www.flexiot.de) is used as middleware for the two example implementations. This IoT platform FlexIOT is based on the research results of the project WOAS [16] and provides a flexible, extensible and easy-to-use kit for the IoT and IIoT. With FlexIOT, SICS controllers can be designed as web-based, and connected and operated with other automation services (HMI, SCADA, etc.).

4.4. Clearing of SICS Services

If you want to offer PLC functions as smart services or control as a service (CaaS) in the context of a business model, you have to create interfaces and make them transparent and open, in order to give third-parties the opportunity to connect their business models to these interfaces

In the IC Industry, this has already been standard for some years and is being used with a strong upward trend. In the classical automation industry, these service models are so far largely unknown or, for various reasons, difficult to implement or even unwanted.

What does this look like for PLCs as smart services, according to the SICS concept?

1. The PLC functionality as a service with disclosed and well-defined interfaces is available with the SICS controllers.

2. A third party may integrate and offer the SISC services in his cloud system or IoT system (e.g., FlexIOT).
3. For implementation as a business model, however, the SICS services must now also be able to be settled appropriately.

The implementation of topic 3 requires special consideration, since previous SaaS systems (including various IoT platforms) charge IC services very differently.

For automation services (control systems, HMI, SCADA, alarm monitoring, etc.), business models make sense from the point of view of future globalized, convertible and digitized production, which can bill individual services on a runtime basis. Since machines and stations normally are not permanently operated with all available functionalities, a billing after runtime significantly improves the cost-efficiency of the user, and also ensures a cost transparency of the services used.

So far, the authors, however, do not know a system in which automation services can be billed on runtime basis. Most billing is done with a flat rate and/or according to the number of integrated devices and/or transmitted data volume. One reason for this is certainly that on one hand, in the usual server-centered systems a customer and service-specific runtime determination within the server is very expensive and on the other hand, the runtime of services in the field of IC plays rather a subordinate role.

For the settlement of the SICS services, a clearing system has therefore, been developed for the FlexIOT platform (FlexIOT Service Portal) with which the service's runtime can be determined and settled, at least for a SICS SMM controller.

The clearing system runs independently of the FlexIOT portal and uses its own database for the billing data. Figure 13 illustrates the basic structure of the interaction between the FlexIOT portal and the clearing system.

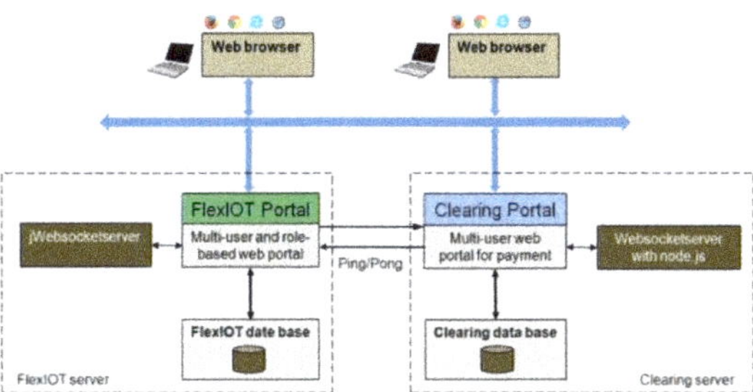

Figure 13. Basic structure of the interaction between the FlexIOT portal and the clearing system.

In RUN mode of the FlexIOT portal, i.e., if a user uses a configured FlexIOT functional system (for example, a SICS control function or an HMI panel), the runtime measurement of the FlexIOT services in the clearing system takes place via a ping-pong mechanism. The services can be analyzed in detail in the clearing portal and settled via PayPal. Detailed information about the clearing of described smart services, you can find in [17]. The clearing portal can be tested online on http://www.flexiot.de:3000 and uses the same user accounts as the FlexIOT portal.

Figure 14 shows a part of the billing/analysis webpage in the clearing system for a sample project of a user in which 5 SICS-SMM controller instances were used for several minutes. A total of 106 process datums from an automated station were connected to the SISC controllers.

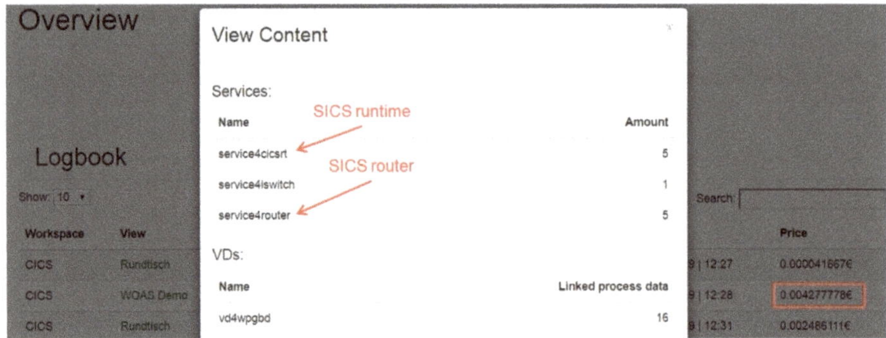

Figure 14. Settlement and analysis in the clearing system for an example with five SISC-SMM controller instances.

5. Discussion

The control programs for the SISC-SMM and SICS-SM prototype were created with the programming system CoDeSys from 3S in the language ST and exported as PLCopen XML programs for execution in the SICS runtime. The connection of the I/Os of the automation devices occurs via WebSocket/OPC or MQTT/Modbus TCP.

In the following, an application example for a SICS-SM and a SICS-SMM will be discussed.

5.1. Application Examples of a SICS-SM Controller

The application example 1 consists of a mounting system for model cars with five assembly/ disassembly stations (Figure 15).

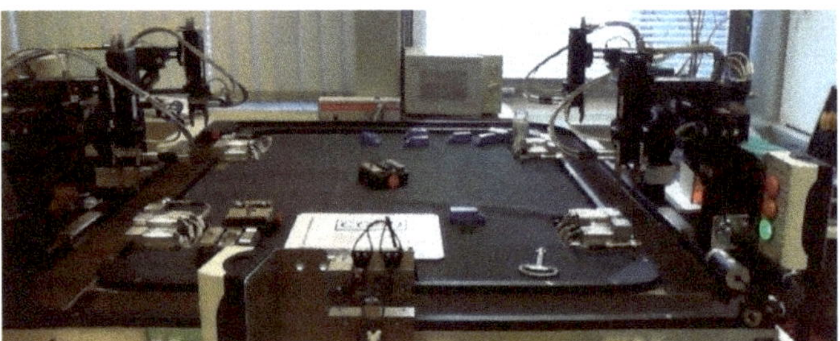

Figure 15. Mounting system for model cars with five assembly/disassembly stations.

In standard mode, the system is operated with five classic PLC controllers (one PLC per station) from Siemens and Phoenix Contact. The average production time for a model car is 37 s.

For the SICS test, all five stations were connected to the Internet via device gateways/web connectors for OPC DA [16]. This means that all 135 process datums of the stations are available in the IP network. The SICS-SM control system for the station was designed in the web browser via the IoT Platform FlexIOT and by use of a special administration tool. A separate SICS controller instance is used for each of the five stations.

With the SICS-SM controller, the production time for a model car increases by approximately 10%. However, this allows the removal of five PLC controllers as hardware and their associated maintenance.

5.2. Application Example of a CICS-SMM Controller

In application example 2, a SICS-SMM controller controls a working cell, consisting of two processing and test stations, and a loading robot. Figure 16 shows the technological structure for this example.

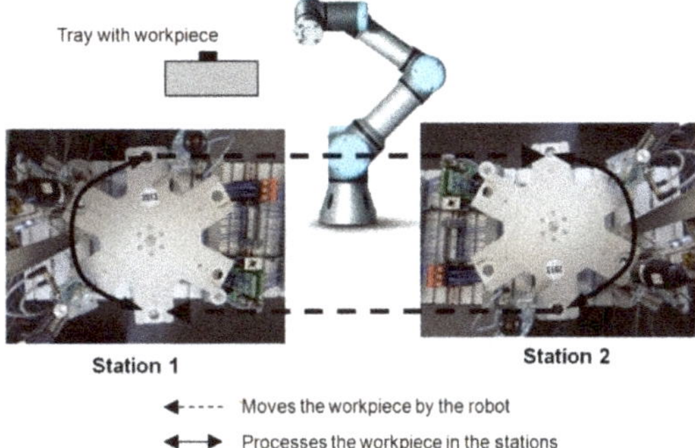

Figure 16. Technological structure of application example 2.

Two SICS controller instances control the two processing stations. Another SICS instance is responsible for coordinating the robot with the two stations. The connection of the respective SICS router instances to the devices of the two stations takes place via a universal gateway as a web connector [18]. The Modbus TCP interface of the robot is connected to the Internet via WebSocket using a device gateway, realized by means of Node-RED. The application was successfully presented at exhibition SPS/IPC/Drives 2016 in Nuernberg (Germany).

In contrast to example 1, the application example with the SICS SMM controller can be configured completely in the IoT platform. Additional administration tools are not required.

Using the clearing portal described above, the runtime of the control instances for the operation of the system can be fine-tuned analyzed and billed.

The possible realization of the further requirements, which are necessary for future industrial 4.0 solutions, was also demonstrated with the application examples.

- Orchestration of heterogeneous industrial control systems: Easy connection with robot control, proprietary I/O modules and servo drive from different manufacturers.
- Fast reconfiguration and agility (plug and work): The activation of completely new control programs and the I/O configuration is possible in a few seconds on the fly. Studies on synchronization and "safe state" issues are still pending.
- Remote control and improved manufacturer service: Remote control is an integral part. All data are already available in the cloud for diagnostic purposes without additional effort.
- Simulation support for planning and process optimization: Simulation and visualization models of the systems can be easily connected via the cloud.
- Application of smartphone and tablet PCs: Through browser technology and responsive design, all devices are ready to be used for operation.

5.3. Realtime Features

A SICS control system uses IP networks for data transmission, regardless of the solution variation. From the perspective of an automation technician, these networks are a priori neither reliable nor deterministic, and are not within the jurisdiction of the respective technical automation solution. Extensive time measurements for different communication structures were, therefore, performed for both SICS prototypes [19].

A practice-oriented method was chosen for the time measurements, which allows direct statements about the reaction time of the described SICS controllers. Figure 17 shows the measurement setup for the examples.

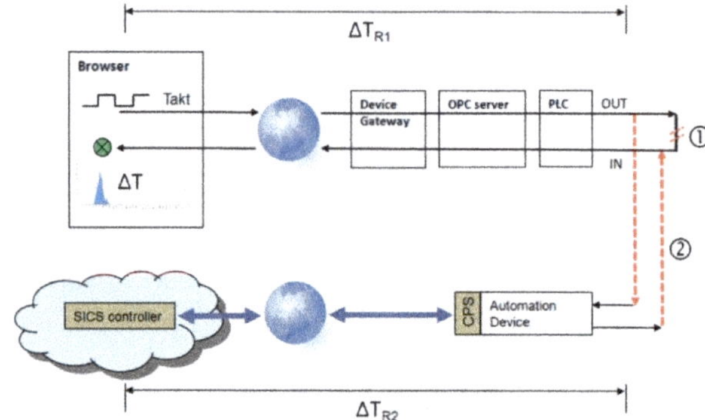

Figure 17. Measurement setup to determine the reaction time of a SICS controller.

In the upper part of the measurement setup ①, an HTML page generates a pulse frequency with a JavaScript generator, which is sent to a PLC a digital signal. This digital output is connected by a short circuit bridge to a PLC input, which sends its value back to the HTML page. A time difference measurement is then carried and recorded in a histogram. This gives the reaction time ΔT_{R1}. Using this measurement method, the time characteristic of the process data transmission is obtained over a longer period of time and also detects the variance of the measurements over a probability curve.

In the second measuring part ②, the short circuit bridge is opened at the PLC and the I/O signal is connected to an automation device (e.g., I/O module), which has a CPS interface and is controlled by a SICS controller. As a result, a histogram is displayed in the browser for the entire reaction time of the measurement setup

$$\Delta T = \Delta T_{R1} + \Delta T_{R2}.$$

The response time of the SICS controller thus results in

$$\Delta T_{R2} = \Delta T - \Delta T_{R1}.$$

Figure 18 shows the reaction time of the SICS controller.

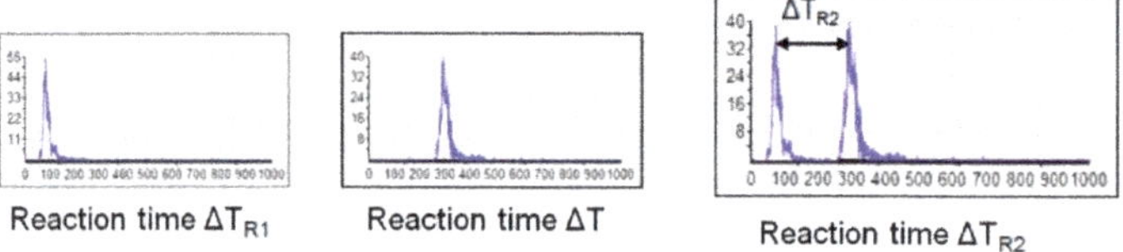

Figure 18. Determination of the reaction time of a SICS controller.

With respect to the real-time features, the following general statement can thus be made:

With a SICS-SM or SICS-SMM solution, response times of about 80–100 ms at a 95% probability can be achieved by a usual Internet connection.

If the SICS controller is operated only on the Intranet, response times of under 40 ms can be achieved.

Altogether, the statement can be made that technical processes with process times of >150 ms (simple assembly processes, temperature and mixing processes, climate and energy processes, etc.) can already be performed successfully from the cloud by means of a SICS.

Compared to a classic PLC, it is currently not possible to guarantee 100% compliance with a specific time requirement for a SISC controller using IP networks and the cloud, but only with a certain probability. This must be taken into account in the respective application.

The majority of the response time is caused by the run times in the network. The actual cycle time for the execution of the control program in the SISC runtime is less than 1 ms. Considering the use of new real-time networks, e.g., time sensitive networking (TSN), is expected to allow SISC controllers in the local or edge computing networks to control drive motors as well.

5.4. Security and Safety

The reliability, data security and machine safety aspects were considered in a study of the operability of a SICS controller. An analysis and determination of the protective measures required for SICS was carried out. In addition, the required methods and measures for securing the process data access to networked sensors/actuators were investigated and tested using the example of encrypted access (HTTPS or WSS).

Critical aspects considered include: Disturbance of the process data communication, disconnection, runtime error in the SICS controller, connection delay and/or connection interruption, breaking into the cloud and modifying the files for SICS services and breaking into the client and/or into the cloud and disruption of the SICS operation.

In general, it can be estimated that there are deficits compared to a conventional PLC, especially with regards to machine safety, but these depend on the respective operating and application environment. However, the use of edge computing or the local connection of the sensors and actuators to a SISC controller can certainly provide workable solutions.

Interested readers are welcome to ask the authors about the topic further documents and research reports (in German).

5.5. Engineering

An important objective in SICS is the use of IEC 61131-3 programs which, as is usual in the classical PLC technology, are created with professional programming systems, such as CoDeSys (3S), PC WORX (Phoenix Contact) or STEP7/TIA Portal (Siemens). The basis for implementing of IEC 61131 programming with SISC is the representation of the control programs in the uniform format PLCopen XML. For the programming systems PC WORX, CoDeSys, STEP7 (TIA Portal) and openPCS, a special research study [20] analyzed the conformity with the PLCopen standard; and the capabilities of those tools to export and import PLC programs as PLCopen-compliant XML programs. The results of the study can be summarized as follows:

- Classic PLC programs exported as PLCopen XML programs by CoDeSys in the notation structured text (ST) or instruction list (IL) exported by PC WORX can form the basis for a SICS control program.

openPCS is not suitable as a programming environment for SICS because of instability during export of a PLCopen XML program. By means of STEP 7 (TIA-Portal), presently it is not possible to export a PLCopen XML according the standard IEC 61131-3.

The exported PLCopen XML programs in the IL or ST notation must be converted into the internal SICS-RT programming language (JavaScript) for executability in the SICS-RT. Two versions were developed and tested:

Version 1:

For a SICS-CM (Figure 6b), the IL program was executed after being translated (transformation of the PLCopen XML program into an internally suitable form) by loading the program into the SICS-RT using an interpreter. The interpreter was an inseparable part of the SICS-RT.

Version 2:

A separate ST compiler was developed for the SICS-SMM and SICS-SM, which compiles the PLCopen XML programs into directly executable JavaScript programs. The ST compiler was based on server-side JavaScript with node.js and functions as a web service for all SICS instances. The compiler is housed in a docker container and can be operated on any server.

Regardless of the programming language notation in the PLCopen XML program, however, there are some limitations to SICS engineering compared to the traditional setup of a PLC program, since the PLC engineering world does not have a common area within the Internet world (at least not right now). Among others, these include:

- Online debugging with PLC programming tools available on the market is not possible. Therefore, additional web-based testing possibilities (e.g., step operation and the output of step commands in the browser console) were introduced into the SICS controllers.
- Absolute addresses must be used for the I/O process data in the program. This can be easily implemented during the creation of the IEC 61131-3 control program
- The use of libraries or several tasks is up to now not possible. Thereto the compiler should be extended.

The lack of online debugging in the source code of the IEC 61131-3 program makes testing a SICS controller more difficult and increases the time required to develop an error-free SICS control program.

In summary, for SICS engineering, the engineering applied to industry, with a slightly higher cost compared to traditional PLC programming, is possible and can be used safely in a transitional period for the first SICS application solutions. In the long term, however, it is essential for smart control services to develop appropriate cloud and web-based engineering tools that fit seamlessly into the globally networked world of Industry 4.0 and the Industrial Internet of Things.

6. Conclusions

Using the concept of smart industrial control services, a new type of industrial control was developed and tested that allows for the complete detachment of control function and associated equipment to globally distributed, cloud-based software control services. A SICS controller is operated by a classic IEC61131-3 control program, thus ensuring the interoperability and industrial compatibility of the control system. The application of the service paradigm for industrial control functions significantly increases flexibility, meets industry 4.0 requirements such as changeability, reconfiguration and autonomy, and enables new business models to lease control functions.

Examples are presented that make clear the distinctions between general and methodological issues of a SICS.

For testing and evaluation, prototypical implementations were deployed for a purely server-based SICS controller and for a mixed client/server-based SICS controller. Both SICS controller types were successfully tested in the context of application scenarios from production automation. An evaluation of the SICS applications showed that simple technical processes with process times of greater than 150 ms can already be controlled reliably over the Internet.

With SICS, previous hardware-oriented and centralized procedures for the control of automated devices, machines and systems (e.g., PLC controllers) can be distributed and used transparently for uncritical real-time conditions (e.g., environmental processes, logistics processes, energy processes, simple assembly processes) through IP-network-distributed software functions.

Among other things, further research work in the project will focus on increased data security in the system through consistent encryption of communication for each SICS component; reduction of the latencies for process data communication between the SICS controller; the automation devices through optimized web protocols; the use of TSN as a deterministic IP network; examination of the practicability (long-term testing) of a SICS control system for climate control and lighting control of an office building in everyday life; investigating the application of block chain and distributed ledger technologies for the identification management in a SISC control system; and last but not least, testing the business model for the leasing of control functions by an IoT platform with integrated smart SICS services.

Author Contributions: R.L. authored the main text and M.S. added some parts (e.g., testing a SICS-SMM controller), gave recommendations and suggestions and re-edited the text.

Funding: A part of this research work was funded by Federal Ministry for Economic Affairs and Energy, grant number IGF 18354N.

Conflicts of Interest: The authors declare no conflict of interest.

References

1. Kagermann, H.; Wahlster, W.; Helbig, J. *Recommendations for Implementing the Strategic Initiative INDUSTRIE 4.0*; Research union: Business-Science; National Academy of Science and Engineering: Munich, Germany, April 2013.
2. WAGO Kontakttechnik GmbH: Effective Upgrade: A Plug-In Makes WAGO Controllers IoT-Ready. Available online: https://www.wago.com/us/plc-mqtt-iot (accessed on 23 April 2019).
3. EU Project: SOCRADES. 2006–2009. Available online: http://www.socrades.net (accessed on 10 September 2019).
4. Colombo, A.W.; Bangemann, T.; Karnouskos, S.; Delsing, J.; Stluka, P.; Harrison, R.; Jammes, F.; Lastra, J.L. *Industrial Cloud-Based Cyber-Physical Systems*; Springer International Publishing: Zug, Swiss, 2014.
5. Microsoft: Introducing DPWS. 2015. Available online: https://msdn.microsoft.com/en-us/library/dd170125.aspx (accessed on 23 June 2018).
6. ISW of the University Stuttgart: Industrial Cloud-Based Control Platform for the Production with Cyber-Physical Systems (piCASSO-in German)—BMBF-Verbundprojekt, Stuttgart. 2013. Available online: http://www.projekt-picasso.de/index (accessed on 10 September 2019).
7. Grischan, E.; Stahl, C.; Greiner, T.; Barth, M.; Gorecky, D. Cloud-Based Automation. *atp ed.* **2015**, *3*, 38–47. (In German)
8. Schmitt, J.; Goldschmidt, T.; Vorst, P. Cloud-Enabled Automation Systems Using OPC UA. *atp ed.* **2014**, *7*, 34–40. [CrossRef]
9. Cristani, M.; Demrozi, F.; Tomazzoli, C. ONTO-PLC: An ontology-driven methodology for converting PLC industrial plants to IoT. *Proced. Comput. Sci.* **2018**, *126*, 527–536. [CrossRef]
10. Langmann, R.; Stiller, M. Control Services from the Cloud According IEC 61131. *atp ed.* **2017**, *4*, 3–15. (In German)
11. Langmann, R.; Meyer, L. Architecture of a Web-oriented Automation System. In Proceedings of the 18th IEEE International. Conference on Emerging Technologies & Factory Automation (ETFA 2013), Cagliari, Italy, 10–13 September 2013.
12. *Competence Center Automation Düsseldorf (CCAD): Description of a Virtual Device*; R&D Document; CCAD: Düsseldorf, Germany, 2013. (In German)
13. Brass, M. Development and Analysis of a Web and Cloud-Based Runtime Environment for Dynamic Control of Automated Station. Master's Thesis, Hochschule Duesseldorf University of Applied Sciences, Düsseldorf, Germany, 2016. (In German).
14. Coppenrath, M. Development, Test and Evaluation of a Server-Based Control Service by Node.js. Master's Thesis, Hochschule Duesseldorf University of Applied Sciences, Düsseldorf, Germany, 2018. (In German).
15. Aschbacher, H: Smart Services—Fachhochschule CAMPUS 02. Available online: http://www.serviceengineering.at/blog/?page_id=274 (accessed on 23 April 2019).
16. Langmann, R. Automation Systems with Web Technology. *atp ed.* **2014**, *10*, 887–897. (In German)
17. Zechel, C. Developing of a Clearing System for Determination, Storing and Analysis of Using Times of Automation Services in an Industry 4.0 Environment. Master's Thesis, Hochschule Duesseldorf University of Applied Sciences, Düsseldorf, Germany, 2017. (In German).
18. Langmann, R. *HTML5-Based Web Connector for OPC (in German)*; A&D-Kompendium 2014/2015; Publish Industry Verlag: Munich, Germany, 2014; pp. 54–56.

19. Braß, M. *Development and Test of an Industry 4.0 Gateway for the Internet Access to Modbus TCP and Proprietary TCP Protocols*; Documentation of Project Work; Hochschule Duesseldorf University of Applied Sciences: Düsseldorf, Germany, 25 January 2016. (In German)
20. Rojas-Peña, L. *IEC 61131-3 Programming in CICS-Problems and Possible Solutions*; Study Report; Hochschule Duesseldorf University of Applied Sciences: Düsseldorf, Germany, 7 October 2015.

Article

Harnessing the Full Potential of Industrial Demand-Side Flexibility: An End-to-End Approach Connecting Machines with Markets through Service-Oriented IT Platforms

Martin Roesch [1,*], Dennis Bauer [2,3], Leon Haupt [4], Robert Keller [5], Thomas Bauernhansl [2], Gilbert Fridgen [4], Gunther Reinhart [1] and Alexander Sauer [3]

[1] Fraunhofer Research Institution for Casting, Composite and Processing Technology IGCV, Am Technologiezentrum 2, 86159 Augsburg, Germany

[2] Fraunhofer Institute for Manufacturing Engineering and Automation IPA, Nobelstrasse 12, 70569 Stuttgart, Germany

[3] Institute for Energy Efficiency in Production EEP, University of Stuttgart, Nobelstrasse 12, 70569 Stuttgart, Germany

[4] FIM Research Center, University of Bayreuth and Project Group Business & Information Systems Engineering of the Fraunhofer FIT, Wittelsbacherring 10, 95444 Bayreuth, Germany

[5] Project Group Business & Information Systems Engineering of the Fraunhofer FIT, Universitätsstr. 12, 86159 Augsburg, Germany

* Correspondence: martin.roesch@igcv.fraunhofer.de; Tel.: +49-821-90678-142

Received: 15 August 2019; Accepted: 5 September 2019; Published: 10 September 2019

Featured Application: Manufacturing companies with high energy consumption can contribute significantly to balancing energy demand and supply while generating monetary benefits for themselves at the same time. In the case of a voltage drop caused by a lack of electricity for example, a market participant, e.g., an aggregator, can request some negative energy flexibility, which is offered by industrial consumers at a specified price. If the offer is accepted, a typical example would be disrupting or delaying an industrial oven process in order to temporarily reduce energy consumption. In doing so, it also has to be ensured that the product quality and other manufacturing objectives, such as logistical deadlines, are not affected.

Abstract: The growing share of renewable energy generation based on fluctuating wind and solar energy sources is increasingly challenging in terms of power grid stability. Industrial demand-side response presents a promising way to balance energy supply and consumption. For this, energy demand is flexibly adapted based on external incentives. Thus, companies can economically benefit and at the same time contribute to reducing greenhouse gas emissions. However, there are currently some major obstacles that impede industrial companies from taking part in the energy markets. A broad specification analysis systematically dismantles the existing barriers. On this foundation, a new end-to-end ecosystem of an energy synchronization platform is introduced. It consists of a business-individual company-side platform, where suitable services for energy-oriented manufacturing are offered. In addition, one market-side platform is established as a mediating service broker, which connects the companies to, e.g., third party service providers, energy suppliers, aggregators, and energy markets. The ecosystems aim at preventing vendor lock-in and providing a flexible solution, relying on open standards and offering an integrated solution through an end-to-end energy flexibility data model. In this article, the resulting functionalities are discussed and the remaining deficits outlined.

Keywords: industrial load management; demand-side management; demand-side response; energy flexibility; IT concept; platform-based ecosystem

---

1. Introduction and Motivation

The mitigation of climate change is without a doubt one of the biggest challenges facing humanity [1]. The effects of global warming are already perceptible, e.g., the global increase in extreme weather events [2]. Climate change will not only cause irreversible damage to our environment, but will also have a big financial impact on the world economy. Thus, global warming of 1.5 degrees centigrade is expected to cause damages of 72 trillion U.S. dollars by 2060 [3]. This leads to growing environmental awareness and puts pressure on political and economic leaders all over the world to increase their efforts for a sustainable energy system, because power generation contributes one third of global emissions [4]. As a consequence, the renewable energy consumption in Europe, for example, has more than doubled in the last twenty years. Nevertheless, there is still a long way to go, as renewables still held a relatively small share of 13.2 percent of the total energy consumption in 2018 [5].

Germany was one of the first industrial countries to initiate an ambitious energy transition and increase the share of electricity from renewable sources up to 80 percent by 2050 [6]. Currently, 38 percent of the electricity is generated from renewable sources [7]. Of this, two thirds are contributed by wind and solar, which are therefore key to achieving the sustainability goals. However, power generation from these sources depends strongly on the instantaneous weather conditions and thus fluctuates significantly. As conventional power plants are easier to control and can be adapted to the actual energy demand, this characteristic presents a new challenge for balancing electricity supply and demand in the power grid. Thus, to match the increased share of wind and solar power, the grid operators are making greater efforts to ensure the power grid stability by temporarily activating reserves or turning off power plants. The costs for such measures have more than doubled in the last five years [8].

Besides the reactive management on the power generation side, the flexible adaption of energy demand is a way to balance the power grid. This mechanism is commonly known as DR and is usually supported by IT. Triggered by external incentives, e.g., price signals, electricity consumers have to reduce their power demand in times of low energy availability and vice versa. Since industry consumes the biggest share of electricity globally, this sector has also the biggest potential for balancing the power grid [9]. However, this calls for a new collaboration approach between industry and energy providers, which requires new mechanisms and interaction types for cost-competitive energy procurement against the background of increasing uncertainty and volatility. To enable industry to adapt its energy consumption actively, the technical and organizational preconditions must be developed into an eligible platform ecosystem, in which all stakeholders can take part.

In this paper, an overview of existing platform ecosystems in this field is given, and corresponding research is initially presented in Section 2. In the next step, a detailed specification analysis summarizes the current conditions of industry and the energy markets. In Section 4, additional requirements are analyzed before the research deficit is derived. The results with regard to addressing the identified deficits are presented in Section 5, and the paper concludes with a summary in Section 6, as well as an outlook on future research in Section 7.

2. State-of-the-Art

2.1. Fundamentals of Demand-Side Management and Demand-Side Response

In the past, changes in the electrical load were matched by controlling the power generation of conventional power plants [10]. Currently, due to the intermittent and hardly controllable nature of renewable energy sources, this control mechanism no longer provides a sufficient option anymore. This trend is described by Papaefthymiou et al. [10] as a "flexibility gap". Commonly, four options are available to reintroduce the necessary flexibility into the system [11,12]:

- Generation: new flexibility on the supply side
- Transmission: flexibility through the expansion of the power grid
- Storage: flexibility through storage

- Sector coupling: flexible conversion of energy between energy sectors
- Consumption: flexibility through DR

Due to the lack of sufficient deployment to expand the supply side [13], the high social costs of grid expansion [14], the still very high cost of electricity storage [15], and slow progress in interconnecting the energy sectors (power-to-gas, electromobility, etc.) [10], DR is a competitive option.

DR is a category of DSM and is generally understood as a generic term for measures that influence the level or timing of power consumption. These serve to adapt the electricity demand to the electricity supply, especially to the current generation. Hence, possible DSM measures mainly concern the short term, but also the long term. A very short-term control power could provide balancing energy within seconds [16]. In the long term (several months to years), DSM will also be allocated to programs that promote energy saving or efficiency measures on the part of consumers [17].

More short-term adjustment effects focus on DR measures, which are understood as a subcategory of DSM. By means of incentive payments or variable electricity prices, DR activities initiated by grid operators and energy providers cause changes in electricity demand [18,19]. Motivated by such price signals, participating electricity consumers autonomously choose to provide flexibility in their energy demand [16]. However, the consumption adjustment covers a period of minutes to hours. The so-called "load control", which includes the load connection, load disconnection, and load shift initiated by the utility company or grid operator, goes one step further with the same term [20]. Especially, in the industrial sector, which holds the largest share of electricity consumption [9], the DR potential can be provided at comparatively low marginal cost [21]. Energy-intensive companies are already using DR, albeit to a small extent [10,22].

2.2. Industrial Energy Procurement

For most companies, electrical energy procurement is the first and sometimes only contact point with energy planning and energy markets. As part of the procurement of electricity, companies usually pursue different and sometimes even conflicting goals. For example, a key objective of procurement can be to ensure financial and demand-driven planning security. This would require a fixed price for a certain planning level, resulting in a higher cost. Another key objective of procurement is to reduce costs by optimum exploitation of price fluctuations [23].

Historically, companies understood electrical energy procurement as a unidirectional process, in which electricity is bought from the market based on a fixed-price model, procuring the necessary amount of energy for a given period of time for a fixed price on a reference date. More advanced procurement models, such as a tranche or portfolio model, in which the procured energy is split up and bought in different tranches, are only attractive for energy-intensive industries, since they require more know-how about energy markets and, therefore, give rise to additional administrative expenses [24].

In recent decades, energy markets have become much more volatile and, therefore, uncertain (see Section 1), exposing companies to much higher risk when procuring energy. This not only leads to an increase of interest in intermediary entities, such as electricity providers and aggregators, by offering electricity procurement solutions [23], but also service providers addressing the need for additional decision support systems. This way, the company outsources the administrative efforts of analyzing markets and wins back planning certainty with regards to the electricity price.

2.3. Requirement Profiles for Industrial Demand-Side Response

In addition to improving the information base (e.g., through forecasting and better optimization) and outsourcing risk of energy procurement to intermediary entities, flexibility through DR plays an increasingly important role [25]. Table 1 shows four requirement profiles for DR together with the required request duration and remuneration possibilities. DR offers an attractive option by being able to react to short-term (minutes to hours) market price signals, and thus, it can be used as an insurance against unforeseen market volatility and even for arbitrage purposes for the intraday- or day-ahead-market (see Nos. 1 and 2, Table 1) [26,27]. Besides the use of DR in the energy-only market, industrial DR plays an important role in balancing power markets, in which loads are offered for compensation in the case of system instability [28]. Moreover, new marketing opportunities, such as opening up special markets for interruptible load, have also been created [29] (see No. 1, Table 1). With increasing deployment of renewable energies, the grid becomes exposed to systemic risks with respect to the security of supply. This is due to a higher possibility of prolonged periods (1–5 days) with an excess or lack of (renewable) energy generation. It is expected that extreme market prices will incentivize the activation of large-scale manufacturing flexibility (see No. 3, Table 1). Moreover, DR can be used to reduce peak load during the peak-time windows, which are prescribed by the system operator. This is already a common practice used by companies in order to reduce network charges and the contracted demand charge tariff (see No. 4, Table 1). Despite the variety of applications for DR, the current rate of demand flexibility realized by companies is far below the level required for the successful realization of the energy transition [12,30].

Table 1. Requirement profiles and potential remuneration for DR.

Requirement Profile	Short Description	Request Duration	Remuneration
1. Short-term load adjustment	Flexibility can compensate short-term fluctuations in generation or demand.	5 min–1 h	Balancing power markets, interruptible load market, intraday-market
2. Load adjustment over several hours	Mismatch of renewable generation and demand lead to significant fluctuations in electricity prices.	3 to 12 h	Day-Ahead-Market, derivatives
3. Reduction/increase of load over several days	Relevance for flexibility over longer periods increases with regard to the security of supply.	1–5 days	Day-ahead-market, derivatives
4. Atypical grid usage	By avoiding grid usage at congested times, the grid and procurement costs can be reduced.	Several hours per day	Reduction of network charges

2.4. Energy Flexibility for Industrial Companies

The identification of flexibility measures, the technical and economic assessment, as well as the subsequent marketing on the energy market present a complex challenge for companies [31]. On the one hand, the different use cases for flexibility are characterized by an individual product design and compensation methods. On the other hand, on the manufacturing level, different machines or manufacturing processes inherit different forms of flexibility measures and therefore a divergent level of flexibility potential. The characteristics are most commonly distinguished by their controllability. Hence, the potential measures are categorized into uncontrollable (no flexibility), curtailable, shiftable, buffered, and freely controllable loads (full flexibility) [32]. Curtailable loads, for example, are those that do not need to recover the curtailed energy once they are reconnected. In contrast to that, shiftable loads are those that can be moved in time. However, the amount of consumed energy does not change; it only gets shifted in the time domain. In addition, the identification and implementation of measures for DR are quite complex. The technical processes define many different boundary conditions and dependencies, such as a minimum or maximum period of interruption due to quality issues, which have to be considered [33]. In addition, DR can also counteract the main objectives of manufacturing, such as throughput or delivery reliability [34]. The high number of dependencies between individual flexibility measures and a large number of manufacturing parameters represents a major challenge in the manufacturing industry. This includes planning horizons, temporal resolutions, and maximum permissible runtimes [35]. At the same time, the complexity of the boundary conditions and the abstraction level increase with higher automation levels [36]. Moreover, data from a wide range of heterogeneous sources on the shop floor (e.g., machine control, manufacturing planning, etc.) is required for industrial DR. As manufacturing

IT platforms have been proven to provide such features, the existing ecosystems in this field are examined below.

2.5. Platform Ecosystems

In recent years, platforms have arisen in many business fields in order to bring together customers and providers and offer innovative services. In this respect, the term platform is used very frequently; however, its meaning is not clear and consistent [37]. In the scientific literature, three perspectives on platforms from the fields of engineering, economics, and organization can be found [38]. Considering this, platforms offer a medium of interaction for business partners based on software, hardware, organizational processes, and standards [39]. Platforms are characterized by a layered architecture [40] and act as a central cornerstone, which supplies core functionalities [41]. In this way, they offer a common access point for users and a set of basic services [42]. Ecosystems and platforms are closely connected with each other because ecosystems are platforms in which numerous firms can contribute components (e.g., apps) to a technical platform [43]. Hence, the term "platform ecosystems" is also frequently used [44,45].

From an economic perspective, platform ecosystems offer several major benefits, especially the co-creation of business value by encouraging complementary invention and exploiting network effects, as well as increasing flexibility [42]. The fact that five out of the ten most valuable brands are built on platform business models demonstrates these outstanding advantages [46]. Regarding platform ecosystems, there are several aspects to consider. Platforms may result in a winner-takes-all situation, where eventually, the free competition of different vendors is eliminated and substituted by a single dominator [47]. However, there are also systems aiming at co-evolution promoting symbiotic situations, which provide benefits for all participants. In this way, the complementors, which contribute, e.g., applications to the platforms are crucial, because they moderate the competition [48]. Therefore, platform users should be able to easily exchange services from different vendors and, in doing so, avoid so-called vendor lock-in. Another characteristic of platform ecosystems is the indirect network effects, where the value of a product increases with the total number of users [42], e.g., platforms with a high number of potential customers are very attractive for vendors. Consequently, the thresholds and obstacles for participants need to be minimized, in order to make it easy for new partners to participate in an ecosystem. This effect was for example studied by [49], where the pricing of gaming hardware, which indicates the entry hurdles for a video game console, was proven to be crucial for the success of vendors. Moreover, platform ecosystems benefit highly from complementary inventions from other partners and thereby value co-creation, as by this means, new and innovative services can be offered [50].

To coordinate the variety of platform participants, a suitable governance approach is also required. The governance of a platform ecosystem is of contingent and dynamic nature and has to be carried out by the platform operator [51]. Thereby, standardization is an important element for governance. Moreover, due to the situational and temporal contingencies, establishing self-selection is very effective [52]. This means the participants self-select the different partner levels and define the appropriate rules for their business.

Overall, platform ecosystems play a very important role in the business world and bring some major advantages. Nevertheless, establishing and running a platform ecosystem includes some challenges. Platform ecosystems are already widespread in the manufacturing sector and energy markets; the existing solutions are outlined below.

2.5.1. Manufacturing IT Platforms

Modern manufacturing systems require multiple IT systems, which operate on different operational levels, to make the complex and numerous processes controllable. A commonly-used reference model to structure these manifold software solutions in a functional and hierarchical manner of five levels is the automation pyramid [53]. As every level may consist of several individual elements, shaping a suitable architecture to achieve an integrated information flow and meet all functional requirements is challenging. In line with the increasing digitization, cyber-physical systems, which merge the physical and virtual world through embedded hardware and software, present a new approach [54–56]. Thus, the conventional pyramid architecture is supplemented or replaced by a decentralized and non-hierarchical structure with increasing agility and flexibility [57]. Nevertheless, an architecture following the automation pyramid is still common in present-day manufacturing system architectures, which may also be indicated by its adaption within the Reference Architectural Model Industry 4.0 [58].

In order to handle increasing complexity and in line with the trend of digitization, manufacturing IT platforms, which cover all levels of the automation pyramid, are commonly applied. In addition, customers and suppliers are able to interact on these platforms, shaping complex ecosystems [59]. Driven by the recent trend towards increasing service orientation, also referred to as XaaS [60], many authors [61–64] and companies have begun to work on service-oriented and often cloud-based platform ecosystems. They focus on the efficient provision of functionality in the form of software services for end users and are, therefore, well suited for applications in manufacturing. Compared to consumer-grade software platforms, industrial-grade manufacturing IT platforms have additional requirements [65,66]. Reliability and security are central. While reliability ensures that the platform is capable of supporting 24/7 manufacturing, information and data security are necessary to protect the intellectual property of the company.

Manufacturing IT platforms are offered by leading equipment and software vendors such as Predix by General Electric (https://www.ge.com/digital/iiot-platform), MindSphere by Siemens (https://siemens.mindsphere.io), IoT Suite by Bosch (https://www.bosch-iot-suite.com), and Axoom by Trumpf (https://marketplace.axoom.com), as well as by new developments in research such as the Industrial Data Space (https://www.internationaldataspaces.org) or Virtual Fort Knox (https://virtualfortknox.de). These platforms aim at supporting manufacturing companies in networking their manufacturing and generate added value through digital services such as process optimization or predictive maintenance. Most existing and commercially available manufacturing IT platforms are tailored to products and services offered by the respective vendor [65,66]. They utilize proprietary interfaces and protocols instead of open ones. As a consequence, neither interaction with external systems nor interoperability with other platform vendors are possible. This causes vendor lock-in effects for the customer, which is one of the largest barriers that companies face when introducing cloud platforms [67,68].

Besides manufacturing IT platforms, EMS are also widely spread in industrial companies [69,70]. It is their task to gather and preprocess data of the current and historical energy consumption, in order to increase transparency and uncover trends or outliers. Thus, the main scope of the EMS is to increase energy efficiency. However, these systems are not able to introduce any control actions, and so far, there is no linkage to the shop floor [68,71]. Consequently, in order to adapt the energy consumption flexibly, EMS needs to be enhanced and linked to the conventional systems for manufacturing management and execution.

2.5.2. Market-Side Platforms and Services

As described in Section 2.2, intermediary entities are of increasing importance for companies. However, not only third-party energy suppliers and aggregators are commissioned. Service providers, e.g., decision support or forecasts, are also at the focus of increasing attention. Companies are focusing on their core business, and thus, on-site support processes are increasingly outsourced to third parties [72]. Additional reasons for outsourcing are the reduction and control of costs, risk minimization, and benefits from the best practice of others.

For energy procurement, plenty of decision support service providers offer a wide range of software solutions. These range from electricity market forecasts to solutions for manufacturing planning optimization, taking into consideration different horizons of electricity market prices. Such service providers are, for example, Kisters (https://www.kisters.de/en), N-SIDE (https://www.n-side.com), DNV-GL (https://www.dnvgl.com/power-renewables/services/index.html), Software AG (https://www.softwareag.com), and Siemens Industry Services (https://www.industry.siemens.com/services). These services are usually limited to decision support and do not interact with markets. Following the process of optimization, a company can generally market the flexibility on its own or via an intermediary, such as a direct marketer or an aggregator [73]. Aggregators provide comprehensive solutions, incorporating energy supply contracts, optimization of manufacturing flexibilities, and the final trading of the same. Examples include Next Kraftwerke (https://www.next-kraftwerke.de) and e2m (https://www.e2m.energy). These solutions build on current market designs and scarcely integrate new potential flexibility markets, such as local flexibility markets. Moreover, the solutions provided with regard to the interfaces, data models, and processes used for communication between the company and aggregators are not standardized.

For this reason, public projects are aiming to establish reference frameworks, which deliver a common standard for systems to ensure interoperability for flexibility trading and integrating all potential flexibility customers. Several publicly-funded research projects are focused on developing open source flexibility platform solutions in order to standardize the processes and protocols for flexibility usage, including DR in the commercial and residential sector. For example, the non-profit organization USEF (https://www.usef.energy) developed an open standard for transparent communication procedures fitting on top of most energy market models, extending existing processes to offer the integration of both new and existing energy markets. It is designed to make flexible energy a tradeable commodity. Thus, its focus lies on the marketing processes for potential customers, such as system operators and balance responsible parties after the flexibility was identified. The OpenADRproject (https://www.openadr.org), on the other hand, very much focuses on the implementation of DR and final execution of the flexibility measure by providing a common language communication data model incorporating different DR programs, such as critical peak pricing for spot markets or fast demand response for ancillary markets.

On the one hand, the USEF and OpenADR approach are making an important contribution to the standardization of interfaces and communication within the flexibility value chain, but on the other hand, they do not provide the IT solutions needed to bring together manufacturing machines and flexibility markets.

2.6. Preliminary Conclusions

It can be summarized that (1) industrial demand flexibility plays a key role in overcoming the current flexibility gap in the electricity sector and (2) demand flexibility has not yet been harvested to its full potential. Moreover, (3) in order to implement DR in industrial companies, the complex impact of single measures on the whole manufacturing system and its interdependencies on all organizational levels have to be considered. (4) A large number of manufacturing IT platforms already exists. They are, however, predominantly designed as closed ecosystems. Proprietary interfaces are used, and there is no interaction with external systems. (5) EMS, however, only focuses on gathering and preprocessing information on the energy consumption and is not capable of executing adapting actions. (6) On the other hand, platforms targeting aggregation and marketing of demand flexibility lack the complexity management required for industrial manufacturing processes and thus only partially address the value chain of DR (see Figure 1).

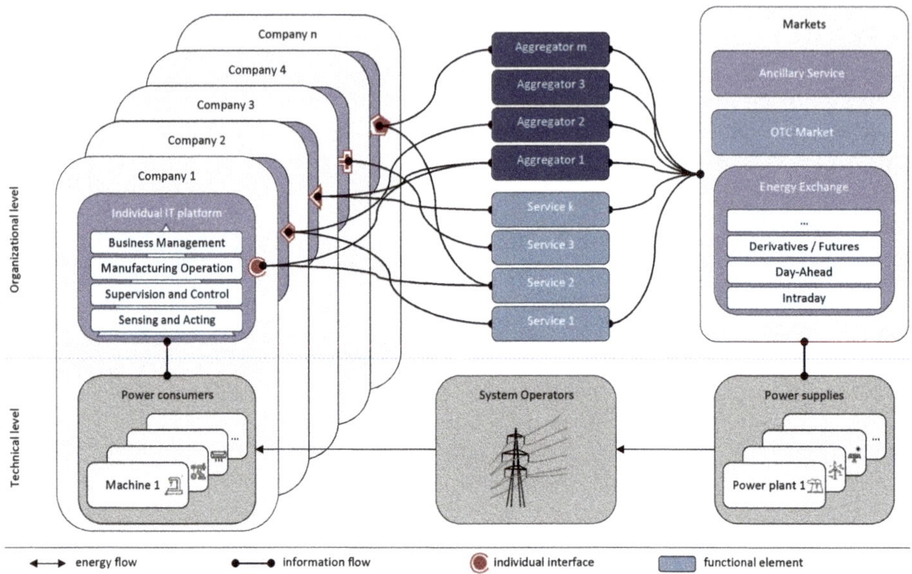

Figure 1. Current landscape of IT platforms, services, aggregators, and energy markets.

3. Industry Specification Analysis

The previous section showed that there remains much unused potential for industrial DR. In order to determine why this potential is currently not being exploited, the existing barriers are identified below. To achieve this, the method of structured interviews and workshops was applied. These were conducted with different companies and stakeholders, which were elicited in focus group interviews comprising representatives of manufacturing companies and the energy market. The information was collected in the context of "SynErgie" (https://www.kopernikus-projekte.de/synergie), a large German research project, and beyond that discussed, filtered, and consolidated in internal workshops with numerous researchers from different disciplines. In doing so, the creation of an integrated approach for industrial DR, which includes all relevant providers, consumers, and other market participants is the central goal [74]. As more than 80 partners from industry, energy markets, research, and society were involved, a substantiated specification could be prepared that records the demands of all relevant stakeholders. The requirements identified by the method described above are summarized and outlined below:

1. Functionality: The automated synchronization of industrial energy consumption and supply requires several different partners and functional levels from the energy market down to single machines to be involved. The increasing number of participating companies and the increasing challenges to balance the power grid make automated synchronization necessary. For this reason, an end-to-end approach is required, which defines the continuous information flow throughout all relevant stages and maps the following three central use cases: procurement of market information (e.g., price forecasts), evaluation of flexibility (e.g., ideal marketing time), and marketing of flexibility (on the appropriate target market). In addition, information on energy consumption needs to be equally considered within manufacturing. Thus, appropriate functionalities throughout all organizational levels of a company are required.

2. Interoperability: Currently, a variety of communication protocols, data models, and IT systems are used in manufacturing. On top of that, energy markets and grid providers apply several interfaces and data models. Consequently, interoperability is another important requirement to enable a high number of heterogeneous companies to take part in a flexible manner. The integration of

existing standards, shaping a transferable and robust architecture, as well as the harmonization of data models are crucial for the development of a new ecosystem in the context of energy flexibility.

3. Free competition: To achieve wide acceptability and to offer incentives for companies to participate in the energy markets, a solution needs to be offered that ensures free competition. Accordingly, obstacles for new market actors need to be minimized, and an open ecosystem must be offered that provides flexible access to the services of different providers. In addition, industrial companies and other participants should be able to extend independently the functionalities of and solutions to their and their customers individual needs.

4. Privacy and security: Privacy and security are key requirements for companies to participate. The detailed energy consumption contains crucial information about a manufacturing company, because for example, the given utilization or the used technology can be derived. Thus, no confidential information about energy consumption, etc., should be provided to external competitors. Besides privacy, security probably plays an even more important role. As energy systems are part of the critical infrastructure, the highest security standards need to be adopted, and given specifications need to be fulfilled, e.g., [75]. In addition, data leaks have to be prevented to ensure that personalized information does not fall into the wrong hands.

5. Credibility and trustworthiness: The pursued solution should offer the possibility to purchase and sell energy flexibility automatically. Therefore, signing legally-binding digital contracts is an additional essential requirement. For this purpose, appropriate services and processes need to be established that at the same time provide proper technical solutions to ensure trustworthiness in terms of traceability and transparency.

6. Market entry threshold: The specification analysis also involved the current market design and resulted in requirements for key changes. In order for demand flexibility to be used to its full extent, sanctioning flexibility by network charges needs to be eliminated. Furthermore, a non-discriminatory access to all flexibility markets needs to be ensured.

7. General architecture requirements: Considering the discussed requirements for IT-based automated industrial DR, additional specifications regarding the architecture can be derived. In particular, modularity and extensibility are very important in order to provide the required reusability, adaptability, and scalability of the solution. These requirements can be summed up with the concept of service-oriented architectures. In addition, near real-time processing, as well as robust and reliable communication flows are essential. Furthermore, different deployment models, e.g., private or public cloud solutions, should be possible.

4. Research Deficit

In the previous two sections, the state-of-the-art was analyzed (see Section 2), and a broad specification analysis (see Section 3) was presented. Based on these findings, the need for action and research deficits are derived below, as the next methodological steps. In their practical form, these deficits represent obstacles to companies making full use of demand flexibility and to marketing it economically. These obstacles need to be addressed in the research in order to provide more demand flexibility in industry, thereby contributing to a successful energy transition. The deficits identified are described below.

4.1. End-to-End Approach

Regarding existing frameworks and standards for market demand flexibility, it must be pointed out that there is no approach covering information flows and automation from machines to energy markets. However, such an end-to-end approach from machines to energy markets is necessary to achieve consistency and interoperability with all technical entities and stakeholders involved.

- Information flows: All bidirectional information flows between machines and energy markets, including every intermediate stage, must be covered to achieve an end-to-end approach. Bidirectionality because demand flexibility must be offered in the markets and, in addition, the purchase signals must subsequently be converted into load profiles within the company and ultimately into control signals for the machines and equipment. The information flows do not only cover company in-house processes as described by the automation pyramid [35,53]. Therefore, the concept of the automation pyramid needs to be broadened to include external processes with regard to marketing demand flexibility (see Nos. 1 and 2, Section 3). In addition, decentralized approaches also need to be considered as these concepts are increasingly common and provide more flexibility for companies (Section 2.5.1).

- Multi-level optimization: In order to regulate complex information flows and control the efficient use of flexibility, different levels of optimization are necessary. These levels decompose the overall optimization problem of efficient flexibility usage into sub-problems considering characteristics such as planning horizons, temporal resolutions, and maximum permissible runtimes. It is important that optimization levels can be implemented dynamically, e.g., depending on the target process and the company's infrastructure, some of the optimization levels may be left out or split, e.g., into sub-optimizations. However, with decomposition, new challenges arise. First, restrictions made at certain levels must be respected by the following levels. Second, the·interaction between various optimization levels needs to be coordinated by a corresponding architecture (see No. 1, Section 3).
- Generic data model: To ensure consistency within the end-to-end approach for information flows and the interaction of different optimization levels, a generic data model covering a wide range of demand flexibility is necessary. However, not every communication between technical entities in the described end-to-end approach needs to be mapped in this data model. Rather, it is a matter of using the data model where it creates added value. Data models for further communication must then be designed in such a way that they can be derived from this generic data model or transferred back to it (see No. 2, Section 3).
- Traceability: While information flows and the data model ensure technical interoperability, the communication within the described end-to-end approach needs to be credible and trustworthy. Therefore, technical solutions meeting these specifications are necessary, e.g., step-by-step traceability of transactions. From a company-side perspective, this is to ensure that commercialized flexibility has really been implemented and the contract fulfilled accordingly. From a market-side point of view, the fulfillment of the contracts is equally important to ensure balanced groups in the energy system (see No. 5, Section 3).
- Encapsulation: Industry and the energy sector are completely different domains with diverse knowledge, methods, and technologies. To realize automated flexibility commercialization, both domains need to be connected and work interlocked. Consequently, approaches to encapsulate both domains without affecting the system's performance are necessary. This includes, but is not limited to, commercializing load profiles without revealing manufacturing secrets and ensuring free competition on markets, etc. (see No. 4, Section 3).

4.2. Company-Side

Even without the use of demand flexibility, industrial manufacturing and the associated supply networks are very complex. Demand flexibility adds another dimension of complexity to this, which is why manifold research deficits can be identified:

- Vendor lock-in: Most existing manufacturing IT platforms are tailored to products and services offered by the respective vendor and lack interoperability with other platform providers or integration of external systems. To prevent vendor lock-in, open platforms with the ability to connect proprietary (e.g., Siemens S7, SAP BAPI, etc.) and open protocols (e.g., OPC-UA, REST, etc.) for hardware and software flexibly are required (see Nos. 2, 3 and 7, Section 3).
- Interoperability: To ensure interoperability, communication must be protocol independent, and the platform must be able to harmonize data models. For platforms to incorporate the described end-to-end approach, this does not only apply to internal interfaces and data models, but also to

external communication. This allows, for example, the vendor-independent integration of PLC by Siemens and Bosch Rexroth with an EMS by econ solutions and an ERP by SAP. Additionally, it should be possible to establish a connection between platforms of different vendors. At the same time, since various stakeholders, vendors, components, and services are involved, the concepts for security and privacy become crucial. (see Nos. 2, 3, 4, and 6, Section 3).

- Energy as a decisive target: Existing platforms do not necessarily consider energy as a decisive target in manufacturing (see Section 2.5.1). Yet, with an increasingly volatile energy supply and the resulting need for greater demand flexibility, energy and its related availability and costs must be taken into account. Therefore, the functionalities of existing platforms need to be enhanced to consider energy aspects and to provide solutions for the synchronization of energy demand and supply (see No. 1, Section 3).

- Technical flexibility assessment: The variety of industry sectors yields a wide range of manufacturing processes with individual flexibility measure patterns. Besides the lack of adequate flexibility products on the energy markets, the technical assessment and integration into a flexibility portfolio within a complete flexibility management approach still requires high individual efforts, resulting in unpredictable project costs for companies. Even though aggregators and other service providers already offer audits to identify flexible loads for potential commercialization, the focus mainly lies on large-scale manufacturing plants, leaving unused potential (see No. 1, Section 3).

- Flexibility management: Due to the synchronization of energy demand and supply, the traditional magic triangle of time, cost, and quality becomes more and more volatile. Demand flexibility offers a possibility for cooperating with the rising importance of energy procurement. Therefore, platforms must enable an integrated management of demand flexibility within the company, integrating energy-related data with other manufacturing data for an adequate technical assessment. Consequently, the acquisition, aggregation, analysis, and optimization of process and manufacturing data are necessary to achieve energy-synchronized control of the systems, plants, and components (see No. 1, Section 3).

- Energy-synchronized control: Covering automated marketing of demand flexibility, functionality at all levels of the automation pyramid needs to be considered. Key features of this energy-synchronized control of manufacturing are the transformation of process data into flexibility measures and the aggregation by combining, splitting, and adapting the flexibility measures for optimized usage. In addition, a communication interface to the energy markets is required, which ideally is implemented using standardized and open protocols to automate the access to different offers of demand flexibility marketing, e.g., day-ahead-market, and information procurement, such as market price forecasts (see Nos. 1 and 2, Section 3).

- Entry hurdles: While the workload of administrators is reduced by the cloud offerings of certain manufacturing IT platforms, there are still significant entry hurdles for users. On the one hand, due to probable vendor lock-in and, on the other hand, due to the availability of functionalities in the field of manufacturing and the necessary effort for connection and usage. For these reasons, platforms should be designed as open development, sales, and operating platforms. Available functionalities can thus be obtained and operated via the platform, while missing functionalities can be developed by the user or software partners. (see Nos. 1, 2, 3, and 6, Section 3).

4.3. Market-Side

The reasons for the slow expansion of industrial DR are manifold: First, complex regulatory frameworks and weak market incentives on the market side; second, the lack of integrated economic assessment of flexibility, from machine to the flexibility market, as part of the manufacturing planning; third, the uncertainty in price forecasts and the resultant risk to manufacturing planning. The resulting research deficits on the market side are outlined below:

- Market design: The step towards trading flexibility from a market and regulatory perspective is subject to a number of obstacles. Three main obstacles to developing an efficient energy market embracing demand flexibility can be identified [76]:

  1. As the energy sector is subject to a complex regulatory system, a high uncertainty exists with regard to the continuity design of energy laws, subsidies, taxes, etc. This uncertainty is currently reflected by the indecision on the part of companies to invest.
  2. Energy market design aims to treat different technologies equally. However, in reality, the nature of flexibility measures (as described in Section 2), dependent on the type of machine or manufacturing process, does not imitate conventional generation schemes. The result is a distortion of flexibility options and technologies.
  3. The complexity of the energy sector is also reflected in the price structure. Daily price fluctuations on the electricity market are only partially visible to consumers. The high fixed cost share (electricity taxes, network charges, etc.) is leveled, the price fluctuations are reduced, and the profit margin of the demand flexibility project is reduced.

  Further research is needed to highlight the inefficiencies and weaknesses of the current market design and to identify solutions for obtaining economically-viable demand flexibility options (see No. 6, Section 3).
- Economic flexibility assessment: Participation in the energy market is attractive when the value of the flexibility measure, i.e., the electricity cost savings that can potentially be achieved through load shifting, is higher than the opportunity costs that the company incurs by making the process more flexible and possibly losing value due to the flexibility measure. If the value of flexibility signaled on the market is below the opportunity cost, a flexibility measure will not be stimulated. Therefore, the internal financial assessment of the flexibility needs to have both information on the manufacturing costs (including all implied potential costs if used) and real-time market prices. The information on the optimization combines the two prices and indicates which flexibilities should be drawn from an economic point of view. Regardless of whether the optimization takes place on the company-side or market-side platform, some kind of data model for communication of flexibilities is used. To ensure consistency in communication while maintaining a high level of functionality and traceability, a standardized description and modeling of industrial demand flexibility is required. (see Nos. 1, 2, and 5, Section 3).
- Energy market forecast: To ensure planning security, the role of forecasting energy market prices is crucial. However, a predictability of more than five days, e.g., the day-ahead-market, due to the intermittent nature of renewable energy sources, is difficult. In some company cases, predictions of more than five days, regardless of the forecast quality, are needed. Currently, plenty of forecast service providers are competing for the growing market of demand flexibility. The forecasting service market is not yet transparent, and therefore, it is not easy to compare different services. This might be suitable for individual service providers; however, it does not contribute to a performance-based free market embracing free competition (see No. 3, Section 3).

5. Concept of the Energy Synchronization Platform

Considering the state-of-the-art (see Section 2), required specifications (see Section 3), and existing research deficits (see Section 4), the ESP as depicted in Figure 2 is proposed. This platform ecosystem enables the industry to participate actively in energy markets through both a faster and more accurate scheduling of its energy consumption (consumer role) and by offering DR potential (supplier role). Therefore, the ESP allows a proactive and continuous synchronization between companies' offer of flexibility and the markets' demand for flexibility, as well as between energy consumption and energy supply. In this way, the platform supports companies in ensuring that only the flexibility for which it is economically viable is marketed. At the same time, the reliability and stability of the power grid are increased from a system perspective.

To fulfill all requirements, especially regarding free competition, the ESP is designed as an ecosystem with extensive functionalities preventing vendor lock-in and ensuring interoperability. Currently, the energy market is decentralized, and there are individual service providers and aggregators with different interfaces, which results in the displayed disadvantages (see Section 4). To overcome this, the ESP aims to bundle all necessary services on one platform and integrate all service providers as complementors. In doing so, the evolution of the ESP should be initiated by the publicly-funded project "SynErgie" (see Section 3) and later be transferred to an independent platform ecosystem with a community as the platform operator. Consequently, the ESP grows the flexibility market by increasing offered and demanded flexibility, as well as by synchronizing both. This networking and communication in particular is expected to have a huge effect on the market growth [42,77].

The technical concept of the ESP was developed with regard to security by design and consists of two logical platform types: the CoP as described in Section 5.1 and the MaP as described in Section 5.2. The division into two logical platforms ensures privacy and security by encapsulating their specific domain knowledge, technologies, and methods while, at the same time, maintaining a safe state without affecting the operation and performance of the overall system. Summarizing, the ESP describes the interaction of several CoPs on a central MaP to carry out transparently IT-supported demand flexibility trading.

Both logical platform types are connected via an interface, allowing the exchange of necessary data for automated DR by a specified data model. This data model is universally valid, as well as generic for various flexibility measures and applicable in all data-handling steps for automated DR from machines to energy markets. Such an EFDM is proposed by [78]. Especially for comprehensive processes, this EFDM can be applied end-to-end. However, this EFDM is not designed as the single source of truth, but much more as a superordinate data model from which further data models can be derived. A prerequisite for this is that the derived data models can be translated back into the generic EFDM at any time. A use case for the derivation of data models is the multi-level optimization of the use of flexibility. In the following, the two logical platform types are described and then merged to form the end-to-end approach.

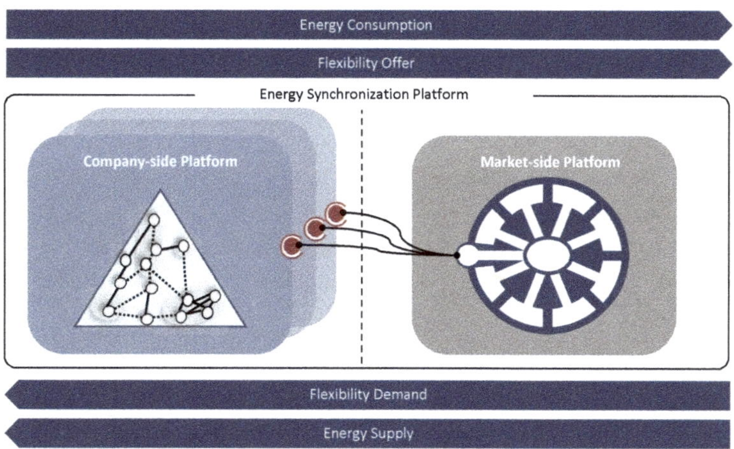

Figure 2. Concept of the energy synchronization platform.

5.1. Company-Side Platform

As part of ESP, a CoP offers the necessary functionalities for the IT connection and control of energy-flexible manufacturing processes and infrastructure in a service-oriented architecture. Therefore, it represents the modular, service-oriented, secure, and externally-encapsulated IT system within a company. Compared to other manufacturing IT platforms, the CoP is characterized by the fact that energy is explicitly taken into account or focused as a decisive target for manufacturing. Consequently, it includes acquisition, aggregation, analysis, and optimization of process and manufacturing data, on the one hand, and energy-synchronized control of the systems, plants, and components, on the other hand. The functionalities represented by services thus cover all levels of the automation pyramid, and information can be distributed in both ways, bottom-up and top-down. Hence, demand-flexible behavior in the future electricity system is enabled for technical entities in manufacturing companies. Thereby, information from both manufacturing and the energy system is brought together. The architecture of the CoP is depicted in Figure 3 and described below.

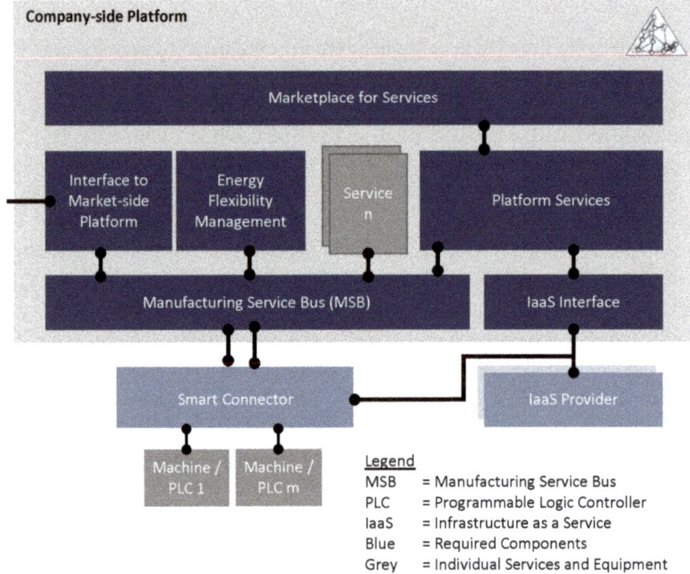

Figure 3. Architecture of the company-side platform.

Most important for ensuring the interoperability and preventing a vendor lock-in is the open, extensible, and modular architecture of the platform itself together with the integration layer MSB. First, the modularity of the platform allows a flexible configuration, including only necessary components (blue in Figure 3), as well as the possible enhancement by individual additional components and services (gray in Figure 3). The development of these additional services is fostered within the ESP's ecosystem through close cooperation between manufacturing companies and software vendors. Second, each component of the platform can be deployed by different operators. This enables deployment models ranging from operating the CoP as a private instance on-premise to a public cloud operated by an independent third party, as well as multiple hybrid forms. Additionally, the deployment model is not set for all times, but can be adjusted at any time in the event of changed boundary conditions, by exchanging or adding individual modules instead of exchanging the entire platform. Companies are able to use a wide range of existing services suitable for their platform without their own extra development effort and often also without a considerable integration effort.

The previously-described modular, service-oriented approach and the consideration of energy as a target variable runs through all components of the platform:

- Marketplace for services: The marketplace for services is comparable to the well-known concept of app stores and represents the web-based point of entry for users. It enables obtaining new services and deploying them automatically in conjunction with the platform services. Due to the simple booking of services on demand, entry hurdles for new participants are reduced. Offering a wide range of services from different independent services vendors, the marketplace ensures free competition.

- Services: In contrast to existing approaches, which do not consider energy as a decisive target for manufacturing [79,80], additional services on the CoP, such as manufacturing planning and control or various optimization services, are designed taking it into account. A typical approach with respect to logistic operating curves in manufacturing planning and control was described in [81]. Since these additional services are highly individual and vary greatly between use cases, they are flexibly orchestrated using the MSB and integrated into the handling of demand flexibility by the energy flexibility management. As complementors can flexibly contribute their services, the ecosystem is able to profit from co-evolution. In addition, the central role of the complementors prevents from winner-takes-all situations (see Section 2.5).
- Platform services: The operational functions of the CoP are summarized in platform services. Among them are identity and permission management, service repository, service life cycle management, as well as service accounting and service monitoring. Platform services are well described in the literature, e.g., in [62,82].
- Interface to the market-side platform: The interface to the market-side platform enables services on the CoP to access data and services on the MaP via a single standardized interface. Services on the CoP benefit from this by only having to implement a single interface instead of multiple interfaces to all requested services on the MaP. Furthermore, this represents a security barrier between companies and energy markets.
- Energy flexibility management: The CoP's hub for aggregating and managing all demand flexibilities and their dependencies is represented by the energy flexibility management. First, it includes an overview of the company's flexibility. Second, it supports the technical assessment of flexibility. Third, the energy flexibility management provides an API to combine, split, and adapt flexibility for optimized usage. However, the optimization is not part of the energy flexibility management and is provided by third party services instead. Furthermore, the energy flexibility management is responsible for controlling the implementation of demand flexibility measures and communication with energy markets. Most likely, the functionality of the energy flexibility management will be integrated or at least closely connected to a company's EMS.
- Manufacturing service bus: An integration layer for manufacturing companies needs to ease reconfiguration, enable a loose coupling, allow for asynchronous communication, and offer standards-based integration [83]. The MSB meets these requirements, as well as additional ones as described in [84]. It provides an abstraction layer for different protocols, which can easily be extended by additional interfaces. The purpose of this abstraction layer is to harmonize data models by allowing individual data objects to be mapped flexibly to each other. This mapping can either be modeled automatically via a self-description or manually. Furthermore, the easy extensibility enables effortless additions of protocols, which can either be proprietary or open. Consequently, the MSB can also be used to translate proprietary protocols (e.g., Siemens S7) to open protocols (e.g., OPC-UA). Summarizing, the MSB ensures interoperability between all components of the CoP in the sense of a close cooperation of independent, heterogeneous systems in order to exchange information efficiently and in a usable manner.
- Smart connector: The smart connector is designed as a bidirectional interface to access machines and their respective PLCs from higher-level IT systems. Therefore, it extends the MSB by translating proprietary PLC protocols to open IT protocols and allows for process data and machine data acquisition. Additionally, the smart connector's machine interface is designed to process energy data and to transfer it to the generic EFDM to model demand flexibility. Moreover, orchestrated by services on the CoP, the smart connector is also capable of an energy-synchronized control of the process, e.g., triggered by price signals.
- IaaS interface: The IaaS interface represents an abstraction layer between the CoP and its underlying infrastructure. It enables the CoP to distribute the deployment of services,

ranging from in-house infrastructure to external cloud infrastructure such as Amazon Web Services and edge devices such as the smart connector.

If a service, e.g., the energy flexibility management, is ordered via the marketplace for services, its deployment and provisioning on an IaaS provider, as well as its orchestration with other services via the MSB are done by the platform services. Furthermore, process data from machines are integrated using the smart connector. Therefore, with the architecture described above, it is possible to automate all activities regarding energy flexibility marketing on the company side by virtualizing flexibility and distributing it via services.

5.2. Market-Side Platform

The MaP represents the second main part of the ESP. In contrast to the CoP, it is designed as a single multi-sided platform solution, connecting and integrating a variety of CoP, existing energy markets (e.g., balancing power market, electricity exchange), and third-party services (e.g., optimization and forecast services, software providers, energy supplier, aggregators). The MaP, which was co-designed by the participating companies, fosters the monetization of industrial flexibility on power markets in a B2B market setting.

The MaP is connected to the CoP via an interface that allows data exchange and should enable automated trading of flexibility in the future. In contrast to the CoP and services, which in principle can occur in an unlimited number on the ESP, only one logical instance of the MaP is provided. Its multi-sided architecture enables direct interactions between two or more distinct sides. Each side is affiliated with the platform [85]. The platform itself consist of various components, such as hardware, software, or service modules with specific arrangements and rules. Figure 4 shows the modular and flexible configuration, featuring two main layers and their respective components.

- Runtime layer: This represents the active component of the MaP, which has the necessary interfaces to the outside. On the one hand, the runtime layer is the basis for system-related and domain-specific services; on the other hand, it also implements the routing to the corresponding services and data.
- Persistence layer: The persistence layer acts as a scalable data management component and allows writing and reading access from the runtime layer. It represents the passive "database" component of the MaP.

Within the layers in Figure 4, components define the functionality of the MaP. The key components are highlighted and explained in the following:

- Portal (access layer): The portal represents the component, which allows users to communicate with the platform through a graphical user interface. Besides basic features, such as registration and login, it provides an input mask for information transfer to other components, API documentation, the possibility to execute test calls, community functionalities (e.g., rating), and the monitoring of offered services. It is connected to the platform services for access management and authentication, as well as the service broker for further access to functionalities.
- Service broker: The service broker acts as a central access point for market participants implementing the API gateway, makes inquiries to registries, and forwards the request of the current market participant according to the response from the registries. The core task of the service broker is to establish contact between flexibility providers and their users by integrating supporting services. As the core of the multi-sided architecture, it is also exposed to threats and requires threat protection against cyber attacks, such as distributed denial of service attacks, SQL injections, etc. Moreover, it is the logical instance that regulates access control policies and enforces policies related to subscriber authentication and authorization of access to services. Thus, the service broker uses services of operational components from platform services.

- Platform services: The operational functions of the MaP are summarized in platform services. They inherit control and contract services, in order to monitor and maintain the platform, but also oversee the access management, which regulates access to sensitive data according to the respective authorization. In order to monitor the access authorizations, the platform services access the system data containing the log, user, and contract data, market services, the service broker, and the access layer.
- Market services: These are the services offered on the MaP by third parties and allow service providers and platform users to analyze what features have been used and what registered users are looking for. It is important to note that market services are only addressed from the service broker. The connection to platform services is for authentication reasons only. Market service-related data are stored within the component service data. This comprises among others the storage of flexibility data in the form of a filterable list, e.g., in the form "market participant X has offered flexibility Y of quantity Z [for time interval T]".
- Customer services: In order to provide an interface to the external services or datasets offered by market participants, but not directly published on the market platform, the customer services complement the service broker by a collection of virtual services that bijectively map to these external services. In this way, the service broker can address the virtual services and datasets and thus communicate with the external ones. It has a connection to custom data and a database to store the required information.

The architecture embraces standardized communication between all user groups, standard data models for the distinct communication of flexibility such as the EFDM, as well as guidelines, which enable an open-integration hub, similar to the CoP. The MaP is not only compatible with the existing power system, integrating APIs with the electricity exchange and the balancing power markets, but also implements solutions for the local marketing of demand (building upon concepts of USEF) in so-called local flexibility markets. The local aspect is especially interesting for system operators, which can utilize the flexibility for grid applications.

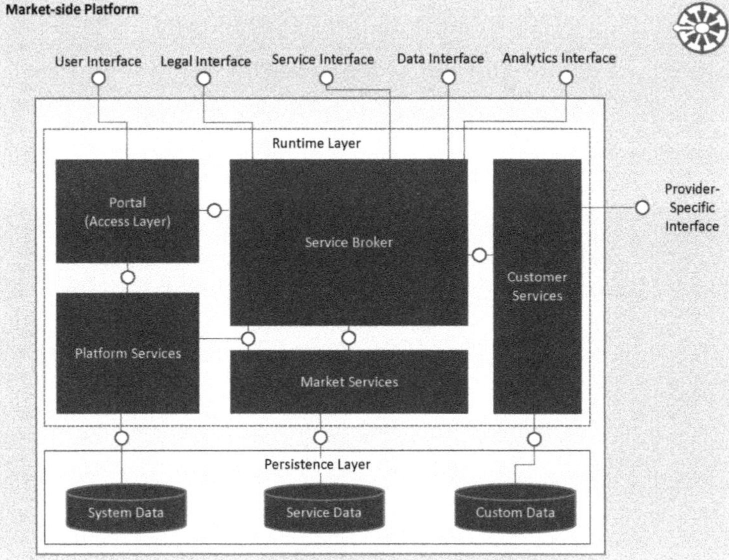

Figure 4. Architecture of the Market-side Platform.

The modular design of the MaP enables a wide range of possibilities and a flexible and company–individual extension of the interaction with regard to the degree to which the company wants to interact with the MaP. While some companies with greater flexibility potential will access several markets simultaneously and without an intermediary agent, other companies just need service providers, such as optimization or forecast services. The low entry hurdles are important to provide non-discriminatory access to all actors in the ecosystem and to foster the acceptance and reduce the threshold for participation. The MaP could be perceived as an app store, in which services can be tried on a low- or non-binding agreement level, aiming to promote excellent services through competition. The MaP not only increases transparency in the complex landscape of the power ecosystem, it also lowers the market entrance barrier for industrial power flexibility providers, while also considering (and maintaining) the regulatory framework, as well as the physical requirements.

5.3. Synthesis to an End-to-End Approach

The ESP is designed as an integrated concept including data, information, and energy flows between machines and energy markets (see Figure 5). At a technical level, the electricity flows from power plants via the system operator's public grid to the power consumers, e.g., manufacturing machines. The information flow at an organizational level is entirely managed by the ESP. The company–individual CoP offers extensive functionalities to control the power consumers and aggregate their energy flexibility potential. Due to the standardized interface between MaP and CoP, companies can easily commercialize their energy flexibility via an aggregator or services on the MaP. Here, the service broker is the central connector between energy markets and manufacturing companies. Thus, vendor lock-in is avoided because companies can easily replace aggregators or apply new services, and free competition is ensured. Beside direct trading on the energy markets, the usage of flexibility lists to reduce local grid congestion and infrastructure investments, as well as the application of optimization tools and forecast services is also possible.

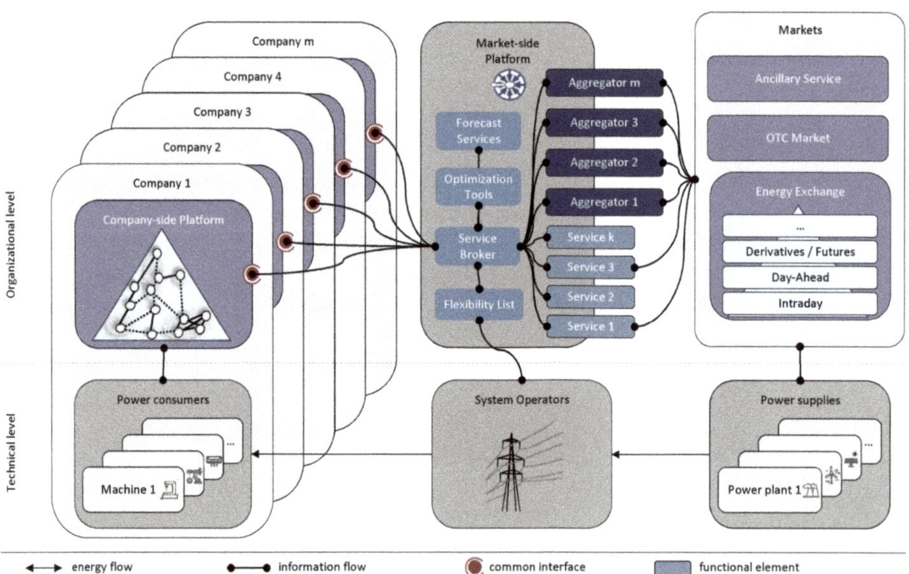

Figure 5. End-to-end approach of the energy synchronization platform integrating IT platforms, services, aggregators, and energy markets.

With respect to the end-to-end approach, processes on both MaP and CoP have been identified and modeled using BPMN. Subsequently, it was possible to design a process map for the ESP (see Figure 6). The process map ranges from the need for flexibility (customer requirements) to the reliable delivery of flexibility (customer satisfaction) and, therefore, matches the goal of the ESP as described in Section 5. Within this process map, processes are divided into comprehensive processes that affect the whole ESP and those that affect only one sub-platform. Besides this classification is based on the affected platform, processes are clustered into core processes, support processes, and management processes [86,87]. Core processes are the processes creating value for the customer, on the ESP ranging from the assessment of flexibility to its delivery and, therefore, fulfill the purpose of the ESP by synchronizing demand and offer of flexibility. It is important to note, that the core process is a cross-platform process. Hence, the successful interplay of MaP and CoP is crucial for achieving the purpose of the ESP. Support processes are necessary to ensure the successful interplay. In Figure 6, three different levels of support process are distinguished. Firstly, overlapping support processes are mainly focused on the cross-platform communication and optimization between the platforms and to third parties. Secondly, MaP and CoP support processes comprise processes that are platform-specific and do not interact with other platforms, e.g., internal management of flexibility data or user authentication. Possibly, the support processes of the ESP constitute the operations with regard to servicing tasks such as IT-management, controlling, accounting, as well as customer relationship management, but also initial processes such as company on-boarding to the ESP. The ESP community is a complementary service, in which ESP participants can exchange best practice and knowledge. Management processes set the framework for core and support processes by deciding a strategy, establishing development processes, ensuring reliable quality management, and providing a financial scope of the ESP.

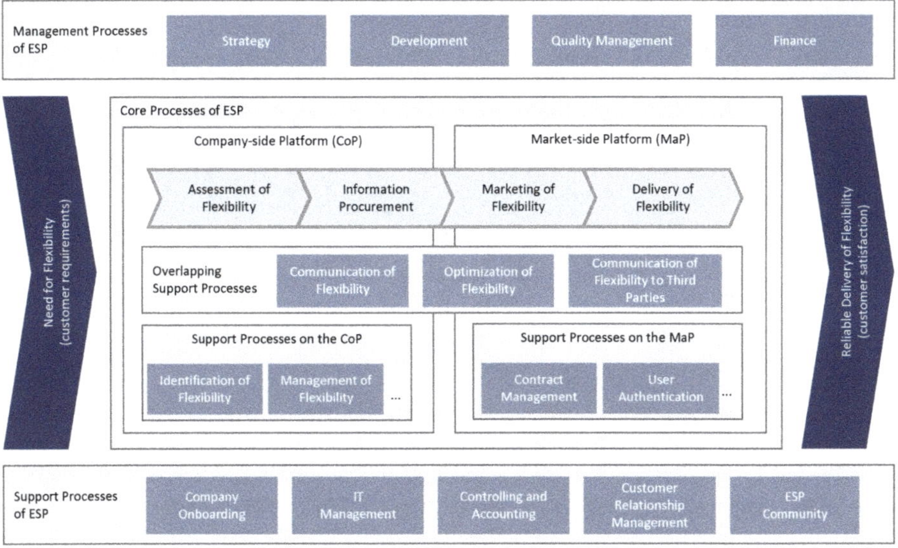

Figure 6. Process map of the energy synchronization platform.

There is much communication within and between the processes. Due to the complex nature of industrial manufacturing, a multi-level optimization is necessary to foster an effective and efficient use of demand flexibility. The resulting sequence is shown in Figure 7. The decomposition into multiple levels (Market-side optimizer, ERP optimizer, MES optimizer, Machine-side optimizer) allows for a specific optimization at each level of the automation pyramid, taking into account characteristics such as planning horizons, temporal resolutions, and maximum permissible runtimes.

Consequently, an optimization at the manufacturing planning level will target different energy markets rather than an optimization at the manufacturing control level. Long-term planning (weeks to months) on the ERP level corresponds to the time horizon of the derivatives market (see 1 in Figure 7). Optimization horizons of MES and machine-side optimizer compare to the one of day-ahead-market. Thus, the prices on the day-ahead-market are already known at noon of the previous day. Consequently, flexible adaption of production processes as a reaction to volatile prices based on this market is no longer necessary from this moment on (see 3 and 5 in Figure 7). The residual flexibility can be traded as reactive flexibility on balancing (control) power markets (see 6 in Figure 7). However, with this distributed approach, the integration of a wide range of optimization services in the described end-to-end approach becomes crucial. A solution considering all restrictions of previous optimization steps was described with detailed information flows in [56]. According to this approach, decisions on classic logistic, as well as energy goals can be made, and this with the necessary foresight. Optimization steps are chronologically ordered (marked by steps 1–7 in Figure 7), and strategic decisions, which are made at a higher level of the automation pyramid, are passed to the lower levels as given requirements. This enables an energy-oriented optimization on all levels of the automation pyramid without violating constraints defined at previous levels and, therefore, supports companies in marketing flexibility in the optimum possible markets.

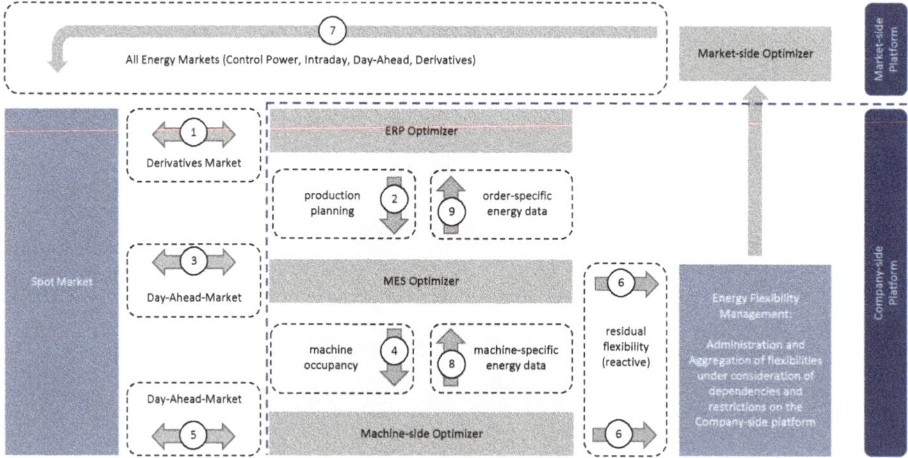

Figure 7. Approach for multi-level optimization on the ESP (adapted from [56].

6. Summary

Industrial DR shows high potential for adapting the energy demand to the increasingly fluctuating power generation of wind and solar power plants and presents an economic option to stabilize the power grid. However, the given state-of-the-art and elaborated specification analysis showed that there are currently some major deficiencies preventing manufacturing companies from participating in the energy markets. Therefore, this paper described an end-to-end approach that encompasses all necessary processes, from single machines to the energy markets, based on a platform ecosystem. Here, complementors can flexibly contribute services to other platform participants. Additionally, some fundamental services are already provided, e.g., energy flexibility management. The complex interaction of processes and services was modeled using BPMN.

The designed platform ecosystem of the ESP incorporated several advantages. First, it presented a single, continuous solution to connect machines to energy markets including all functionalities such as management, optimization, and marketing of flexibility. Second, as complementors can contribute services, co-evolution and co-creation were ensured. Consequently, comprehensive and innovative functionalities can be offered. Third, due to the architecture of the ESP and the interoperability of its components and services, companies were prevented from vendor lock-in effects, and free competition between different complementors on the ESP was ensured. In a strictly regulated market such as the energy system, this is an innovative approach that has not yet been attempted. Fourth, entry hurdles for new participants were minimized by the modularity of the platform and the different deployment models of the CoP, which were well-suited to companies of all sizes. This enabled an increased provision of flexibility. Fifth, considering security aspects in the design and operation of the ecosystem was proven as a key factor for its acceptance by companies.

Therefore, the ESP's ecosystem allows industry to automate DR with a reduced effort. Consequently, this study delivered an important contribution to the successful realization of the German and global transition of power grids and the increasing integration of fluctuating renewable energies.

7. Outlook and Remaining Deficit

As summarized in Section 6, the identified research deficits were all addressed in this paper. However, there are still some weaknesses in the developed ESP ecosystem in its current state, which are currently being investigated by the authors and will be addressed in future studies:

- Governance: The concept of governance for the invented platform ecosystem is one of the most significant remaining deficits. The governance approach should be mainly based on two elements, introducing standards and a community. To enable external access and contributions, a strict standardization of all processes, services, interfaces, data models, and communication flows is necessary. The BPMN documentation presents a first step. Nevertheless, this needs to be further developed by standardizing a reference architecture. In addition, the ecosystem requires a platform carrier to coordinate future extensions, maintenance, regulation, and safeguard ongoing operations. Since the literature indicates self-selection as the most effective approach for platform ecosystems, a central task is to build a broad community, where all participants and complementors of the platform can participate.
- Additional services: The implementation and marketing of energy flexibility in manufacturing companies are complex. Therefore, the existing services on CoP and MaP are not yet all-encompassing and cannot fulfill all requirements in every use case. Consequently, some further extensions and additional services need to be implemented, e.g., for advanced price and signal predictions, aggregation of flexibility measures, hierarchical optimization at the different operational levels, evaluation costs for energy flexibility at the manufacturing level, risk assessment of flexibility measures, etc.
- Information procurement: While the existing energy flexibility management based on the EFDM is very well suited to the applications of assessing and marketing flexibility, there is a gap with respect to information procurement, e.g., market price predictions. Therefore, an approach for flexibly connecting services on the CoP with MaP has to be developed. The challenge is to make it possible to use more than a single data model such as the one described for flexibility measures, but rather a wide range of data models must be translated without affecting the operation of the existing components.
- Security: Security by design has been the main way of integrating security aspects so far. In future research, a detailed security analysis needs to be conducted in which feared events are identified and relevant counter measures derived. In addition, standardized security requirements for any kind of interface and service need to be defined. Thus, security aspects were mainly incorporated for the encapsulation of companies and energy markets. A serious threat, however, is the deliberate disruption of the energy system through manipulation of the energy markets via the ESP. As part of a critical infrastructure [75], this aspect must, therefore, also be taken into account when extending the security concept.

- Credibility and trustworthiness: The elaborated architecture and information flow guarantee high standards of credibility and trustworthiness. However, in the recent past, there has been an increase of new distributed ledger-based technologies (e.g., block chain), which show promising results in this field. Their application should be examined and implemented to further enhance the capability of the solution.

In addition, it is essential to further apply the elaborated architecture with its components and services in a wide range of industrial usage cases. This will provide several insights into how the concept can be further developed for wider industrial applicability. Subsequently, the ESP must be prepared for roll-out and scale-up to provide significant energy flexibility potential.

Author Contributions: M.R., D.B., and L.H. contributed equally to this work. They worked jointly on the conceptualization, investigation of the research subject, the design of the general structure, and wrote the manuscript. R.K. assisted in aligning the research to an appropriate methodology and reviewed the manuscript. G.R. supervised the work of M.R. A.S.and T.B. supervised the work of D.B. G.F. supervised the work of L.H. and R.K. All authors and supervisors provided critical feedback and helped shape the research, analysis, and manuscript.

Funding: This research was funded by the German Federal Ministry of Education and Research (BMBF), Grant No. 03SFK3G1.

Acknowledgments: The authors gratefully acknowledge the financial support of the Kopernikus project "SynErgie" by the Federal Ministry of Education and Research (BMBF) and the project supervision by the project management organization Projektträger Jülich (PtJ). The authors thank all participants in SynErgie Cluster III for the fruitful discussions.

Conflicts of Interest: The funders had no role in the design of the study; in the collection, analyses, or interpretation of data; in the writing of the manuscript; nor in the decision to publish the results.

Abbreviations

The following abbreviations are used in this manuscript:

API	Application Programming Interface
BPMN	Business Process Model and Notation
CoP	Company-side Platform
DR	Demand-Side Response
DSM	Demand-Side Management
EFDM	Energy Flexibility Data Model
EMS	Energy Management Systems
ERP	Enterprise Resource Planning
ESP	Energy Synchronization Platform
IaaS	Infrastructure as a Service
IT	Information Technology
MaP	Market-side Platform
MES	Manufacturing Execution System
MSB	Manufacturing Service Bus
OPC-UA	Open Platform Communications Unified Architecture
OpenADR	Open Automated Demand Response
OTC	Over The Counter
PLC	Programmable Logic Controllers
REST	Representational State Transfer
USEF	Universal Smart Energy Framework
XaaS	Everything as a Service

References

1. United Nations. *Transforming Our World: The 2030 Agenda for Sustainable Development*; United Nations: New York, NY, USA, 2015.
2. Mann, M.E.; Rahmstorf, S.; Kornhuber, K.; Steinman, B.A.; Miller, S.K.; Coumou, D. Influence of Anthropogenic Climate Change on Planetary Wave Resonance and Extreme Weather Events. *Sci. Rep.* **2017**, *7*, 1–10. [CrossRef] [PubMed]
3. Channel, J.; Churmi, E.; Ngyen, P.; Prior, E.; Syme, A.; Jansen, H.; Rahbari, E.; Morse, E.; Kleinman, S.; Kruger, T. *Energy Darwinism II: Why a Low Carbon Future Doesn't Have to Cost the Earth*; Citi GPS: London, UK, 2015.
4. International Energy Agency. *Key World Energy Statistics 2018*; International Energy Agency: Paris, France, 2018.
5. Bertram, R.; Primova, T.; Herbert, J.; Bulantova, K.; Metaxa, K.; Ugryn, K.; Walsh, M. *Energy Atlas: Facts and Figures About Renewables in Europe 2018*; Heinrich Böll Foundation: Berlin, Germany, 2018.
6. Federal Ministry for Economic Affairs and Energy. *Energiekonzept für Eine Umweltschonende, Zuverlässige und Bezahlbare Energieversorgung*; BMWi: Berlin, Germany, 2010.
7. Umweltbundesamt. *Erneuerbare Energien in Deutschland: Daten zur Entwicklung im Jahr 2018*; Umweltbundesamt: Dessau-Roßlau, Germany, 2019.
8. Bundesnetzagentur für Elektrizität, Gas, Telekommunikation, Post und Eisenbahnen. *Quartalsbericht zu Netz- und Systemsicherheitsmaßnahmen: Gesamtjahr und Viertes Quartal 2018*; Bundesnetzagentur für Elektrizität, Gas, Telekommunikation, Post und Eisenbahnen: Bonn, Germany, 2019.
9. European Environmental Agency. *Final Energy Consumption by Sector and Fuel*; European Environmental Agency: Kopenhagen, Denmark, 2018.
10. Papaefthymiou, G.; Haesen, E.; Sach, T. Power System Flexibility Tracker: Indicators to track flexibility progress towards high-RES systems. *Renew. Energy* **2018**, *127*, 1026–1035. [CrossRef]
11. Lund, P.D.; Lindgren, J.; Mikkola, J.; Salpakari, J. Review of energy system flexibility measures to enable high levels of variable renewable electricity. *Renew. Sustain. Energy Rev.* **2015**, *45*, 785–807. [CrossRef]
12. Müller, T.; Möst, D. Demand Response Potential: Available when Needed? *Energy Policy* **2018**, *115*, 181–198. [CrossRef]
13. Lacal Arantegui, R.; Jäger-Waldau, A. Photovoltaics and wind status in the European Union after the Paris Agreement. *Renew. Sustain. Energy Rev.* **2018**, *81*, 2460–2471. [CrossRef]
14. Battaglini, A.; Komendantova, N.; Brtnik, P.; Patt, A. Perception of barriers for expansion of electricity grids in the European Union. *Energy Policy* **2012**, *47*, 254–259. [CrossRef]
15. Lund, H.; Østergaard, P.A.; Connolly, D.; Ridjan, I.; Mathiesen, B.V.; Hvelplund, F.; Thellufsen, J.Z.; Sorknæs, P. Energy Storage and Smart Energy Systems. *Int. J. Sustain. Energy Plan. Manag.* **2016**, *11*, 3–14. [CrossRef]
16. Palensky, P.; Dietrich, D. Demand Side Management: Demand Response, Intelligent Energy Systems, and Smart Loads. *IEEE Trans. Ind. Inf.* **2011**, *7*, 381–388. [CrossRef]
17. Feuerriegel, S.; Neumann, D. Measuring the financial impact of demand response for electricity retailers. *Energy Policy* **2014**, *65*, 359–368. [CrossRef]
18. Albadi, M.H.; El-Saadany, E.F. A summary of demand response in electricity markets. *Electr. Power Syst. Res.* **2008**, *78*, 1989–1996. [CrossRef]
19. Markle-Huss, J.; Feuerriegel, S.; Neumann, D. Decision model for sustainable electricity procurement using nationwide demand response. In Proceedings of the 49th Annual Hawaii International Conference on System Sciences, Koloa, HI, USA, 5–8 January 2016; pp. 1010–1019. [CrossRef]
20. Jazayeri, P.; Schellenberg, A.; Rosehart, W.D.; Doudna, J.; Widergren, S.; Lawrence, D.; Mickey, J.; Jones, S. A Survey of Load Control Programs for Price and System Stability. *IEEE Trans. Power Syst.* **2005**, *20*, 1504–1509. [CrossRef]
21. Steurer, M. Analyse von Demand Side Integration im Hinblick auf eine Effiziente und Umweltfreundliche Energieversorgung. Ph.D. Thesis, University of Stuttgart, Stuttgart, Germany, 2017.
22. Bertsch, J.; Fridgen, G.; Sachs, T.; Schöpf, M.; Schweter, H.; Sitzmann, A. *Ausgangsbedingungen für die Vermarktung von Nachfrageflexibilität: Status-Quo-Analyse und Metastudie*; Bayreuther Arbeitspapiere zur Wirtschaftsinformatik; University of Bayreuth: Bayreuth, Germany, 2017; Volume 62.

23. Lübbecke, M.; Koster, A.M.; Letmathe, P.; Madlener, R.; Peis, B.; Walther, G. *Operations Research Proceedings 2014: Selected Papers of the Annual International Conference of the German Operations Research Society (GOR), RWTH Aachen University, Germany, 2–5 September 2014*; Springer: Cham, Switzerland, 2016.

24. Maier, F.; Belhassan, H.; Klempp, N.; Koetter, F.; Siehler, E.; Stetter, D.; Wohlfrom, A. Decision Support for Structured Energy Procurement. In Proceedings of the 6th International Conference on Smart Cities and Green ICT Systems-Volume 1, Porto, Portugal, 22–24 April 2017; pp. 77–86. [CrossRef]

25. Rackow, T.; Kohl, J.; Canzaniello, A.; Schuderer, P.; Franke, J. Energy Flexible Production: Saving Electricity Expenditures by Adjusting the Production Plan. *Procedia CIRP* **2015**, *26*, 235–240. [CrossRef]

26. Zafirakis, D.; Chalvatzis, K.J.; Baiocchi, G.; Daskalakis, G. The value of arbitrage for energy storage: Evidence from European electricity markets. *Appl. Energy* **2016**, *184*, 971–986. [CrossRef]

27. Fridgen, G.; Häfner, L.; König, C.; Sachs, T. Providing Utility to Utilities: The Value of Information Systems Enabled Flexibility in Electricity Consumption. *J. Assoc. Inf. Syst.* **2016**, *17*, 537–563. [CrossRef]

28. Fridgen, G.; Keller, R.; Thimmel, M.; Wederhake, L. Shifting load through space—The economics of spatial demand side management using distributed data centers. *Energy Policy* **2017**, *109*, 400–413. [CrossRef]

29. Koliou, E.; Eid, C.; Chaves-Ávila, J.P.; Hakvoort, R.A. Demand response in liberalized electricity markets: Analysis of aggregated load participation in the German balancing mechanism. *Energy* **2014**, *71*, 245–254. [CrossRef]

30. Dranka, G.G.; Ferreira, P. Review and assessment of the different categories of demand response potentials. *Energy* **2019**, *179*, 280–294. [CrossRef]

31. Strbac, G. Demand side management: Benefits and challenges. *Energy Policy* **2008**, *36*, 4419–4426. [CrossRef]

32. Meeus, L.; Hancher, L.; Azevedo, I.; He, X.; Keyaerts, N.; Glachant, J.M. *Shift, Not Drift: Towards Active Demand Response and Beyond (Topic 11)*; Publications Office: Luxembourg, 2013.

33. Graßl, M.; Reinhart, G. Evaluating Measures for Adapting the Energy Demand of a Production System to Volatile Energy Prices. *Procedia CIRP* **2014**, *15*, 129–134. [CrossRef]

34. Roesch, M.; Berger, C.; Braunreuther, S.; Reinhart, G. Cost-model for Energy-oriented Production Control. In Proceedings of the 2018 IEEE International Conference on Industrial Engineering and Engineering Management (IEEM), Bangkok, Thailand, 16–19 December 2018; pp. 158–162. [CrossRef]

35. VDI. *VDI 5600-1: Manufacturing Execution Systems (MES)*; Beuth Verlag: Berlin, Germany, 2016.

36. Sauter, T.; Soucek, S.; Kastner, W.; Dietrich, D. The Evolution of Factory and Building Automation. *IEEE Ind. Electron. Mag.* **2011**, *5*, 35–48. [CrossRef]

37. Thomas, L.D.W.; Autio, E.; Gann, D.M. Architectural Leverage: Putting Platforms in Context. *Acad. Manag. Perspect.* **2014**, *28*, 198–219. [CrossRef]

38. Gawer, A. Bridging differing perspectives on technological platforms: Toward an integrative framework. *Res. Policy* **2014**, *43*, 1239–1249. [CrossRef]

39. Tilson, D.; Sorensen, C.; Lyytinen, K. Change and Control Paradoxes in Mobile Infrastructure Innovation: The Android and iOS Mobile Operating Systems Cases. In Proceedings of the 2012 45th Hawaii International Conference on System Science, Maui, HI, USA, 4–7 January 2012; pp. 1324–1333. [CrossRef]

40. Parker, G.; van Alstyne, M.; Jiang, X. Platform Ecosystems: How Developers Invert the Firm. *MIS Q.* **2017**, *41*, 255–266. [CrossRef]

41. Tiwana, A.; Konsynski, B.; Bush, A.A. Research Commentary—Platform Evolution: Coevolution of Platform Architecture, Governance, and Environmental Dynamics. *Inf. Syst. Res.* **2010**, *21*, 675–687. [CrossRef]

42. Keller, R. Cloud Networks as Platform-Based Ecosystems: Detecting Management Implications for Actors in Cloud Networks. Ph.D. Thesis, University of Bayreuth, Bayreuth, Germany, 2019.

43. de Reuver, M.; Sørensen, C.; Basole, R.C. The Digital Platform: A Research Agenda. *J. Inf. Technol.* **2018**, *33*, 124–135. [CrossRef]

44. Ozalp, H.; Cennamo, C.; Gawer, A. Disruption in Platform–Based Ecosystems. *J. Manag. Stud.* **2018**, *55*, 1203–1241. [CrossRef]

45. Inoue, Y.; Tsujimoto, M. New market development of platform ecosystems: A case study of the Nintendo Wii. *Technol. Forecast. Soc. Chang.* **2018**, *136*, 235–253. [CrossRef]

46. Forbes. *The World's Most Valuable Brands*; Forbes Media: New York, NY, USA, 2018.

47. Eisenmann, T.R. *Winner-Takes-All in Networked Markets*; Harvard Business School: Boston, MA, USA, 2007; pp. 806–131.

48. Inoue, Y. Winner-Takes-All or Co-Evolution among Platform Ecosystems: A Look at the Competitive and Symbiotic Actions of Complementors. *Sustainability* **2019**, *11*, 726. [CrossRef]

49. Clements, M.T.; Ohashi, H. Indirect Network Effects and the Product Cycle: Video Games in the U.S., 1994–2002. *J. Ind. Econ.* **2005**, *53*, 515–542. [CrossRef]

50. Ceccagnoli, M.; Forman, C.; Huang, P.; Wu, D.J. Cocreation of Value in a Platform Ecosystem! The Case of Enterprise Software. *MIS Q.* **2012**, *36*, 263. [CrossRef]

51. Wareham, J.D.; Fox, P.B.; Cano Giner, J.L. Technology Ecosystem Governance. *Org. Sci.* **2014**, *25*, 1195–1215. [CrossRef]

52. Huber, T.L.; Kude, T.; Dibbern, J. Governance Practices in Platform Ecosystems: Navigating Tensions Between Cocreated Value and Governance Costs. *Inf. Syst. Res.* **2017**, *28*, 563–584. [CrossRef]

53. IEC. *IEC 62264-1: Enterprise-Control System Integration-Part 1: Models and Terminology*; IEC: Geneva, Switzerland, 2013.

54. Acatech. *Cyber-Physical Systems: Driving Force for Innovation in Mobility, Health, Energy and Production*; Acatech-National Academy of Scvience and Engineering: Berlin, Germany, 2011.

55. Körner, M.F.; Bauer, D.; Keller, R.; Rösch, M.; Schlereth, A.; Simon, P.; Bauernhansl, T.; Fridgen, G.; Reinhart, G. Extending the Automation Pyramid for Industrial Demand Response. *Procedia CIRP* **2019**, *81*, 998–1003. [CrossRef]

56. Seitz, P.; Abele, E.; Bank, L.; Bauernhansl, T.; Colangelo, E.; Fridgen, G.; Schilp, J.; Schott, P.; Sedlmeir, J.; Strobel, N.; et al. IT-based Architecture for Power Market Oriented Optimization at Multiple Levels in Production Processes. *Procedia CIRP* **2019**, *81*, 618–623. [CrossRef]

57. Monostori, L.; Kádár, B.; Bauernhansl, T.; Kondoh, S.; Kumara, S.; Reinhart, G.; Sauer, O.; Schuh, G.; Sihn, W.; Ueda, K. Cyber-physical systems in manufacturing. *CIRP Ann.* **2016**, *65*, 621–641. [CrossRef]

58. Zezulka, F.; Marcon, P.; Vesely, I.; Sajdl, O. Industry 4.0–An Introduction in the phenomenon. *IFAC-PapersOnLine* **2016**, *49*, 8–12. [CrossRef]

59. Tsujimoto, M.; Kajikawa, Y.; Tomita, J.; Matsumoto, Y. A review of the ecosystem concept—Towards coherent ecosystem design. *Technol. Forecast. Soc. Change* **2018**, *136*, 49–58. [CrossRef]

60. Duan, Y.; Fu, G.; Zhou, N.; Sun, X.; Narendra, N.C.; Hu, B. Everything as a Service (XaaS) on the Cloud: Origins, Current and Future Trends. In Proceedings of the 2015 IEEE 8th International Conference on Cloud Computing, New York, NY, USA, 27 June–2 July 2015; pp. 621–628. [CrossRef]

61. Ren, L.; Zhang, L.; Tao, F.; Zhao, C.; Chai, X.; Zhao, X. Cloud manufacturing: from concept to practice. *Enterp. Inf. Syst.* **2015**, *9*, 186–209. [CrossRef]

62. Stock, D.; Stöhr, M.; Rauschecker, U.; Bauernhansl, T. Cloud-based Platform to Facilitate Access to Manufacturing IT. *Procedia CIRP* **2014**, *25*, 320–328. [CrossRef]

63. Holtewert, P.; Wutzke, R.; Seidelmann, J.; Bauernhansl, T. Virtual Fort Knox Federative, Secure and Cloud-based Platform for Manufacturing. *Procedia CIRP* **2013**, *7*, 527–532. [CrossRef]

64. Huang, B.; Li, C.; Yin, C.; Zhao, X. Cloud manufacturing service platform for small- and medium-sized enterprises. *Int. J. Adv. Manuf. Technol.* **2013**, *65*, 1261–1272. [CrossRef]

65. Bauer, D.; Stock, D.; Bauernhansl, T. Movement Towards Service-orientation and App-orientation in Manufacturing IT. *Procedia CIRP* **2017**, *62*, 199–204. [CrossRef]

66. Schott, P.; Ahrens, R.; Bauer, D.; Hering, F.; Keller, R.; Pullmann, J.; Schel, D.; Schimmelpfennig, J.; Simon, P.; Weber, T.; et al. Flexible IT platform for synchronizing energy demands with volatile markets. *IT-Inf. Technol.* **2018**, *60*, 155–164. [CrossRef]

67. Hülsbömer, S.; Rozsa, A.; Schonschek, O.; Thomas-Ißbrücker, T. *Studie Cloud Security 2019*; IDG Business Media GmbH: München, Germany, 2019.

68. Keller, R.; König, C. A Reference Model to Support Risk Identification in Cloud Networks. In Proceedings of the International Conference on Information Systems (ICIS 2014): Building a Better World through Information Systems, Auckland, New Zealand, 14–17 December 2014.

69. Federal Office for Economic Affairs and Export Control. *Der Markt für Energiemanagement-Systeme in Kleinen und Mittleren Unternehmen*; Federal Office for Economic Affairs and Export Control: Eschborn, Germany, 2017.

70. Kahlenborn, W.; Kabisch, S.; Klein, J.; Richter, I.; Schürmann, S. *Energy Management Systems in Practice: ISO 50001: A Guide for Companies and Organisations*; BMU: Berlin, Germany, 2012.

71. Sauer, A.; Weckmann, S.; Zimmermann, F. *Softwarelösungen für das Energiemanagement von Morgen: Eine Vergleichende Studie*; University of Stuttgart: Stuttgart, Germany, 2016.

72. Fontana, M.E.; Aragão, J.P.S.; Morais, D.C. Decision support system for outsourcing strategies. *Prod. Eng.* **2019**, *22*, 832. [CrossRef]

73. Siano, P. Demand response and smart grids—A survey. *Renew. Sustain. Energy Rev.* **2014**, *30*, 461–478. [CrossRef]

74. Seifermann, S.; Abele, E.; Bauernhansl, T.; Brecher, C.; Franke, J.; Herrmann, C.; Putz, M.; Reinhart, G.; Thiede, S.; Zaeh, M. Energy Flexibility in Manufacturing: The Kopernikus-Project SynErgie. In Proceedings of the CIRP 2017 General Assembly, STC-A Meeting, Lugano, Switzerland, 20–26 August 2017.

75. Federal Ministry of the Interior. *Cyber Security Strategy for Germany*; Federal Ministry of the Interior: Berlin, Germany, 2011.

76. Ländner, E.M.; Märtz, A.; Schöpf, M.; Weibelzahl, M. From energy legislation to investment determination: Shaping future electricity markets with different flexibility options. *Energy Policy* **2019**, *129*, 1100–1110. [CrossRef]

77. Gilder, G. Metcalf's Law and Legacy. *Forb. ASAP* **1993**, *152*, 158–166.

78. Schott, P.; Sedlmeir, J.; Strobel, N.; Weber, T.; Fridgen, G.; Abele, E. A Generic Data Model for Describing Flexibility in Power Markets. *Energies* **2019**, *12*, 1893. [CrossRef]

79. Neugebauer, R.; Putz, M.; Schlegel, A.; Langer, T.; Franz, E.; Lorenz, S. Energy-Sensitive Production Control in Mixed Model Manufacturing Processes. In *Leveraging Technology for a Sustainable World*; Dornfeld, D.A., Linke, B.S., Eds.; Springer: Berlin/Heidelberg, Germany, 2012; pp. 399–404.

80. Reinhart, G.; Geiger, F.; Karl, F.; Wiedemann, M. Handlungsfelder zur Realisierung energieeffizienter Produktionsplanung und -steuerung. *ZWF Z. für Wirtsch. Fabr.* **2011**, *106*, 596–600. [CrossRef]

81. Pfeilsticker, L.; Colangelo, E.; Sauer, A. Energy Flexibility—A new Target Dimension in Manufacturing System Design and Operation. *Procedia Manuf.* **2019**, *33*, 51–58. [CrossRef]

82. Schel, D.; Bauer, D.; Vazquez, F.G.; Schulz, F.; Bauernhansl, T. IT Platform for Energy Demand Synchronization Among Manufacturing Companies. *Procedia CIRP* **2018**, *72*, 826–831. [CrossRef]

83. Mínguez, J. A Service-Oriented Integration Platform for Flexible Information Provisioning in the Real-Time Factory. Ph.D. Thesis , Unversity of Stuttgart, Stuttgart, Germany, 2012.

84. Schel, D.; Henkel, C.; Stock, D.; Meyer, O.; Rauhöft, G.; Einberger, P.; Stöhr, M.; Daxer, M.A.; Seidelmann, J. Manufacturing Service Bus: An Implementation. *Procedia CIRP* **2018**, *67*, 179–184. [CrossRef]

85. Hagiu, A.; Wright, J. Multi-Sided Platforms. *Int. J. Ind. Org.* **2015**, *43*, 162–174. [CrossRef]

86. Porter, M.E. *Competitive Advantage: Creating and Sustaining Superior Performance*; Free Press: New York, NY, USA, 1985.

87. Rosing, M.V.; Scheer, A.W.; Scheel, H.V. *The Complete Business Process Handbook: Body of Knowledge from Process Modeling to Bpm: Volume I*; Morgan Kaufmann: Waltham, MA, USA, 2015.

Article

Technology Portfolio and Role of Public Research Institutions in Industry 4.0: A Case of South Korea

SungUk Lim [1] and Junmo Kim [2,*]

[1]   Department of Industrial Engineering, Dae-Jin University, 1007 Hoguk-Ro, Gyeonggi, Pocheon 11159, Korea
[2]   Department of Public Admin, Konkuk University, Neung Dong Ro 120, Gwang-Jin-Gu, Seoul 05029, Korea
*   Correspondence: junmokim@konkuk.ac.kr or junmokim@unitel.co.kr

Received: 7 June 2019; Accepted: 21 June 2019; Published: 28 June 2019

Featured Application: Management and Public Policy oriented research on the realignment of research organizations.

Abstract: The 4th industrial revolution has been a hot topic in various societies for several overlapping reasons. It may be a huge wave for researchers to navigate through. In this context, research institutions are not different from major industrial sectors, in that both consider the 4th revolution a major turning point as well as a threat. Today's industries and research institutions are knowledge-intensive in nature. Consequently, their potential for survival depends on scientific and technological aspects as well as their organizational dimension. This study analyzes 25 major public research institutions in South Korea, located in the DaeDuk area, based on their technological capability for organizational and expert evaluation. It also proposes a matching scheme between research institutions and research topics related to the 4th industrial revolution.

Keywords: 4th industrial revolution; industry 4.0; AHP; QFD; matching

---

1. Introduction

The 4th industrial revolution has been a hot topic in various societies for several overlapping reasons. It may be a huge wave for researchers to navigate through. In this context, research institutions are not different from major industrial sectors, in that both consider the 4th revolution a major turning point as well as a threat [1,2]. Today's industries and research institutions are knowledge-intensive in nature. As a result, their potential for survival depends on scientific and technological aspects as well as their organizational dimension [3,4]. This implies that research institutions' organizational and manpower strategy is one of the key aspects in understanding such institutions [5,6]. If the characteristic of the 4th industrial revolution can be understood as increasing connectiveness with information technology, countries that are aiming, by certain degrees, at reconfiguration of research institutions, whether they are public or private in their origin, may find similar problem definitions and solutions. This research intended to find a clue with a case study of research institutions in Korea with potential implications for other countries.

In this context, this study analyzes 25 major public research institutions located in the DaeDuk area in South Korea, based on their technological capability for organizational as well as expert evaluation. The study also proposes a matching scheme between research institutions and research topics related to the 4th industrial revolution.

2. Literature Review

2.1. The Industry 4.0 Context and Product/Technology Life Cycle Theory

Industry 4.0 has been a keyword not only in business circles, but also in general media and academic circles [7]. Industry 4.0, or the 4th industrial revolution, has been on the agenda, despite the fact that the concept has often been either overstated or misrepresented to encompass diverse meanings [8]. Instead of analyzing the concept in depth, this study focuses on its core element: the connectivity of things [1]. The use of Information Technology (IT) has made this connectivity possible. This implies that, if there had been an alternative for the connectivity, it could be a medium other than IT.

While connectivity is the core element of the 4th industrial revolution [9], technology life cycle and product life cycle theories still hold their validity even for Industry 4.0 [10]. Proposed by Vernon in 1966, the product life cycle theory shows that there are product flows among country groups based on how early a country can initiate a new movement. Advanced industrial nations start the waves, which then flow into less developed countries. The technology life cycle theory follows the same logic, but focuses on technology flow [11] (Figure 1).

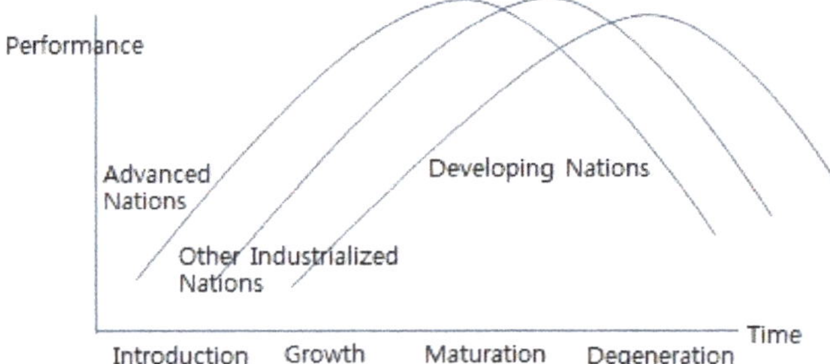

(Source: Raymond Vernon 1966: "International Investment and International Trade in the Product Cycle" The Quarterly Journal of Economics MIT Press/adapted and re-drawn from the original article)

Figure 1. A typical technology life cycle model.

2.1.1. Differing Temperature Between Country Groups

In the Industry 4.0 context, both the theories still make sense. For a country like the Philippines, the accepted meaning of the 4th industrial revolution is far different from that of the European countries [12]. The Philippines is still going through the so-called IT revolution or the 3rd industrial revolution by participating in the field of IT-related manufacturing and increasing its presence in the field. The case would be the same, if not much more intensive, in Vietnam [13], for example, where a good portion of Samsung smartphones has been assembled.

Based on this logic [14], it would be reasonable to infer that the 4th industrial revolution will take the form of a series of waves, as predicted by the technology life cycle or product life cycle theory. Consequently, the requirements for adapting to the 4th industrial revolution will vary across countries [15,16], and so will the roles of universities and research organizations.

2.1.2. "Leave us Alone" (Pure Research) vs. Entrepreneurial Research Model

There have been successful cases of transformation involving research communities, universities, and firms. Cases from the Basque area of Spain are typical examples. The entrepreneurial research model, called the CIC marGUNE (the Co-operative Research Centre into High-performance Manufacturing), showed that technological research institutions and universities worked well with regional industries with the sponsorship of the government for upgrading industrial manufacturing processes [17].

Compared to the traditional "stand-alone" type research, which will find its final application fields at the terminal stage, the Basque model shows a concurrent collaboration among players in research and production [18]. This can be viewed as a harbinger for the forthcoming "4th industrial revolution" era spirit of collaboration. Additionally, cases like the Basque have an implication for other countries and regions that are seriously looking at the transformation of research institutions and industries into the 4th industrial revolution.

2.2. Changing Role of Public Research Institutions

2.2.1. Changing Role of Public Research Institutions

The technology life cycle theory, as explained in the previous section, has provided insights for understanding the dynamics between countries in different groups. Like other developing countries, South Korea also had its own needs for technology development. The first type of need was related to the high costs of technology transfer from advanced nations. Vernon's international division of labor, which covers the technology life cycle, does not sufficiently explain the cost side of international technology transfer [19,20]. This does not mean that the theory is deficient; rather, it is reasonable to argue that the cost side was treated as hidden [21]. Thus, it would also be sensible to infer that if the costs of technology transfer exceed the traditional level, a country will have greater incentives for technology development [22,23].

The second type of need may arise from the potential restrictions on buyers that constrain their use of the technology. If the costs of bearing the restrictive clauses outweigh the costs of indigenous development, a technology-importing country is likely to opt for indigenous development.

It was in this context that public research institutions were established for supporting technology development in South Korea in the 1970s. The roles of public research institutions have changed over time with economic growth, allowing for a generalization of their roles with potential applications in other countries.

Column B in Table 1 shows the typical area in which public research institutions in South Korea have been engaged in since the 1970s. With increased economic development, the role of and expectations from public research institutions have also been strengthened.

Table 1. Stages of technology development cooperation.

Pre-Competition			Competition Stage				
R&D Cooperation			Technical Cooperation			Manufacturing/Marketing Cooperation	
A	B	C	D	E	F	G	H
Univ. centered research funded by private entities	Gov't-industry cooperation project in which univ and public research institutions participate	arrangement based on private firms	a venture capital investment within a big company	a non-equity technological arrangement between firms on selective areas	inter-firm technical arrangement which includes a multiple types of cross licensing of technology	a joint venture or a comprehensive R&D, manufacturing, and marketing consortium	a licensing agreement or marketing arrangement

(Source: Adapted from OECD Science, Technology and Industry Outlook 2012, 2014).

2.2.2. Performance Review Requirements

Performance evaluation of public research institutions had been based mainly on performance review of their research projects until the 1980s. Starting from the 1990s, a comprehensive watershed came under the governance of public research organizations for two reasons [24,25]. The first was the expansion of government budget allocated to different research functions, under which public organizations could receive direct or indirect funds by participating in specific research projects. This brought in some sort of scrutiny in the management of the research budget. Second, by the time this reform was initiated, there were 22 government-funded research institutions, and the central government wanted to raise their budgetary allocation. The approach of project-based systems (PBS) was introduced in 1996 because of these two reasons.

Under the PBS, the Ministry of Finance would allocate funds to each public research institution for paying a share of salaries (usually between 30% and 80%) and for covering research budget, which had direct costs and overhead cost elements. After introducing the PBS, the government constituted a governing body named National Science & Technology Research Council for the 22 government-funded research institutions; the Council still exists, albeit with some modifications in terms of its structure and functions. In December 2005, the government promulgated a new law for national research and development projects, titled Government Performance and Review Act. The Act designated the jurisdiction of the evaluation of the public research organizations at the Ministry of Finance, which was formerly at the National Science and Technology Research Council. One peculiar feature was that after the Ministry of Finance assumed this control, the emphasis was more on general management evaluation than on research-related evaluation.

Reflecting the Ministry of Finance's lack of attention to research institutions, in 2013, the government transferred its control tower function of the evaluation of research institutions to the Ministry of Science and Technology.

2.2.3. New Roles in the 4th Industrial Revolution

With the 4th industrial revolution becoming a hot topic in policy circles, the role of public research institutions in South Korea became an issue in 2018. The Ministry of Science and Technology asked public research institutions to prepare their own "Role & Responsibility" (R&R) statements, which the institutions duly completed in 2018. The Ministry of Finance then introduced a real challenge, by proposing to link these institutions to their new roles in the context of the 4th industrial revolution [26–28]. This study identifies a way to enhance the unique contribution of public research institutions to the waves of the 4th industrial revolution.

3. Methodology

Data and Methodology

In order to fulfill the aim of the study, this research employed an extensive data of the 25 public research institutions in Korea during summer and fall of 2018. As for details, first, this research acquired organizational, man power-related data, and pay structure of the institutions. Second, this study acquired performance-related data that were published to have validity. Third, by processing the two data sets, this research produced a AHP survey text, which has generated new data to be analyzed in the research. Then, this study utilized three types of methods. First, a three-stage analytical hierarchy process (AHP) technique was employed to set up an in-depth analysis. Second, an organizational analysis of all the 25 public research institutions was performed by using their manpower-related data. Third, this study utilized the quality function deployment (QFD) analysis to match the potential roles of research institutions with topics under the 4th industrial revolution. Lastly, this study linked the AHP and QFD scores to derive implications for the institutions and their roles in the context of the 4th industrial revolution.

The AHP analysis (Table 2) was carried out to assess the existing data. While it is possible to gather and analyze institution-level performance data, it is not easy to evaluate articles and patents in pure numbers. The method will urge experts to assign weight to individual institutions by looking at "all possible" performance data of the institutions. The AHP analysis was conducted with 25 experts who have been active in the research community for more than 20 years after their Ph.D. This study conducted a three-stage AHP survey to obtain relative weights of factors, which will be multiplied to organizational performance data from the 25 public institutions. AHP participants were asked to give weights at the three levels from the first stage to the third so that their averaged weighting was to be applied to the existing performance data of the 25 institutions. The first round produced a result of scientific/engineering excellence (0.669), excellence in operation/management (0.126), and efficiency in manpower and social values (0.205), which was then divided to be narrowed down to the next two stages.

Table 2. AHP weighting elements.

1st Stage	2nd Stage	3rd Stage
Scientific/Engineering Excellence	Necessity for research Levels of research objectives Excellence in research	Articles/patens per researcher Profitability of outcomes Growth potentials o research Spin-off potentials
Excellence in Operation/Management	Innovation strategy R&D effectiveness Validity of Research Efficiency in budget	R&D intensity R&D manpower R&D cooperation R&D investment Organizational efficiency R&D budget efficiency
Efficiency in Man power & social values	Socioeconomic perspective Public interest perspective Future change Adapatability Infra structure	

QFD is a quality management technique used to match consumer demand requirements and product specifications, and later, to check the affinity between different categories. In this study, QFD was utilized to link research institutions and their potential contributions to research topics related to the 4th industrial revolution.

After completing both the AHP and QFD analyses, we plotted their results to link research institutions and their potential contribution to the 4th industrial revolution. The focus was on finding those institutions that scored high on both analyses. These organizations can serve as the mediating entity for collaborative research in the 4th industrial revolution.

4. Findings

4.1. Organizational Dynamics Analysis

We aimed to forecast manpower changes in five years from 2019 to 2024. We found that approximately 21% of researchers are retiring due to age limits. This had tremendous implications in the functioning of a research institution. Currently, most of the 25 public research organizations had a reverse pyramid structure, where the composition of senior researchers takes the "top heavy" part. This also implies that in the next five years, if no special changes happen, younger Ph.Ds. will occupy the top positions, which will bring the organizational shape back to the "normal" pyramid type until the same organizations turn into a reverse pyramid shape in the future. This point necessitates a retooling of manpower structure in these institutions in the 4th industrial revolution era so that research organizations may not naturally follow the cyclical changes from normal pyramid to reverse pyramid shape. In fact, empirical findings have confirmed the point.

Another way of interpreting the implication for the organizational shape is based on the data this study analyzed. While it is difficult to determine who—Senior researchers or junior researchers—Contribute more to organizational performance in a cold context, because different ways of proving will always override the opposite hypothesis, Table 3 shows an indirect implication.

Table 3. Comparison between senior and junior researchers.

Composition		Number of Articles	Number of Patents
Senior Level	Pearson Coefficient	0.298	0.914
Junior Ph.D., level	Pearson Coefficient	0.27	0.495
Lower level	Pearson Coefficient	0.202	0.610
Senior Level Annual salary/man	Pearson Coefficient	−0.26	−0.154
Junior Ph.D., level Annual salary/man	Pearson Coefficient	−0.262	−0.244
Lower level Annual salary/man	Pearson Coefficient	−0.157	−0.203

The upper part of Table 3 shows that seniority matters, because senior researchers excelled in both articles and patents. However, considering the bottom part of the table expressed in "pure" money terms, there was no difference across the three groups. It could mean that we need researchers from all three age brackets to achieve better results. This coincides with the academic finding of the "I" shape organization model for research institutes.

4.2. AHP Analysis Using the Pre-Announced Performance Data

Table 4 presents a partial example of preannounced performance data of the institutions. While it is seemingly organized data-wise, it is, in fact, not feasible to derive implications, because the value or impact of articles and patents of each organization, on average, is underrepresented. That is why this study employs the AHP-type value-assessing method.

Table 4. Institutional performance data.

	Excellence of Research						Potential for Market Success
Results	1. Articles, Patents			2.Profitability Of Research Outcomes	Growth Potential ff Research Outcomes		Business Feasibility
	Academic Articles per Researcher	Patents Application per Researcher	Patents Registration per Researcher	Running Patents(B)	Tech Transfer Contracts per Researcher		Paid Tech Transfer Contracts per Researcher
Mean	0.57	0.17	0.15	0.65	0.03		0.03
Min	0.00	0.00	0.00	0.00	0.00		0.00
Median	0.34	0.14	0.13	0.52	0.03		0.02
Max	2.03	0.59	0.37	2.48	0.11		0.11

	Effectiveness of research related resources		Validity of research activities		Efficiency in Budget		Efficiency in Manpower management	
	R&D intensity	R&D Manpower proportion	R&D cooperation	R&D Investment intensity	Budget efficiency	R&D Budget efficiency	Man power efficiency	
Results	Degrees Of patent use	Total number of R&D related personnel	# of Cooperative Research Projects Across institutions Per researcher	Amount of Sustainably Secured Research Budget Per researcher	Ratio of Total Current operating Budget over Total Budget	Total Direct Research Budget Per researcher	% of Full time emploees	Salary Per person
Average	26.7%	71.3%	0.005	135	6.8%	180	78.4%	71
Min	12.2%	43.0%	0.000	35	2.4%	84	61.8%	55
Median	27.1%	71.7%	0.004	132	6.3%	169	78.6%	71
Max	53.9%	87.0%	0.018	231	13.2%	480	95.3%	92

By multiplying the AHP results of weighting to the existing data, we obtained a performance portfolio of the 25 institutions (Table 5).

Table 5. AHP results.

Grouping	Ranking	Converted Scores
A Group	1–9	94.69–81.32
B Group	10–16	78.94–71.10
C Group	17–21	68.53–60.51
D Group	22–25	59.85–54.47

4.3. QFD Results

With the QFD analysis (Tables 6 and 7), a matching between the 4th industrial revolution-related research topics and research institutions was made. It can be understood that the higher the number, the more the relatedness of the institution to the research topic. In Table 6, research organizations such as the KIST, ETRI, and KIMM are listed as the key institutions to play the central role in research related to the 4th industrial revolution.

Table 6. Matching between institutions and topics related to the 4th industrial revolution (QFD analysis).

	1 (KIST)	2 (GTC)	3 (KBSI)	4 (NFRI)	5 (KASI)	6 (KRIBB)	7 (KISTI)	8 (KIOM)	9 (KITECH)	10 (ETRI)	11 (NSRI)	12 (KICT)	13 (KRRI)	14 (KRISS)	15 (KFRI)	16 (WIKIM)	17 (KIGAM)	18 (KIMM)	19 (KIMS)	20 (KARI)	21 (KIER)	22 (KERI)	23 (KRICT)	24 (KIT)	25 (KAERI)	RELATED-NESS
Big Data	3		1				9																			13
Communications										9																9
A.I.							1			3												3				7
Autonomous vehicle									1									3								4
Drones)									1									3	1	1						6
Health						1		3							1	1						1	1	1		9
Smart city	1											9	3													13
Virtual reality										3																3
Robots			3						3									9	3							18
Semiconductor	9									3									3							15
material	3								3	1									1							8
Medicine						9		3															3			15
energy	1	1		1																	9		3		9	24
Fitness	17	1	4	1	0	10	10	6	8	19	0	9	3	0	1	1	0	15	8	1	9	4	7	1	9	

One possible interpretation comes from the very nature of the 4th industrial revolution, where different fields are expected to come together for integration of oriented research fields; this clearly testifies that IT, machinery, and an interdisciplinary institution (KIST) could be the best fit.

This fitness relationship between research institutions and research topics related to the 4th industrial revolution is presented in detail in Figure 2. Taking AHP scores and QFD scores as the horizontal and vertical axes, respectively, Figure 2 shows that the institutions that have obtained high scores in both criteria can be regarded as the center for collaborative research in the context of the 4th industrial revolution. These institutions were KIST, KIMM, and ETRI, all of which can be characterized as institutions with integration role as one of their key functions in their original mission statement-based tasks.

Table 7. Index of research institutions.

1	KIST (Korea Institute of Science and Technology)
2	GREEN TECH CENTER (Green Technology Center)
3	KBSI (Korea Basic Science Institute)
4	NFRI (National Fusion Research Institute)
5	KASI (Korea Astronomy and Space Science Institute)
6	KRIBB (Korea Research Institute of Bioscience & Biotechnology)
7	KISTI (Korea Institute of Science and Technology Information)
8	KIOM (KOREA INSTITUTE OF ORIENTAL MEDICINE)
9	KITECH (Korea Institute of Industrial Technology)
10	ETRI (Electronics and Telecommunications Research Institute)
11	NSRI (National Security Research Institute)
12	KICT (Korea Institute of Civil Engineering and Building Technology)
13	KRRI (Korea Railroad Research Institute)
14	KRISS (Korea Research Institute of Standards and Science)
15	KFRI (Korea Food Research Institute)
16	WIKIM (World Institute of Kimchi)
17	KIGAM (Korea Institute of Geoscience and Mineral Resources)
18	KIMM (Korea Institute of Geoscience and Mineral Resources)
19	KIMS (Korea Institute of Materials Science)
20	KARI (Korea Aerospace Research Institute)
21	KIER (Korea Institute of Energy Research)
22	KERI (Korea Electrotechnology Research Institute)
23	KRICT (Korea Research Institute of Chemical Technology)
24	KIT (Korea Institute of Toxicology)
25	KAERI (Korea Atomic Energy Research Institute)

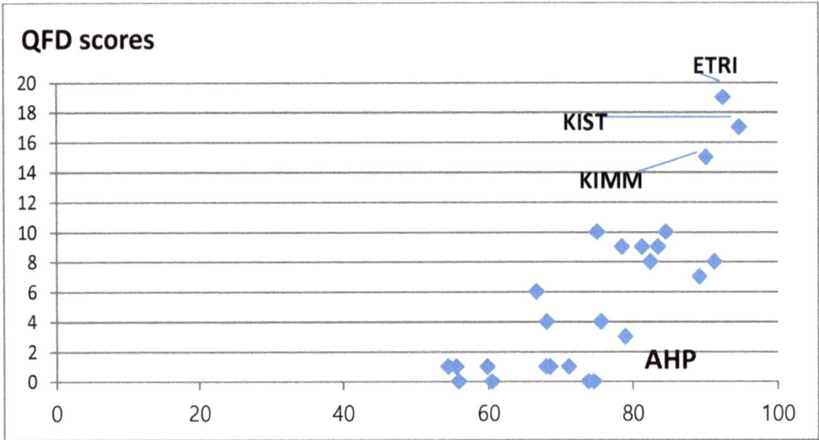

Figure 2. Fitness relationship between research institutions and the research topic of the 4th industrial revolution.

5. Discussion

As shown in this paper, the 4th industrial revolution is presenting a huge challenge as well as an opportunity to different levels of organizations including research institutions [29,30]. This paper, noting the salience, tried to capture the changing momentum of the 25 public research institutions in South Korea. In doing so, for going beyond the already existing performance measures, this study utilized the AHP expert survey to plot organizational performance and used organizational analysis to find inflows and outflows of manpower in the institutions.

This study found several key elements that are applicable to other countries too. First, independent to the research institutions' scientific and technological capabilities at the current period, the organizations can be transformed by just changing their manpower structure, as if a machine undergoes a major refit.

● This study suggests that there should be a non-conventional recruitment process in the future following the "I" shape organization structure. Knowing the characteristic of the 4th industrial revolution which emphasizes connectiveness with information technology, countries that are aiming, by certain degrees, at reconfiguration of research institutions, whether they are public or private in their origin, may find similar problem definitions and solutions that have been described in this research. In addition, many analytical tools seem to rearrange those institutions, which this research tried to overcome by using expert level AHP analysis.

Second, through the QFD analysis of matching, the study found that research topics related to the 4th industrial revolution can be unevenly distributed among institutions, while most of them can still contribute to the analysis. This implies that the 4th industrial revolution, due to its nature, predesignates the kind of institutions that can function as integration control towers.

● In other words, research on the 4th industrial revolution can be carried out by establishing a key institution, which will lead a group of participating research institutions. For example, in an autonomous vehicle case, it would be reasonable to find that either an IT area institution or a machine-based institution takes the leading role as a control tower, while other participating institutions follow the lead. This, in some sense, perfectly dovetails with the nature of the 4th industrial revolution, which needs connectedness between and among fields.

Third, although this study tried to present a ranking-based result from the AHP, it did not intend to foster a relative ranking system; rather, the intention was to augment the capability of interpreting the existing performance data of research institutions [31,32].

● As more and more research institutions and company research units are being retooled to meet the requirements of the 4th industrial revolution [33,34], this study tried to shed light on a new avenue to design interdisciplinary research programs [35,36], which can be applied in other contexts too.

Author Contributions: J.K. designed the overall research plan and methodology, sampling framework for the research institutions, and wrote the manuscript. S.L. performed AHP and QFD analysis. Both authors together prepared the AHP questionnaire and indepth interviews of researchers.

Funding: This research received no external funding.

Conflicts of Interest: The authors declare no conflict of interest.

References

1. Schwab, K. The Fourth Industrial Revolution. World Economic Forum Public Discussions on the 4th Industrial Revolution Include the Following Forums. Available online: https://www.brookings.edu/blog/brown-center-chalkboard/2017/04/11/surfing-the-4th-industrial-revolution-artificial-intelligence-and-the-liberal-arts/ (accessed on 20 May 2019).
2. West, D.M. *The Future of Work*; The Brookings Institution: Washington, DC, USA, 2018.
3. Moretti, E. *The New Geography of Jobs*; First Mariner Books; Mariner Books: New York, NY, USA, 2013.
4. Womack, J.P.; Jones, D.T.; Roos, D. *The Machine That Changed the World: The Story of Lean Production*; Rawson Associates: New York, NY, USA, 1990.
5. Brynjolfsson, E. *The Second Machine Age: Work, Progress, and Prosperity in a Time of Brilliant Technologies*, 1st ed.; W. W. Norton & Company: New York, NY, USA, 2016.
6. Palley, T. *From Crisis to Stagnation: The Destruction of Shared Prosperity and the Role of Economics*; Cambridge University Press: Cambridge, UK, 2012.
7. Boyer, R. The present crisis. A trump for a renewed political economy. *Rev. Political Econ.* **2013**, *25*, 1–3. [CrossRef]
8. Duperrin, B. Industry 4.0: The Dark Side of the Digital Enterprise. Available online: http://www.duperrin.com/english/2016/08/25/industry-4-0-digital-enterprise/ (accessed on 6 May 2019).

9. Mainstream, B. *Industry 4.0, The New Industrial Revolution: How Europe Will Succeed*; Roland Berger: Munich, Germany, 2014; Available online: http://www.rolandberger.com/media/pdf/ (accessed on 6 May 2019).

10. Moretti, E. *The New Geography of Jobs*; Mariner Books: Wilmington, DE, USA, 2013.

11. Schumpeter, J.A. *The Theory of Economic Development*; Oxford University Press: New York, NY, USA, 1912.

12. Koch, M. *Roads to Post-Fordism: Labor Markets and Social Structures in Europe*; Routledge: Abingdon-on-Thames, UK, 2017.

13. Galbraith, J.K. *History of Economics: The Past as the Present*, 1st ed.; David & Charles: Exeter, UK, 1987.

14. Ross, A. *The Industries of the Future*; Reprint edition; Simon & Schuster: New York, NY, USA, 2017.

15. Kelly, K. *The Inevitable: Understanding the 12 Technological Forces That Will Shape Our Future*; Penguin Books: London, UK, 2017.

16. Piore, M.J.; Sabel, C. *The Second Industrial Divide*; Basic Books: New York, NY, USA, 1984.

17. Maidagan, X.; López de Lacalle, L.N.; Sánchez, J.A.; Lamiki, A.; Maidagan, E.; Cabanes, S. Entrepreneurial research model, CIC marGUNE: A case study. *J. Sci. Ind. Res.* **2008**, *67*, 272–276.

18. Barrio, H.G.; Moran, I.C.; Ealo, J.A.; Barrena, F.S.; Beldarrain, T.O.; Zabaljauregui, Mi.C.; Zabala, A.M.; Arriola, P.A.; Lopez de Lacalle, L.N. A reliable machining process by means of intensive use of modelling and process monitoring: Approach 2025. *DYNA* **2018**, *93*, 689–696.

19. Robinson, J. *Essays in the Theory of Economic Growth*; Macmillan: London, UK, 1962.

20. Minsky, H.P. *Stabilizing an Unstable Econom*; McGraw-Hill: New York, NY, USA, 2008.

21. Kim, J. Manufacturing or service? Market saturation and cycles of over-investment as a clue to future of service economies. *Technol. Forecast. Soc. Chang.* **2010**, *78*, 1345–1355. [CrossRef]

22. Vernon, R. The product cycle Hypothesis in a new international environment. *Oxf. Bull. Econ. Stat.* **1979**, *41*, 255–267. [CrossRef]

23. Vernon, R.; Wells, L.T. International trade and International investment in the product life cycle. *Q. J. Econ.* **1966**, *81*, 190–207. [CrossRef]

24. Hill, C.W.L. *International Business Competing in the Global Marketplace*, 6th ed.; McGraw-Hill: New York, NY, USA, 2007; p. 168.

25. Boyer, R. Is a finance-led growth regime a viable alternative to Fordism? A preliminary analysis. *Econ. Soc.* **2000**, *29*, 111–145. [CrossRef]

26. Shoujun, L. Economic crisis and regulation theory: review of international conference of research & regulation 2015. *World Rev. Political Econ.* **2016**, *7*, 145–153.

27. Kim, J. Are countries ready for the new meso revolution? Testing the waters for new industrial change in Korea. *Technol. Forecast. Soc. Chang.* **2018**, *132*, 34–39. [CrossRef]

28. Minsky, H.P.; Whalen, C.J. *Economic Insecurity and the Institutional Prerequisites for Successful Capitalism*; Working Paper No. 165; Jerome Levy Economics Institute: Annandale-On-Hudson, NY, USA, 1996.

29. Piketty, T. *Capital in the 21st Century*; Belknap Press, An Imprint of Harvard University Press: Cambridge, MI, USA, 2014.

30. Nelson, R.R. Research on productivity growth and productivity differences: Dead ends and new departures. *J. Econ. Lit. Am. Econ. Assoc.* **1981**, *19*, 1029–1064.

31. Nelson, R.R. *The Sources of Economic Growth*; Harvard University Press: London, UK, 1996.

32. Baumol, W.J.; Nelson, R.R.; Wolff, E.N. *Convergence of Productivity: Cross-National Studies and Historical Evidence*; Oxford University Press: Oxford, UK, 1994.

33. Sabel, C.; Zeitlin, J. Historical alternatives to mass production: Politics, markets and technology in nineteenth-century industrialization. *Past Present* **1985**, *108*, 133–176. [CrossRef]

34. Kim, J.; Yoo, C.W. Why 'design' does not work well for cluster policy: With the implications for Science and Technology (S&T) manpower policy. *Int. J. Technol. Manag.* **2007**, *38*, 321–338.

35. Audretsch, D.B. Technological regimes, industrial demography and the evolution of industrial structures. *Ind. Corp. Chang.* **1997**, *6*, 49–82. [CrossRef]

36. Piore, M.J. Beyond Markets: Sociology, street-level bureaucracy, and the management of the public sector. *Regul. Gov.* **2011**, *5*, 145–164. [CrossRef]

Review

Digital Manufacturing Platforms in the Industry 4.0 from Private and Public Perspectives

Jon Kepa Gerrikagoitia [1,*], Gorka Unamuno [1], Elena Urkia [1] and Ainhoa Serna [2,*]

[1]   IDEKO, ICT and Automation Research Group, Arriaga 2, 20870 Elgoibar, Spain
[2]   Digital Manufacturing Group, Universidad de Deusto, Pozoa Kalea, s/n, 01013 Vitoria-Gasteiz, Araba, Spain
*   Correspondence: jkgerrikagoitia@ideko.es (J.K.G.); aserna@deusto.es (A.S.)

Received: 21 June 2019; Accepted: 19 July 2019; Published: 23 July 2019

Abstract: The fourth industrial revolution is characterized by the introduction of the Internet of things (IoT) and Internet of Services (IoS) concepts into manufacturing, which enables smart factories with vertically and horizontally integrated production systems. The main driver is technology, as Industry 4.0 is a collective term for technologies and concepts of value chain organization. Digital manufacturing platforms play an increasing role in dealing with competitive pressures and incorporating new technologies, applications, and services. Motivated by the difficulties to understand and adopt Industry 4.0 and the momentum that the topic has currently, this paper reviews the concepts and approaches related to digital manufacturing platforms from different perspectives: IoT platforms, digital manufacturing platforms, digital platforms as ecosystems, digital platforms from research and development perspective, and digital platform from industrial equipment suppliers.

Keywords: digital manufacturing; smart factory; Industry 4.0; digital platforms

---

1. Introduction

The Industry 4.0 concept has gained a global context beyond the German Industrie 4.0 initiative and suggests a perspective of the fourth industrial revolution that brings ICT (Information and Communication Technologies) into manufacturing that aims at the development of smart factories with fully integrated production systems. Among the different Industry 4.0 definitions, there is a common understanding that Cyber-Physical Systems (CPS), Internet of Things (IoT), Big Data Analytics, and Internet of Services (IoS) are the main components that embody Industry 4.0 [1,2]. Besides the technological perspective, Industry 4.0 can be understood as a collective term for technologies and concepts of value chain organization [3]. As a result, in industries worldwide, highly flexible processes that can be changed quickly enable individualized mass production [4].

Within modularly structured Smart Factories, Cyber Physical Systems (CPS) monitor physical processes, create a virtual representation of the physical world, and make decisions. The traditional structure of the automation pyramid and the IT/OT (Information Technologies/Operational Technologies) frontiers are blurring, CPS enable new means of communication and cooperation among devices, production assets and information systems in an orchestrated and decentralized way in real time. Via the Internet of Services (IoS), both internal and cross organizational services are offered and used by stakeholders of the value chain.

Different countries and regions have designed their own programs to achieve this fourth revolution. For example, the German government and the European Union (EU) promote the Industry 4.0 program [5], while in the United States, the Smart Manufacturing Leadership Coalition (SMLC) is the main initiative [6]. Other important manufacturing countries such as Japan [7] and Korea [8] have also established national programs on smart manufacturing.

Economic expectations are also important, as PwC estimated a global investment in technologies for industry digitization in 2020 of more than $900 billion annually. It also reported that 72% of the companies that took part in the survey expected to have achieved "advanced levels of digitization" by 2020 [9]. There is a growing awareness in business communities regarding the market opportunity for platform suppliers based on the level of investment.

Focusing on Europe, in 2016, the European Commission started the Digitizing European Industry initiative (DEI) aimed to reinforce the EU's competitiveness in digital technologies. The European Commission strategy defines four pillars: Digital Innovation Hubs, regulatory framework, skills, and Digital Platforms [10]. The EU has launched several calls in the Horizon 2020 program to advance in the development of digital industrial platforms like DT-ICT-07 2018–2019 and 2019–2020 with a budget over €100 M. Digital platforms for manufacturing play a key role in addressing competitive pressures and integrating new technologies, apps and services. The challenge is to make full use of new technologies that enable manufacturing businesses, particularly mid-caps and small and medium-sized enterprises (SMEs) to meet the requirements of evolving supply and value chains. Besides innovation and research actions there are also coordination and support Activities in order to cross-fertilize the industrial platform communities, facilitating the adoption of digital technologies from ongoing and past research projects to real-world use cases and encouraging the transfer of skills and know-how between industry and academia.

The first steps towards the digitization of the European industry were made in 2013 by an Industry 4.0 working group that published a report [5], which provided the vision, the integration features, the priority areas for action, and the example applications for the fourth industrial revolution. Nevertheless, there are still problems for its correct implementation and Industry 4.0 is tackling adoption issues argued by the literature, such as that Industry 4.0 is a complex concept which is not clearly understood yet by companies' managers, the lack of a detailed roadmap, the need of strategic guidance and clear implementation details, the large IT investments required in order to achieve a vertical and horizontal integration, the uncertainty about the outcomes, and the maturity and capability of the companies [11–15].

The digital manufacturing platform scenario is complex and uncertain, as the main players and roles are still being shaped. Trying to foresee market scenarios, in December 2016. The Economist compared two platforms, the General Electric (GE) Predix and Siemens Mindsphere in order to evaluate the likelihood of finally dominating the industrial Internet. It found that it is unlikely that a single platform will reach complete dominance and highlighted the significance of an open strategy [16].

Motivated by the difficulties to understand and adopt Industry 4.0 and the momentum that the topic has currently, this paper reviews the concepts and approaches related to digital manufacturing platforms from a qualitative research methodology considering the period of the last five years, when it all started. The structure of the paper contains an introduction, the followed methodology, the results of the research classified in subsections as explained in the methodology, a conclusion section that states from a qualitative perspective the outcome of the review work, and the references.

The relevance and contribution of the paper relies to a great extent in the proposed review methodology approach that follows the view and motivation of a research organization specialized in manufacturing technologies and digital platforms that is part of a machine tool builder industrial group. Thus, the whole lifecycle of research, knowledge transfer to industry, funding sources and ROI (Return on Investment) for industrial partners are present in the rationale of the methodology.

The findings of the research work show the broad and complex scope of digital manufacturing platforms following the motivation and view of the authors. The relationship between private (IoT platform vendors, manufacturing equipment suppliers, and machine tool builders) and public stakeholders (European Commission, Public Private Partnerships, etc.) in the strategy of digitizing the European industry contributes to build a global vision towards addressing future challenges posed by the need to create new business models based on data economy and the growth of digital ecosystems fostered by digital manufacturing platforms.

2. Literature Review

The current concept of a digital manufacturing platform has had an evolution since its roots in the sixties when the concept of manufacturing systems came up with advances in computing capability. Since then, it has attracted the attention of researchers who reported their findings in the literature from different perspectives and viewpoints in the past years. This section will review this evolution covering the different terms, names, and features related with digital manufacturing platforms.

Chryssolouris et al. [17] describe the scope of digital manufacturing and show the evolution of information technology systems in manufacturing, outlining their characteristics and future trends. Digital manufacturing and digital factory concepts in the pre-industry 4.0 era focus mainly on PLM (Product Lifecycle Management) technologies as computer-aided design, engineering, process, product data and life-cycle management, simulation and virtual reality, process control, shop floor scheduling, decision support, decision making, manufacturing resource planning, enterprise resource planning, logistics, supply chain management, and e-commerce systems.

As results of digitization advancement, manufacturing system controls have to deal with materials and machines and integration issues that started to come up in manufacturing, as machines and devices in a manufacturing process were no longer isolated but parts of a system, where all the components could be effectively coordinated. To handle integration issues, computer-integrated manufacturing system (CIMS) are starting to be widely adopted by companies. In this context Chen et al. [18] study the perspectives and enablers of integrated and intelligent manufacturing. Increasing opportunities were opened by Internet of Things (IoT) and CPS technologies, which enabled integration to be made wider and more open, comprising three levels of integration in manufacturing—vertical integration, horizontal integration, and end-to-end integration [5].

According to Chen et al. [18] intelligent manufacturing platforms are the enablers to implement intelligent manufacturing technologies. The point of view of industries that are preparing to develop cloud computing platforms based on IoT motivated by business is explained. Predix [19], ThingWorx [20], and Siemens [21] software platforms are described as the main references by the authors. In addition, the authors remark that "digital twins" are a significant feature of all such platforms to allow for the prediction of future conditions of productive assets.

Alcácer et al. [22] show an approach of Industry 4.0 for manufacturing systems based on the Smart Factory concept. There is a clear focus on enabling technologies for Smart Factory, such as IoT, IoS, systems integration, and the Cyber-Physical Production System (CPPS). The presented Smart Factory concept relies heavily on distributed computing as a core concept of Industry 4.0, as opposed to the most common manufacturing environments that are centralized. Authors explain the connection between technologies and standards with the role of RAMI 4.0 and its importance in leading the growth of CPPS [23].

Zhong et al. [24] present a review of intelligent manufacturing in the context of Industry 4.0. The authors present the major advances in manufacturing technologies as intelligent manufacturing, IoT enabled manufacturing, and cloud manufacturing. They define intelligent manufacturing (also known as smart manufacturing) as a broad concept of manufacturing, with the purpose of optimizing production and product transactions by making full use of advanced information and manufacturing technologies [25]. Intelligent manufacturing system (IMS) are considered to be the next-generation manufacturing system by adopting new models, new forms, and new methodologies to transform the traditional manufacturing system into a smart system. Authors remark the importance of service-oriented architecture (SOA) via the Internet to that end, providing collaborative, customizable, flexible, and reconfigurable services to end-users. Moreover, the authors highlight the essential role of AI (Artificial Intelligence) in an IMS by providing features such as learning, reasoning, and acting in a human-machine cooperation context. IMS shape an ecosystem where manufacturing elements are involved with organizational, managerial, and technical implications.

Another technological aspect is cloud manufacturing [26] that refers to an advanced manufacturing model under the support of cloud computing, IoT, virtualization, and service-oriented technologies.

It covers the extended whole life cycle of a product, from its design, simulation, manufacturing, testing, and maintenance, aiming to provide on-demand manufacturing services can from the cloud.

De Reuver eta al. [27] approach the study of digital platforms and examine the ecosystems that surround them. The authors state that digital platforms have a transformative and disruptive impact on organizations and their business models to the extent that platforms change the power structure and the relationship between participants in the ecosystem. The way service providers and device manufacturers strategize in a platform environment is discussed based on prior ecosystemic thinking work [28], taking into account that organizations are not isolated anymore, and value is co-created and co-delivered by multiple contributing entities.

3. Methodology

The present research work follows a qualitative approach focusing on the broad concept of the digital manufacturing platform from the perspective of a research organization specialized in manufacturing technologies and digital platforms that is part of a machine tool builder industrial group. Thus, research and industry views are included in the review paper as they complement each other to reach research and business goals.

In order to cover the technological perspective, the first step was to understand which were the technology enablers that make possible the development of digital manufacturing platforms. The essential features of an IoT platform were well known considering the authors' background and the search was focused on the major IT (Information Technology) players that provide application enablement platforms for horizontal domains covering connectivity, data acquisition, data storage, device monitoring, data analysis, and data visualization capabilities. As the scientific literature in this field is not extensive so far, the data sources were mainly IT vendors' web sources and white papers combined with benchmarking articles among different solutions. Small and niche IoT platform vendors were also included in the search and additional sources, such as the Hannover Messe Fair from 2015 to 2018, were considered as well. The next step of the review method was to focus and search for manufacturing domain platforms and specifically the concept of digital manufacturing platforms. The sources were both research papers and documents from the European Commission, as the influence of policy making with the DEI (Digitization of the European Industry) strategy has had a remarkable impact in the reviewed topic. Following the impact, the next stage was to obtain the research initiatives and the main outcomes of the DEI regarding digital manufacturing platforms. In this case the main information sources were web sources containing European Commission documents and European research projects. The research and search was focused on the European Research and Innovation calls under the Horizon 2020 programs related to digital manufacturing. The last stage of the method was to include the strategy and approach carried out by industrial equipment suppliers and machine tool builders. The data sources used for that purpose were based on a competitive intelligence tool, INNGUMMA (Elgoibar, Spain), that allows to identify and monitor competitors under specific observation parameters.

4. Results

4.1. IoT Platforms

An IoT platform is a middleware between the IoT devices and IoT gateways on one hand and applications on the other hand. The IoT platform enables the building of applications and are also called Application Enablement Platforms or AEPs. The essential features and capabilities of an IoT platform are on the level of connectivity and network management, device management, data acquisition, security, event processing, monitoring, analysis, visualization, integration, storage, and application enablement.

There are several vendors with different architectures, ways of connecting and managing IoT devices, possibilities to manage and analyze data, capabilities to build applications, and options to leverage IoT in a meaningful way for any given IoT use case in any given context—consumer applications, enterprise IoT applications, and Industrial IoT or Industry 4.0. In the end, IoT is part of an integrated approach to leverage data from devices and assets. There are hundreds of players in the market and although the IoT platforms have many functions in common there are differences in the offerings with sometimes very different features [28].

The main players in the IoT platform market are:

Microsoft Azure IoT [29]: The Microsoft Azure IoT offers device monitoring, rules engine, device shadowing, and identity registry. Upon these basic services, Azure IoT incorporates several existing products such as Stream Analytics, Power BI, IoT Hub, notifications hub, and some pre-packed machine learning. In addition, Azure Digital Twins allows to create digital models of any physical environment; including places, things, and people.

Oracle Internet of Things (IoT) Cloud Service [30] is a managed Platform as a Service (PaaS) cloud-based offering that allows to connect devices to the cloud, analyze data from those devices in real time, and integrate data with enterprise applications, web services, or with other Oracle Cloud Services, such as Oracle Business Intelligence Cloud Service.

Google Cloud IoT Core [31]: Google Cloud IoT Core is a fully manageable IoT platform. This platform is labeled as a major rival against the other similar platforms, since it mainly concentrates on intelligence. To achieve this intelligence, it utilizes ad-hoc queries using Google Big Query and Cloud Functions workflows. Thus, the devices can automate changes based on real-time events, data visualizations are done using Google Data Studio, and machine learning is done with Cloud Machine Learning Engine.

IBM Watson IoT [32]: This cloud IoT platform enables users to connect their devices and the IoT device data into a repository from where this cloud IoT helps them gain insight into an IoT network to not just improve their operations, but also launch various new business models. The users of this cloud platform receive real-time data exchange, data storage, device management, and secure communications.

AWS IoT Core [33]: This cloud IoT platform helps turn cars, sensory grids, and turbines to "smart" objects by helping in connecting and managing the sensors on these objects. The AWS IoT Core provides a secure device gateway, device shadows, device SDK, a registry for recognizing the different devices, a message broker, and rules engine that would evaluate the inbound messages.

Bosch IoT Suite [34]: This cloud platform is one with services designed to meet the requirements of every IoT project. The platform was initially designed and built to provide the IoT solution developers flexibility and ease to perform their daily tasks.

Above the mentioned IoT platforms there are large generic IoT cloud platforms from vendors such as IBM, Google, Amazon, Microsoft, and more. The so-called IoT network providers' platforms, such as AT&T and Orange Business Services, as well as Telefónica or Vodafone are another category of IoT platforms.

4.2. Digital Manufacturing Platforms

In the early stages of the digitization of the industry there were Remote Machine Monitoring systems (RMMS) that are machine manufacturer software products designed to allow their clients to monitor their shop floor equipment [35]. DMG Mori Seiki has been a pioneer with an RMMS solution called Mori Net that allowed customers to monitor their DMG Mori Seiki machines over a local network or the Internet. For non DMG MORI machines they developed a solution named Messenger that is based on MTConnect, a standard for accessing machine tool data [36,37].

The advent of the Internet of Things in the industry sector pushed the adoption of sensor-based information collection to address their key problems related primarily to machine downtimes and process delays. This way, machine monitoring evolves towards condition monitoring, which is the practice of monitoring electrical equipment, usually with external sensors, in order to gather the required data for diagnosis. To achieve this goal, data acquisition systems and data loggers are used to monitor all kinds of industrial equipment and devices. A 2017 Research & Markets magazine study provides that the market for condition monitoring is almost $11 billion for factories and OEMs. Dell, Caterpillar, Microsoft, IBM, General Electric, and Siemens, among others, are industry leaders named in the study.

Besides condition monitoring, more types of services and requirements, such as preventive maintenance, runtime and uptime measurement, energy monitoring or performance tracking are coming up and as results Digital Manufacturing platforms are being shaped to cover a broader scope.

In this broader scope, the European Factories of the Future Research Association (EFFRA) that is a non-for-profit, industry-driven association is performing an important role in the digitization of manufacturing industry and digital manufacturing platforms. As an example of the activity under EFFRA's umbrella, the "Connected Factories" CSA (Coordination and Support Action) establishes a structured overview of available and upcoming technological approaches and best practices [38]. The project identifies present and future needs, as well as challenges, of the manufacturing industries. The digitization of manufacturing connects people, devices, machines and enterprises, and includes concepts such as 'Industrial Internet', 'digital manufacturing platforms', and the 'Internet of Things' (IoT). Moreover, the Connected Factories explores pathways to the digital integration and interoperability of manufacturing systems and processes and the benefits this will bring. Hence, Connected Factories will enhance the awareness among companies of the use of digital technologies in the manufacturing sector, but also provide them with expertise to make informed choices about technology and business models. Connected Factories will improve companies' understanding of the use of digital techniques in the manufacturing industry, and also provide them with expertise to make informed technology and company model decisions.

Digital manufacturing platforms allow the provision of manufacturing services in a broad sense [39]. Digital platforms provide services that can be used for data collection, storage, processing, and delivery. These data describe the whole context which includes the product that is being manufactured, the manufacturing process, the production assets, the worker, and the entire value network. In general, the digital platform for manufacturing can provide any "digital" extension of functionalities for physical assets, through the adoption of ICT technologies. Digital platforms play a crucial role in enabling the application scenarios of digital manufacturing [40]. All services are aimed at optimizing manufacturing from different perspectives such as efficiency, availability, quality, performance, flexibility, etc.

Digital platforms can be on premise, in the cloud or in a hybrid architecture. Nevertheless, the thrust into a productive environment includes the need for agreements on industrial communication interfaces and protocols, common data models, semantic models and the interoperability of data. RAMI 4.0 is a framework that will help accomplish this task [41]. RAMI 4.0 is a three-dimensional layer model that compares the life cycles of products, factories, machinery or orders with the hierarchy levels of Industry 4.0. The model divides existing standards into manageable parts, integrates different user perspectives, and provides a common understanding of Industry 4.0 technologies, standards, and use cases.

Digital Platforms as Ecosystems

A digital manufacturing platform is part of a layered architecture that integrates a set of functions or software services that can be implemented by different technologies using interfaces and making the data available to be consumed by third party applications [39]. For example, a platform could make available operational state and machining process data provided by a machine tool to be used in business intelligence applications that provide production or OEE (overall equipment efficiency) insights. Platforms can be understood as operating systems that offer a set of applications as services. These services make shop floor data (machines, products, operators, …) accessible to other software applications (production planning, operation and process, quality management, maintenance, troubleshooting, energy management, etc.). The services usually will be exposed using IT flavor open standard interfaces, such as API Rest or OT standards such as OPC-UA. This way, an ecosystem of application developers can be fostered.

The digital manufacturing platform includes three characteristics. First, the community aspect that embodies an ecosystem of users in a social network connected to each other. In this community there is a marketplace where the offer and the demand can be matched. Some users are service providers, their raw material is data, and the offered product is usually a software app as a value-added data-driven service. The value creation relies on a solid technology infrastructure and this is the second aspect. The infrastructure aspect of the digital manufacturing platform is the enabler for users and partners to develop apps and create value added data-driven services. The ability to develop and deploy software apps in the platform is a core issue in order to develop a growing ecosystem of data-driven service consumers and producers. The infrastructure is the basic layer to boost the digital economy in manufacturing and to do so it is mandatory to be an open infrastructure that is able to integrate and unlock technologies and systems. The last aspect is the data role. Data is the raw material of digital manufacturing platforms provided by enterprise management systems, industrial assets, devices, and sensors and has to be exchanged, accessed, and processed in a proper way. The outcome of the process will be produced and consumed by allowed platform users.

Moreover, the ecosystem of digital manufacturing platforms is composed by four types of players. These are the owners of platforms in charge of the governance, the providers who are the interface with users, the producers who create their offerings, and consumers.

As the perception of date value is gaining importance in the global value creation, IDS (international data spaces) is devoted to forming the basis for data ecosystems and market places based on the principles of trust and data-sovereignty, which is guaranteed for data creators with respect to who is using their data, for how long, for which application, how many times, and according to which terms and conditions [42].

4.3. Digital Platforms from R&D Perspective

Digital manufacturing platform related initiatives have been fostered by Public-Private Partnerships (PPPs) at European level. Factories of the Future (FoF) (discrete manufacturing) and Sustainable Process Industry through Resource and Energy Efficiency (SPIRE) PPP (industrial processing) are two PPPs that explicitly address manufacturing/production [43].

Under the FoF PPP, a set of ten projects and one coordination and support action, called "Connected Factories" were started in autumn 2016 that develop reference implementations of platforms in a multi-sided market ecosystem and include user-driven proof-of-concept demonstrations and validation in several different scenarios [40]. The Connected Factories project was launched in the same call as the Factories of the Future FoF-11-2016 research and innovation projects, where six of them were focusing on digital platforms for factory automation (AUROWARE, DISRUPT, DAEDALUS, FAREDGE, SAFIRE and scalABLE4.0) and four projects were focusing on supply chain and logistics (COMPOSITION, DIGICOR, NIMBLE, vf-OS).

The digital manufacturing platform concept in its wider scope covers whole RAMI space and the mentioned projects fill different gaps within the reference architecture proving a set of building blocks for that purpose. With this complementary and incremental build up strategy, there are three other relevant projects (Industrial Data Space, ARROWHEAD, and Productive 4.0). Figure 1 shows the place that each project fills in the two-dimensional cross-section table of the RAMI 4.0 framework.

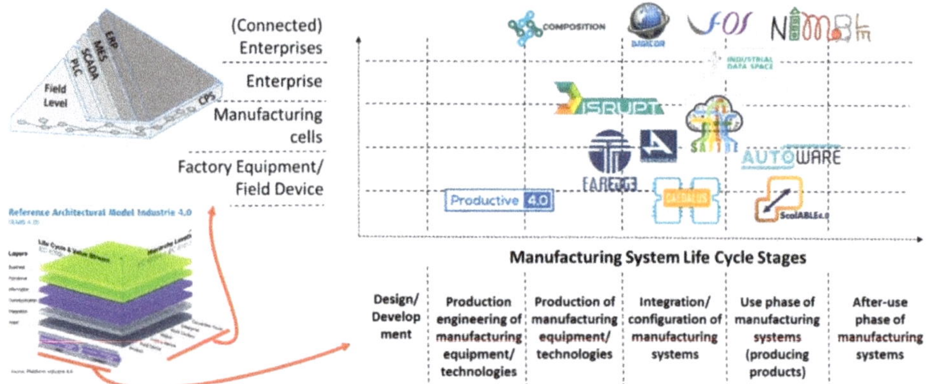

Figure 1. Positioning of digital platform related research and development (R&D) projects on RAMI 4.0. Source Reference [43] (p. 21).

In addition, DT-ICT-07 2018, the H2020 ICT call focused on Digital Manufacturing Platforms for Connected Smart Factories began three projects in January 2019 aimed to develop orchestrated open platforms ecosystem, atomized components, and digital enablers to meet the Industry 4.0 ZDM and lot-size-1 challenges.

4.4. Digital Platform from Industrial Equipment Suppliers

In addition to the efforts by leading IT vendors, industrial conglomerates (Siemens, GE, etc.), and the R&D initiatives, machine tool builders aim to transform their businesses digitally. For instance, by making use of data deriving from the machine tools they build, they develop predictive and prescriptive solutions for customers, improving machining performance, health and safety, energy-efficiency, business domain integration and so on. Their offer includes HMI software, production management software, machine and shop floor monitoring software, technical assistance software, etc. Usually, technology consultancy services with a global offer are necessary to be able to define the solution with the client. Therefore, industrial equipment suppliers are partnering with IT and consultancy companies.

Leading machine tool builders are investing more and more in digital platforms to provide a comprehensive solution to their customers:

DMG Mori [44] is a pioneer when it comes to digitization in machine tool construction. Under the key phrase "Integrated Digitization", DMG Mori with CELOS [45] is realizing a consistent strategy—starting with CELOS Machine, via CELOS Manufacturing up to the Digital Factory. DMG Mori started with CELOS, an operating and control system based on applications. Supported in ADAMOS [46], CELOS can become an open network and a digital market for the machine construction industry. ADAMOS relies on Microsoft Azure infrastructure.

Homag [47] machines have been connected to the plant level for a long time, they even have their own MES system. They started with their own digital platform that connected them to the cloud. Now, it is an open platform (Tapio) integrated in ADAMOS. The HomagGroup offers its customers solutions for digitized production with a complete software platform. They offer many services in the field of machines and production facilities, in addition to the corresponding control software.

Trumpf [48]: Stands for the digital business model supported in Truconnect as a smart factory platform that includes a comprehensive portfolio of consultancy, software and hardware resources. Axoom [49] provides a cloud monitoring platform for Trumpf machines but it is not exclusive to competitors.

Bosch [50]: Bosch Connected Industry bundles software and services for Industry 4.0 in a comprehensive portfolio called Nexeed.

Siemens [51]: Mindsphere is the cloud-based, open IoT operating system from Siemens that connects products, plants, systems, and machines, enabling you to harness the wealth of data generated by the Internet of Things (IoT) with advanced analytics. Mindsphere runs on top of Microsoft Azure or AWS infrastructures.

Fanuc [52]: FIELD (FANUC Intelligent Edge Link & Drive system) is the FANUC's open platform system that gives machine tool builders, robot manufacturers, and sensor and peripheral device manufacturers the freedom to develop their own applications. The target of the FIELD system connects each device within a factory, but also allows the flexibility to connect to upper host systems, such as ERP (Enterprise Resource Planning), SCM (Supply Chain Management) and MES (Manufacturing Execution Systems).

Schaeffler [53]: Schaeffler focuses its products in the digital world. Schaeffler has incorporated sensors, actuators, and control units with embedded software into these products. As a result, it is possible for these parts to collect and process valuable data on the condition of a machine and then convert this data into added-value services. With IBM as strategic partner, Schaeffler provides a digital platform for processing large amounts of data, generating valuable insight to transform operations. The Schaeffler cloud is a platform for end users to securely and reliably access data from their machines and equipment.

5. Conclusions

The review paper shows the concept and scope of digital manufacturing platforms from different perspectives combining technology and policy making, industry, and academia and the stakeholders that are involved on them as ecosystem.

The development of digital manufacturing platforms is in an early stage but supported in a mature IoT ground. Due to the broad scope of the concept, it has required the definition and development of a reference implementation, RAMI 4.0. In the current platform building context, it is not a matter of making choices for platform adopters but planning an incremental roadmap towards digital transformation. In this sense, the openness of the technological architecture is a must where state-of-the-art technologies regarding IoT, Artificial Intelligence, robotics, cloud or Big Data will be reused and integrated with interfaces described via open specifications. Platforms should aim for openness, avoiding lock-ins, preventing dominant positions of individual players, and compliance with standards and regulation.

Moreover, the openness of the digital manufacturing platform is a major issue as an enabler of digital ecosystems to become an AEP (application enablement platform).

It is remarkable that the role of the major IaaS (Infrastructure as a Service) providers is becoming more and more vertical or domain-oriented. The role of big players, such as Amazon or Microsoft, has been the provision of IoT and IT infrastructures with pay-per-use business models so far. Nowadays, these players are moving towards PaaS (Platform as a Service) services in manufacturing. This movement is being carried out accompanied by reference OEMs of prioritized industrial sectors.

In spite of the relevant advances achieved so far, there is still a lot to do in order to connect to additional services according to the 'plug-and-play' philosophy and considering the multi-sided ecosystem of service providers, platform providers and manufacturing companies, mechanisms for the commercial or open-source provision of the digital services through appropriate marketplaces, modularity of existing or in-development platforms of covering different "regions" of the RAMI framework, legacy system integration, overcoming semantic barriers, considering requirements of specific manufacturing sectors (process industry, consumer goods, capital equipment, etc.), etc.

The benefits of the fourth industrial revolution must be monetized for companies, to the extent that technology advances become reality. The definition and support of new business models based on data will be the next big challenge in relation to digital platforms. All these issues outline future work in digital manufacturing platforms.

Author Contributions: Conceptualization, J.K.G. and E.U.; Methodology, J.K.G. and A.S.; Investigation, J.K.G., E.U., A.S. and G.U.; Resources, J.K.G., E.U., A.S. and G.U.; Data Curation, J.K.G. and E.U.; Writing-Original Draft Preparation, J.K.G. and E.U.; Writing-Review & Editing, A.S. and E.U.; Supervision, J.K.G.

Funding: This research received no external funding.

Conflicts of Interest: The authors declare no conflict of interest.

References

1. Wang, K. Intelligent Predictive Maintenance (IPdM) system—Industry 4.0 scenario. *WIT Trans. Eng. Sci.* **2016**, *113*, 259–268.
2. Muller, J.M.; Buliga, O.; Voigt, K.I. Fortune favors the prepared: How SMEs approach business model innovations in Industry 4.0. *Technol. Forecast. Soc. Chang.* **2018**, *132*, 2–7. [CrossRef]
3. Hermann, M.; Pentek, T.; Otto, B. Design principles for industrie 4.0 scenarios. In Proceedings of the IEEE 2016, 49th Hawaii International Conference on System Sciences (HICSS), Koloa, HI, USA, 5–8 January 2016; pp. 3928–3937.
4. Thoben, K.D.; Wiesner, S.; Wuest, T. Industrie 4.0 and smart manufacturing-a review of research issues and application examples. *Int. J. Autom. Technol.* **2017**, *11*, 4–16. [CrossRef]
5. Kagermann, H.; Helbig, J.; Hellinger, A.; Wahlster, W. *Recommendations for Implementing the Strategic Initiative INDUSTRIE 4.0: Securing the Future of German Manufacturing Industry, Final Report of the Industrie 4.0 Working Group*; Forschungsunion: Berlin, Germany, 2013.
6. Smart Manufacturing Leadership Coalition. Implementing 21st Century Smart Manufacturing, 2011, Workshop Summary Report. Washington, DC, USA. September 2010. Available online: https://www.controlglobal.com/assets/11WPpdf/110621_SMLC-smart-manufacturing.pdf (accessed on 13 May 2019).
7. Fujii, T.; Guo, T.; Kamoshida, A. A Consideration of Service Strategy of Japanese Electric Manufacturers to Realize Super Smart Society (SOCIETY 5.0). In *Knowledge Management in Organizations, Proceedings of the International Conference on Knowledge Management in Organizations, Zilina, Slovakia, 6–10 August 2018*; Springer: Berlin/Heidelberg, Germany, 1842; pp. 634–645.
8. Park, J.; Lee, J. Presentation on Korea smart factory program. In Proceedings of the International Conference on Advances in Production Management Systems, Tokyo, Japan, 7–9 September 2015; pp. 5–9.
9. Geissbauer, R.; Vedso, J.; Scharauf, S. PWC. 2016 Global Industry 4.0 Survey, Industry 4.0: Building the Digital Enterprise. Available online: https://www.pwc.com/gx/en/industries/industries-4.0/landing-page/industry-4.0-building-your-digital-enterprise-april-2016.pdf (accessed on 3 April 2019).
10. European Commission. Pillars of the Digitising European Industry Initiative. 2018. Available online: https://ec.europa.eu/digital-single-market/en/pillars-digitising-european-industry-initiative (accessed on 3 April 2019).
11. Schumacher, A.; Erol, S.; Sihn, W. A Maturity Model for Assessing Industry 4.0 Readiness and Maturity of Manufacturing Enterprises. *Procedia CIRP* **2016**, *52*, 161–166. [CrossRef]
12. Liao, Y.; Deschamps, F.; Loures, E.D.F.R.; Ramos, L.F.P. Past, present and future of Industry 4.0—A systematic literature review and research agenda proposal. *Int. J. Prod. Res.* **2017**, *55*, 3609–3629. [CrossRef]
13. Theorin, A.; Bengtsson, K.; Provost, J.; Lieder, M.; Johnsson, C.; Lundholm, T.; Lennartson, B. An event-driven manufacturing information system architecture for Industry 4.0. *Int. J. Prod. Res.* **2017**, *55*, 1297–1311. [CrossRef]
14. Qin, J.; Liu, Y.; Grosvenor, R. A categorical framework of manufacturing for industry 4.0 and beyond. *Procedia CIRP* **2016**, *52*, 173–178. [CrossRef]
15. Erol, S.; Jäger, A.; Hold, P.; Ott, K.; Sihn, W. Tangible Industry 4.0: A scenario-based approach to learning for the future of production. *Procedia CIRP* **2016**, *54*, 13–18. [CrossRef]
16. The Economist. Siemens and General Electric Gear up for the Internet of Things. 2016. Available online: http://www.economist.com/news/business/21711079-american-industrial-giant-sprinting-towards-its-goal-german-firm-taking-more (accessed on 11 February 2019).

17. Chryssolouris, G.; Mavrikios, D.; Papakostas, N.; Mourtzis, D.; Michalos, G.; Georgoulias, K. Proceedings of the Institution of Mechanical Engineers, Part B: Journal of Engineering Manufacture. *Digit. Manuf. Hist. Perspect. Outlook* **2009**, *223*, 451–462.

18. Chen, Y. Integrated and Intelligent Manufacturing: Perspectives and Enablers. *Engineering* **2017**, *3*, 588–595. [CrossRef]

19. GE Digital. *Predix: The Industrial Internet Platform*; White Paper; General Electric Company: Boston, MA, USA, 2016.

20. The ThingWorx IoT Technology Platform [Internet]. Needham: PTC. Available online: https://www.thingworx.com/platforms/ (accessed on 02 June 2019).

21. Feuer, Z.; Weissman, Z. Smart Factory—The Factory of the Future [Internet]. Sunnyvale: LinkedIn. Available online: https://www.linkedin.com/pulse/smart-factory-future-zvi-feuer?articleId=8390740796107302304 (accessed on 19 December 2016).

22. Alcácer, V.; Cruz-Machado, V. Scanning the Industry 4.0: A Literature Review on Technologies for Manufacturing Systems. *Eng. Sci. Technol. Int. J.* **2019**, *22*, 899–919. [CrossRef]

23. Liu, C.; Xu, X. Cyber-physical Machine Tool—The Era of Machine Tool 4.0. *Procedia CIRP* **2017**, *63*, 70–75. [CrossRef]

24. Zhong, R.Y.; Xu, X.; Klotz, E.; Newman, S.T. Intelligent Manufacturing in the Context of Industry 4.0: A Review. *Engineering* **2017**, *3*, 616–630. [CrossRef]

25. Kusiak, A. Encyclopedia of Sustainable Technologies. *Smart Manuf.* **2017**, *7543*, 417–427. [CrossRef]

26. Wu, D.; Greer, M.J.; Rosen, D.W.; Schaefer, D. Cloud manufacturing: Strategic vision and state-of-the-art. *J. Manuf. Syst.* **2013**, *32*, 564–579. [CrossRef]

27. De Reuver, M.; Sørensen, C.; Basole, R.C. The digital platform: A research agenda. *J. Inf. Technol.* **2018**, *33*, 124–135. [CrossRef]

28. I-scoop. IoT Platform Definitions, Capabilities, Selection Advice and Market. Available online: https://www.i-scoop.eu/internet-of-things-guide/iot-platform-market-2017-2025/ (accessed on 11 May 2019).

29. Azure IoT. Available online: https://azure.microsoft.com/es-es/overview/iot (accessed on 17 June 2019).

30. IoT-Enable Your Business Applications. Available online: https://cloud.oracle.com/en_US/iot (accessed on 17 June 2019).

31. Cloud IoT Core. Available online: https://cloud.google.com/iot-core (accessed on 20 June 2019).

32. The Internet of Things Delivers the Data. AI Powers the Insights. Available online: https://www.ibm.com/internet-of-things (accessed on 20 June 2019).

33. AWS IoT Core. Available online: https://aws.amazon.com/es/iot-core (accessed on 19 June 2019).

34. Bosch IoT Suite. Available online: https://www.bosch-iot-suite.com (accessed on 17 June 2019).

35. Mori, M.; Fujishima, M.; Komatsu, M.; Zhao, B.; Liu, Y. Development of Remote Monitoring and Maintenance System for Machine Tools. *CIRP Ann.* **2008**, *57*, 433–436. [CrossRef]

36. Mori, M.; Fujishima, M. Remote monitoring and maintenance system for CNC machine tools. In Proceedings of the 8th CIRP Conference on Intelligent Computation in Manufacturing Engineering, Ischia, Italy, 18–20 July 2012.

37. Edrington, B.; Zhao, B.; Hansel, A.; Mori, M.; Fujishima, M. Machine Monitoring System Based on MTConnect Technology. *Procedia CIRP* **2014**, *22*, 92–97. [CrossRef]

38. EFFRA, The European Factories of the Future Research Association. Connected Factories. Available online: https://www.effra.eu/connectedfactories-project (accessed on 18 May 2019).

39. EFFRA, The European Factories of the Future Research Association. Digital Manufacturing Platforms. Available online: https://www.effra.eu/digital-manufacturing-platforms (accessed on 18 May 2019).

40. BMWi, Federal Ministry for Economic Affairs and Energy. Benefits of Application Scenario Value-Based Service. 2017. Available online: https://www.plattformi40.de/PI40/Redaktion/DE/Downloads/Publikation/benefits-application (accessed on 19 June 2019).

41. VDMA. RAMI 4.0 and Industrie 4.0 Components. 2016. Available online: https://industrie40.vdma.org/en/viewer/-/v2article/render/15557415 (accessed on 19 June 2019).

42. International Data Spaces. Available online: https://www.internationaldataspaces.org/ (accessed on 20 June 2019).

43. European Union, Brussels. Digitising European Industry: Working Group 2—Digital Industrial Platforms. 2017, p. 21. Available online: https://ec.europa.eu/futurium/en/implementing-digitising-european-industry-actions/report-wg2-digital-industrial-platforms-final (accessed on 19 June 2019).
44. European Commission. Digitising European Industry. Working Group 2. Digital Industrial Platforms. 2017. Available online: https://ec.europa.eu/futurium/en/system/files/ged/dei_wg2_final_report.pdf (accessed on 19 June 2019).
45. DMG MORI. Available online: https://es.dmgmori.com/productos/digitization (accessed on 19 June 2019).
46. ADAMOS. Available online: https://www.adamos.com (accessed on 4 June 2019).
47. HOMAG. Available online: https://www.homag.com/ (accessed on 4 June 2019).
48. TRUMPF. Available online: https://www.trumpf.com/es_ES/productos/fabricainteligente/ (accessed on 3 June 2019).
49. AXOOM. Available online: https://axoom.com/ (accessed on 9 June 2019).
50. BOSCH. Nexeed Software and Services for Production and Logistics. 2018. Available online: https://assets.bosch.com/media/en/global/products_and_solutions/connected_products_and_services/industy_40/bosch-connected-industry-brochure.pdf (accessed on 4 June 2019).
51. SIEMENS. This Is MINDSPHERE! Available online: https://new.siemens.com/global/en/products/software/mindsphere.html (accessed on 14 April 2019).
52. FANUC. Available online: https://www.fanuc.co.jp/en/product/field/index.html (accessed on 9 June 2019).
53. SCHAEFFLER. Available online: http://www.schaeffler.com/content.schaeffler.com/en/innovation/digitization/digitization.jsp (accessed on 19 June 2019).

Article

Implementation of R&D Results and Industry 4.0 Influenced by Selected Macroeconomic Indicators

Jena Švarcová [1,*], Tomáš Urbánek [2], Lucie Povolná [1] and Eliška Sobotková [1]

[1] Department of Economics, Tomas Bata University in Zlín, Zlín 76001, Czech Republic; povolna@utb.cz (L.P.); sobotkova@utb.cz (E.S.)
[2] Department of Statistics and Quantitative Methods, Tomas Bata University in Zlín, Zlín 76001, Czech Republic; turbanek@utb.cz
* Correspondence: svarcova@utb.cz

Received: 15 March 2019; Accepted: 24 April 2019; Published: 5 May 2019

Featured Application: An economic recession can destroy the potential of INDUSTRY 4.0 projects and R&D projects. Current EU grant support, tax relief, and other specific factors appear to be more important for the development of new R&D projects in the Czech Republic than the effects of economic recession.

Abstract: Successful timing of INDUSTRY 4.0 projects in businesses can be disrupted by the coming of a recession. The authors assume a close link between INDUSTRY 4.0 and research and development (R&D) projects. R&D projects are statistically internationally monitored and have a significant impact on European Union economic policies. This article explores the impact of the two economic recessions in 2009 and 2012–2013 on the number of R&D entities and human resources involved in R&D in the Czech Republic. The method of multivariate statistics with dummy variables was used. Research has shown that different sectors (business sector, government sector, higher education sector, and non-profit sector) show a different development of the number of R&D entities in times of economic crisis. The research findings indicate that current European Union grant support, tax relief, and other specific factors appear to be more important for the development of R&D projects in the Czech Republic than the effects of economic recession. In terms of longer time horizons, however, the effects of the business cycle cannot be ignored. In order to predict economic development, enterprises and other subjects can use leading macroeconomic indicators.

Keywords: INDUSTRY 4.0; economic recession; research and development indicators

---

1. Introduction

A number of authors, such as Weyer et al. [1] or Wang et al. [2], recognize INDUSTRY 4.0 as one of the major factors in improving manufacturing processes, increasing productivity and economic performance. At the same time, an economic crisis may be one of the most important factors that could lead to the failure of promising INDUSTRY 4.0 projects, as Povolná and Švarcová [3] point out. Pavelková et al. [4] examines economic performance in the Czech Republic (CR) in the dominant automotive sector in the pre-crisis, crisis, and post-crisis periods. There is clear pressure on the introduction of INDUSTRY 4.0 and research and development (R&D) knowledge. Statistical data in the CR shows that the number of subjects actively involved in R&D did not decline during two recessions in 2009 and 2012–2013, but even increased in selected sectors (the business sector). This is a completely untypical phenomenon. Usually, in times of recession, gross domestic product (GDP) and the number of economically active entities decrease in the economy under review. Macroeconomic performance (measured by GDP) and number of economically active entities increase at a time of boom. This article tested hypotheses of whether the recessions affect the number of subjects actively involved in R&D,

both in the business sector and in the government sector, universities, and non-governmental non-profit organizations in the CR. At the same time, the development of workforces in R&D workplaces is examined. The results show that the impact of economic recession on the R&D entities in the CR is not significant nowadays and the economic crisis does not affect the number of subjects evolving INDUSTRY 4.0 and R&D. Subjects of R&D in Europe are much more influenced by other factors, for instance grants to support R&D from European Union funds. The Czech Republic is very closely linked to the German economy, which is the main driver of the European Union (EU) economy, so it can be expected that the results in the Czech Republic will be similar to those in Germany. At the same time, it should be emphasized that the support system for R&D projects is uniform across the EU, so similar impacts can be expected in other EU countries. It is a stimulus for economic policy and state support and transnational grant programs (such as EU programs), to increase the supply of subsidies, grants, and tax relief in times of economic crisis. Active support for R&D can, thus, be one of the counter-cyclical measures to reduce the negative effects of economic crises.

This paper is arranged as follows. We first introduce some literature review and related work in Section 2. Section 3 explains the materials and methods. Section 4 describes the results. Section 5 presents discussion and conclusions in detail, and Section 6 highlights actual INDUSTRY 4.0 successes in Central Europe, brings final conclusions of the paper, provides corresponding policy suggestions, and describes the limitations.

2. Literature Review

Conceptual framework questions: Is the INDUSTRY 4.0 phenomenon beneficial in both developed and emerging economies? Which areas does INDUSTRY 4.0 focus on? Does INDUSTRY 4.0 have negative impacts? What is the macroeconomic significance of INDUSTRY 4.0 for the Czech Republic, and in a wider international context? Compared to other countries in the world, do the Czech Republic and its business partner Germany need to revitalize the Information and Communication Technologies (ICT) industry and support INDUSTRY 4.0? Can new technologies be the only source of economic growth or is human resource development and its long-term impact important? In the short term, it is important for the economy to monitor the signals of coming economic recessions, which are provided by the statistical indicators IFO Business Climate Index (leading indicator for economic activity in Germany prepared by the Ifo Institute for Economic Research in Munich, Germany) and ZEW indicators (The ZEW – Leibniz Centre for European Economic Research in Mannheim is an economic research institute and monthly publish their forecasts for macroeconomic indicators). Can we use R&D data to describe the INDUSTRY 4.0 phenomenon?

Dalenogare et al. [5] highlights evidence of INDUSTRY 4.0's benefits, not only in advanced economies, but also in emerging economies. In their extensive study, they examined and discussed the contextual conditions of the Brazilian industry that may require a partial implementation of the INDUSTRY 4.0 concepts created in developed countries.

Diez-Olivan et al. [6] concludes that one of the main objectives of data science in the context of INDUSTRY 4.0 is to effectively predict abnormal behavior in industrial machinery, tools, and processes, so as to anticipate critical events and damage that eventually cause significant economic losses and safety issues. Many authors focus on the review of essential standards and patent landscapes for the Internet of Things (IoT) and industrial IoT, such as Xia et al. [7] and Trappey et al. [8]. Some authors emphasize the relevance of visual computing for manufacturing processes, for example, Lee et al. [9] focus on service innovation and smart analytics using big data as a key enabler for next generation advanced manufacturing. Rajput and Singh [10] show that the IoT ecosystem and IoT big data are the most influential IoT enablers that help industry practitioners effectively implement INDUSTRY 4.0.

Kovacs [11], in his article, The Dark Corners of INDUSTRY 4.0, emphasizes that INDUSTRY 4.0 processes have not only positive economic impacts but also negative side-effects on the environment and society. It is a serious topic that needs to be addressed so that negative impacts are minimized.

Implementation of INDUSTRY 4.0, taking into account influences on the environment and society, provides immense opportunities for the realization of sustainable manufacturing [12].

The macroeconomic significance of INDUSTRY 4.0 is being studied both in the context of the Czech Republic and in a wider international context. The initial conditions of aspects of process management in the context of company strategies in Czech enterprises are evaluated by Tuček [13]. Pavelková et al. [4] examines economic performance in the Czech Republic in the dominant automotive sector in the pre-crisis, crisis, and post-crisis periods. There is clear pressure on the introduction of INDUSTRY 4.0 and R&D knowledge. Increasing the productivity and economic performance of companies using INDUSTRY 4.0 is becoming an important part of the competitive struggle and more companies are trying to use R&D results in business processes. Kang et al. [14] brings comparisons between Germany, the United States, and Korea, and talks about support from the economic policies of the states. INDUSTRY 4.0 is also an important growth factor in China. Li [15] notes not only the technological but also sociological context of these changes, and finds an upward trajectory in China in manufacturing capability development, research and development commitment, and human capital investment. It is very important to see the connections between R&D, INDUSTRY 4.0, the development of human capital, and economic political strategies of individual states. No less important is the monitoring of the links between science and research, innovation, INDUSTRY 4.0, and the development of the workforce to meet these strategies.

Min et al. [16] focuses on a very broad innovation framework in the US and the smartization strategy in Japan. The authors show that developed countries are pushing nation-wide innovation strategies. Similarly, China is pursuing the Made in China 2025, and Korea has announced the Manufacturing Industry Innovation 3.0 strategy. Min et al. [16] provides a comparative study on industrial spillover effects among Korea, China, the USA, Germany, and Japan, especially the spillover effect of the Information and Communication Technology (ICT) industry and equipment (the foundation of smart manufacturing through convergence with the ICT industry). Practical implications of their findings are that Germany needs to revitalize the ICT industry to strengthen its manufacturing industry. It is a very important finding with many implications for the Czech Republic, because Czech industry is an important subcontractor to the German industry. The connection of Czech companies to German industry has been the basis for the growth of the Czech GDP in the last 25 years. There is transfer know-how in subcontracting chains, which enhances the development of workforce skills. At the same time, there is a very strong pressure to increase production efficiency and reduce costs, which is manifested by the pressure to increase labor productivity, the introduction of robotics, and the successful implementation of INDUSTRY 4.0 in the Czech Republic.

Macroeconomics captures economic growth through production functions, where the basic variables are human resources and capital, taking into account that capital is financial, and also includes the impact of new technologies and processes, including INDUSTRY 4.0. However, it would be a fundamental mistake to separate it from human resources, in their quantitative and, above all, qualitative assessment (qualifications, personalities responsive to change, teamwork, etc.). Most of the theoretical work, however, does not take into account the required growth of labor force qualification for INDUSTRY 4.0. For example, Grassetti and Hunanyan [17] used a neoclassical one-sector growth model with differential savings, while assuming a Kadiyala production function that shows a variable elasticity of substitution symmetric with respect to capital and labor. Authors declare that thanks to the proposed methodology, the government can select a proper economic policy to reduce production costs without decreasing the capitalization trend of the economy. The theory, however, does not correspond to the issue of investment in education and the growth of labor force skills needed to grow productivity and the whole economy. The causes of economic crises are perceived as external factors of the growth model.

Some authors are looking for the causes of cyclical macroeconomic behavior. Devezas and Corredine [18] focused on the effective causality of long-term macroeconomic rhythms, commonly referred to as long waves or Kondratieff waves. The authors have demonstrated that the unfolding of each structural cycle of a long wave is controlled by two parameters: the diffusion-learning rate δ and the aggregate effective generation tG, whose product is maintained in the interval $3 < \delta tG < 4$ (deterministic chaos) of social systems. For the development of INDUSTRY 4.0, it is particularly important to define the basic variables that the authors have included in their model and which show the importance of the human factor for long-term development. Dosi et al.'s [19] model results show that seemingly more rigid labor markets and labor relations are conducive to coordinating successes, leading to higher and smoother growth. Emphasis on the development and organization of the workforce can be crucial for the development of INDUSTRY 4.0.

On the other hand, for businesses that are using INDUSTRY 4.0 technologies and practices, it is important to anticipate the near-term (not long-term) macroeconomic development of the economy. For Czech companies, it is mainly macroeconomic developments in Germany, but also in many other European countries and the world. The anticipation of close macroeconomic developments is the responsibility of a number of institutions. For a short time horizon (typically several months), the statistical authorities perform standardized statistical surveys—purchasing managers' expectations, but also IFO Index, ZEW indicators, and more. Homolka and Pavelková [20] examined the predictive power of the ZEW sentiment indicator in the case of the German automotive industry, which is crucial for the Czech economy. The ZEW Indicator of Economic Sentiment is a leading indicator of the German economy, similar to the IFO index. All of these indicators can reduce the uncertainty of businesses that want to invest in R&D and INDUSTRY 4.0, but fear that their investment can be destroyed by a future economic crisis.

INDUSTRY 4.0 is a very new phenomenon that does not go beyond this decade. Macroeconomic research of this phenomenon, therefore, faces insufficient methodological definition and coverage of data by international and national statistical institutions. The definition of Research and Development (R&D) is much older and more sophisticated, as developed by the Organization for Economic Cooperation and Development OECD in 1963 in its first Frascati Manual, which started the process of institutional R&D surveys. Nowadays, R&D surveys regularly monitor, among other things, expenditure on ICT equipment, software, biotechnology, and nanotechnology, funded from both private and public sources. Subsequently, there has been an effort to link the R&D methodological system with the System of National Accounts (SNA), which has been methodically managed by the United Nations (UN) since 1947. In the latest version of SNA 2008, both systems were already compatible and usable for the analyses of Gross Domestic Product (GDP). SNA and GDP are important for analyzing periods of recession or boom periods. Some authors (e.g., Monsori [21]) see a very close relationship between INDUSTRY 4.0 and R&D. In this article, the concept of R&D, which is more methodically covered, has been used to research the new INDUSTRY 4.0 phenomenon. However, it would certainly be worth exploring the characteristics and degree of overlap of the two important concepts in the future.

3. Materials and Methods

Research Question 1 (RQ1): Is the number of economic subjects investing in R&D in the Czech Republic in times of economic crisis (recession) decreasing?

Research Question 2 (RQ2): Do the economic sectors of the Czech Republic (business, government, university, and non-governmental non-profit sectors) develop equally, or do they each have specific conditions and are affected by specific factors?

Research question 3 (RQ3): Can human resources be a limiting factor for the development of R&D and INDUSTRY 4.0?

Data: Data on subjects investing in R&D results, including INDUSTRY 4.0 technologies and methods, were taken from the Czech Statistical Office [22–26]. Supporting the introduction of R&D results into production and business processes is part of the Czech Republic's economic policy strategy. Companies in the Czech Republic interested in obtaining scientific grants and subsidies from the economic policy of the state are voluntarily registered with CZ-NACE 72 Research and Development (R&D in the technical industry has a number CZ-NACE 72.19.2). The Czech Classification of Economic

Activities CZ-NACE is derived from the International Standard Industrial Classification of All Economic Activities (ISIC). The Czech Statistical Office sends to these companies the Annual Report on R&D VTR 5-01 [24–26]. It is the collection of internationally comparable data on human and financial resources in the field of research and development in the Czech Republic. This collection of statistics is in line with Decision No. 1608/2003/EC of the European Parliament and of the Council on a methodology for statistics on science and technology for the European Union, within the Europe 2020 Strategy. The collection monitors all R&D subjects (CZ-NACE 72), regardless of whether R&D is their main or secondary economic activity. The Czech Statistical Office uses the method of combining exhaustive and sample surveys [24–26].

Method: The method of multivariate statistics with dummy variables was used to analyze the impact of economic contractions on the number of R&D subjects. Multiple regression is an extension of the bivariate linear regression firms [27], predicting a variable from another's scores. Dummy variables (nominal variables coded 1 = "recession", 0 = "non-recession") were used for the purpose of identifying the influence of the economic cycle-phase recession.

Procedure: Firstly, Model 1 was created and tested; the model included R&D entities from all four sectors of the Czech economy surveyed (business, governmental, university, and non-profit sectors). Secondly, different conditions (especially legal and economic) and different factors of influence (in particular taxes, grants and subsidies) in individual sectors were expected. Therefore, Model 2 was subsequently created and this model examined each sector separately. The results of both models were then compared and discussed. Subsequently, data on the development of human resources in R&D in the Czech Republic and their possible impact on the development of INDUSTRY 4.0 were presented and discussed.

4. Results

The results are presented in connection with the research questions (RQ1–RQ3).

4.1. Research Questions 1 and 2

4.1.1. Recession and Non-Recession Model 1—All Sectors Together

Table 1 describes the number of Czech subjects reporting R&D results over the years 2007 to 2017. The recession periods are based on the definition of declining GDP and are marked. The economic crises in 2009 and 2012–2013 were accompanied not only by a decline in GDP, but also by a sharp decline in the production performance of CZ-NACE C Manufacturing in the Czech Republic (The Czech Classification of Economic Activities CZ-NACE is derived from the International Standard Industrial Classification of All Economic Activities).

Table 1. Numbers of Research and Development (R&D) subjects in the Czech Republic in sectors 2007–2017 [24–26].

Sector	2007	2008	2009	2010	2011	2012	2013	2014	2015	2016	2017
Business enterprise	1736	1766	1872	2107	2237	2312	2299	2368	2387	2355	2628
Government	224	222	223	219	209	195	199	195	196	195	199
Higher education	186	185	187	193	202	203	208	213	228	227	229
Non-gov-non-profit	58	60	63	68	72	68	62	64	59	53	58
GDP [1] real	3964	4070	3874 [2]	3962	4033	4001 [2]	3981 [2]	4089	4307	4412	4601

[1] Database of National Accounts. Czech Statistical Office (CZSO) in billion Czech crowns (CZK) [23]. [2] Recessions (decline in GDP).

The number of subjects of the Czech Republic declaring the results of R&D in the years 2007 to 2017 raises from 2021 to 3114 subjects [24–26]. The total number of all entities in the Czech Republic increased from 570,000 to 668,000 subjects in the same period (the non-profit sector consisted of

120–150,000 subjects, the government sector had 18,000 subjects, and the business sector had developed from 409,000 to 502,000 subjects) [23].

Table 2 shows descriptive statistics of all four sectors of the Czech economy in terms of the number of entities that declare R&D results.

Table 2. Descriptive statistics of each sector.

Sector	n	Mean	Sd	Median	Min	Max	Se
Business enterprise s.	11	2187.91	285.03	2299	1736	2628	85.94
Government s.	11	206.91	12.64	199	195	224	3.81
Higher education s.	11	205.55	17.03	203	185	229	5.14
Non-gov. non-profit s.	11	62.27	5.5	62	53	72	1.66

As can be seen for business enterprise sector the numbers of subject is much higher than for other sectors. This is the main reason why we use the logarithm function in our first model (Model 1) to eliminate the different levels of measurement. The fact that government and higher education sectors have very similar mean and median number of subjects is also interesting.

The following overview shows the results of the bivariate linear regression forecasting variable from other scores. Dummy variables (nominal variables coded 1 = "recession", 0 = "non-recession") were used for the purpose of identifying the influence of economic cycle phase recession.

Model 1 is given in Equation (1):

$$\log(\#subjects) = \beta_{0,j} + \beta_{1,j}\,(t) + \beta_{2,j}\,(recession) + \epsilon_{i,j}, \tag{1}$$

where #subjects denote the number of subjects, j is the index of each sector.

The logarithm of the number of R&D entities working in all sectors (business, government, higher education, and non-profit) is determined by the year and whether or not there has been a recession. Results of Model 1 (all sectors together) are presented in Table 3.

Table 3. Results of linear mixed-effect Model 1.

	Value	Std. Error	DF	t-Value	p-Value
intercept	5.557	0.6861	38	8.09836	0.0000
t	0.00954	0.0131	38	0.7309	0.4693
recession	0.00585	0.0184	38	0.318373	0.7519

Table 3 shows results of linear mixed effect model. As can be seen the only statistically significant variable is the intercept with the value of 5.557. When the p-value for t and recession variable are greater than 0.05 we can conclude that these two variables are not statistically significant. Interestingly, the t variable was observed to be insignificant, which can be firstly seen as counterintuitive. However, from Table 1 we can see that the number of subjects rises only for two sectors (Business enterprise sector and Higher education sector) over time, as can be seen the number of subjects in the government sector decreases. This is the main reason why the time variable is insignificant if we analyze all four sectors together.

4.1.2. Recession and Non-Recession Model 2—Each Sector Separately

Model 2 works with the same data as Model 1, but it examines each sector of the Czech economy separately. Therefore, it is no longer necessary to use the logarithm function.

Model 2 (separate results for each sector) is given in Equation (2):

$$\#subjects = \beta_0 + \beta_1\,(t) + \beta_2\,(recession) + \epsilon_i, \tag{2}$$

where #subjects are number of subjects.

Results of Model 2 (Business enterprise sector) are presented in Table 4.

Table 4. Model 2 results for business enterprise sector.

Predictors	Estimates	CI	p
(Intercept)	1685.42	1547.18–1823.67	<0.001
t	82.01	62.95–101.08	<0.001
Recession	38.18	−97.19–173.55	0.596
Observations	11		
R^2/adjusted R^2	0.899/0.874		

The results of linear regression (Table 4) show that the p-value is lower than 0.05 for intercept and time variable. As can be seen, each year the number of subjects in this sector rises by a value of 82. It also can be seen that the dummy variable (recession) is insignificant. The R^2 of this model is 0.89, which can be interpreted as this model explaining 89.9% of variation.

Results of Model 2 (Non-government non-profit sector) are presented in Table 5.

Table 5. Model 2 results for non-governmental non-profit sector.

Predictors	Estimates	CI	p
(Intercept)	64.52	56.74–72.30	<0.001
t	−0.48	−1.56–0.59	0.403
Recession	2.39	−5.23–10.01	0.556
Observations	11		
R^2/adjusted R^2	0.142/−0.073		

The results of linear regression (Table 5) show that the p-value is lower than 0.05 for the intercept only. This model has very poor explanatory power. It can be seen from the data that this sector has a very subtle number of subject fluctuations, which can be explained by the intercept itself. The result shows that the number of subjects can be predicted by intercept, which is 64 subjects with no growth or descent annually.

Results of Model 2 (Government sector) are presented in Table 6.

Table 6. Model 2 results for Government sector.

Predictors	Estimates	CI	p
(Intercept)	229.07	220.85–237.29	<0.001
t	−3.47	−4.60–−2.34	<0.001
Recession	−4.89	−12.94–3.16	0.268
Observations	11		
R^2/adjusted R^2	0.819/0.774		

The results of linear regression (Table 6) show that the p-value is lower than 0.05 for the intercept and time variable. As can be seen, each year the number of subjects in this sector decreases by a value of 3. It also can be seen that the dummy variable (recession) is insignificant. The R^2 of this model is 0.82, which can be interpreted as this model explaining 82% of variation.

Results of Model 2 (Higher education sector) are presented in Table 7.

The results of linear regression (Table 7) show that the p-value is lower than 0.05 for the intercept and time variable. As can be seen each year, the number of subjects in this sector rises by a value of 5. It also can be seen that the dummy variable (recession) is insignificant. The R^2 of this model is 0.96, which can be interpreted that this model explains 96% of variation.

Table 7. Model 2 results for higher education sector.

Predictors	Estimates	CI	p
(Intercept)	176.97	172.23–181.71	<0.001
t	4.95	4.29–5.60	<0.001
Recession	−4.01	−8.65–0.63	0.129
Observations	11		
R^2/adjusted R^2	0.967/0.959		

4.2. Research Questions 3

4.2.1. Human Resources Working in R&D in the Czech Republic

The focus of this research is also on human resources for R&D and INDUSTRY 4.0 projects in relation to the period of recession or non-recession. Table 8 shows an overview of the development of R&D personnel by sector and field of science. Recession periods are highlighted.

Table 8. R&D personnel by sector of the Czech Republic by fields of science 2007–2017 [24–26].

Sector	2007	2008	2009 [1]	2010	2011	2012 [1]	2013 [1]	2014	2015	2016	2017
Business enterprise	30,640	31,660	32,375	34,658	37,437	41,445	44,255	48,194	49,252	51,069	55,232
Science	4513	4226	4675	5872	7156	8494	9810	11,558	11,812	12,148	13,361
Technical sciences	23,830	25,114	25,279	26,063	27,085	29,079	31,614	32,673	33,697	34,262	37,727
Medical sciences	744	801	820	1196	1174	1156	993	1022	1080	1212	1168
Agricultural sciences	1256	1243	1230	1255	1305	1622	1151	1262	1302	1302	1131
Social Sciences	271	242	343	246	686	1077	666	1666	1347	2116	1818
Humanities	26	35	29	27	29	16	21	14	14	28	27
Government	15,470	15,559	15,402	15,029	15,313	15,482	15,996	16,177	16,705	16,615	17,941
Science	8276	8544	8586	8532	8576	9165	9464	9544	9517	9686	10,094
Technical sciences	464	557	484	413	414	411	488	492	559	554	604
Medical sciences	2195	2124	2168	1948	2000	1931	1935	1802	1980	1605	2029
Agricultural sciences	1239	1136	956	1089	1107	829	842	993	1225	1211	1319
Social Sciences	1029	893	820	777	874	874	758	785	943	973	1112
Humanities	2267	2305	2388	2270	2342	2272	2509	2562	2481	2586	2783
Higher education	26,735	26,993	27,694	27,844	29,149	30,301	32,173	32,680	33,891	31,915	34,234
Science	2887	3431	3286	3788	5818	5864	6512	6804	6743	6528	7517
Technical sciences	8573	8463	8492	8304	7436	8947	8308	8955	9530	9118	9383
Medical sciences	6347	6439	7386	6648	6773	6172	7340	7149	7361	6677	7213
Agricultural sciences	2353	2356	2657	2538	2003	2166	2276	2097	1927	1861	1996
Social Sciences	4574	4367	3393	3487	4959	4597	5447	5365	5550	5194	5509
Humanities	2001	1937	2480	3079	2160	2555	2290	2310	2780	2537	2616
Non-gov. non-profit s.	236	296	317	372	384	300	290	302	280	276	327
Science	26	29	74	103	84	61	55	73	79	62	68
Technical sciences	69	93	28	49	56	47	40	51	23	36	59
Medical sciences	5	16	11	6	23	10	3	1	-	-	8
Agricultural sciences	13	25	24	16	12	8	4	10	4	5	5
Social Sciences	115	106	156	146	174	155	183	166	174	165	179
Humanities	7	27	24	52	36	19	5	1	1	8	8

[1] Recessions.

The business enterprise sector has a very strong predominance of R&D personnel in the field of technical sciences. This is fundamentally different from other sectors, specifically the university sector, which has a high share of technical scientists, but a high proportion of workers are also seen in the fields of natural sciences and medicine. This comparison shows the strong unilateral orientation of CZ-NACE C Manufacturing in the business sector, which also directs the current INDUSTRY 4.0 projects in the Czech Republic.

Figure 1 shows the international comparison of the number of R&D personnel in selected countries of the world during 2005–2015 [24]. Data on human resources working in R&D in the United States was not available.

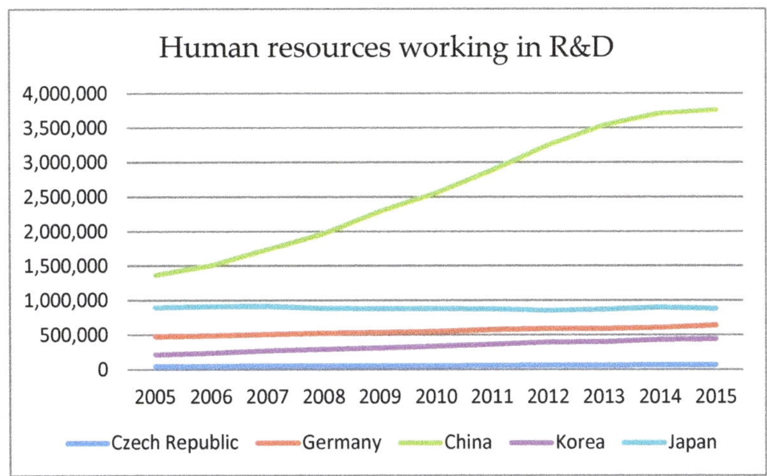

Figure 1. Human resources working in R&D—international comparison for the period 2005–2015 [24].

Figure 1 shows an international comparison of R&D personnel development numbers in the Czech Republic and Germany, showing that numbers are increasing, which gives good conditions for the development of INDUSTRY 4.0 projects. Compared with China, however, the growth rate is relatively slow, which may worsen future competitiveness.

4.2.2. Future Human Resources for R&D in the Czech Republic

The future competitiveness of the Czech Republic will be influenced mainly by the number of graduates of higher education institutions working in the area of R&D, especially in the field of technical sciences. Table 9 shows current vacancies for graduates.

Table 9. The statistics of vacancies for university graduates and high school graduates in the Czech Republic on the date 18 January 2019 [28].

Field	Total Number of Vacancies for University [1] Graduates	Total Number of Vacancies for School [2] Graduates
Administration	67	2150
Transport	9	994
Finance	38	1352
Information Technology	347	854
Culture and Sport	23	290
Management	16	357
Trade and Tourism	65	4553
Defense and Security	3	720
Services	0	1351
Building Industry	44	1634
Science and Research	144	98
Education	378	351
Operations	256	10,131
Health Care	1112	1912
Agriculture and Forest	26	239
Law	44	8

[1] University (college) graduates. [2] Jobs for applicants with completed secondary education.

Table 9 shows current vacancies for university graduates, where the R&D area in the Czech Republic is not large. For the further development of INDUSTRY 4.0, these vacancies will need to be expanded and more attractive to young people.

5. Discussion and Conclusions

INDUSTRY 4.0 is a very important topic that can be explored from a variety of perspectives. In the introduction to this article at least some of them have been highlighted. The research has revealed a number of definitions of INDUSTRY 4.0, but the authors have not found a precise definition of the relationship of INDUSTRY 4.0 to another important topic, namely the development of science and research. The authors of this article perceive a close relationship between INDUSTRY 4.0 and R&D, but this relationship itself is not the subject of research. The authors see it as important to seek a more precise definition of this relationship in future research.

A very interesting topic is the process of implementing INDUSTRY 4.0 (or R&D projects) in relation to the macroeconomic cycle stage. The arrival of an economic recession especially affects the introduction of new projects in companies, as Pavelková has stated [4]. Homolka and Pavelková [20] examined the predictive power of the ZEW sentiment indicator in the case of the German automotive industry, which is important for the Czech economy. Leading macroeconomic indicators can reduce the uncertainty of businesses that want to invest in R&D and INDUSTRY 4.0, but there is still fear that their investment could be destroyed by an economic crisis. The authors of this article based their work on these assumptions and focused on the question of whether the evolution of the business cycle measured by GDP at constant prices affects the numbers and selected indicators of R&D results, both in the business sector and in the government, universities, and non-governmental non-profit organization sectors.

The research was divided into three research questions, of which the first two research questions examined the number of R&D entities in the Czech Republic during 2007–2017. The research horizon included two periods of economic recession, in 2009 and during 2012–2013, but the research is not robust enough to be considered representative. It is possible to discuss whether R&D entities voluntarily register with CZ NACE 72 due to their research activities or due to subsidies and grants. On the other hand, the methodology used for monitoring R&D subjects is internationally recognized and widely implemented.

The method of multivariate statistics with dummy variables was used to analyze the impact of economic recession on the number of R&D subjects. The results (Table 3) show that the recession variable in Model 1 is insignificant. This can have at least three possible explanations. The first explanation is that recession does not have any explanatory power. The second explanation is that recessions are relatively very rare events and there is not enough data to show their explanatory power. The third explanation is that each sector (business, governmental, university, and non-profit sectors) has different conditions (especially legal and economic) and different factors of influence (in particular taxes, grants, and subsidies). This idea is supported by the fact that almost in every year (no matter if recession strikes or not) the number of subjects rising or descending is continuous. Therefore, Model 2 was subsequently created, and this model examined each sector separately.

Results for Model 2 are presented separately from each sector (Tables 4–7). The results of linear regression for the business sector (Table 4) show that the p-value is lower than 0.05 for the intercept and the time variable. As can be seen, each year the number of subjects in this sector rises by a value of 82. It also can be seen that the dummy variable (recession) is insignificant. The R^2 of this model is 0.89, which can be interpreted that this model explains 89.9% of variation. The results of linear regression show that the p-value is lower than 0.05 for intercept only. The result of the linear regression for the non-governmental non-profit sector (Table 5) has very poor explanatory power. This sector has very subtle number of subject fluctuations, which can be explained by the intercept itself. The result shows that the number of subjects can be predicted by the intercept, which is 64 subjects, with no growth or descent annually. The results for the governmental sector (Table 6) show that the number of subjects in this sector decreases by a value of 3 each year. It can also be seen that the dummy variable (recession) is insignificant. The R^2 of this model is 0.82 which can be interpreted as this model explaining 82% of the variation. This result shows that the government sector is not affected either by economic recession or by grants and subsidies, and is steadily declining. This is surprising and alarming. On the contrary the results of the higher education sector (Table 7) show that the p-value is lower than 0.05 for the intercept and time variable. As can be seen, each year the number of subjects in this sector rises by a value of 5. Although a significant number of the universities in the Czech Republic are public, it is positive that the number of R&D subjects is steadily increasing. Here you can see the positive impact of grants, especially from European Union funds. It also can be seen that the dummy variable (recession) is insignificant. The R^2 of this model is 0.96, which can be interpreted as this model explaining 96% of the variation.

The results of both models show that the impact of recession (dummy variable) is not significant. The growth of the number of R&D entities in the CR is much more influenced by other factors (mainly tax relief and grants supporting R&D from EU funds).

Subsequently, data on the development of human resources in R&D in the Czech Republic and their possible impact on the development of INDUSTRY 4.0 (Research question 3) were presented and discussed. Table 8 presents an overview of the development of R&D personnel by sector and field of science. The business enterprise sector in the Czech Republic has a very strong predominance of R&D personnel in the field of technical sciences. This is fundamentally different from other sectors. Specific is the university sector, which has a high share of technical scientists, but a high proportion of workers are also seen in the fields of natural sciences and medicine. This comparison shows the strong unilateral orientation of Manufacturing in the business sector, which also directs the current INDUSTRY 4.0 projects in the Czech Republic. Strong pressure on increasing the number of R&D personnel, however, is not reflected in the currently offered vacancies for university graduates (Table 9). For the further development of INDUSTRY 4.0 projects in the Czech Republic, there is a need to increase and make attractive the qualified job positions that are important for these projects.

Min et al. [16] emphasizes that Germany needs to revitalize the ICT industry to strengthen its manufacturing industry. The Czech industry is an important subcontractor for the German industry. The development of INDUSTRY 4.0 projects in the Czech Republic could boost the competitiveness of both the Czech Republic and Germany, but it is conditioned by a change in the structure of the supply of qualified positions in the necessary structure and relevant demand from university graduates.

The research findings of this article also showed that current EU grant support and tax relief appear to be more important factors for the development of new R&D projects in the Czech Republic than the effects of the economic recession. In terms of longer time horizons, however, the effects of the business cycle cannot be ignored, and their impact on the future development of INDUSTRY 4.0 may be strengthened over time.

6. Conclusions

The current INDUSTRY 4.0 projects in the Czech Republic bring a number of cutting-edge solutions that can be used globally. There are also research institutions dealing with advanced manufacturing in cooperation with companies, which is a step closer towards smart factories. A few of these are listed below.

- The Research Center of Manufacturing Technologies (RCMT) at the Czech Technical University cooperates with industry, especially on topics of advanced simulation models, virtual prototyping and virtual testing, development of advanced feed drive control techniques and vibration suppression methods, advanced monitoring and diagnostics of machine tool condition, multi-axis machining technology, etc. Introducing R&D results helps the industry address its challenges. Machine tools can use, for example, a model to predict micromilining cutting forces, which estimates the tool's deflection and the real tool-path during the micromilling process [29]. Manufacturing also needs methods to monitor tool wear; for instance, one of such methods uses the application of a ceramic piezoelectric sensor mounted on the tool holder in the turning machine to monitor vibration signals due to the flank wear progression [30]. A very important objective, not only in the automotive industry, is to achieve a good surface quality directly from machining without any additional manual work, especially with use of advanced high-strength steels (AHSS). For instance, López de Lacalle's technological model of the milling process estimates values of cutting forces and can offer manufacturers a reduction of production and lead times [31].

- Many companies have their own research facilities, for instance the company Prusa Research focuses on manufacturing of 3D printers and is a global leader in its category thanks to the innovation of using a full metal nozzle and PCB heated bed [32]. Smart factories are in the process of automation and robotization of their production.

- Hybrid manufacturing technologies are used, for instance, the technology developed by Kovosvit and RCMT, which enables manufacturing with additive technology and welding of various combinations of materials, welding of functional surfaces, parts, and details, repairs, creation of full parts with internal channels, shell parts, and hollow parts, all in combination with machining. The rate of growth of parts of different steels is in the range of 0.2 to 1.0 kg/h [32].

- The remote diagnostic is useful and popular in smart factories, for instance, Wikov launched the remote diagnostic tool for online monitoring of gearbox condition. This is a system for the complete driveline and enables optimization of the maintenance plan for maximum availability and minimum downtime. Various data, such as vibration, temperature, speed, pressure, and other parameters are monitored and postprocessed. Outputs are accessible in real time via a web-based interface. The software's advanced algorithms can detect gear-teeth and bearing damages at a very early stage, and thus prevent major gearbox damage [32].

There are many fields in which IoT and monitoring or high-tech approaches are in use. Multiple IoT networks are already being constructed and are ready for use in the country.

The results of both models show that the impact of economic recession on the R&D entities in the CR is not significant nowadays. These entities in Europe are much more influenced by other factors, for instance, grants to support R&D from EU funds, such as Horizon 2020 [32]. The number of R&D subjects in the Czech Republic grew in some sectors (business enterprise sector) in the monitored period of two economic recessions. Usually, in a period of economic recession, the decline in economic performance is accompanied by a decline in the number of economically active entities. From this perspective, it is possible to supplement the scientific literature with findings from this paper.

The added value of Models 1 and 2 is seen primarily in the business sector, which is sufficiently large and dynamically evolving. The dummy variable is not significant in model 2 (Table 4, business enterprise sector), so it can be said that the number of R&D subjects in the business sector is not dependent on the recession. Other sectors are small in number of entities, for example, non-profit non-governmental sector, which has about 150,000 subjects, but only about 60 entities declare R&D results. The study wanted to show the difference in business sector development from other sectors. The ability of the business sector to develop R&D projects, even in times of economic recession, could be used in economic policy. The effective direction of grants in this area could then accelerate the faster return of the economy to economic growth.

This study has its limitations, mainly in the area of available data, because the recession is a rare phenomenon and we have only three data points for each sector (data on the number of R&D entities are collected only once a year; Report VTR 5-01). On the other hand, this phenomenon is very interesting from perspectives of both scientific literature and business, so it would be beneficial to develop this research and perform calculations on a representative set of data.

The other limitation of this research is that in this article, the concept of R&D, which is more methodically covered, has been used to research the new INDUSTRY 4.0 phenomenon. However, it would certainly be worth exploring the characteristics and degree of overlap of the two important concepts.

Author Contributions: Conceptualization, J.Š. and T.U.; methodology, J.Š.; software, T.U.; validation, L.P., and E.S.; formal analysis, T.U.; investigation, J.Š.; resources, L.P.; data curation, J.Š. and E.S.; writing—original draft preparation, J.Š.; writing—review and editing, L.P. and E.S.; visualization, T.U.; supervision, J.Š.; project administration, L.P.; funding acquisition, L.P.

Funding: This research was funded by the Internal Grant Agency of FaME TBU IGA/FaME/2018/001 "Leading indicators in the buying behavior of companies in B2B markets" and RO/2018/08 "Research on qualitative and quantitative changes in demand on the Czech labor market with the introduction of INDUSTRY 4.0".

Acknowledgments: The authors wish to thank the Internal Grant Agency of FaME TBU IGA/FaME/2018/001 "Leading indicators in the buying behavior of companies in B2B markets" and RO/2018/08 "Research on qualitative and quantitative changes in demand on the Czech labor market with the introduction of INDUSTRY 4.0".

Conflicts of Interest: The authors declare no conflict of interest.

References

1. Weyer, S.; Schmitt, M.; Ohmer, M.; Gorecky, D. Towards Industry 4.0-Standardization as the crucial challenge for highly modular, multi-vendor production systems. *Ifac-Papersonline* **2015**, *48*, 579–584. [CrossRef]
2. Wang, S.; Wan, J.; Zhang, D.; Li, D.; Zhang, C. Towards Smart Factory for Industry 4.0: A Self-organized Multi-agent System with Big Data Based Feedback and Coordination. *Comput. Netw.* **2016**, *101*, 158–168. [CrossRef]
3. Povolná, L.; Švarcová, J. The Macroeconomic Context of Investments in the Field of Machine Tools in the Czech Republic. *J. Compet.* **2017**, *9*, 110–122. [CrossRef]
4. Pavelková, D.; Homolka, L.; Knápková, A.; Kolman, K.; Pham, H. EVA and key performance indicators: The case of automotive sector in pre-crisis, crisis and post-crisis periods. *Econ. Sociol.* **2018**, *11*, 78–95. [CrossRef] [PubMed]
5. Dalenogare, L.S.; Benitez, G.B.; Ayala, N.F.; Frank, A.G. The expected contribution of Industry 4.0 technologies for industrial performance. *Int. J. Prod. Econ.* **2018**, *204*, 383–394. [CrossRef]
6. Diez-Olivan, A.; Del Ser, J.; Galar, D.; Sierra, B. Data Fusion and Machine Learning for Industrial Prognosis: Trends and Perspectives Towards Industry 4.0. *Inf. Fusion* **2019**, *50*, 92–111. [CrossRef]
7. Xia, F.; Yang, L.T.; Wang, L.; Vinel, A. Internet of things. *Int. J. Commun. Syst.* **2012**, *25*, 1101–1102. [CrossRef]
8. Trappey, A.J.; Trappey, C.V.; Govindarajan, U.H.; Chuang, A.C.; Sun, J.J. A review of essential standards and patent landscapes for the Internet of Things: A key enabler for Industry 4.0. *Adv. Eng. Inf.* **2017**, *33*, 208–229. [CrossRef]
9. Lee, J.; Kao, H.A.; Yang, S. Service innovation and smart analytics for industry 4.0 and big data environment. *Procedia CIRP* **2014**, *16*, 3–8. [CrossRef]
10. Rajput, S.; Singh, S.P. Identifying Industry 4.0 IoT enablers by integrated PCA-ISM-DEMATEL approach. *Manage. Decis.* **2018**. [CrossRef]
11. Kovacs, O. The Dark Corners of Industry 4.0—Grounding Economic governance 2.0. *Technol. Soc.* **2018**, *55*, 140–145. [CrossRef]
12. Stock, T.; Seliger, G. Opportunities of sustainable manufacturing in industry 4.0. *Procedia CIRP* **2016**, *40*, 536–541. [CrossRef]
13. Tuček, D. New Strategy for Business Process Management—Quantitative Research in Czech Republic. *Int. Adv. Econ. Res.* **2017**, *23*, 131–132. [CrossRef]
14. Kang, H.S.; Lee, J.Y.; Choi, S.; Kim, H.; Park, J.H.; Son, J.Y.; Kim, B.H.; Noh, D.S. Smart manufacturing: Past research, present findings, and future directions. *Int. J. Precis. Eng. Manuf.-Green Technol.* **2016**, *3*, 111–128. [CrossRef]
15. Li, L. China's manufacturing locus in 2025: With a comparison of "Made-in-China 2025" and "Industry 4.0". *Technol. Forecast. Soc. Chang.* **2018**, *135*, 66–74. [CrossRef]
16. Min, Y.K.; Lee, S.G.; Aoshima, Y. A comparative study on industrial spillover effects among Korea, China, the USA, Germany and Japan. *Ind. Manag. Data Syst.* **2018**, *119*, 454–472. [CrossRef]
17. Grassetti, F.; Hunanyan, G. On the economic growth theory with Kadiyala production function. *Commun. Nonlinear Sci. Numer. Simul.* **2018**, *58*, 220–232. [CrossRef]
18. Devezas, T.C.; Corredine, J.T. The biological determinants of long-wave behavior in socioeconomic growth and development. *Technol. Forecast. Soc. Chang.* **2001**, *68*, 1–57. [CrossRef]

19. Dosi, G.; Pereira, M.C.; Roventini, A.; Virgillito, M.E. When more flexibility yields more fragility: The microfoundations of Keynesian aggregate unemployment. *J. Econ. Dyn. Control* **2017**, *81*, 162–186. [CrossRef]

20. Homolka, L.; Pavelková, D. Predictive power of the ZEW sentiment indicator: Case of the German automotive industry. *Acta Polytech. Hung.* **2018**, *15*, 161–178. [CrossRef]

21. Monostori, L. Cyber-physical Production Systems: Roots, Expectations and R&D Challenges. *Procedia CIRP* **2014**, *17*, 9–13. [CrossRef]

22. Obst, C.; Hein, L.; Edens, B. National Accounting and the Valuation of Ecosystem Assets and Their Services. *Environ. Resour. Econ.* **2015**, *64*, 1–23. [CrossRef]

23. Database of National Accounts. CZSO. Available online: http://apl.czso.cz/pll/rocenka/rocenka.indexnu?mylang=EN (accessed on 21 January 2019).

24. Research and Development. CZSO. Available online: https://www.czso.cz/csu/czso/statistika_vyzkumu_a_vyvoje (accessed on 22 January 2019).

25. Výkaz VTR 5-01. CZSO. Available online: http://apl.czso.cz/pll/vykazy/pdf113?xvyk=2414&cd=0 (accessed on 22 January 2019).

26. VTR 5-01 Roční Výkaz o Výzkumu a Vývoji. CZSO. Available online: https://www.czso.cz/csu/vykazy/vtr-5-01-rocni-vykaz-o-vyzkumu-a-vyvoji_psz_2019 (accessed on 22 January 2019).

27. Cooper, D.R.; Schindler, P.S. *Business Research Methods*, 12th ed.; McGraw-Hill Higher Education: New York, NY, USA, 2013; pp. 547–549, ISBN-13: 9781259070952.

28. The Ministry of Labour and Social Affairs. Available online: https://portal.mpsv.cz/sz/obcane/vmjedno (accessed on 18 January 2019).

29. Uriarte, L.; Azcárate, S.; Herrero, A.; Lopez de Lacalle, L.N.; Lamikiz, A. Mechanistic modelling of the micro end milling operation. *J. Eng. Manuf.* **2008**, *222*, 23–33. [CrossRef]

30. Ahmad, M.A.F.; Nuawi, M.Z.; Abdullah, S.; Wahid, Z.; Karim, Z.; Dirhamsyah, M. Development of Tool Wear Machining Monitoring Using Novel Statistical Analysis Method, I-kaz™. *Procedia Eng.* **2015**, *101*, 355–362. [CrossRef]

31. López de Lacalle, L.N.; Lamikiz, A.; Muñoa, J.; Salgado, M.A.; Sánchez, J.A. Improving the high-speed finishing of forming tools for advanced high-strength steels (AHSS). *Int. J. Adv. Manuf. Technol.* **2006**, *29*, 49–63. [CrossRef]

32. Industry 4.0 in the Czech Republic. CzechInvest. Available online: https://www.czechinvest.org/getattachment/27479be7-854a-4720-8258-e9143cc2d22c/Industry-4-0-in-the-Czech-Republic (accessed on 3 March 2019).

 *applied sciences*

Article

Industrial Cyber-Physical System Evolution Detection and Alert Generation

Aitziber Iglesias [1,*], **Goiuria Sagardui** [2] and **Cristobal Arellano** [1]

[1]   Ikerlan Technology Research Centre, Big Data Architectures, 20500 Arrasate, Spain; carellano@ikerlan.es
[2]   Mondragon Unibertsitatea, Information Systems-HAZI-ISI, 20500 Arrasate, Spain; gsagardui@mondragon.edu
*   Correspondence: aiglesias@ikerlan.es

Received: 12 February 2019; Accepted: 11 April 2019; Published: 17 April 2019

Abstract: Industrial Cyber-Physical System (ICPS) monitoring is increasingly being used to make decisions that impact the operation of the industry. Industrial manufacturing environments such as production lines are dynamic and evolve over time due to new requirements (new customer needs, conformance to standards, maintenance, etc.) or due to the anomalies detected. When an evolution happens (e.g., new devices are introduced), monitoring systems must be aware of it in order to inform the user and to provide updated and reliable information. In this article, CALENDAR is presented, a software module for a monitoring system that addresses ICPS evolutions. The solution is based on a data metamodel that captures the structure of an ICPS in different timestamps. By comparing the data model in two subsequent timestamps, CALENDAR is able to detect and effectively classify the evolution of ICPSs at runtime to finally generate alerts about the detected evolution. In order to evaluate CALENDAR with different ICPS topologies (e.g., different ICPS sizes), a scalability test was performed considering the information captured from the production lines domain.

Keywords: Cyber-Physical Systems (CPS); scalability test; Internet of Things (IoT)

---

1. Introduction

Nowadays, Industrial Cyber-Physical Systems (ICPSs) play an important role in the current trend of automation in manufacturing as Industry 4.0 is increasingly gaining strength [1–3]. ICPSs are "physical, biological, and engineered systems whose operations are monitored, coordinated, controlled, and integrated by a computing and communication core" [4]. An ICPS is composed of different types of devices, i.e., actuators, displays, and sensors. Data from the devices is monitored in order to transform information that is visualized by the user to monitor the industrial domain [2,5,6].

In the automotive domain, manufacturing production lines are based on press machines (press lines). A press line is composed of different machines (e.g., press machine and furnace) and every machine within the press line is composed of different devices. Additionally, each device is able to send different attributes where the information about that device is reflected. Thus, the press line composition in addition to the devices within each machine depends on a customer's real needs, i.e., depending on the production line, one machine or another is introduced in the press line. Note that the characteristics of each machine are variable, e.g., different types of furnace exist or different press machine sizes exist. Additionally, devices within each machine are variable, and they also depend on the customer needs [5]. These makes each press line different from each other, i.e., different ICPS topologies exist. Thus, a monitoring solution in Industry 4.0 receives large amounts of data coming from heterogeneous and distributed devices, which implies that the monitoring system must be scalable enough to respond to different ICPS topologies. These data are being used to identify anomalies during operation [7,8].

Additionally, after analyzing the industrial domain, we realized that they required a continuous renovation, known as retrofitting, i.e., new devices can be introduced, removed, or modified depending on the customer needs (e.g., new elements of an industrial domain need to be monitored due to customer requirements, so new devices must be inserted). Users can make decisions that impact the business [9]. Furthermore, ICPS devices are intelligent; they are able to change the data architecture depending on the status of the machine that is being monitored. Note that a concrete machine within a press line can be composed by more that 500 devices [10]. Therefore, being aware of what is happening in the industrial domain is important. When an evolution happens, many people need to be alerted in order to detect anomalies to reduce system downtime, since a monitoring system is supported by different user roles that are interested in different information. Therefore, ICPSs evolve throughout their lifetimes [11], and managing the variability is crucial in Industry 4.0 [1,2,10], as the data captured from the ICPS are converted into information for decision-making. Therefore, managing the variability requires capabilities, posing additional challenges for monitoring ICPSs [1,11,12]. As a part of the monitoring logic, it is necessary to communicate the occurred evolution to all users who are supervising the ICPS that is being monitored. Thus, having updated and reliable information at any time allows users to make decisions. Therefore, monitoring solutions in a dynamic context, such as Industry 4.0, need to be flexible to identify and integrate ICPS evolutions rapidly to meet the requirements.

Although the identified issues are motivated by a press line domain, we notice that the ICPS evolution is something known in the literature [13,14], and considering the domain analysis performed in our previous work [5], we realized that ICPS evolution besides the different ICPS topologies are not specific problems for press lines. In the literature, many authors consider software evolution [14–17], but very few of them consider software and hardware evolution [18,19] even though those who consider runtime variability do not consider uncertainties; hence, as far as we know, no one in the literature has give the response to detecting ICPS evolution in order (1) to have the traceability of what has happened in an ICPS over time; (2) to classify the occurred evolution in order to communicate immediately to the users to avoid any bad decisions; and (3) to give a solution which considers different ICPS topologies.

That is why, considering the importance of (1) the awareness of the status of ICPS, (2) the existence of different user roles with each one working with different data, and (3) the need to communicate changes immediately in order to make decision-making more effective, we propose a system that can detect and classify the evolution of an ICPS in a fast and efficient manner. To support those challenges, we present CALENDAR, a **C**yber-physic**AL** syst**Em** evolutio**N** **D**etection and **A**lert gene**R**ation system, that is capable of detecting and effectively classifying the evolution of ICPSs at runtime. CALENDAR compares the data received in time Qt, with the data received in the previous time (Qt-1). By comparing this information, CALENDAR detects changes in an ICPS in a structured way. CALENDAR is able to identify and classify ICPS evolution and then generates user alerts. Moreover, considering our solution needs to respond to different ICPS topologies (e.g., different ICPS sizes), we have performed a scalability test: (I) to prove the validity of our solution in growing ICPSs sizes, (II) to check the performance based on different types of evolution, and (III) with different press line sizes.

The rest of the article is structured as follows: In Section 2, the use case based on press lines is explained. Section 3 introduces the problem statement and an overview of the monitoring ICPS in Industry 4.0. In Section 4, the CALENDAR module for monitoring systems is explained. In Section 5, the scalability evaluation of CALENDAR is performed followed by the related work in Section 6. Finally, we conclude the article in Section 7.

2. Use Case: Press Line Domain

One of our partners designs and manufactures mechanical and hydraulic press machines, complete stamping systems, transfer presses, robotic lines, etc. Considering the manufacturing production lines designed and developed by our partner are based on press machines, we refer to them as press lines.

The automotive world is a sector that is in constant movement and where technological developments require a continuous technological renovation. The Hot Stamping of Boron Steels is a recent creation technology that is settling in the sector and which the processes are in constant evolutions, changes, and improvements. For example, a hot forming manufacturing line for boron steels consists of 3 fundamental machines, each one tied to the other:

- Destacker: It is the component responsible for (1) unstacking previously cut formats and (2) introducing the format in the furnace.
- Furnace: Inside this component, the material remains for a minimal time until it reaches a completely austenitic structure and finally achieves a diffusion of the coating in the substrate. Currently, our partner used different furnace types: (I) Roller furnaces, (II) Multilevel furnaces, and (III) Furnace "carousels".
- Press Machine: Once the format is heated, the press machine changes the shape of a workpiece with pressure. The main characteristic of this machine is that, unlike the trajectory that is necessary in the forming of cold steels, in the Hot Stamping, the press has to approach the mold as quickly as possible.

Notice that different press lines exist. Depending on the customer's needs, the quantity and type of machines that constitute a press line are different, e.g., three different types of furnaces are used by our partner; thus, depending on customers needs, one or the other would be used. In turn, each machine within the press line is composed of different devices. These devices are also variable; they depend on customers needs, since the customer is the one who decides what to monitor inside the press lines, e.g., the temperature of the clutch break inside the press machine.

In order to collect quantitative information in addition to getting information about their daily work, we conducted interviews with our industrial partner. For example, we realized that three different sizes of press machines can be used inside a press line: large, medium, and small (see Table 1). Though the number of devices is incremental to the size of the machine, the incidences (i.e., machine breakdowns) occurring per week are similar in all press machine sizes and are resolved in 1 to 2 days. Additionally, due to machine maintenance, retrofitting, etc., a press machine can evolve, i.e., new devices can be introduced or existing ones can be removed or replaced, and these changes affect between 40% and 59% of the machine.

Table 1. The characteristics of a Press Machine.

Press Machine Characteristics	Product Scale		
	Small	Medium	Large
Average number of devices per Press Machine	20 to 49	50 to 99	>500
Percentage of affected devices when the Press Machine evolves (added, removed, or modified)?	40% to 59%	40% to 59%	40% to 59%

The data captured for each press machine varies, e.g., depending on installed devices. A customer decides what s/he wants to monitor, and the customers' requirements keep on evolving, resulting in several types of changes. Thus, the variability within a press machine exists, since depending on its purpose, the customer may decide what s/he wants to monitor.

Therefore, in a press line, several machines can be found, each tied to one another. Each machine has a different objective, and therefore, the characteristics of each one are different. Additionally, note that each machine (e.g., press machine), as such, can be different (e.g., devices can be from different providers and the mechanism of the machine can be different). This implies that, in the same press line, variability exists. Therefore, each machine can evolve, since each machine is independent. This evolution is known as retrofitting. New requirements usually have an impact on the devices inside each machine, having to insert new devices (e.g., new elements of an industrial domain need to be monitored due to customer requirements, so new devices must be inserted), remove existing ones (e.g., due to an anomaly, the device is damaged and must be removed). or modifying them (e.g., a device is updated and is now able to send more data that was not previously considered). Additionally, note that every device can send different attributes. These ones can also vary depending on the state of the ICPS, i.e., some of the devices located in the ICPS are intelligent; they are able to change the data architecture depending on the status of the machine that is being monitored. Device information is then sent to the users, so they can make decisions, e.g., repairing a device, since it is not working properly, or do predictive maintenance because the Remaining Useful Lifetime (RUL) is approaching to zero (predictive maintenance [20]).

In spite of this, note that the ICPS is composed of different machines with different characteristics. In the same manner that devices inside the machines can evolve, the press line itself can evolve due to customers requirements, i.e., new machines can be introduced inside the press line. At the same time, it is necessary that none of these machines stop, since that would bring negative consequences to the production (e.g., loss of money).

In addition, in this particular use case explained above, between 30 and 50 people are needed to support the proper functioning of the press line. Notice that the quantity of people would depend on each press line to supervise.

Therefore, being aware of what is happening in the industrial domain is crucial. When an evolution happens, many people need to be alerted in order to detect anomalies to reduce system downtime. In Table 2 is shown the different roles that the users have in order to support a press line. Thus, depending on the user role, the interest of the users in terms of data is different. In spite of that, all of them need to be aware of what is happening in the ICPS that is being monitored. This results in the following conclusions: (1) a solution able to automatically detect ICPS evolution is necessary. (2) Alerting each user about the evolution is necessary to be aware of the status of ICPS, as it helps in making decisions. (3) The presented solution needs to be scalable in order to give responses to different ICPS sizes.

Table 2. The user roles for press line supervision.

Role	Definition
Operator	Controls the operation of the press line.
Analytical Manager	Analyzes the historical data in order to to find machine patterns or trends.
Domain expert	Analyzes at runtime the raw data of a specific device or group of devices to detect any malfunction or anomaly.
Technical Assistant	Provides technical assistance, i.e., people in charge to solve any incidence that can occurred as fast as possible
Assistance management	If an incident cannot be solved by the Technical Assistants, a more exhaustive assistance has to be planned. Thus, in that case, the issue will be transferred to the Assistance Management in order to solve the incidence.

3. Problem Statement and Solution Overview

After analyzing the press line domain, we discovered that monitoring their ICPSs is necessary. Motivated by our industrial partner, in this section, we explain the problem statement which is (Section 3.1) followed by the solution overview in Section 3.2, where the given solution is provided.

3.1. Problem Statement

Different ICPS sizes exist and are being supported by different user roles. That means that not everyone is working or is interested in the same information. Additionally, the ICPS can evolve over time, for example, when an anomaly occurs, the devices need to be repaired; this means that there are often changes in the ICPS itself in order to continue operating normally. The ICPS can also evolve due to business reasons, i.e., new machines need to be introduced in the press line in order to adapt the product to the business demands. This means that an ICPS is not static and can evolve over time and that, as many people are supporting the ICPS, it is necessary to inform them about the occurred changes.

Although the motivation of the problem comes from the press line domain, notice that it is not an isolated case; the evolution of an ICPS already appears in the literature [13,14]. Moreover, taking into account the domain analysis performed in our previous work [5], we realized that the evolution of an ICPS besides the different ICPS topologies, are issues that also occur in automated warehouse domain, besides in a catenary-free tram, an intelligent elevator, or wind turbine domains [21].

Note that an ICPS is composed of different devices, and these can have logic or physical distributions. Every device is able to send different attributes (e.g., temperature and pressure). Additionally, as discussed in our previous work [5], a device is associated with an agent. This agent can be intelligent, i.e., depending on what happens in the industrial environment, the information to be sent may be different, e.g., alerts. Hence, an intelligent device which is associated with an agent is able to start sending a new attribute that, in a previous state, was not sent. Thus, considering that an ICPS can be composed of different machines and each one can be composed of more than 500 devices, having control of all agents is not feasible. This causes the proper distribution of the data to change. Thus, every ICPS has a concrete distribution, either logical or physical, in addition to the fact that each device can send different attributes. Thus, either the attributes or the distribution of the ICPS can evolve, i.e., in an ICPS, structural changes can occur.

In spite of this, it should be considered that self-adaptation is important when talking about ICPS [12]. These means that different self-adaptation levels exist when an ICPS evolves: (1) sensor or hardware level, (2) software monitoring level, and (3) data visualization level. However, as Schütz et al. remark [13], the reconfiguration is not available at a sensor level. However, the monitoring software [5] and the information visualization [10] do need to be adapted. This is crucial in Industry 4.0 [1,2,10], as the data captured from the ICPS are converted into information for decision-making. Thus, once the data is received and structured, we propose to identify the evolution by comparing the data of two-time instances. A dataset comparison is widely used to predict future trends [22,23], but as far as we know, it has not been used to identify ICPS evolution.

Considering the importance of (1) the awareness of the status of ICPS, (2) the existence of different user roles with each one working with different data, and (3) the need to communicate changes immediately in order to make decision-making more effective, we propose a system that can detect and classify the evolution of an ICPS in a fast and efficient manner. To support those challenges, we present CALENDAR, a Cyber-physicAL systEm evolutioN Detection and Alert geneRation system.

3.2. Solution Overview

Figure 1 illustrates an overview of the monitoring of an ICPS and how an evolution can be detected by CALENDAR. CALENDAR is capable of detecting additions, removals, and modifications on an ICPS (e.g., integration of a new device) immediately (next time the data is received).

The monitoring system is composed of three subsystems: (1) the data management subsystem, responsible for capturing data from an ICPS and building the corresponding Data Models; (2) CALENDAR, responsible for analyzing the evolution occurred in the ICPS; and (3) the visualization subsystem, which is responsible for communicating to the user the evolution in a proper manner.

The data management subsystem uses Data Models to create a snapshot of the ICPS at a specific moment in time. Thanks to it, both the structure of the ICPS and the data are captured. Thus, from the received data, the Data Collector extracts a specific Data Model that saves all the information received from the ICPS at a given time as explained in our previous work [5].

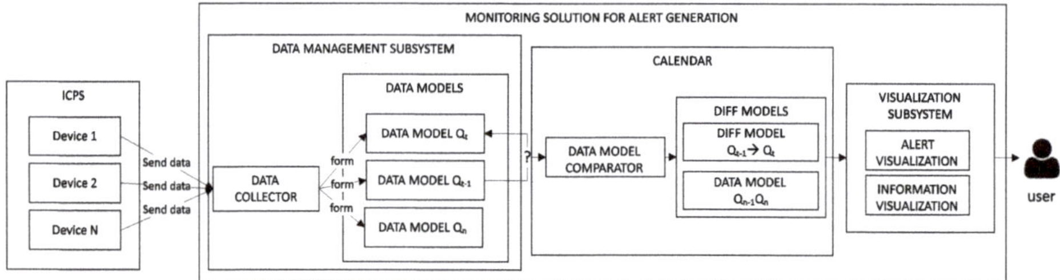

Figure 1. An overview of monitoring Industrial Cyber-Physical Systems (ICPSs): the captured data and detected evolution to alert users.

Once the Data Model is formed by the Data Collector, CALENDAR starts working. It is important to know that CALENDAR is suitable once the evolution has occurred. CALENDAR is a reactive system, i.e., reacts to changes that have happened; it does not anticipate them.

Inside CALENDAR, the Data Model Comparator analyzes if any change has happened since the last time the data was received. In the case that any evolution occurred, CALENDAR is responsible for classifying the occurred evolution in a Diff Model.

ICPS evolves over time, so detecting changes in the structure such as the addition or removal of devices are critical to providing the user with the right information to make decisions. CALENDAR is responsible for detecting the evolution in an ICPS and its classification. For each Data Model, CALENDAR compares the current instance (Data Model Qt) with the previous instance in time (Data Model Qt-1) to identify the evolution of an ICPS. Diff Models are used to classify changes and to alert the user on the instat that evolution occurs through the visualization subsystem. In this way, the user is fully informed of what is happening and can, therefore, be supported in decision-making.

Thus, if the ICPS can be represented by the Data MetaModel, CALENDAR is able to analyze at runtime if any changes have occurred. Hence, CALENDAR is able to detect any evolution which can be represented by the Data MetaModel (presented in Section 4.1). For example, imagine that due to the intelligence of a device, this one starts sending a new attribute which was not previously represented in the Data Model. In this particular scenario, CALENDAR is able to detect a new attribute at runtime in addition to classifying it.

The visualization subsystem, already developed and evaluated in Reference [10], is capable of visualizing both the information (information visualization) and the alerts (alert visualization) to communicate changes to the user. Moreover, the subsystem is also responsible for managing the invalid visualizations, i.e., if the ICPS evolves (e.g., a device is removed), the visualization fails due to the fact that the information to be displayed has disappeared. Thus, the visualization subsystem is able to manage those situations so that the visualization is adequate in addition to informing users about the occurred changes.

CALENDAR ensures the detection of evolution in an ICPS. An ICPS faces changes frequently, and controlling them is necessary for users to make valid decisions. In the following sections, we detail CALENDAR and provide an evaluation that shows its applicability in real scenarios.

4. CALENDAR

In this section, we focus on CALENDAR, i.e., a system that compares the current instance (Data Model Qt) with the previous instance in time (Data Model Qt-1) to identify the evolution of an ICPS.

The objective of CALENDAR is to detect an ICPS evolution. Once an evolution occurrs, our solution is able to identify at any level, the additions, removals, or modifications by comparing Data Models in subsequent timestamps.

In Figure 2, the real scenario of monitoring an ICPS is presented, which was developed with SpringBoot. As mentioned above, the Data Collector, which is based on Kafka, is a distributed streaming platform that is responsible for generating the Data Models, taking into account the data received from the ICPS. Once the Data Model is created, it is stored by the Data Collector in a NoSQL database (Elasticsearch), so in this manner, CALENDAR can then use any stored Data Model.

Once the Data Model is stored and a new one is detected, CALENDAR starts working. CALENDAR is composed of two main components: (I) Data Model Comparator, the component responsible for comparing two subsequent Data Models, and (II) Diff Models, the instance of Diff MetaModel responsible classifying the occurred evolution.

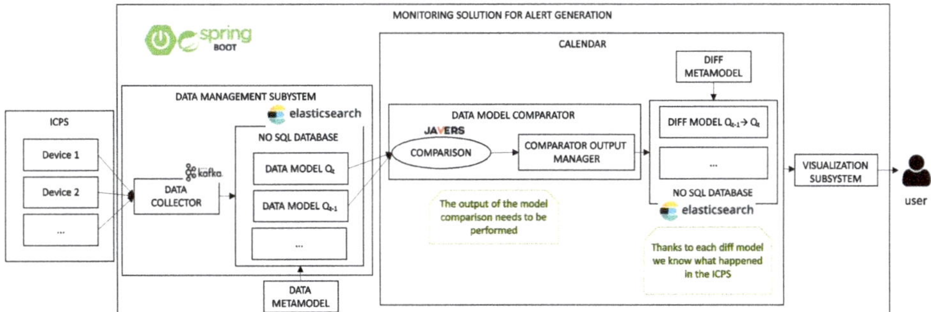

Figure 2. A real scenario of monitoring ICPS, detecting the evolution, and alerting users.

In this section, first, we present the Data Models that CALENDAR is going to use in order to detect ICPS evolution (Section 4.1). We present the characteristics of the Data Model in order to define what kind of evolution will be able to detect. Then, in Section 4.2, how CALENDAR is able to compare two Data Models using the Data Model Comparator component is presented. Finally, in Section 4.3, the Diff Model where the result of the Data Model Comparator classification is explained.

4.1. Data MetaModel

The Data MetaModel is the artifact that allows a representation of both the data and structure at once. It has a tree structure to facilitate the evolution detection [24] and contains seven levels [5] that represent the logical and physical structures of different ICPSs. The Data MetaModel is a combination of two different standards (i.e., IEC 62264 and IEC 61850). In order to be valid this Data MetaModel in different ICPSs, the following requirements need to be considered:

- **Quantity of levels:** The Data Model conformed by the Data MetaModel will always be composed of 7 levels, i.e., there cannot be a branch containing only 5 of them.
- **ICPS representation:** The Data MetaModel has a hierarchical structure. This implies that a node of the Data Model cannot depend on several nodes, i.e., a single node contains a single parent node.
- **Physical/logical structure:** Even if in an ICPS, a node can communicate with other nodes, in our Data MetaModel, the relation between nodes is not reflected. Each node is independent of the rest. If we wanted to reflect the relation between nodes, another model must be used.
- **Atomic values:** Although the Data MetaModel supports complex data structures, we do not focus on the analysis of these complex data. That is why, the Data MetaModel is designed for atomic data values, i.e., simple data (e.g., Boolean, Integer, and String).

In addition, it is important to notice that these Data MetaModels are also valid for monitoring multiple ICPSs, i.e., multiple press lines. More information about the Data MetaModel is provided in Reference [5]. In the next table, information about each level of the Data MetaModel is provided:

Table 3. The level types of the Data MetaModel.

Level Type	Descriptions
Enterprise	ICPS identification, name
Site	Geographical or physical distribution of the ICPS
Area	Logical distribution inside the ICPS
Logical device	Devices description
Logical node	Device identification
Data	Information description that the devices send
Data attribute	Concrete information that the device sends inside the data.

Thus, the Data Collector generates a Data Model that conforms to this Data MetaModel in each timestamp. Once the Data Model is captured, the Data Model Comparator analyzes if any change has happened since the last time data was received.

4.2. Data Model Comparator

For each Data Model, the Data Model Comparator compares the current instance (Data Model Qt) with the previous instance in time (Data Model Qt-1) to identify the evolution of an ICPS. To perform the comparison between the two Data Models, CALENDAR uses Javers. Javers is a library able to compare complex structures and to detect changes. Javers' output is not enough for our purpose so we have post-processed the results of Javers.

Notice that Javers does not take into account the hierarchical dependencies between nodes. However, the dependencies in an industrial environment are something necessary because it is valuable to visualize the result in a simple and meaningful way to the user in order to help him/her make decisions. That is why we need to post-process the Javers result. For example, in the press line domain due to business strategy, imagine it is necessary to remove Zone B. In the upper part of Figure 3, the Data Model before an evolution occurred (Qt-1) is shown, e.g., the Press Machine product line of Mexico is composed by two areas. However, in the lower part of the figure, the Data Model after an evolution (Qt) is presented, e.g., Zone B is removed from the Press Machine product line. Therefore, all nodes that depend on that zone are removed (e.g., Machine 1). When CALENDAR compares Qt with Qt-1 using Javers, this one detects a change for each modified, added, or removed node. In this case, Javers generates 1001 alerts when Zone B node is removed using a Json Object format or 1502 alerts using a Json Array format. This quantity of alerts do not facilitate the task to the user when data is represented. Even if the example given is due to a business strategy, note that, as mentioned in Section 3, many reasons can trigger the evolution of an ICPS, e.g., devices' intelligence itself can cause changes in the Data Model. Additionally, it is important to notice that many people are supporting the monitoring system and that not all of them are interested in the same data or information [10]. In spite of that, all of them need to be informed about the occurred evolution.

Thus, we concluded that the comparison provided by Javers is format-dependent, i.e., the format of the text model impacts the result. If Json arrays [25] are used, the order is taken into account. Thus, when removing node Zone B, the nodes on the right are marked as modified, i.e., Zone C is marked as modified, ascending the number of alerts to 1502. Instead, if we use Json Objects [25], when the order of a node changes (due to an addition or removal), it is not marked as a change.

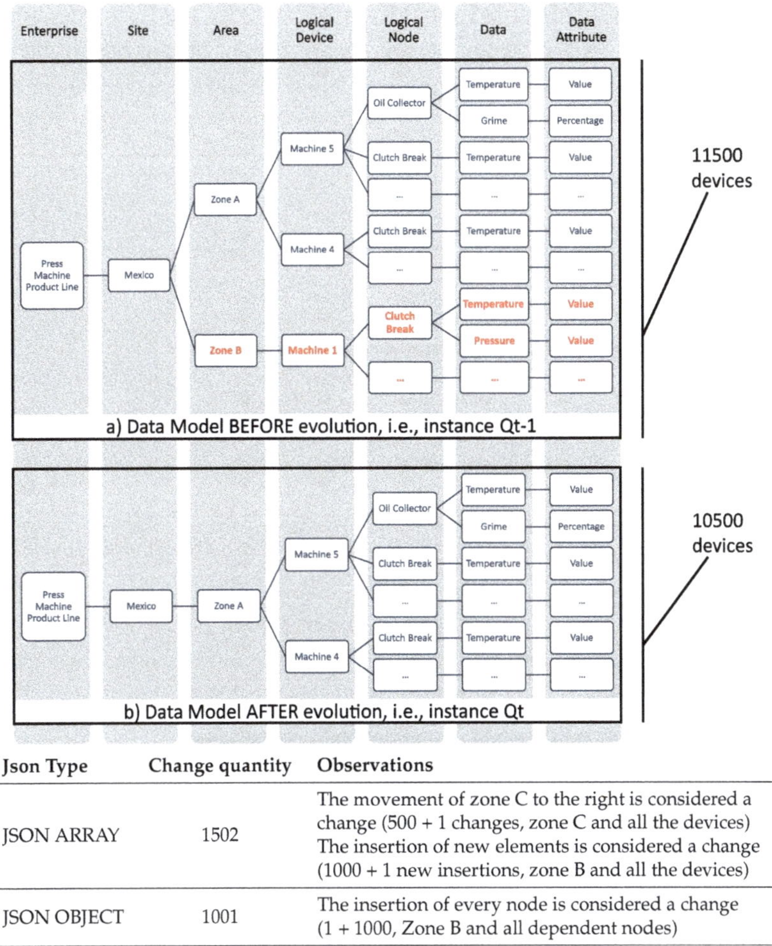

Json Type	Change quantity	Observations
JSON ARRAY	1502	The movement of zone C to the right is considered a change (500 + 1 changes, zone C and all the devices) The insertion of new elements is considered a change (1000 + 1 new insertions, zone B and all the devices)
JSON OBJECT	1001	The insertion of every node is considered a change (1 + 1000, Zone B and all dependent nodes)

Figure 3. Text model comparator operations depending on the input.

To overcome this limitation, besides using Json Objects, we post-process the Javers result and use a model to manage the post-procesed output, which only generates the necessary alerts for the user (Diff Model); in this particular case, we will only generate a unique alert, i.e., Zone B is removed.

4.3. Diff MetaModel

In Figure 4, a diagram of the process for creating the concrete Diff Model is presented. The process starts when a new Data Model is received, on the right side of Figure 5, the Unified Modeling Language (UML) diagram of the Data MetaModel is shown. In that moment, CALENDAR gets the received Data Model and the previous instance, i.e., Qt and Qt-1. Then, the Data Models are compared by the Data Model Comparator using Javers as explained in the previous section. If the result is not empty, CALENDAR gets the result of Javers and treats each action differently (new, remove, or change). This is because the information to be saved depends on the type of change that occurred. Once the attributes are mapped, the information is classified based on the Data MetaModel types (see Table 3). This manner simplifies to identify where the evolution has occurred. Then, if elements are added or removed, it is necessary to delete the unnecessary information despite duplicated information as mentioned above (e.g., from 1001 alerts to 1, i.e., we do not take into account nodes below Zone B).

Finally, all the information is set to create the Diff Model. Once the process is finished, the created Diff Model is stored in a NoSQL database in order to keep track of the traceability of the occurred evolutions over time.

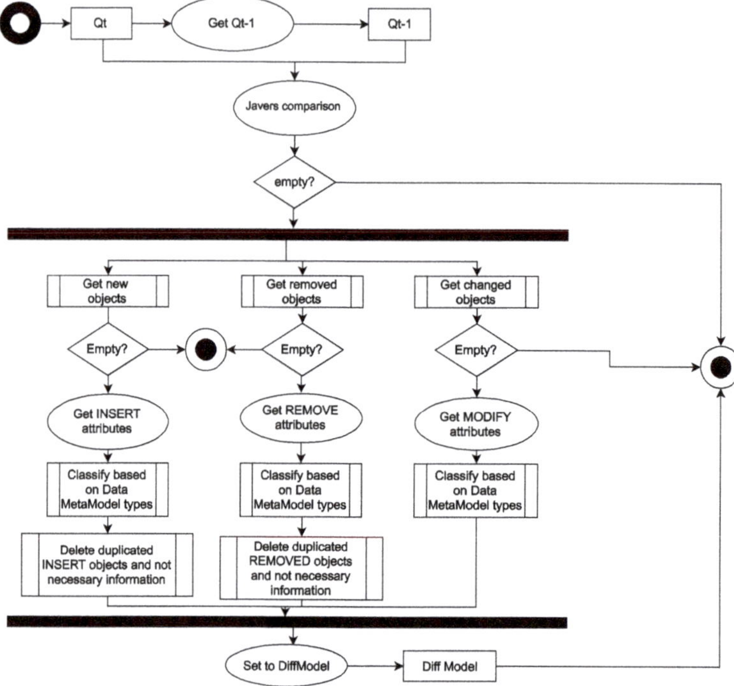

Figure 4. The Diff Model creation a diagram.

The Diff MetaModel contains information about the types of changes, the level in which the changes have occurred and the nodes affected by the changes. The level type is related to the level of the Data MetaModel and describes in which level of the Data Model the changes occurred (e.g., site). Therefore, the Diff Model can have a maximum of seven level types, one for each Data Model level. That is, all the identified changes in a level (e.g., area) are grouped. This classification facilitates communication among users.

Inside each level, the metamodel considers three types of changes that occur when an ICPS evolves (see Figure 5).

- ADD: all the new nodes that do not exist in the previous instant.
- REMOVE: all the nodes that have disappeared at the previous instant.
- MODIFY: if the node exists, but a change has happened in it.

To reduce the alerts of removal or addition explained in the previous section (i.e., *"delete duplicated INSERTED/REMOVED objects and unneccesary information"*; see functions of Figure 4), e.g., from 1001 alerts to 1 alert, a FatherChildNode is used to group the affected nodes, see Figure 5. The FatherChildNode is an instance of a DataModelChild, that is, a subtree with the node changed and its children and contains the nodes affected by an addition and removal. Note that the FatherChildNode does not need to contain the seven levels; despite this, the Data Model will always be composed of seven levels as mentioned in Section 4.1.

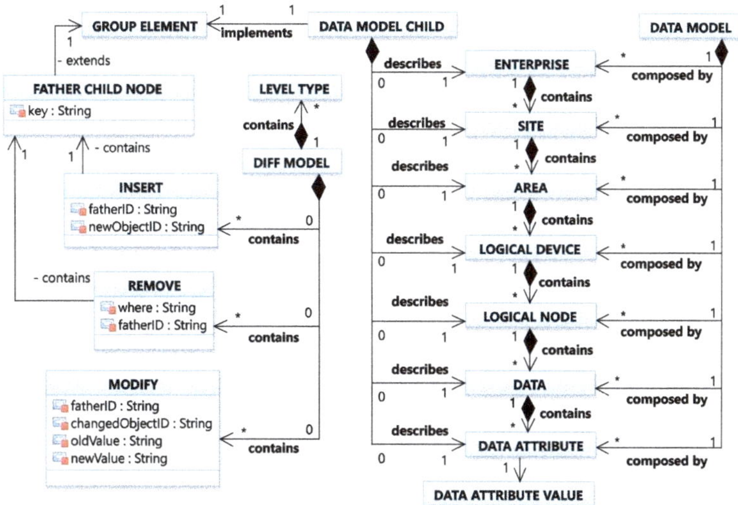

Figure 5. The Diff MetaModel structure for alerting users about ICPS evolution.

When a node is removed or inserted, a FatherChildNode is saved that extends from GroupElement i.e., the GroupElement describes the FatherChildNode, since it describes the DataModelChild. For example, Figure 3 only shows the first five levels (enterprise, site, area, logical device, and logical node); in a real scenario, it would be composed of seven levels. When Zone B is removed, the FatherChildNode would be composed of Zone B and its children, i.e., DataModelChild only contains five levels (area, logical device, logical node, data, and data attribute); the enterprise and site will not be saved. In this manner, we reduce the number of alerts. A description of each variable in the Diff MetaModel is presented in Table 4:

Table 4. A description of the information saved in each Diff Model.

Variable	Description
father_ID	The identifier of the node from which an element has been added, deleted, or changed.
newObjectID	The identification of the newly inserted node
where	A path pointing to the entire chain from the enterprise to the newly inserted object.
FatherChildNode	The tree that depends on the inserted or deleted node. This tree does not necessary have seven levels; it will depend on the fatherID level.
changedObjectID	The identification of the changed node
oldValue	The value previously held by that node
newValue	The value currently held by the node

A Diff Model that conforms to a Diff MetaModel is created automatically from Javers output every timestamp, and then, it is stored in Elasticsearch. Thus, it is possible to classify all the changes so that they can be presented easily to the user. Thanks to this metamodel, the number of alerts are reduced and we avoid redundant information because we just save the deepest node that is changed, excluding all below nodes.

5. CALENDAR Evaluation: Scalability Test

Despite ICPS evolution, different ICPS Data Model sizes exist, e.g., a press machine can be composed of 50 or more than 1000 devices, this being one of the machines within the press line. In addition, a customer can also be interested in monitoring multiple press lines; thus, the size of the environment can be huge. In order to ensure the usability of CALENDAR in different ICPS sizes, we provide a scalability test.

The data to perform the scalability test were created in a random way, since we had no access to the real data to perform the scalability test. Thus, considering that the Data Collector generates every Data Model that conforms to the Data MetaModel, we simulate these Data Models randomly. In this way, we can also conclude that CALENDAR is applicable for other domains that use the Data MetaModel as a base, as long as the scenario to be monitored complies with the Data MetaModel characteristics presented in Section 4.1.

All the experiments have been performed on a laptop with a CPU Intel (R) Core (TM) i7-4600U CPI @ 2.10GHz CPU. In addition, the computer in use had a 16 GiB memory with a 64-bit operating system (Linux) and a 500-GB disk (HDD). The correctness of the Diff Models was tested manually with a smaller Data Model. Java Microbenchmark Harness (JMH) (https://www.baeldung.com/java-microbenchmark-harness) was used to execute an accurate microbenchmark in an automatic way, i.e., measure the average time that CALENDAR takes to run and the confidence interval of the average.

In order to get reliable numbers, each query was processed 200 times for each evaluation case and Java Virtual Machine was restarted for each execution for each test.

Considering that changes can occur in any of the Data Model levels but mostly that the evaluation occurs at the devices (i.e., Logical Node), in the evaluation, we simulate changes in this level. For the evaluation, as mentioned above, we randomly generated Data Models with 50, 100, 500, 1000, 5000, and 10,000 devices. Each Data Model was cloned and the changes were inserted: an addition, removal, modification, and random changes (addition, modification, and removal). Finally, we established different percentages of changes (from 20% to 100%). Note that 100% means an addition of 10,000 devices in the largest model or a modification of the devices. In the case of removal, we skipped the removal of 100%, as it would result in an invalid Data Model. The scalability test results are reflected in Appendix A and Table A1.

In this evaluation, we address "Which is the performance of CALENDAR?" To do so, we distinguish three different configuration factors (F) that may impact

- **F1 → Type of change:** We measure how the type of change (addition, modification, or removal) impacts the performance (execution time).
- **F2 → Percentage of devices changed:** We measure how the number of devices changed impacts the performance (execution time).
- **F3 → Size of the Data Model:** We measure how the increasing size of the Data Model impacts the performance (execution time).

5.1. Discussion

Considering all the results obtained, the following section discusses the factors F1, F2, and F3 as well as a joint analysis of them all. The results have been evaluated against the following quantitative metrics: AVG: Average Execution Time in milliseconds (ms) and CI: Confidence Interval (ms).

5.1.1. F1: Type of Change

In Figure 6, we show the different ICPS Data Models separated by the different changes, i.e., inserted, removed, or modified. The type of change seems to affect our system. That is, the detection of adding a device is not the same as detecting that a device is removed or modified. Taking, for example, the Data Model of 50 devices, the adding devices average is 21.0176 ms ± 0.738 ms, removing devices is 12.30825 ms ± 0.3645 ms, and modifying devices is 16.2452 ms ± 0.5166 ms. In the case where all kind of changes are made in the same Data Model (insert, remove, and modify), the time needed is 16.1666 ms $\pm$ 0.6328 ms. The difference between removing and inserting devices (maximum and minimum execution time) is about 71% for this Data Model.

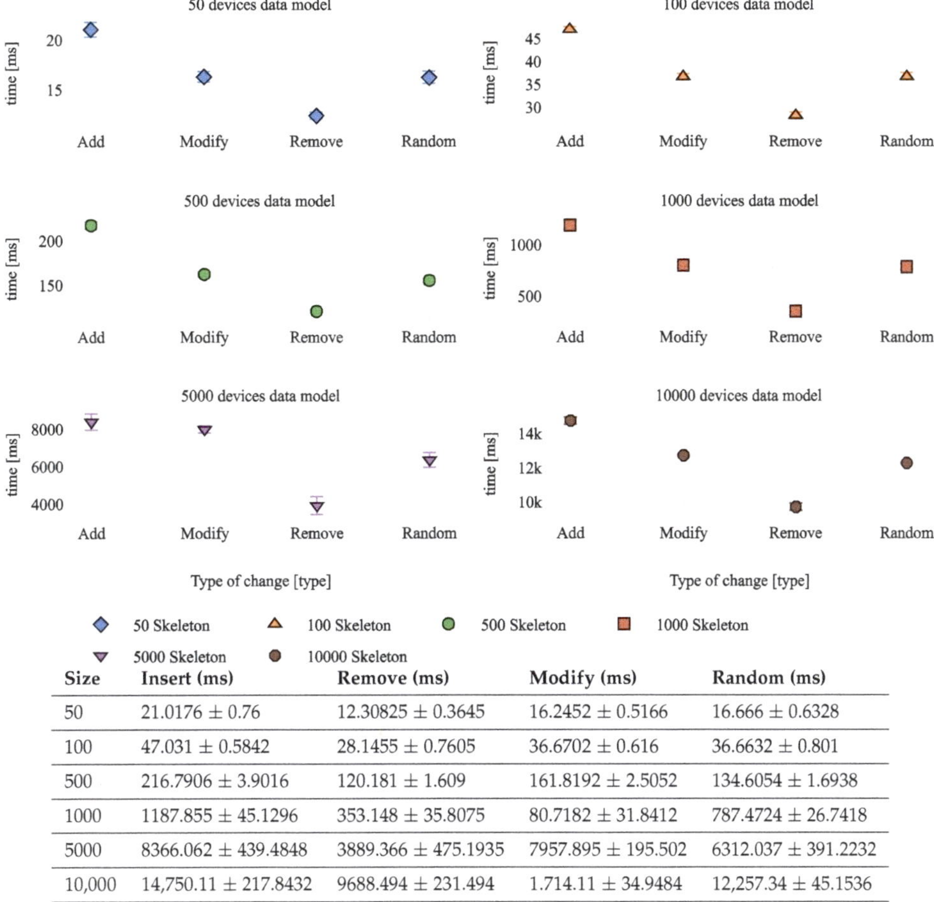

Size	Insert (ms)	Remove (ms)	Modify (ms)	Random (ms)
50	21.0176 $\pm$ 0.76	12.30825 $\pm$ 0.3645	16.2452 $\pm$ 0.5166	16.666 $\pm$ 0.6328
100	47.031 $\pm$ 0.5842	28.1455 $\pm$ 0.7605	36.6702 $\pm$ 0.616	36.6632 $\pm$ 0.801
500	216.7906 $\pm$ 3.9016	120.181 $\pm$ 1.609	161.8192 $\pm$ 2.5052	134.6054 $\pm$ 1.6938
1000	1187.855 $\pm$ 45.1296	353.148 $\pm$ 35.8075	80.7182 $\pm$ 31.8412	787.4724 $\pm$ 26.7418
5000	8366.062 $\pm$ 439.4848	3889.366 $\pm$ 475.1935	7957.895 $\pm$ 195.502	6312.037 $\pm$ 391.2232
10,000	14,750.11 $\pm$ 217.8432	9688.494 $\pm$ 231.494	1.714.11 $\pm$ 34.9484	12,257.34 $\pm$ 45.1536

Figure 6. The execution time averages in different ICPS Data Model sizes.

Considering the differences between the highest and lowest execution times of all Data Models (see Table 5), we can conclude that it does not have a relation with the size of the Data Model. However, in all the cases, the difference between the maximum and the minimum is 50% up to 220%.

Table 5. The max and min execution time difference percentages.

	50	100	500	1000	5000	10000
The difference between insert and remove	71%	67%	80%	224%	115%	52%

The type of change to detect and classify the evaluation affects the result. Thus, detecting removals is less expensive than detecting insertions in a Data Model. In the same way, the execution time for modifying devices is between adding and deleting. In the case of random changes, the maximum (insert) and minimum (remove) time are compensated, and therefore, the average time achieved is more or less in the middle.

5.1.2. F2: Percentage of Changes

This factor evaluates whether the percentage change affects the execution time, i.e., if with the same Data Model (e.g., 1000 devices), changing 20% (e.g., 200) or 100% (e.g., 1000) of devices influences the time required (execution time) for detecting changes. The results are shown in Figure 7.

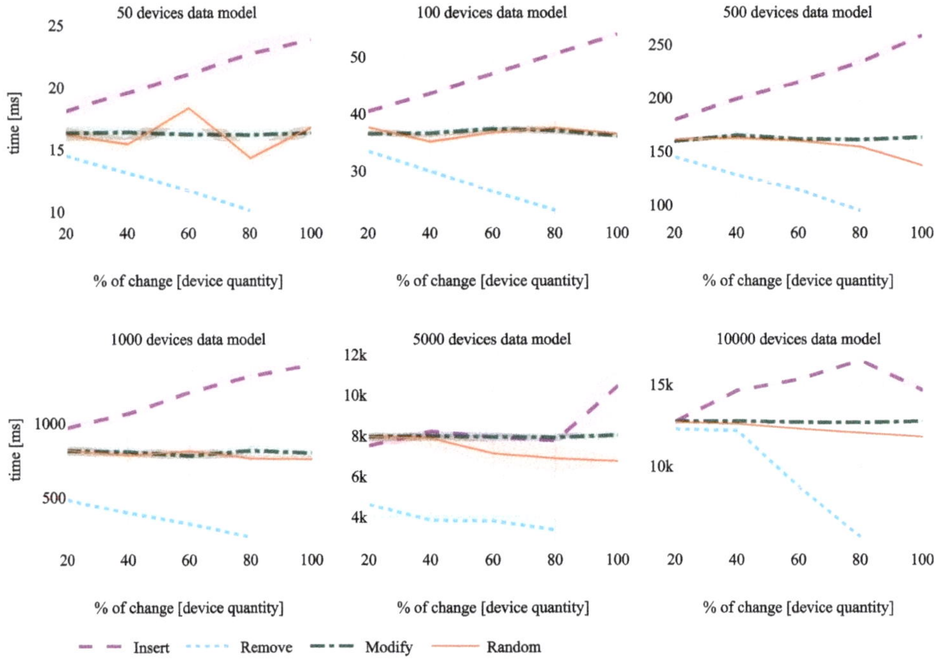

Figure 7. The increase in the percentage of changes in different ICPS Data Model sizes.

Considering the different changes that can occur, we observe that

- **Inserting:** The higher the percentage of change, the higher the execution time. In the smallest Data Model (50 devices), the execution time has an increase of 41.89%. With a bigger Data Model (e.g., 10000), the difference is 46.49%. Considering all the results (see Table 6), more or less of the difference between the minimum and maximum execution time when the percentage of change changes is between 40% and 60%. Therefore, as it is reflected in Figure 7, the growth of time is linear to the percentage of change.
- **Removing:** The lower the percentage of change made, the longer the execution time. In addition, the time decreases linearly. The execution time required for removing 20% in a 50 device Data Model is 14.462 ms, but in 80%, it is 10.06 ms. A difference of 52.86% exists, reaching 143.98% in the case of 1000 devices.
- **Modifying:** Increasing the percentage of change does not have a negative impact on time. The major percentage of change occurs with 1000 Data Model sizes, i.e., 0.013%. That is why we can concluded that the trend is constant.
- **Random:** This scenario is similar to modifying devices, i.e., increasing the percentage of change does not have a negative impact on the time, making the trend rather constant. Moroever, this is a case that depends a lot on the changes that have taken place.

Therefore, in a common scenario where we do not control over what is going on, we can assume that the percentage of change does not have a negative impact on the execution time needed. The times for adding and removing are compensated for each other, leaving a rather stable time when making random changes.

Table 6. The difference between the minimum (remove) and maximum (insert) execution times.

	50	100	500	1000	5000	10000
% of ADD	41.89	37.27	49.21	56.20	63.51	46.49
% of MODIFY	7.19	7.02	7.40	12.57	6.44	0.60
% of REMOVE	52.86	52.00	56.70	143.96	84.63	120.71
% of RANDOM	40.92	12.08	7.71	15.49	134.10	8.19

5.1.3. F3: Size of the Data Model

In order to see if the ICPS Data Model size affects the execution time needed, we extracted the values of the random test case (see Appendix A and Table A2). As we have concluded in factor F2, in a real scenario, we do not know about the change that is going to happen, and it has also been shown that time is more or less constant in terms of the number of changes that have occurred. In this manner, we contemplate all the cases without focusing on a single change. As it is shown in Figure 8, the larger the ICPS Data Model, the longer the execution time is. For detecting 100% of changes in a 1000 device Data Model, i.e., 1000 changes, CALENDAR needs an average of 754.174 ± 736.894 ms in contrast in the 5000 device Data Model detecting 20% of changes, i.e., for the same quantity of changes (1000), the time needed is bigger (7948.296 ± 7740.268 ms). For detecting the same quantity of changes, 546% more time is needed, i.e., equivalent to 6.6% of seconds. Looking at the graph, we know that this is not an isolated case; it is something that occurs if we compare different Data Model sizes. That means that the Data Model size, i.e., the input, impacts the output. The larger the size of the Data Model, the longer the time needed to calculate the differences even though the number of changes is the same.

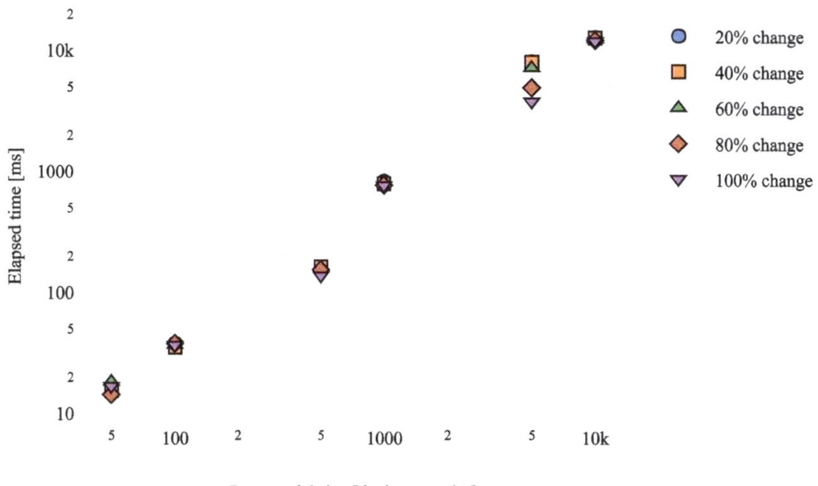

Figure 8. The time need for detection versus ICPS Data Model sizes (the axes are in logarithmic scale).

In order to see the trend that our system has, we calculated the average needed for each ICPS Data Model size (considering random changes results). Analyzing the results, we can observe that they tend in a potential way, which can be represented as follows: $y = cx^a$.

Thanks to a linear regression, i.e., a mathematical model used to approximate the relationship of dependence between a dependent variable (time) and the independent variables (quantity of devices), we obtain values $a = 1.28852412$ and $c = 0.08764925519477608$.

Thus, we obtain the equation $f(x) = 0.08764925519477608x^{1.28852412}$, which is represented in Figure 9 and shows the relation between the ICPS Data Model size and the execution time.

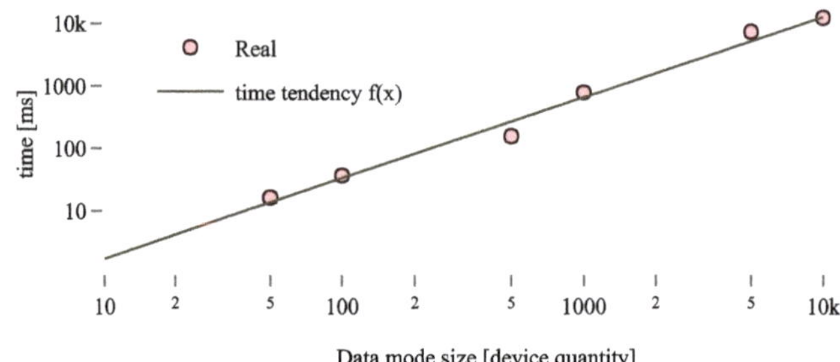

Figure 9. The time growth trend when we increase the number of devices (the axes are in logarithmic scale).

Henceforth, using this solution with a high quantity of devices could be a problem, since if CALENDAR needs to be used in an ICPS with many devices, it is necessary to provide a solution to the scalability problem found in order to reduce the execution time. If a fast response is needed when the number of devices is high, this module would not be able to give a fast enough response to the user. For example, in a small (50 Data Model size) scenario, 0.016 ± 0.0006 s are needed, but in a bigger one (5000 Data Model size), 6.312 ± 0.3912 s are needed. Usually, the latency of an industrial monitoring system is about 2000 ms. That is why our system is not profitable enough in real-time big Data Model scenarios.

5.1.4. Factor Analysis (F1, F2, and F3)

Considering all factors, we realize that, with small Data Models, the average time that CALENDAR needs for communicating alerts is smaller. In the same manner, taking into account Figure 9, we see that the behavior of the small Data Models is smoother than that of the larger ones as the trends are clearer. Thus, currently, CALENDAR is able to give responses to small ICPSs as long as it meets the customer's requirements, i.e., the latency is adequate.

In industrial scenarios in which monitoring has a latency that does not support our solution, we propose to split the Data Model file into different files, making a comparison in each of the files. The following chart shows (Figure 10) the relation between the file size and the execution time, where it is shown that the smaller the file, the smaller the execution time.

Taking into account the results of Figure 10, we propose a division of the Data Models into sub-Data Models. Because there are no data dependencies between the tasks to be parallelized, we can use the parallel computing theory to decrease the response time. Dividing the model, we get that each sub-Data Model is smaller and, thus, that the individual execution time needed will be smaller. In the same way, several sub-Data Models would be executed at the same time, so we would lower the total execution time. However, it is necessary to consider that a Deviation Time (det) exists, since time is needed to split the Data Model in sub-Data Models and then the result needs to be joined, i.e., the different sub-Diff Models need to be converted into a unique Diff Model to finally transfer the information to the user. In Figure 11 is shown an activity diagram where, using the parallel computing theory, we can reduce the execution time in order to give a response to the problem founded.

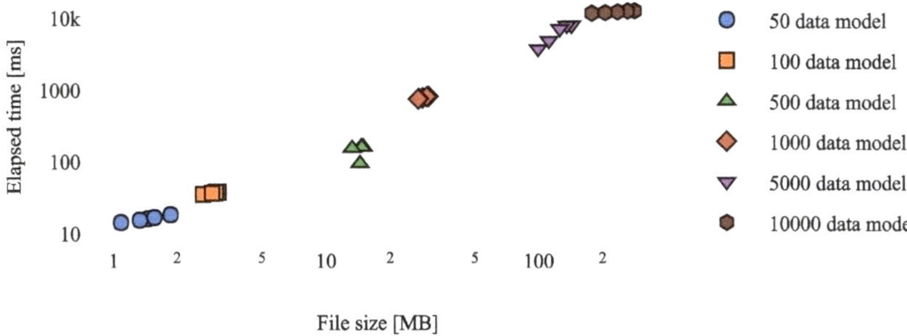

Figure 10. The relation between different Data Model file sizes and the elapsed execution time (the axes are in logarithmic scale).

Note that, for dividing the Data Model into sub-Data Models, we need to calculate the optimal division value, i.e., the value that instructs in how many sub-Data Models we should separate the model. To do so, we need to consider: (I) the time tendency formula (Formula 2) and (II) the deviation time that the proposal will have (Figure 11). Thus, by developing this improvement and performing a scalability test, we will be able to calculate the deviation time. Once we have the deviation time, we will be prepared to calculate the corresponding function to obtain the optimal division value.

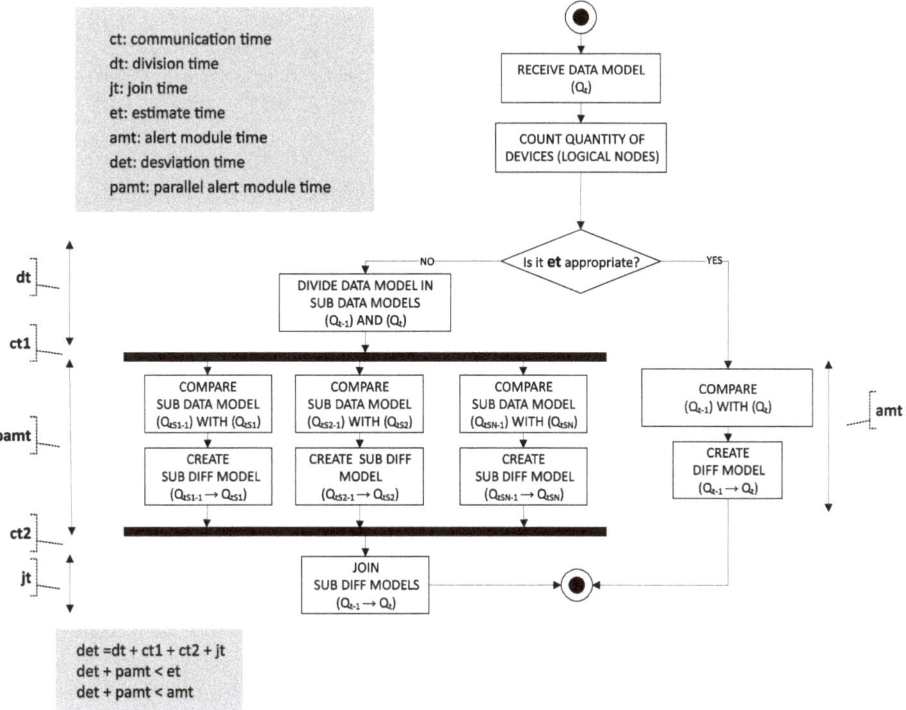

Figure 11. An activity diagram for reducing the alert module result time.

We make use of the parallel computing theory; dividing our Data Model in sub-Data Models and calculating the optimal division value, we will generate a model capable of dividing the Data Model in an efficient manner with the aim of parallelizing the comparison and, thus, reducing time. Furthermore, in Big Data scenarios, thanks to the Map & Reduce programming models, there would be no problem in parallelization. Thanks to the Map functions, the differences between the sub-Data Models will be founded, and then, the Reduce functions will join the results. Notice that this is only a hypothesis, which will need to be evaluated in the future.

5.2. Threats to Validity

As for internal validity, not all test cases are analyzed, i.e., we only evaluated the variability on the device level (Logical Node), and this may be a threat at the time of conclusion. Although, considering the personal interviews with our industrial partner, the most variable part has been evaluated (device variability), it would be interesting to examine other industrial domains to consider that it is applicable to them.

As for the external validity, an evaluation is not performed in a real environment, i.e., the computer used has fewer resources than a possible industrial PC. It is, therefore, appropriate to perform this test in a real environment in order to obtain more realistic results.

6. Related Work

There are some proposals in the literature that address monitoring solutions for ICPSs [19,21]. ICPS monitoring solutions are used in different domains such as traffic control and safety, manufacturing, or energy conservation [21,26]. In the same manner, some authors propose monitoring ICPSs to detect attacks that can affect the systems [27] or even for, storage data, data analysis, and the use of machine learning techniques to automatically update ICPS functionalities [28]. Despite that, different authors manage ICPS variability regarding the software [14,15], and a few of them consider hardware variability [18,19]. Hernandez and Reiff-Marganiec [18] propose a framework where smart objects start working in a autonomous way from a passive position to an active one. Unlike our system, these smart objects only consider variability at the Cyber Layer not in the Physical Layer. Chen et al. [19] recognizes variability at the Cyber and Physical Layers, but it is not able to manage the variability at runtime.

Therefore, ICPSs evolve throughout their lifetimes [11], but the previous presented solutions do not consider the traceability and the communication of the evolution at runtime. Note that managing the variability is crucial in Industry 4.0 [1,2,10], as the data captured from the ICPS are converted into information for decision-making. Note that the ICPS evolution includes not only software but also the addition, removal, and replacement of already installed devices [10,19], as ICPSs and their operating environment are highly dynamic due to, e.g., a deployment of autonomous devices. Therefore, flexibility and adaptation are the two required capabilities posing additional challenges for monitoring ICPSs [1,11,12].

Thus, even if the evolution of an ICPS is something known in the literature [13,14], as far as we know, no one in the literature has given a response to detecting ICPS evolution in order (1) to have the traceability of what has happened in an ICPS over time and (2) to communicate the evolution immediately to the users to avoid any bad decision. Additionally, note that the data structure changes when the ICPS evolves; that implies that new data need to be captured by the monitoring system in addition to being visualized as new information to the user. It is, therefore, necessary to manage the evolution, since different user roles exist and all must be informed.

Thus, in our solution, the data captured from the monitoring system is structured with tree models, since according to Reference [24], using a tree model structure facilitates the detection of an evolution. Tree models have been used in medicine (mutation tree) to identify the mutation of atoms, molecules, particle, etc. which are reflected in trees [29]. Melnik et al. use the comparison of models to then match the two trees in order to turn them into one [5]. Thus, tree-shaped models are recognized when it is necessary to detect evolution. Additionally, with a tree-shaped model, the structure beside the information is be transferred [30].

Thanks to tree models, the detection of an evolution is more immediate. This allows users to report what has happened at all times. For that, different methods for the comparison exist, which are used for tree-shaped models [24] such as SiDiff, UMLDiff, WinDiff, WinMerge, and SMDiff. Some are text comparators like SiDiff, and others are graphic comparators (UM-LDiff, WinDiff, WinMerge, and SMDiff). Although the graphics buyers show you the result visually, they are not suited for runtime use and our need is to find such changes at runtime. WinDiff (https://windiff.waxoo.com/) and WinMerge (http://winmerge.org/?lang=es) are line-based tools, i.e., changes will not be expressed at logic issues. Trip-wire (https://github.com/Tripwire/tripwire-open-source) is able to detect changes, but it is not able to detect which are the changes, and Remedy (http://www.bmcsoftware.es/it-solutions/remedy-itsm.html) is able to detect changes in the structure but not on the values. In our case, the whole architecture needs to be analyzed, i.e., the ICPS structure and the corresponding values. Cobana et al. and Weaver et al. have selected ad hoc solutions. The former one proposes an algorithm to compares XML files [31], and in our case, the information has a JSON format. In addition, the solution is for websites, and the movements inside the same father are considered changes, i.e., the format of the XML impacts the result; hence, it is not a valid solution for us. The latter instead [30], even if they are able to detect a change in a concrete moment in time, do not retain the traceability of changes. In addition, the output script must be saved and processed for it to be interpretable.

Inspired on these works, our proposal (CALENDAR) uses a tree-shaped model structure as input (i.e., the Data Model is the output of the monitoring system) for a comparison and detection of evolution. Each detected evolution is stored in a specific model (Diff model) in order to have the traceability of what has happened during the whole useful life of an ICPS. Among all the comparison methods available for tree-shaped mode, in our case, we have selected Javers, as in our previous work [10], it is shown that it is a model text comparator suitable for JSON format (tree-model structure) received from the ICPS.

The novelty of this paper lies in the following which, as far as we know, has not been addressed in the literature: (1) CALENDAR helps users to be informed immediately. (2) Thanks to CALENDAR, you can keep track of all changes that have happened over time, and thus, users can avoid any bad decisions. (3) With CALENDAR, we are able to classify the occurred evolution, reducing the quantity of alerts. (4) With the union of CALENDAR and the contribution published previously [10], the users will be able to receive graphical alerts.

7. Conclusions and Future Work

This paper presents CALENDAR, a Cyber-physicAL systEm evolutioN Detection and Alert geneRation System, in order to detect structural changes and to classify them in an efficient way when the ICPS evolves. Different changes (insertion, removal, or modification) can occur when an ICPS evolves; hence, it is necessary to alert the user in order to make correct decisions. For that, the ICPS is monitored and a data model is formed. The data model captures the structure and the information of the ICPS at every timestamp, and it has a tree structure. Then, in CALENDAR, a model text comparator is used to compare the data model received in a time (Qt), with the data model received in the previous time (Qt-1). The problem is that these comparators (e.g., Javers) do not take into account the dependencies between nodes. Due to this problem, the number of generated alerts grows, so Diff Models are used to classify these alerts, reducing the number of alerts and communicating the evolution to the user in a more direct way, avoiding redundant information.

Despite ICPSs evolution, different ICPS exist, i.e., a press line can be composed by different machines which are composed by different devices. For example, in a press line, a common machine is the press machine which can be composed of 50 or more than 1000 devices. Thus, considering that a press line is composed by different machines, the quantity of devices to monitor is high. In order to ensure the usability in different ICPSs, we provide an evaluation (scalability test) to evaluate different factors: (1) if the time needed to detect ICPS evolution is different depending on the occurred changes (insert, remove, or modify); (2) if the time needed to detect ICPS evolution is the same when varying the percentage of changes in the same tree structure; and (3) if the time needed to detect ICPS evolution is the same when the number of devices is different.

After performing the evaluation, we discuss the obtained results, considering the factors. We concluded that detecting a device addition is more costly than detecting eliminations, but they compensate each other when we analyze joint tests. The cost of detecting modifications is similar. In the same manner, when we analyze different percentage changes, we realize that, in a real scenario where different changes can occur even if the percentage of change changes, the execution time is mainly constant. Finally, when analyzing different data model sizes and when the number of devices increased, the execution time of CALENDAR increases.

Thus, we raise a problem, since the execution time increases sup-linearly as we increase the ICPS data model size. That is why, in the future, we would like to improve CALENDAR in order to reduce the response time, since our hypothesis says that, if we decrease the size of the input files and use parallel computing theory, we will be able to decrease the total execution time, enabling new real time scenarios.

Author Contributions: Conceptualization, A.I., G.S., and C.A.; data curation, A.I.; software, A.I., G.S., and C.A.; writing—original draft, A.I.; writing—review and editing, G.S. and C.A.

Funding: This work received funding from the Basque Government through the Elkartek program under the TEKINTZE project (Grant agreement No. KK-2018/00104)

Appendix A

Table A1. The obtained results from the data model comparator. The average time (AVG) and confidence interval (CI) of the execution time needed (ms).

	50 Devices							
	INSERT		MODIFY		REMOVE		RANDOM	
	AVG	CI $\pm$	AVG	CI $\pm$	AVG	CI $\pm$	AVG	CI $\pm$
20	18.06	0.773	16.288	0.602	14.462	0.443	16.155	0.627
40	19.553	0.508	16.347	0.457	13.062	0.478	15.403	0.525
60	21.026	0.589	16.183	0.556	11.649	0.228	18.28	0.747
80	22.654	1.086	16.127	0.45	10.06	0.309	14.265	0.763
100	23.795	0.734	16.281	0.518	-	-	16.73	0.502
	100 Devices							
	INSERT		MODIFY		REMOVE		RANDOM	
	AVG	CI $\pm$	AVG	CI $\pm$	AVG	CI $\pm$	AVG	CI $\pm$
20	40.398	0.645	36.449	0.561	33.38	0.611	37.586	0.618
40	43.483	0.529	36.547	0.674	29.791	0.91	35.09	1.005
60	47.029	0.555	37.289	0.629	26.35	0.823	36.723	0.639
80	50.377	0.492	37.03	0.611	23.061	0.698	37.501	1.235
100	53.868	0.7	36.036	0.605	-	-	36.416	0.508

Table A1. *Cont.*

	500 Devices							
	INSERT		MODIFY		REMOVE		RANDOM	
	AVG	CI ±	AVG	CI ±	AVG	CI ±	AVG	CI ±
20	179.135	2.836	159.188	2.234	144.423	1.828	161.05	2.366
40	198.863	3.486	164.873	3.699	127.603	1.774	161.808	2.217
60	214.576	3.57	161.443	2.364	114.108	1.576	159.476	2.093
80	233.163	4.82	160.746	2.112	94.59	1.258	154.081	1.793
100	258.216	4.841	162.846	2.117	-	-	36.612	1.965

	1000 Devices							
	INSERT		MODIFY		REMOVE		RANDOM	
	AVG	CI ±	AVG	CI ±	AVG	CI ±	AVG	CI ±
20	966.694	42.941	812.452	31.71	479.819	60.566	816.526	34.539
40	1062.361	41.325	805.063	34.992	390.658	39.605	781.418	23.924
60	1206.35	32.811	778.48	28.555	315.706	38.158	810.628	41.136
80	1312.73	56.793	812.255	37.455	226.409	4.901	759.616	16.87
100	1391.139	51.778	795.341	26.494	-	-	754.174	17.28

	5000 Devices							
	INSERT		MODIFY		REMOVE		RANDOM	
	AVG	CI ±	AVG	CI ±	AVG	CI ±	AVG	CI ±
20	7504.666	253.046	7948.106	192.028	4599.335	601.986	7948.296	208.028
40	8207.304	137.2	7983.115	199.595	3825.186	376.088	7871.655	223.714
60	7932.337	251.416	7937.625	195.618	3794.048	401.019	7123.38	570.081
80	7751.169	133.73	7906.775	186.239	3338.893	521.681	4872.885	694.389
100	10,434.83	1422.032	8013.856	204.03	-	-	3743.969	259.904

	10,000 Devices							
	INSERT		MODIFY		REMOVE		RANDOM	
	AVG	CI ±	AVG	CI ±	AVG	CI ±	AVG	CI ±
20	12,723.95	383.693	12,734.7	42.86	12,254.94	33.845	12,671.83	43.162
40	14,616.28	82.877	12,754.4	35.86	12,168.72	402.94	12,565.88	52.22
60	15,310.77	97.376	12,685.11	28.597	8709.968	363.877	12,277.84	34.849
80	16,461.61	85.731	12,666.93	28.371	5620.353	52.583	11,991.42	67.946
100	14,637.95	439.539	12,729.39	39.054	-	-	11,779.71	27.591

Table A2. The Data Model sizes in MegaBytes (MBs) for comparison.

Size	Qt-1 (MB)	Qt (kb)					
		Change	20%	40%	60%	80%	100%
50	1.515	Remove	1.212	911	609	307	–
		Insert	1.811	2.105	2.399	2.698	2.993
		Modify	1.513	1.506	1.500	1.496	1.494
		Random	1.453	1.328	1.862	1.082	1.558
100	3.024	Remove	2.420	1.818	1.212	609	–
		Insert	3.614	4.208	4.795	5.391	5.979
		Modify	3.017	3.007	3.000	2.990	3.026
		Random	3.136	2.655	3.035	3.057	2.930
500	15.106	Remove	12.083	9.066	6.043	3.024	–
		Insert	18.057	21.018	23.980	26.924	29.877
		Modify	15.065	15.022	14.974	14.928	14.888
		Random	14.891	14.592	14.692	13.246	14.385
1000	30.212	Remove	24.161	18.127	12.086	6.045	–
		Insert	36.119	42.028	47.941	53.837	59.751
		Modify	30.115	30.036	29.944	29.846	29.770
		Random	30.266	28.287	29.701	27.107	27.071
5000	147	Remove	117	88.4	59	29.5	–
		Insert	176	205	234	263	292
		Modify	147	146	146	145	145
		Random	143	136	126	112	99.6
10000	294	Remove	235	176	118	58.9	–
		Insert	294	411	469	295	585
		Modify	294	293	192	291	290
		Random	282	262	235	206	178

References

1. Leitão, P.; Colombo, A.W.; Karnouskos, S. Industrial automation based on cyber-physical systems technologies: Prototype implementations and challenges. *Comput. Ind.* **2016**, *81*, 11–25. [CrossRef]
2. Lee, J.; Bagheri, B.; Kao, H.A. A Cyber-Physical Systems architecture for Industry 4.0-based manufacturing systems. *Manuf. Lett.* **2015**, *3*, 18–23. [CrossRef]
3. Ruppert, T.; Jaskó, S.; Holczinger, T.; Abonyi, J. *Enabling Technologies for Operator 4.0: A Survey; Appl. Sci.* **2018**, *8*, 1650. [CrossRef]
4. Rajkumar, R.; Lee, I.; Sha, L. ; Stankovic, J. Cyber-physical systems: The next computing revolution. In Proceedings of the 47th Design Automation Conference, Anaheim, CA, USA, 13–18 June 2010; pp. 731–736.
5. Iglesias, A.; Lu, H.; Arellano, C.; Yue, T.; Ali, S.; Sagardui, G. Product Line Engineering of Monitoring Functionality in Industrial Cyber-Physical Systems: A Domain Analysis. In Proceedings of the 21 st International Systems and Software Product Line Conference, Sevilla, Spain, 25–29 September 2017; pp. 195–204.
6. Lee, J.; Kao, H.A.; Yang, S. Service Innovation and Smart Analytics for Industry 4.0 and Big Data Environment. *Procedia Cirp* **2014**, *16*, 3–8. [CrossRef]

7. Bagozi, A.; Bianchini, D.; Antonellis, V.D.; Marini, A.; Ragazzi, D. Big Data Summarisation and Relevance Evaluation for Anomaly Detection in Cyber Physical Systems. In Proceedings of the On the Move to Meaningful Internet Systems. OTM 2017 Conferences—Confederated International Conferences: CoopIS, C&TC, and ODBASE 2017, Rhodes, Greece, 23–27 October 2017; Springer International Publishing: New York, NY, USA, 2017; pp. 429–447.

8. Stojanovic, L.; Dinic, M.; Stojanovic, N.; Stojadinovic, A. Big-data-driven anomaly detection in industry (4.0): An approach and a case study. In Proceedings of the 2016 IEEE International Conference on Big Data, BigData 2016, Washington DC, USA, 5–8 December 2016; pp. 1647–1652.

9. De, S.; Zhou, Y.; Larizgoitia Abad, I.; Moessner, K. *Cyber–Physical–Social Frameworks for Urban Big Data Systems: A Survey*; Multidisciplinary Digital Publishing Institute: Basel, Switzerland, 2017; Volume 7, p. 1017.

10. Iglesias, A.; Arellano, C.; Yue, T.; Ali, S.; Sagardui, G. Model- Based Personalized Visualization System for Monitoring Evolving Industrial Cyber-Physical System. In Proceedings of the 25th Asia-Pacific Software Engineering Conference, APSEC 2018, Nara, Japon, 4–7 December 2018.

11. García-Valls, M.; Perez-Palacin, D.; Mirandola, R. Time-Sensitive Adaptation in CPS through Run-Time Configuration Generation and Verification. In Proceedings of the IEEE 38th Annual Computer Software and Applications Conference, COMPSAC 2014, Vasteras, Sweden, 21–25 July 2014; pp. 332–337.

12. Gerostathopoulos, I.; Bures, T.; Hnetynka, P.; Keznikl, J.; Kit, M.; Plasil, F.; Plouzeau, N. Self-adaptation in software-intensive cyber-physical systems: From system goals to architecture configurations. *J. Syst. Softw.* **2016**, *122*, 378–397. [CrossRef]

13. Schütz, D.; Wannagat, A.; Legat, C.; Vogel-Heuser, B. Development of PLC-Based Software for Increasing the Dependability of Production Automation Systems. *IEEE Trans. Ind. Inform.* **2013**, *9*, 2397–2406. [CrossRef]

14. Bougouffa, S.; Meßzmer, K.; Cha, S.; Trunzer, E.; Vogel-Heuser, B. Industry 4.0 interface for dynamic reconfiguration of an open lab size automated production system to allow remote community experiments. In Proceedings of the 2017 IEEE International Conference on Industrial Engineering and Engineering Management (IEEM), Singapore, 10–13 December 2017; pp. 2058–2062.

15. Tang, H.; Li, D.; Wang, S.; Dong, Z. CASOA: An Architecture for Agent-Based Manufacturing System in the Context of Industry 4.0. *IEEE Access* **2018**, *6*, 12746–12754. [CrossRef]

16. Aicher, T.; Spindler, M.; Fottner, J.; Vogel-Heuser, B. Analyzing the industrial scalability of backwards compatible intralogistics systems. *Prod. Eng.* **2018**, *12*, 297–307. [CrossRef]

17. Mauro, J.; Nieke, M.; Seidl, C.; Yu, I.C. Context-aware reconfiguration in evolving software product lines. *Sci. Comput. Program.* **2018**, *163*, 139–159. [CrossRef]

18. Hernandez, M.E.P.; Reiff-Marganiec, S. Towards a Software Framework for the Autonomous Internet of Things. In Proceedings of the 4th IEEE International Conference on Future Internet of Things and Cloud, FiCloud 2016, Vienna, Austria, 22–24 August 2016; pp. 220–227.

19. Chen, Z.; Zhang, X.; He, K. Research on the Technical Architecture for Building CPS and Its Application on a Mobile Phone Factory. In Proceedings of the 5th International Conference on Enterprise Systems, Beijing, China, 22–24 September 2017; pp. 76–84.

20. Lei, Y.; Li, N.; Guo, L.; Li, N.; Yan, T.; Lin, J. Machinery health prognostics: A systematic review from data acquisition to RUL prediction. *Mech. Syst. Signal Process.* **2018**, *104*, 799–834. [CrossRef]

21. Iglesias, A.; Iglesias-Urkia, M.; López-Davalillo, B.; Charramendieta, S.; Urbieta, A. TRILATERAL: Software Product Line based multidomain IoT artifact generation for Industrial CPS. In Proceedings of the 7th International Conference on Model-Driven Engineering and Software Development, Modelsward 2019, Prague, Czech Republic, 20-22 February, 2019.

22. Huang, L.; Gromiha, M.M.; Ho, S. iPTREE-STAB: interpretable decision tree based method for predicting protein stability changes upon mutations. *Bioinformatics* **2007**, *23*, 1292–1293. [CrossRef] [PubMed]

23. Lee, J.H.; Ahn, C.W. An Evolutionary Approach to Driving Tendency Recognition for Advanced Driver Assistance Systems. In Proceedings of the MATEC Web of Conferences, EDP Sciences: 8th International Conference on Computer and Automation Engineering (ICCAE 2016), Melbourne, Australia, 3–4 March 2016; Volume 56, p. 02012.

24. Sebastiani, M.; Supiratana, P. Tracing the Differences on an Evolving Software Model. In *Proceedings of the IRCSE '08: IDT Workshop on Interesting Results in Computer Science and Engineering*; Mälardalen University: Västerås, Sweden, 2008.

25. Bo, Y. Querying JSON Streams. Ph.D. Thesis, Uppsala University, Uppsala, Sweden, 2010.

26. Lee, E.A. Cyber Physical Systems: Design Challenges. In Proceedings of the 11th IEEE International Symposium on Object-Oriented Real-Time Distributed Computing (ISORC), Orlando, FL, USA, 5–7 May 2008; pp. 363–369.

27. Pasqualetti, F.; Dörfler, F.; Bullo, F. Attack Detection and Identification in Cyber-Physical Systems. *IEEE Trans. Autom. Control* **2013**, *58*, 2715–2729. [CrossRef]

28. Niggemann, O.; Biswas, G.; Kinnebrew, J.S.; Khorasgani, H.; Volgmann, S.; Bunte, A. Data-Driven Monitoring of Cyber-Physical Systems Leveraging on Big Data and the Internet-of-Things for Diagnosis and Control. In Proceedings of the 26th International Workshop on Principles of Diagnosis (DX-2015) Co-Located with 9th IFAC Symposium on Fault Detection, Supervision and Safety for Technical Processes (Safeprocess 2015), Paris, France, 31 August–3 September 2015; pp. 185–192.

29. Chao, H.; Li, H.; Wang, T.; Li, X.; Liu, B. An accurate algorithm for computing mutation coverage in model checking. In Proceedings of the 2016 IEEE International Test Conference, ITC 2016, Fort Worth, TX, USA, 15–17 November 2016; pp. 1–10.

30. Weaver, G.A.; Smith, S.W.; Bobba, R.B.; Rogers, E.J. Re-engineering Grep and Diff for NERC CIP. In Proceedings of the 2012 IEEE Power and Energy Conference at Illinois, Champaign, IL, USA, 24–25 February 2012; pp. 1–8.

31. Cobena, G.; Abiteboul, S.; Marian, A. Detecting Changes in XML Documents. In Proceedings of the 18th International Conference on Data Engineering, San Jose, CA, USA, 26 February–1 March 2002; pp. 41–52.

Article

In-Line Dimensional Inspection of Warm-Die Forged Revolution Workpieces Using 3D Mesh Reconstruction

Daniel Mejia-Parra [1,2], Jairo R. Sánchez [2,*], Oscar Ruiz-Salguero [1], Marcos Alonso [3], Alberto Izaguirre [4], Erik Gil [5], Jorge Palomar [5] and Jorge Posada [2]

[1] Laboratory of CAD CAM CAE, Universidad EAFIT, Cra 49 no 7-sur-50, Medellín 050022, Colombia; dmejiap@eafit.edu.co (D.M.-P.); oruiz@eafit.edu.co (O.R.-S.)

[2] Vicomtech, Paseo Mikeletegi 57, Parque Científico y Tecnológico de Gipuzkoa, 20009 Donostia/San Sebastián, Spain; jposada@vicomtech.org

[3] Computational Intelligence Group, CCIA Department, UPV/EHU, Paseo Manuel Lardizabal 1, 20018 Donostia/San Sebastián, Spain; malonso117@ikasle.ehu.es

[4] CIS & Electronics Department, University of Mondragon, Loramendi Kalea 5, 20500 Mondragon, Spain; aizagirre@mondragon.edu

[5] GKN Driveline Legazpi S.A., Calle Urola 10, 20230 Legazpi, Spain; Erik.Gil@gkndriveline.com (E.G.); Jorge.Palomar@gkndriveline.com (J.P.)

* Correspondence: jrsanchez@vicomtech.org; Tel.: +34-943-309-230

Received: 12 February 2019; Accepted: 12 March 2019; Published: 14 March 2019

Abstract: Industrial dimensional assessment presents instances in which early control is exerted among "warm" (approx. 600 °C) pieces. Early control saves resources, as defective processes are timely stopped and corrected. Existing literature is devoid of dimensional assessment on warm workpieces. In response to this absence, this manuscript presents the implementation and results of an optical system which performs in-line dimensional inspection of revolution warm workpieces singled out from the (forming) process. Our system can automatically measure, in less than 60 s, the circular runout of warm revolution workpieces. Such a delay would be 20 times longer if cool-downs were required. Off-line comparison of the runout of T-temperature workpieces ($27\,°C \leq T \leq 560\,°C$) shows a maximum difference of 0.1 mm with respect to standard CMM (Coordinate Measurement Machine) runout of cold workpieces (27 °C), for workpieces as long as 160 mm. Such a difference is acceptable for the forging process in which the system is deployed. The test results show no correlation between the temperature and the runout of the workpiece at such level of uncertainty. A prior-to-operation Analysis of Variance (ANOVA) test validates the repeatability and reproducibility (R&R) of our measurement system. In-line assessment of warm workpieces fills a gap in manufacturing processes where early detection of dimensional misfits compensates for the precision loss of the vision system. The integrated in-line system reduces the number of defective workpieces by 95%.

Keywords: in-line dimensional inspection; warm forming; 3D mesh reconstruction; optical system; revolution workpiece

---

1. Introduction

In the context of warm forming of motorcar parts, current production lines of stub axles process around 1200 pieces per hour. The tools used to form these parts are constantly subjected to high structural and thermal stresses [1–3], requiring continuous monitoring and dimensional assessment of the produced parts for process and quality control.

In the case of forged revolution workpieces, the produced parts are not final product, requiring subsequent machining operations. The assessment of the punch orientation with respect to the forming matrix orientation in the forging process is crucial since a severe misalignment between the punch press and the forming matrix axes disables the posterior machining process, resulting in a scrapped part. The circular runout [4] of the forged revolution workpiece indicates the deviation between the punch orientation and the forming matrix axis.

Standard tools for dimensional assessment of these workpieces rely on contact between the probe and the measured workpiece. Such is the case of Coordinate Measurement Machines (CMMs), which provide highly accurate measurements [5]. However, dimensional assessment with standard CMMs (and contact methods in general) is not convenient due to (1) the high temperatures directly affect (or even damage) the probe and, (2) long measurement times for the cooled-down workpieces. Consequently, a delay of nearly 20 min between the production of a single part and its dimensional assessment (including its cooling down, transportation to the metrology office and measuring times) arises. Such time delay translates into an uncertainty in the quality control process of approximately 400 potential defective workpieces (worst case scenario) for each measurement.

This manuscript presents an optical (i.e., contact-avoiding) system for in-line dimensional assessment of warm forming of revolution workpieces. Our system can continuously measure the circular runout of the parts at around 600 °C in less than 60 s per part. Results from experiments conducted in this manuscript show no temperature vs. runout correlation for the system uncertainty level (0.1 mm). The system is integrated and deployed in a warm-die forge industry, in the supply chain of world-class auto makers. Such a system allows continuous monitoring for quality and process control of the production line, providing an early detection mechanism of manufacturing failures which reduces the number of potential defective parts from 400 to 20 (95%) between consecutive measurements. This dimensional assessment for warm workpieces fills a gap in warm-die manufacturing processes in which the advantage of early detection of process bias compensates for the disadvantage of precision loss regarding higher-precision mechanisms (such as contact-based CMM).

The deployed system improves on the classic approaches for dimensional assessment in warm-die industry. Using technologies from Visual Computing and Industry 4.0 [6], the system allows in-line visual assessment of warm workpieces, either by metrologists, engineers, or operators. Furthermore, the increased cadence of the measurements (from 1 measurement every 20 min to 1 measurement per minute) improves the efficiency in product quality and process control, and leaves open future lines of dimensional assessment focused on data analytics.

The remainder of this manuscript is organized as follows: Section 2 reviews the relevant literature. Section 3 describes the developed system. Section 4 presents and discusses the results. Section 5 concludes the manuscript and introduces what remains for future work.

2. Literature Review

In the automotive and aeronautic industry, dimensional inspection of manufactured parts requires high precision methods to assess the quality of the final product. Currently, CMMs are one of the most common tools used to inspect forged workpieces due to their high precision [5]. However, CMMs are not suitable for in-line dimensional inspection of warm-die manufacturing parts due to: (1) their contact-based nature requires the workpieces to be in a cooled state to avoid damaging the measuring probe and, (2) taking measurements with the probe is highly time-consuming (even with cold workpieces). Visual computing provides contact-avoiding technologies and methodologies for Reverse Engineering and dimensional inspection which improve the productivity and efficiency of such CAD CAM CAE processes [6]. While sacrificing accuracy to some extent, optical scanners have become an alternative for dimensional inspection of different kinds of warm-die manufacturing processes [5,7]. Table 1 presents a comparative of standard CMMs vs. optical scanner dimensional assessment.

Table 1. Comparison between standard contact-based Coordinate Measurement Machines (CMMs) vs optical scanners for dimensional assessment.

CMM	Optical Scanner
Highly accurate measurements [5].	Less-accurate measurements [5].
Data collection relying on probe vs. piece contact.	Contact probe vs. piece not required.
Technician assistance required for definition of piece feature coordinate systems [8].	Technician assistant required for point sample vs. B-Rep (i.e., CAD model) registration [9,10].
Time-consuming data acquisition protocol [8].	Real-time data acquisition and post-processing of the digitized mesh (triangulation, mesh registration, feature extraction) [11,12].
Inherently sparse point samples, conducted according to discrete trajectories. Analytic form fitting needed as a consequence [8].	Dense point samples. Both mesh computation and analytic form fitting possible [9].
Competing equipment precision at the cost of off-line measurements [3].	Accurate measurement systems for in-line dimensional assessment [3].
Requires specific clamps for each reference model, introducing additional complexity in the management of measuring resources.	Allows the use of fixed universal setups for many different workpiece references.

2.1. Off-Line Dimensional Inspection in Warm-Die Manufacturing

In warm-die manufacturing, constant monitoring of forming tools is crucial for the quality control of produced parts. Forming tools such as punches, are subjected to high structural and thermal stresses that limit their lifetime [1]. Refs. [13,14] analyze the progressing wear of forging tools and forging defects by monitoring volume changes in the manufactured workpieces using 3D mesh reconstruction. In addition, the use of optical scanners allows the integration of numerical methods (such as Finite Element Analysis) in the dimensional inspection pipeline to quantify the thermal and structural damage of the forging tool [1,3]. Other dimensional inspection methods in warm forming include computed tomography [7,15], thermographic assessment [16], ultrasonic assessment [17], liquid penetrant testing [18], among others.

2.2. In-Line Dimensional Inspection

All the previously presented methods only execute off-line measurements on cooled-down workpieces. Measurements directly performed on warm and hot workpieces have been rarely reported [1]. Their main shortcoming is that high temperatures affect the measurements of contact-based methods while the strong radiation affects the optical equipment, thus reducing the quality of the captured images [19]. The spectrum selective method presented in [20] filters specific wavelengths from the captured images to allow the reconstruction of the hot parts. Ref. [21] integrates a specific wavelength and power laser beam with surface fitting to measure the length and diameter of hot cylindrical workpieces. Refs. [22,23] presents an in-line measurement system which uses two-dimensional laser range sensors (TLRS) coupled with servo motors for 3D reconstruction of hot cylindric workpieces. As an alternative to optical scanners, ref. [24] presents a vision system with two cameras that capture and process the hot workpiece without requiring any laser beams. These in-line approaches for hot dimensional assessment have been developed for workpieces with non-complex geometries (such as cylindrical workpieces).

It is worth mentioning in-line dimensional inspection approaches for cooled-down workpieces. Point cloud filtering [11] and accelerated mesh registration algorithms [12] have been developed to allow real-time inspection using 3D optical scanners and mesh reconstruction. Applications of these algorithms for in-line dimensional inspection of cooled-down workpieces include flatness inspection of rolled parts [25], inspection of large parts [26], and inspection of generic parts using geometric features [9,27].

2.3. Conclusions of the Literature Review

CMMs have been used in the warm forming industry due to their high measurement precision [5]. However, they are not suitable for in-line processes due to (1) the high temperatures of the workpiece affecting or even damaging the measuring probe and, (2) the long measurement times of the probe even for cooled-down workpieces. Other alternatives for dimensional inspection include optical scanner technologies, which do not rely on contact with the workpiece. However, the radiation due to the high temperatures can affect the data acquired by the scanners [19]. Current literature for in-line dimensional inspection of warm workpieces is very limited [1] and accounts only for very simple geometries (such as cylindrical workpieces) [20,21,23,24].

Responding to the current state of the art, we present an optical system for in-line dimensional assessment of forged still-warm workpieces. Whereas previous cold-state methods would require approximately 20 min for the assessment of a single workpiece, our system spends less than 60 s per part. In-line assessment of these warm (approx. 600 °C) workpieces fills a gap in manufacturing processes in which early detection of an inherent planning, design, or manufacturing error is more important than the higher precision obtained with standard cold-state measurement methods. The system is implemented and deployed in a global automotive part maker plant, where the number of defective workpieces is reduced by a 95% with respect to previous dimensional assessment methods (i.e., cold-state CMM).

The implemented system executes 3D scanning, mesh registration and comparison (against a CAD database) of the geometry of a forged still-warm workpiece. The system is capable of in-line measurements of circular runout of revolution warm workpieces, singled out from the forming process. Contrast test against cold-state CMM measurements show that the warm-workpiece measurement is good enough for the manufacturing plant in which the system is deployed (error below 0.1 mm for parts as long as 160 mm). The temperature-vs.-runout analysis shows no correlation between these two variables at such level of uncertainty. A prior-to-operation ANOVA test with cold workpieces validates the repeatability and reproducibility (R&R) of our measurement system.

3. Methodology

Given an input reference CAD model $\mathcal{C}$ and the triangular mesh $M = (X, T)$ of a scanned workpiece, the objective of the optical system is to compute key dimensional measurements on M with respect to $\mathcal{C}$. In the case of revolution workpieces, the circular runout dimension $\Delta\Phi$ measures how much a circular feature oscillates when the workpiece is rotated around the revolution (datum) axis $A = (\vec{v}, a_0)$ [8]. Such a dimension is crucial to assess the quality of the process and the produced parts in the production line.

The optical system for dimensional inspection has been designed as a process of two phases (Figure 1). In the first phase, the metrologist defines the parameters of the reference CAD model $\mathcal{C}$ required for the dimensional inspection of all workpieces of such reference. In the second phase, the system in-line and automatically estimates the revolution axis A and the circular runout $\Delta\Phi$ of each workpiece M. The operator in the production line is immediately provided with the results, with no intervention of the metrologist. The following sections describe the process in detail.

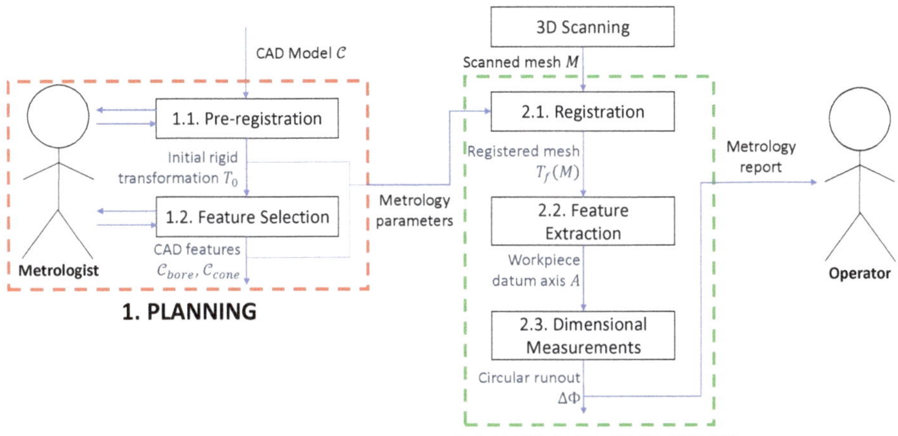

Figure 1. Workflow of the reverse engineering system for dimensional inspection. The system provides the dimensional inspection results to the operator directly in the production line.

3.1. Planning for the Dimensional Inspection

The first phase of the optical system consists in the definition of the dimensional inspection parameters for a given reference CAD model $\mathcal{C}$. The metrologist defines the features of interest in $\mathcal{C}$, which are worth of early assessment in warm workpieces. This phase takes about 5 min, but is performed only once per CAD reference.

3.1.1. Mesh Pre-Registration

To compare the scanned mesh M with the reference CAD model $\mathcal{C}$, it is crucial that both of these surface representations share the same coordinate system $W = \{\vec{w}_x, \vec{w}_y, \vec{w}_z; \vec{p}_w\}$. If W and W_M are the coordinate systems of $\mathcal{C}$ and M, respectively, the objective is to compute a rigid transformation $T_0 \in SE(3)$ such that $T_0(W_M) \approx W$.

To compute T_0, the developed system uses an alignment-of-correspondences algorithm [28]. Let $\{p_0, p_1, p_2\} \subset \mathcal{C}$ and $\{q_0, q_1, q_2\} \subset M$ be three non-collinear points sampled from the reference CAD and the workpiece mesh, respectively. The alignment-of-correspondences algorithm computes the rigid transformation T_0 that minimizes the distance between the two sets of points:

$$T_0 = \arg\min \sum_{i=0}^{2} \|p_i - T_0(q_i)\| \tag{1}$$

$$s.t. \qquad T_0 \in SE(3)$$

where $SE(3) = SO(3) \times \mathbb{R}^3$ is the special Euclidean group (group of all rigid transformations in $\mathbb{R}^3$).

In Equation (1), p_i, q_i are corresponding points in the CAD and the mesh, respectively. These points are interactively selected by the metrologist as illustrated in Figure 2. This pre-registration is performed only once per CAD reference $\mathcal{C}$.

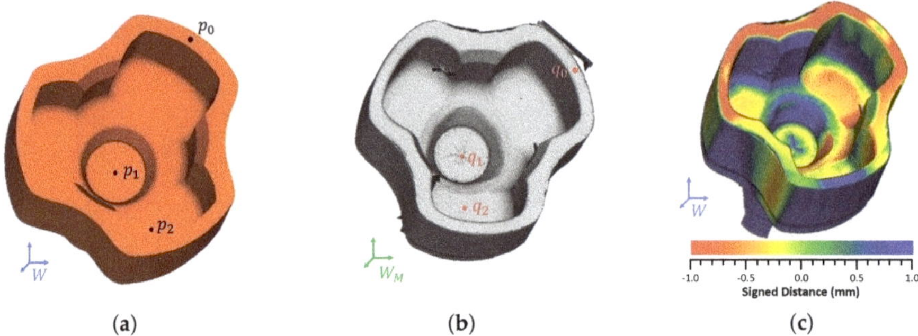

Figure 2. User-assisted alignment of correspondences [28]. The metrologist selects 3 corresponding points in both the CAD (orange) and a scanned mesh (gray). (**a**) CAD reference and its coordinate system W; (**b**) Scanned mesh and its coordinate system W_M; (**c**) Alignment of corresponding points [28].

3.1.2. Feature Selection

As mentioned starting Section 3, the metrologist interactively selects the different CAD features (FACEs) from $\mathcal{C}$ associated with the workpiece revolution axis $A = (\vec{v}, a_0)$ (Figure 3a). In this case, the metrologist selects the CAD FACEs $\mathcal{C}_{bore}$ (blue) of the cylindric surface which dictate the rotation of the workpiece. The axis vector of $\mathcal{C}_{bore}$ defines the theoretical revolution axis vector $\vec{v}$ (green).

On the other hand, the metrologist must define the axis point a_0 as a reference point. In the context of stub axle forming, the point is computed as follows:

1. The metrologist selects the CAD FACEs $\mathcal{C}_{cone}$ (red) corresponding to a conical surface at the bottom of the punch zone of the workpiece (Figure 3a).
2. The metrologist defines the datum diameter $d > 0$. In this case, the metrologist defines d as the diameter of the supporting fixture for the machining of the workpiece (after it has been formed).
3. The point at the revolution axis A where the surface $\mathcal{C}_{cone}$ attains the diameter d is the theoretical axis point a_0. The diameter d is measured perpendicular to the axis A (see Figure 3b). The point a_0 is a reference point for machining operations (after the piece has been formed).

The features $\mathcal{C}_{bore}$ and $\mathcal{C}_{cone}$ (Figure 3a) have been chosen for the definition of the reference axis A due to two main reasons:

1. $\mathcal{C}_{bore}$ is the part of the punch that suffers less wearing since the direction of the compression load during the forging process is parallel to its axis. This fact makes this geometry more stable from a dimensional assessment perspective.
2. The surfaces $\mathcal{C}_{bore}$ and $\mathcal{C}_{cone}$ are the same surfaces used to hold the workpiece during the posterior machining process. In this way, the algorithm uses the same coordinate system that will be used in the next step of the manufacturing process. In addition, it can be said that any possible registration error induced by the tool wearing is not relevant given that the subsequent machining process will use the same defective geometries to establish its reference frame.

The runout height $h > 0$ is the distance from a_0 along the axis A (Figure 3b) where the circular runout is measured. This height h is manually defined by the metrologist.

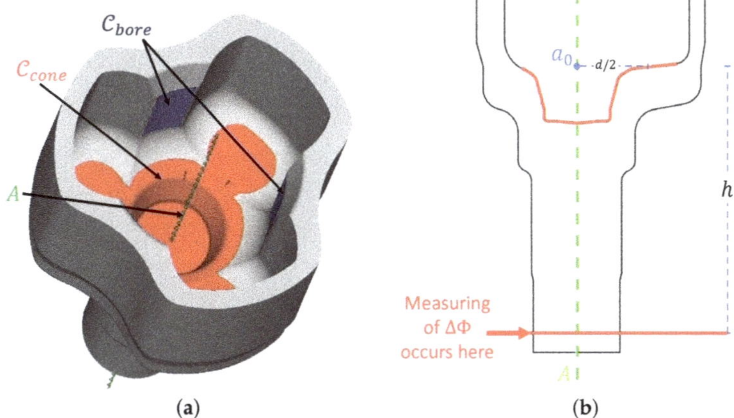

Figure 3. Cylindrical C_{bore} (blue) and conical C_{cone} (red) features on the CAD reference used to compute the revolution axis A (green). (**a**) CAD features C_{bore}, C_{cone} and A; (**b**) Reference axis point a_0 defined where C_{cone} achieves an specific diameter d.

3.2. In-Line Dimensional Inspection

After the planning has been carried for a given reference C, the automatic inspection for every workpiece M related to that reference is automatically performed. The following sections detail the 3D scanning of the warm workpiece, the registration of the mesh regarding the reference CAD C, the computation of the revolution axis A (datum) on M based on the CAD features, and finally the calculation of the circular runout $\Delta\Phi$ of the workpiece.

3.2.1. 3D Scanning System

Figure 4 presents the setup for the 3D scan of the warm workpiece.

Laser triangulation is used to reconstruct the surface. Two independent laser line projectors impact the workpiece inner (punch zone) and outer (forming matrix zone) sides, respectively. Since the surface emits light in the red spectrum due to the high temperatures of the workpiece, the two lasers are chosen to work in the blue spectrum.

The first laser is placed above the workpiece, with an elevation near 45 degrees with respect to the plane that supports the workpiece. This laser allows to scan the inner (punch zone) surface of the axle. This laser is observed by 2 cameras since the geometry of such surface is self-occluding. The cameras must be as close to the workpiece as the heat emitted by it permits (approx. 500 mm), in accordance to the operating temperature prescribed for them.

The second laser is positioned underneath the plane supporting the workpiece (elevation near -45 degrees). The external surface (forming matrix zone) of the workpiece is scanned from below, through slots of the supporting table. The laser projection on such surface is captured by a single camera since no occlusions occur.

Since the laser projections do not lie on a plane parallel to their respective camera plane, Scheimpflug adapters [29] are incorporated in all the cameras to fix their plane of focus. In addition, each camera has an interference filter which allows it to only see light near the laser wavelength ($450\,\text{nm} \pm 25\,\text{nm}$, see Table 2).

A 3-grip system holds the workpiece from the shaft. The grip system is made of steel with ceramic coating to stand the high temperatures. The 3-grip system is mounted on a rotating disk such that the workpiece is rotated 360 degrees around its revolution axis during the scanning. The cameras capture a static image of the workpiece at each pulse of the encoder. Thus, the reconstruction is performed with 360 images per camera.

Table 2. Properties of the laser line projectors used to irradiate the warm workpiece surface.

Property	Value
Power	20 mW
Wavelength	450 nm

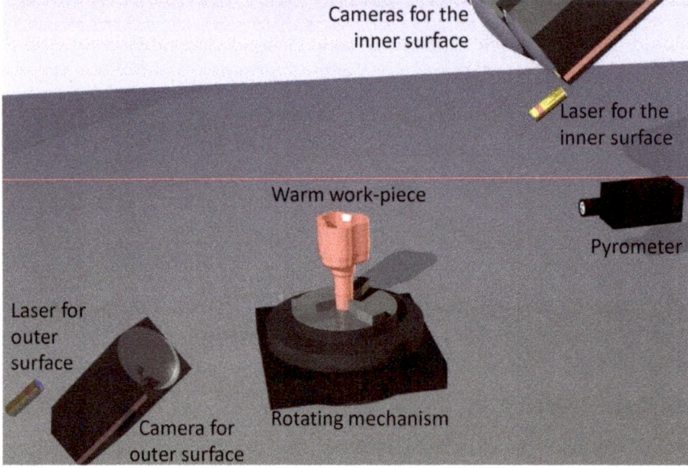

Figure 4. Setup of the 3D optical system designed to digitize the warm workpiece.

The 3D mesh reconstruction from the acquired images is executed with HALCON [30]. The acquisition and reconstruction of the warm workpiece takes about 5 s (average). Finally, a pyrometer is used to track the average temperature of the workpiece during the 3D reconstruction.

The full setup has been installed on a foam cushion layer to isolate the sensors form the vibrations induced by the forging presses.

3.2.2. Device Calibration

The calibration of the scanner involves the characterization of the laser triangulators and their relative positions regarding the axis of the rotating disk. The scanner has a total of two triangulators, with one and two cameras respectively (see Figure 4). Assuming that the reference system of the machine is in the center of the rotating disk, the calibration involves the estimation of the three camera poses.

Both laser projectors have been mechanically positioned and aligned so that their intersection coincides with the axis of the rotating disk. During the construction of the scanner this alignment is verified with a gauge specifically designed for this task, and it does not need further adjustments. This alignment ensures that the points reconstructed by the cameras belong to the plane XZ of the reference system, which is defined by the rotation axis and the two laser lines.

The pose of the cameras is estimated using a calibration object with a hollow revolution geometry, as shown in Figure 5. Such a calibration object has been measured using a CMM by a metrology laboratory certified by an ENAC (National Accreditation Entity).

During the calibration process the object rotates around the axis of the turn table while the cameras observe the projections of the lasers on its surface. Each camera captures an image for each pulse of the encoder, which has been set to 360 pulses per complete revolution. From these images the intersection of the different segments (laser projections) are obtained.

Since the geometry of the calibration object is known, it is possible to establish 3D-2D point matches that relate points from a common reference frame with their corresponding observations in the camera images. In this way, it is possible to obtain the pose of the three cameras solving the homography matrices induced by the sets of correspondences [31].

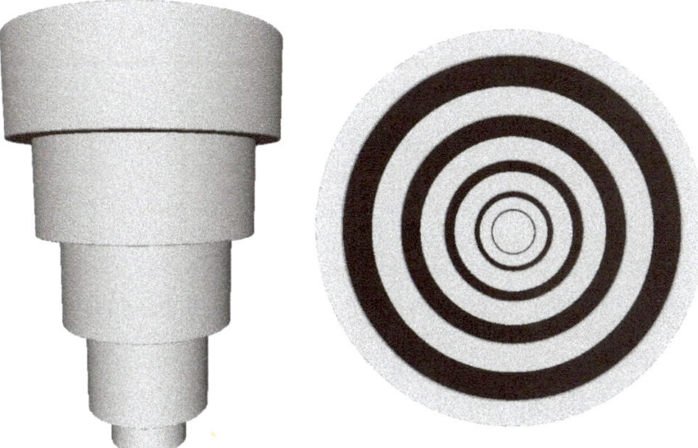

Figure 5. Front and top views of the calibration object geometry.

Experimentally, it has been found that the residual error of the homography evaluated in the intersection points after the calibration is under 0.01 mm. Figure 6 shows some dimensions measured in the calibration object. As it can be deduced from the results, the uncertainty of the scanner is better (less deviation) in the central area of the scanning volume. This effect can be attributed to the fact that the images of the cameras have better focus quality in this area, even after adding the Scheimpflug adapters.

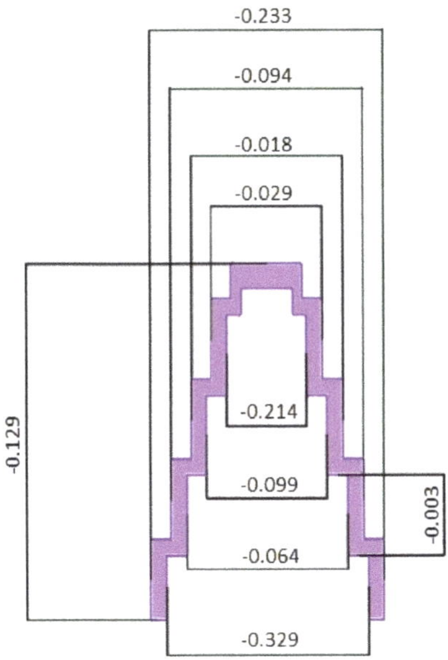

Figure 6. Deviations (mm) after calibration in a cross-section of the scanned pattern (purple), through the *XZ* plane.

3.2.3. Mesh Registration

It is imperative for our system that the scanned triangular mesh M is represented in the same coordinate system W of the CAD reference $\mathcal{C}$. The mesh registration process computes the rigid transformation $T_f \in SE(3)$ that maps the mesh coordinate system W_M to the global coordinate system W i.e., $T_f(W_M) = W$. The rigid transformation T_f is computed by minimizing the distance between the transformed mesh and the reference CAD model:

$$T_f = \arg\min \sum_{x_i \in M} d(T_f(x_i), \mathcal{C})$$
$$s.t. \quad T_f \in SE(3) \tag{2}$$

where $d(T_f(x_i), \mathcal{C})$ is the closest distance from the point $T_f(x_i)$ to the reference $\mathcal{C}$. To solve the minimization problem in Equation (2), the Iterative Closest Point (ICP) algorithm is implemented [32]. The ICP algorithm, transforms the previously defined minimization problem into the following equivalent one:

$$T_{icp} = \arg\min \sum_{x_i \in M} d((T_{icp} \circ T_0)(x_i), \mathcal{C})$$
$$s.t. \quad T_{icp} \in SE(3) \tag{3}$$

where $T_f = T_{icp} \circ T_0$. To avoid local minima, the ICP algorithm requires an initial solution T_0 such that $T_0(W_0)$ is close to the optimal solution W. T_0 has been previously computed in the planning step (Equation (1)).

In optical-based dimensional inspection, selection of reference geometries for mesh registration is crucial for adequate estimation of datums and measurements [9,10]. Therefore, the registration of the scanned mesh M is performed using only the punch zone of the workpiece, which characterizes the revolution axis A. Figure 7 plots the results of the registration process. The colormap shows the distance from the scanned mesh M to the reference CAD $\mathcal{C}$, with green zones indicating closeness between the models ($d(x_i, \mathcal{C}) \approx 0$ mm), and red and blue zones indicating remoteness ($d(x_i, \mathcal{C}) > 0.5$ mm).

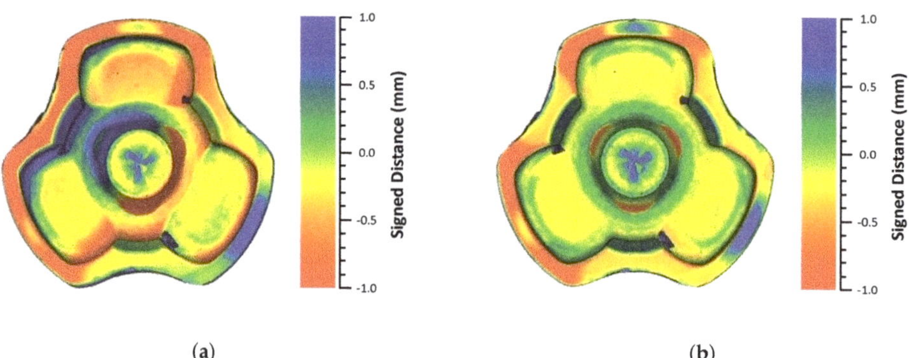

(a)   (b)

Figure 7. Mesh registration results. The colormap shows the signed distance from the scanned mesh M to the CAD model $\mathcal{C}$. (a) Initial guess from np-align (Figure 2c); (b) Iterative Closest Point (ICP) mesh registration.

To save computational expenses, the distance $d(T_f(x_i), C)$ is computed by previously meshing the CAD reference. This is done because we have observed from our experiments that computing point-to-CAD distance is more time-consuming than point-to-mesh distance.

3.2.4. Feature Extraction

After the mesh registration of M has been computed, the system proceeds to extract the mesh features required for the dimensional assessment. To compute the revolution axis, we need to first extract the sub-mesh $M_{bore} \subset M$ which corresponds to the cylindric surface that dictates the rotation of the workpiece. Such cylindric surface has been already identified by the metrologist in the CAD model during the planning phase (Figure 3a). M_{bore} is computed by extracting the mesh points close to the corresponding CAD feature C_{bore}:

$$M_{bore} = \left\{ x \in M \quad | \quad d(x, C_{bore}) < \varepsilon \quad \wedge \quad \cos^{-1}(\vec{n}(x) \cdot \vec{u}(x)) < \theta \right\} \tag{4}$$

where $\varepsilon > 0, 0° \leq \theta \leq 180°$ are a threshold distance (mm) and a threshold angle (degrees), respectively, $\vec{n}(x)$ is the vector normal to the surface at $x \in M$, and $\vec{u}(x)$ is a vector pointing to the theoretical revolution axis $A = (\vec{v}, c_0)$, defined as follows:

$$\vec{u}(x) = ((x - c_0) \cdot \vec{v})\vec{v} - (x - c_0) \tag{5}$$

The term $\cos^{-1}(\vec{n}(x) \cdot \vec{u}(x))$ is introduced in Equation (4) to filter mesh noise and improve the estimation of the revolution axis on the scanned mesh. From our experiments, we have found that the threshold values $\varepsilon = 0.5\,\text{mm}$ and $\theta = 10°$ produce good results, considering the thermal contraction of the workpiece, mesh noise, etc.

The vector $\vec{v}$ is computed by fitting a cylinder to the mesh M_{bore}. The RANSAC algorithm from the Point Cloud Library (PCL) [33] is used to perform the surface fitting.

An approach similar to the previous one is used to calculate the reference axis point a_0. The cone surface feature $M_{cone} \subset M$ is computed by extracting the mesh points close to the corresponding CAD feature C_{cone}:

$$M_{cone} = \{x \in M \quad | \quad d(x, C_{cone}) < \varepsilon\} \tag{6}$$

We have found in our experiments that fitting a cone surface to M_{cone} produces highly unstable results. Instead of fitting the analytical surface, the developed system computes a cylinder $S_{\vec{v}, d/2}$ with axis vector $\vec{v}$ and cylinder radius $d/2$. This cylinder is then intersected with M_{cone}, which produces a polyline Q (Figure 8a). Finally, the intersection between $\vec{v}$ and the plane that contains the polyline Q (in a least-squares sense), is the point a_0 (Figure 8b).

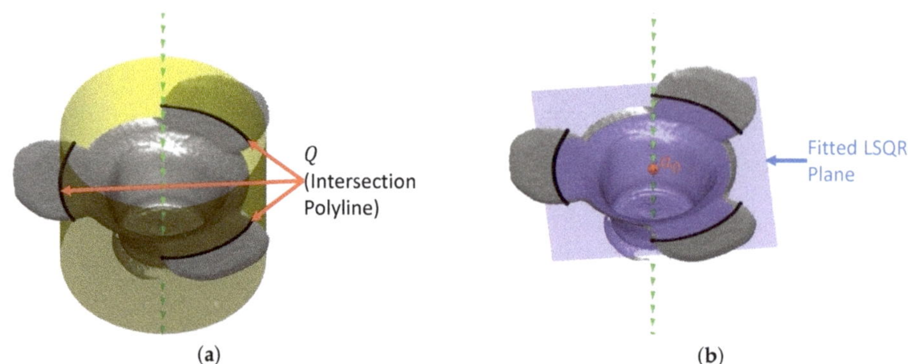

Figure 8. Computation of the reference point a_0 on $\mathcal{C}_{cone}$. (**a**) Mesh-cylinder intersection; (**b**) Plane-axis intersection.

3.2.5. Dimensional Measurements

After the workpiece revolution axis A has been calculated, the circular runout can be computed on M. Given a circular feature $P \subset M$, perpendicular to the revolution axis A (i.e., $P \perp A$), the circular runout of P with respect to A measures how much the feature P oscillates when the workpiece is rotated around the axis A [8].

The following steps describe the system's approach to compute the circular runout of the workpiece:

1. Compute the plane T_h with normal vector $\vec{v}$ and pivot point $a_0 + h\vec{v}$ (the parameter h has been defined already by the metrologist in the planning step, Section 3.1.2). See Figure 9a,b.
2. Compute the circular feature P defined as:

$$P = M \cap T_h \tag{7}$$

3. Filter outliers by removing all points in P whose distance to the theoretical section is greater than a given threshold.
4. Compute the inscribed circle B_{small} and circumscribed circle B_{large} of P with center A and respective radii r_{small}, r_{large} (Figure 9c). $0 < r_{small} \leq r_{large}$.
5. Compute the circular runout $\Delta\Phi$ defined as [8]:

$$\Delta\Phi = r_{large} - r_{small} \tag{8}$$

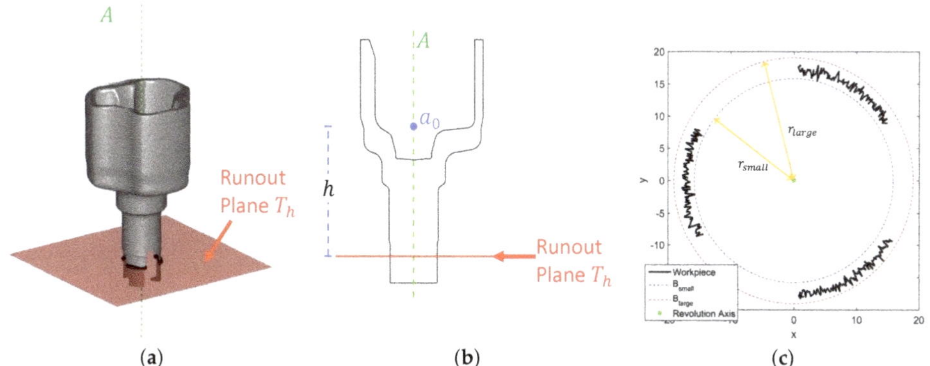

Figure 9. Calculation of the workpiece runout $\Delta\Phi$ regarding the revolution axis A. $\Delta\Phi = r_{large} - r_{small}$. (**a**) Runout plane T_h perpendicular to the revolution axis A; (**b**) The runout plane T_h is defined at a height h from the reference point a_0; (**c**) r_{small} and r_{large} radii computed at the plane T_h.

Before the computation of the circular runout, our system performs circle fitting on the feature P using RANSAC. Such a fitting improves the robustness of the runout estimation by filtering noise and outliers from the scanned mesh.

It is worth noting that the runout deviation includes the eccentricity of both axes and roundness deviations of the measured circle. This is an expected behavior following the standards for geometrical dimensioning and tolerancing [8].

4. Results

Section 4.1 presents and discusses the application of the system to assess the runout of a scanned workpiece at different temperatures. Section 4.2 shows the results of prior-to-operation testing the system using an ANOVA R&R test on cold-state workpieces. Section 4.3 discusses the deployment of the developed system in the automotive manufacturing plant. Finally, Section 4.4 discusses the application of the system in the context of Visual Computing and Industry 4.0 technologies.

4.1. Warm-Workpiece Measurements

In the manufacturing line, each workpiece leaves the forming press at nearly 800 °C. Each part is then left to be cooled naturally by air convection, which takes around 60 min. For this section, two different workpieces are measured continuously during the cool-down. The height of each workpiece is 161.63 mm, and the runout height h (from the axis point a_0 to the runout plane—see Figure 9b) is 64 mm. The objective of this test is to evaluate the accuracy of the system (agreement with the CMM result). Figure 10 plots the runout measurements at different temperatures for the two different workpieces. The workpieces have been measured in the temperature range 27 °C $\leq$ T $\leq$ 560 °C. Each workpiece runout also has been measured with a CMM after cool-down (27 °C). The CMM value is used as ground-truth for assessment purposes. The CMM value obtained for the first workpiece is 0.66 mm. Figure 10a shows that our system measurements deviate in less than 0.1 mm from the CMM measurement. The CMM value obtained for the second workpiece is 0.8 mm. Similar to the first workpiece, our system measurements deviate in less than 0.1 mm from the CMM measurement (Figure 10b). It is worth noting that this deviation is dependent on the height h, increasing as longer workpieces are measured (and decreasing for shorter ones). Such a deviation (vs. height) is small enough for the dimensional assessment purposes of the forging process in which the system is deployed.

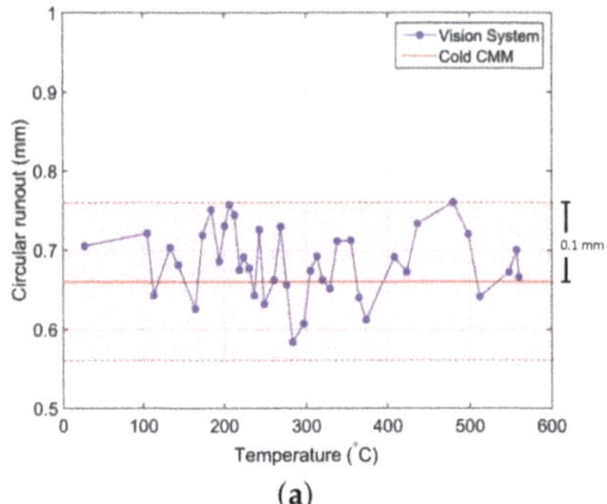
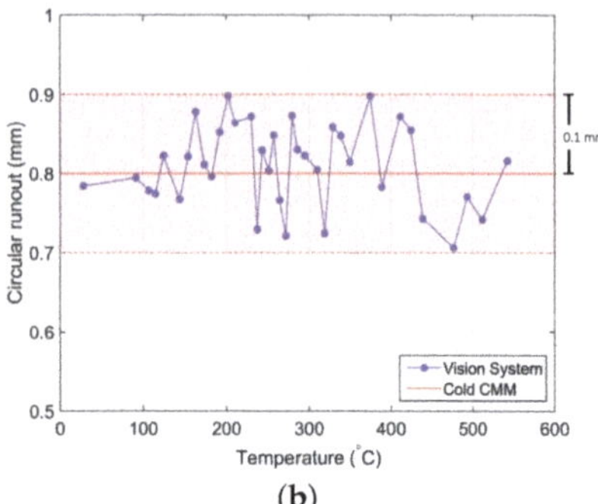

(a) **(b)**

Figure 10. Measurements of 2 warm workpieces until cool-down (27 °C≤ T ≤ 560 °C). Results of our measurement system (blue line) do not deviate more than 0.1 mm from Coordinate Measurement Machine (CMM) measurements (red line). (**a**) Workpiece 1. CMM runout = 0.66 mm; (**b**) Workpiece 2. CMM runout = 0.8 mm.

In Figure 10, there is no apparent correlation between the runout and the temperature of the workpiece at this scale of uncertainty (0.1 mm). Consequently, assessment of the workpiece runout can be performed in our system without the necessity of a correction due to thermal contraction. A more robust study on this matter for this measurement (and other dimensional measurements on the workpiece) is out of the scope of this manuscript, and it is left for future work.

4.2. ANOVA Gauge Repeatability and Reproducibility (R&R) Test

To assess the robustness of the implemented system, a prior-to-operation ANOVA Gauge R&R test is executed. The ANOVA test is performed with cold-state workpieces, which is outside of normal operations. The purpose of the test is to assess the precision (repeatability and reproducibility) of the system but not its accuracy (agreement with the real result). For the control testing, $a = 3$ different workpieces (Figure 11) are measured $m = 10$ times by $b = 3$ different operators, resulting in a sample of 90 measurements. Table 3 presents the measurement results of each experiment.

Figure 11. Cold workpieces used to run the ANOVA R&R test. The three workpieces share the same CAD reference.

Table 3. Runout results (mm) of our system for 3 different cold workpieces (Figure 11), measured 10 times by 3 different operators.

	Workpiece 1			Workpiece 2			Workpiece 3		
	Op 1	**Op. 2**	**Op. 3**	**Op. 1**	**Op. 2**	**Op. 3**	**Op. 1**	**Op. 2**	**Op. 3**
Msh 1	0.71	0.60	0.77	0.78	0.82	0.74	0.79	0.84	0.89
Msh 2	0.63	0.71	0.72	0.79	0.85	0.87	0.77	0.81	0.81
Msh 3	0.64	0.61	0.65	0.81	0.79	0.82	0.80	0.79	0.83
Msh 4	0.70	0.65	0.56	0.79	0.80	0.80	0.77	0.80	0.77
Msh 5	0.65	0.74	0.65	0.79	0.83	0.74	0.79	0.86	0.78
Msh 6	0.69	0.68	0.60	0.84	0.84	0.83	0.82	0.76	0.85
Msh 7	0.68	0.67	0.67	0.85	0.80	0.82	0.84	0.77	0.80
Msh 8	0.70	0.74	0.72	0.75	0.80	0.81	0.76	0.78	0.76
Msh 9	0.61	0.67	0.65	0.81	0.78	0.74	0.82	0.81	0.78
Msh 10	0.72	0.64	0.58	0.75	0.77	0.81	0.84	0.79	0.79
Mean	0.67	0.67	0.66	0.80	0.81	0.80	0.80	0.80	0.80
Std. Dev.	0.03	0.04	0.06	0.03	0.03	0.04	0.03	0.03	0.04
Max-Min	0.11	0.14	0.20	0.10	0.09	0.13	0.08	0.10	0.12

Table 4 shows the ANOVA results for the conducted experiments. The degrees of freedom (DOG) for each sum of squares are defined as:

$$\mathrm{DOG}_{\mathrm{operator}} = b - 1 = 2$$
$$\mathrm{DOG}_{\mathrm{workpiece}} = a - 1 = 2$$
$$\mathrm{DOG}_{\mathrm{interaction}} = (a - 1) * (b - 1) = 4 \tag{9}$$
$$\mathrm{DOG}_{\mathrm{vision\_system}} = ab * (m - 1) = 81$$

The test shows that the variable Workpiece is statistically significant for the computed runout (p-value < 0.05). This means that our system can difference each workpiece from the others due to the inherent manufacturing process bias. On the other hand, the operator and the interaction (operator/workpiece) are not statistically significant (p-value > 0.05), which means that the operator is not a significant source of error for our measurement system (reproducibility).

Table 4. Analysis of Variance (ANOVA) table.

Source of Variability	Degrees of Freedom	Sum of Squares	Mean Square	*F* Statistic	*p*-Value
Operator	2	0.0010	0.0005	1.2282	0.3838
Workpiece	2	0.3575	0.1787	438.5964	0.0000
Interaction	4	0.0016	0.0004	0.2419	0.9137
Vision System	81	0.1364	0.0017		
Total	89	0.4966			

Table 5 presents the Gauge Repeatability & Reproducibility (GRR) [34] results for the executed experiments. The variations induced by the different operators account only for the 0.04% of the total variance of the data, which is an indicator of the reproducibility (less variation equals to more reproducibility) of the developed system. Similarly, the variation introduced by the vision system accounts for the 22.07% of the total variance of the data, which is the indicator of the repeatability (less variation equals to more repeatability) of the system. The manufacturing process accounts for the rest of the variance (77.89%). The Gauge R&R (repeatability + reproducibility) accounts for the 22.11% of the total variance. Such Gauge Repeatability and Reproducibility (GRR) variation is acceptable for the warm forge plant in which the system is deployed (GRR < 25%), where the early detection of dimensional misfits in warm workpieces compensates for the precision loss of the measuring system.

Table 5. Gauge Repeatability and Reproducibility (GRR) table.

Source	Variance	% of Total Variance
Operators (Reproducibility)	0.0000	0.04%
Vision System (Repeatability)	0.0017	22.07%
Gauge R&R (GRR)	0.0017	22.11%
Interaction	0.0000	0.00%
Workpieces	0.0059	77.89%
Total	0.0076	100.00%

Figure 12 presents a boxplot of the data collected in Table 3. Each box represents the 10 measurements taken by an operator. The central marker shows the median of the samples, the bottom and top edges of the box indicate the 25th and 75th percentiles, respectively. The whiskers are the maximum and minimum obtained runout values.

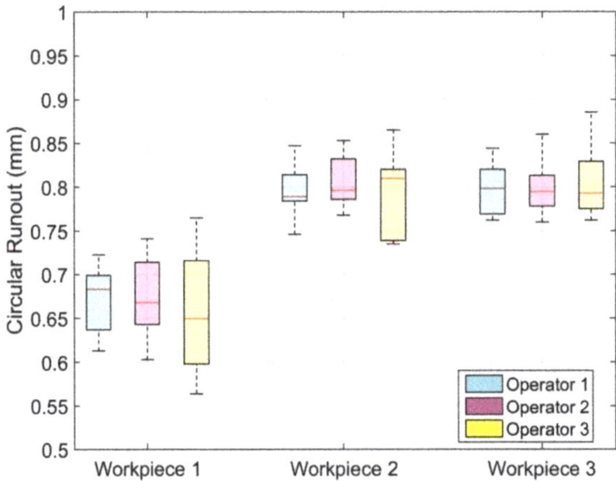

Figure 12. Boxplot of the runout measurements presented in Table 3.

As previously discussed, the manufacturing process induces most of the variance in the form of workpieces with three different runouts (0.67 mm, 0.8 mm and 0.8 mm, respectively). On the other hand, the variance induced by each operator and our measuring system is less significant (largest standard deviation—0.06 mm—in workpiece 1, operator 3).

4.3. Deployment

The metrology system has been deployed in the shop floor of an automotive warm forge of motorcar stub axles, next to one of the press lines. The press line forges around 1200 workpieces per hour. As shown in Figure 13, in this initial setup an operator places manually the warm workpiece into the scanner directly from the press ramp. In the future this operation will be assumed by a robot arm to obtain higher measuring cadences.

To avoid damages to the optical equipment, all the devices are protected with a metal cover. The chassis has been mounted on a foam cushion layer to absorb the vibrations caused by the press.

The measurement software is connected to the Manufacturing Execution System (MES) of the factory. This integration allows automatic loading of the measurement program for the reference that is being manufactured, as well as other data, such as the manufacturing order, the operator name, etc. This data is stored in the report for each workpiece along with the calculated measurements for traceability purposes.

The implemented system automatically assesses the runout of the revolution-like workpieces in less than 60 s. Current measurements are performed with the warm workpiece at approximately 600 °C. The system works as an early detection mechanism for manufacturing and process failures, mainly due to deviations between the forging punch axis and the forming matrix axis. The previous dimensional assessment mechanism (cold-state CMM) used to take only one measurement for every 400 fabricated parts (i.e., every 20 min). On the contrary, our system performs a measurement every 20 fabricated parts (i.e., every minute). Therefore, the system improves the previous measurement method by reducing the number of defective workpieces by around an estimated 95%.

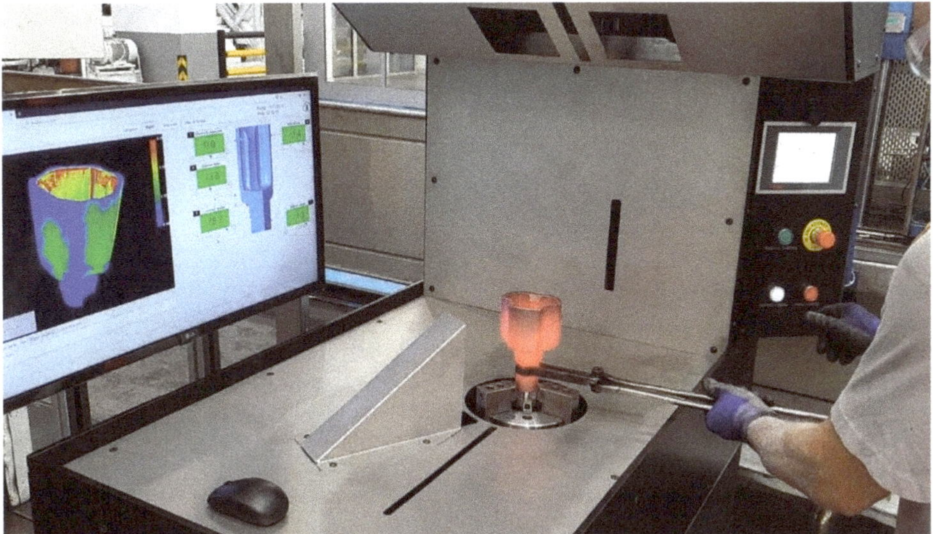

Figure 13. Deployment of the optical system into the production line of warm forming of motorcar stub axles. The temperature of the workpiece is approximately 600 °C.

4.4. Industry 4.0 and Visual Computing

The use of optical systems and visual computing technologies has become an important factor for the improvement of recently developed and classic manufacturing processes [6]. In the context of Industry 4.0, the deployed system improves the classic approaches for dimensional assessment in the warm-die forge industry as follows:

1. Our system performs measurements directly on warm workpieces. Such approach changes the classic scheme for dimensional assessment, which demands workpieces in cold state, limiting in-line metrology application in the warm-die and hot-die forge industry.
2. Thanks to Visual Computing and Industry 4.0 technologies, the developed system can perform fast dimensional measurements on warm workpieces. Over standard cold-state measurement methods, our measurement system reduces the time required to process a warm part by a factor of 95% (from 20 min to 1 min per part).
3. As already mentioned, the Visual Computing technologies provide a framework that allows deployment of the measurement system directly in the manufacturing line. Consequently, the efficiency of the process and product control highly increases as measurements and lines of action can be performed in-line.
4. The deployed system results are shown in a display using a web report tool with visual feedback about the dimensional quality of the measured workpiece. Thanks to the use of such web

technologies [27], the report becomes available in real time to any computer of the factory and any member of the manufacturing plant, including operators, metrologists, and engineers.

5. The visual feedback provided by the visual computing techniques allows easier understanding and more intuitive dimensional assessment of scanned workpieces [9], in contrast to standard CMM numerical data.

6. The automation of the process, together with the high cadence of data acquisition and the aid of web reporting tools, enable a global perspective of the manufacturing process in the context of data analytics. However, such approach is out of scope of the current manuscript, and it is left for future work.

5. Conclusions

This manuscript presents the implementation and deployment of an optical system for automatic in-line dimensional inspection of revolution warm workpieces. The circular runout of warm-forged revolution workpieces is critical as a severe misalignment between the punch press and the forming matrix axes disables the posterior machining, resulting in a scrapped part. The system splits the inspection in two steps: (1) the dimensional assessment planning, performed only once by the metrologist, off-line the production, and (2) the in-line automated dimensional inspection. The developed system automatically assesses, in less than 60 s, the circular runout of the workpiece, whose temperature nears 600 °C. Our prior-to-operation test results show that the measurements of the developed system for warm workpieces (27 °C $\leq$ T $\leq$ 560 °C) deviate less than 0.1 with respect to the standard CMM measurements of the cooled-down workpieces, for workpieces as long as 160 mm. In addition, the temperature-vs.-runout analysis shows no correlation between these two variables at such level of uncertainty. The measuring system repeatability and reproducibility (R&R) has been validated with an ANOVA test. This assessment of dimensions in warm workpieces fills a gap in processes in which the advantage of early detection of an inherent planning, design, or manufacturing error compensates for the disadvantage of precision loss due to the cooling of the workpiece.

Our system has been deployed by an automotive part manufacturer in a warm forming production line of stub axles, working as early detection of dimensional misfits. This early detection reduces the time needed to detect a defective part from 20 min to 1 min. Since the forge is a highly repetitive manufacturing process, when a defective part is found, all the pieces between the last correct part and the defective one are systematically scrapped. Thus, considering the production cadence, the number of parts that are scrapped each time a defective part is detected has been reduced from 400 to 20 (95%).

Future work concerns: (1) A warm-workpiece assessment method for data which are highly sensitive to cooling effects, which accounts for the heat loss effects on the workpiece geometry. (2) The integration of a robot arm for automatic placement of forged workpieces, to increase the measurement efficiency. (3) Metrological certification of the equipment by an ENAC accredited laboratory.

Author Contributions: D.M.-P. and J.R.S. conceptualized and implemented the Computational Geometry and Dimensional Inspection methodology. M.A. and A.I. designed, implemented and executed the system calibration and the Computer Vision system for data acquisition. E.G. and J.P. designed the methodology and experiments, and supervised the the Dimensional Inspection of warm-forge parts. O.R.-S. and J.P. supervised the Stochastic Computational Geometry aspects and applications of this research. All the authors contributed in the writing of this manuscript.

Funding: This research was partially supported by the Basque Government under the grant Basque Industry 4.0.

Acknowledgments: We want to thank our colleagues from Sariki Metrología S.A. whose knowledge of dimensional inspection has been very valuable during this research.

Conflicts of Interest: The authors declare no conflict of interest.

Abbreviations

The following abbreviations are used in this manuscript:

CMM	Coordinate Measurement Machine.
ANOVA	ANalysis Of VAriance.
R&R	Repeatability and Reproducibility.
FACE	A connected region on a parametric surface in $\mathbb{R}^3$.
$\mathcal{C}$	Boundary representation (CAD model) of the reference geometry. $\mathcal{C} \subset \mathbb{R}^3$ is a 2-manifold, represented as set of BODY, LUMPSs, FACEs, LOOPs, EDGEs, and VERTICES.
M	Triangular mesh $M = (X, T)$ of the scanned workpiece. $X = \{x_0, x_1, \ldots, n\}$ and $T = \{t_0, t_1, \ldots, n\}$ are the points (geometry) and triangles (topology) of the mesh, respectively. $M \subset \mathbb{R}^3$ is a 2-manifold.
A	Revolution (datum) axis $A = (\vec{v}, a_0)$ of the workpiece M. The vector $\vec{v} \in \mathbb{R}^3$ defines the direction of the axis and $a_0 \in \mathbb{R}^3$ is a point lying on the axis.
$\Delta\Phi$	Circular runout (mm) of the scanned workpiece M. $\Delta\Phi \geq 0$ measures how much a cylindrical feature oscillates when rotated around the revolution axis A.
h	Height $h > 0$ (mm) where the circular runout $\Delta\Phi$ is measured in the workpiece. This height is measured from a_0, in the direction of $\vec{v}$.
W	Reference coordinate system $W = \{w_x, w_y, w_z; p_w\}$. W is the coordinate system of $\mathcal{C}$ and the coordinate system of M after mesh registration.
W_M	Coordinate system of M before mesh registration.
$SE(3)$	Special Euclidean group. Group of all rigid transformations in $\mathbb{R}^3$. $SE(3)$ is composed by all the possible rotation matrices and all possible translations in $\mathbb{R}^3$, i.e., $SE(3) = SO(3) \times \mathbb{R}^3$.
T_0	Initial rigid transformation $T_0 \in SE(3)$ that approximately maps W_M to W.
T_{icp}	Rigid transformation $T_{icp} \in SE(3)$ that maps $T_0(W_M)$ to W. T_{icp} is the result of registering the mesh M to the reference $\mathcal{C}$.
T_f	Rigid transformation $T_f \in SE(3)$ that maps W_M to W. $T_f = T_{icp} \circ T_0$.
M_{bore}	Cylindrical surface $M_{cylinder} \subset M$ whose axis vector is the revolution axis vector of the workpiece $\vec{v}$.
M_{cone}	Conical surface $M_{cone} \subset M$ used to compute the axis reference point a_0.
C_{bore}	Subset of FACES $C_{bore} \subset C$ which define a cylindrical surface in the CAD reference. These set of faces are used to extract M_{bore} from M.
C_{cone}	Subset of FACES $C_{cone} \subset C$ which define a conical surface in the CAD reference. These set of faces are used to extract M_{cone} from M.
ε	Distance threshold (mm) used to extract the mesh features M_{bore}, M_{cone}.
d	Datum diameter $d > 0$ (mm). The point a_0 is located on the plane where the conical surface M_{cone} attains the diameter d.
P	Circular feature $P \subset M$ where the circular runout $\Delta\Phi$ is measured with respect to A. P defines a polyline perpendicular to the axis A ($P \perp A$).
T_h	Plane $T_h \subset \mathbb{R}^3$ used to extract P from M. The plane T_h has normal $\vec{v}$ and pivot point $a_0 + h\vec{v}$.
T	Temperature of the workpiece (°C).
a	Number of workpieces for the ANOVA test.
b	Number of operators for the ANOVA test.
m	Number of measurements for the ANOVA test.

References

1. Hawryluk, M.; Ziemba, J. Possibilities of application measurement techniques in hot die forging processes. *Measurement* **2017**, *110*, 284–295. [CrossRef]
2. Gronostajski, Z.; Kaszuba, M.; Hawryluk, M.; Zwierzchowski, M. A review of the degradation mechanisms of the hot forging tools. *Arch. Civ. Mech. Engi.* **2014**, *14*, 528–539. [CrossRef]
3. Hawryluk, M.; Gronostajski, Z.; Kaszuba, M.; Polak, S.; Widomski, P.; Ziemba, J.; Smolik, J. Application of selected surface engineering methods to improve the durability of tools used in precision forging. *Int. J. Adv. Manuf. Technol.* **2017**, *93*, 2183–2200. [CrossRef]

4. International Organization for Standardization. *ISO 1101:2017 Geometrical Product Specifications (GPS)—Geometrical Tolerancing—Tolerances of Form, Orientation, Location and Run-Out*; Standard, International Organization for Standardization: Geneva, Switzerland, 2017.

5. Hawryluk, M.; Ziemba, J.; Sadowski, P. A Review of Current and New Measurement Techniques Used in Hot Die Forging Processes. *Meas. Control* **2017**, *50*, 74–86. [CrossRef]

6. Posada, J.; Toro, C.; Barandiaran, I.; Oyarzun, D.; Stricker, D.; de Amicis, R.; Pinto, E.B.; Eisert, P.; Döllner, J.; Vallarino, I. Visual Computing as a Key Enabling Technology for Industrie 4.0 and Industrial Internet. *IEEE Comput. Graph. Appl.* **2015**, *35*, 26–40. [CrossRef] [PubMed]

7. Gapinski, B.; Wieczorowski, M.; Marciniak-Podsadna, L.; Dybala, B.; Ziolkowski, G. Comparison of Different Method of Measurement Geometry Using CMM, Optical Scanner and Computed Tomography 3D. *Procedia Eng.* **2014**, *69*, 255–262. [CrossRef]

8. Henzold, G. 18—Inspection of Geometrical Deviations. In *Geometrical Dimensioning and Tolerancing for Design, Manufacturing and Inspection*, 2nd ed.; Butterworth-Heinemann: Oxford, UK, 2006; pp. 160–254. [CrossRef]

9. Sánchez, J.R.; Segura, Á.; Barandiaran, I. Fast and accurate mesh registration applied to in-line dimensional inspection processes. *Int. J. Interact. Des. Manuf.* **2018**, *12*, 877–887. [CrossRef]

10. Minetola, P. The importance of a correct alignment in contactless inspection of additive manufactured parts. *Int. J. Precis. Eng. Manuf.* **2012**, *13*, 211–218. [CrossRef]

11. Shi, Q.; Xi, N. Automated data processing for a rapid 3D surface inspection system. In Proceedings of the 2008 IEEE International Conference on Robotics and Automation, Pasadena, CA, USA, 19–23 May 2008; pp. 3939–3944. [CrossRef]

12. Zhu, L.; Barhak, J.; Srivatsan, V.; Katz, R. Efficient registration for precision inspection of free-form surfaces. *Int. J. Adv. Manuf. Technol.* **2007**, *32*, 505–515. [CrossRef]

13. Gronostajski, Z.; Hawryluk, M.; Kaszuba, M.; Widomski, P.; Ziemba, J. Application of the reverse 3D scanning method to evaluate the wear of forging tools divided on two selected areas. *Int. J. Autom. Technol.* **2017**, *18*, 653–662. [CrossRef]

14. Hawryluk, M.; Ziemba, J. Application of the 3D reverse scanning method in the analysis of tool wear and forging defects. *Measurement* **2018**, *128*, 204–213. [CrossRef]

15. Jung, K.H.; Lee, S.; Kim, Y.B.; Ahn, B.; Kim, E.Z.; Lee, G.A. Assessment of ZK60A magnesium billets for forging depending on casting methods by upsetting and tomography. *J. Mech. Sci. Technol.* **2013**, *27*, 3149–3153. [CrossRef]

16. D'Annibale, A.; Ilio, A.D.; Trozzi, M.; Bonaventura, L. The Use of Infrared Thermography for Maintenance Purposes in the Production Process of Components for Automotive Alternators. *Procedia CIRP* **2015**, *38*, 143–146. [CrossRef]

17. Fendt, K.T.; Mooshofer, H.; Rupitsch, S.J.; Ermert, H. Ultrasonic Defect Characterization in Heavy Rotor Forgings by Means of the Synthetic Aperture Focusing Technique and Optimization Methods. *IEEE Trans. Ultrason. Ferroelectr. Freq. Control* **2016**, *63*, 874–885. [CrossRef]

18. Reddy, K.A. Non-Destructive Testing, Evaluation Of Stainless Steel Materials. *Mater. Today Proc.* **2017**, *4*, 7302–7312. [CrossRef]

19. Dworkin, S.; Nye, T. Image processing for machine vision measurement of hot formed parts. *J. Mater. Process. Technol.* **2006**, *174*, 1–6. [CrossRef]

20. Jia, Z.; Wang, B.; Liu, W.; Sun, Y. An improved image acquiring method for machine vision measurement of hot formed parts. *J. Mater. Process. Technol.* **2010**, *210*, 267–271. [CrossRef]

21. Zhang, Y.-C.; Han, J.-X.; Fu, X.-B.; Zhang, F.-L. Measurement and control technology of the size for large hot forgings. *Measurement* **2014**, *49*, 52–59. [CrossRef]

22. Du, Y.D. Simple three-dimensional laser radar measuring method and model reconstruction for hot heavy forgings. *Opt. Eng.* **2012**, *51*, 021118. [CrossRef]

23. Du, Z.; Wu, Z.; Yang, J. 3D measuring and segmentation method for hot heavy forging. *Measurement* **2016**, *85*, 43–53. [CrossRef]

24. Liu, W.; Jia, Z.; Wang, F.; Ma, X.; Wang, W.; Jia, X.; Song, D. An improved online dimensional measurement method of large hot cylindrical forging. *Measurement* **2012**, *45*, 2041–2051. [CrossRef]

25. Molleda, J.; Usamentiaga, R.; García, D.F.; Bulnes, F.G. Real-time flatness inspection of rolled products based on optical laser triangulation and three-dimensional surface reconstruction. *J. Electron. Imaging* **2010**, *19*, 031206. [CrossRef]

Appl. Sci. **2019**, *9*, 1069

26. Babu, M.; Franciosa, P.; Ceglarek, D. Adaptive Measurement and Modelling Methodology for In-line 3D Surface Metrology Scanners. *Procedia CIRP* **2017**, *60*, 26–31. [CrossRef]

27. Mejia, D.; Sánchez, J.R.; Segura, A.; Ruiz-Salguero, O.; Posada, J.; Cadavid, C. Mesh Segmentation and Texture Mapping for Dimensional Inspection in Web3D. In Proceedings of the 22nd International Conference on 3D Web Technology (Web3D '17), Brisbane, Queensland, Australia, 5–7 June 2017; ACM: New York, NY, USA, 2017; pp. 3:1–3:4. [CrossRef]

28. Horn, B.K.P. Closed-form solution of absolute orientation using unit quaternions. *J. Opt. Soc. Am. A* **1987**, *4*, 629–642. [CrossRef]

29. Sun, C.; Liu, H.; Jia, M.; Chen, S. Review of Calibration Methods for Scheimpflug Camera. *J. Sens.* **2018**, *2018*, 3901431. [CrossRef] [PubMed]

30. Steger, C.; Ulrich, M.; Wiedemann, C. *Machine Vision Algorithms and Applications*, 2nd ed.; Wiley-VCH: Weinheim, Germany, 2017.

31. Hartley, R.I.; Zisserman, A. *Multiple View Geometry in Computer Vision*, 2nd ed.; Cambridge University Press: Cambridge, UK, 2004; ISBN 0521540518.

32. Besl, P.J.; McKay, N.D. A method for registration of 3-D shapes. *IEEE Trans. Pattern Anal. Mach. Intell.* **1992**, *14*, 239–256. [CrossRef]

33. Rusu, R.B.; Cousins, S. 3D is here: Point Cloud Library (PCL). In Proceedings of the 2011 IEEE International Conference on Robotics and Automation, Shanghai, China, 9–13 May 2011; pp. 1–4. [CrossRef]

34. McNeese, B. ANOVA GAGE R&R—Part 2. 2012. Available online: https://www.spcforexcel.com/knowledge/measurement-systems-analysis/anova-gage-rr-part-2 (accessed on 4 February 2019).

Article

Fabric Defect Detection Using L0 Gradient Minimization and Fuzzy C-Means

Huanhuan Zhang *, Jinxiu Ma, Junfeng Jing and Pengfei Li

College of Electronics and Information, Xi'an Polytechnic University, Xi'an 710048, China
* Correspondence: zhanghuanhuan0557@163.com; Tel.: +86-152-2989-4027

Received: 8 July 2019; Accepted: 20 August 2019; Published: 26 August 2019

Abstract: In this paper, we present a robust and reliable framework based on L0 gradient minimization (LGM) and the fuzzy c-means (FCM) method to detect various fabric defects with diverse textures. In our framework, the L0 gradient minimization is applied to process the fabric images to eliminate the influence of background texture and preserve sharpened significant edges on fabric defects. Then, the processed fabric images are clustered by using the fuzzy c-means. Through continuous iterative calculation, the clustering centers of fabric defects and non-defects are updated to realize the defect regions segmentation. We evaluate the proposed method on various samples, which include plain fabric, twill fabric, star-patterned fabric, dot-patterned fabric, box-patterned fabric, striped fabric and statistical-texture fabric with different defect types and shapes. Experimental results demonstrate that the proposed method has a good detection performance compared with other state-of-the-art methods in terms of both subjective and objective tests. In addition, the proposed method is applicable to industrial machine vision detection with limited computational resources.

Keywords: fabric defect detection; LGM; FCM; image smoothing

---

1. Introduction

Fabric defect detection plays a crucial role in the automatic inspection in textile production processes. However, traditional fabric defect defection is often dependent on human inspection, and quality controls often rely to experience of specialized workers. It is noted that the human workers are prone to fatigue and boredom due to the repetitive nature of their tasks [1]. Thus, the human inspection involves limitations in terms of accuracy, coherence, and efficiency when detecting defects. Since the fabric textures are so complicated (including plain weave fabric, knitted fabric, twill fabric, laces, and pattern fabric) [2], the fabric colors are variable, and the contrasts between fabric defects and background are low, generalized defect detection exploration is highly challenging.

Currently, automated defect detection methods based on machine vision have drawn much attention. Gaussian mixture entropy modeling [3] and wavelet transform [4] were used to detect defect in simple plain and twill fabric images via transformation and reconstruction processes. However, most of these methods designed for the simplest plain and twill fabrics, which cannot be effectively applied on complicated patterns fabric, such as the dot-patterned fabric, star-patterned fabrics and statistical-texture fabrics. The entropy-based automatic selection of the wavelet decomposition level (EADL) [5] method and the automatic band selection method [6] achieved defect defection in statistical and structural textures. Bollinger bands (BB) [7] and image decomposition (ID) methods [8] have been shown to perform robustly for dot-patterned, star-patterned and box-patterned fabrics. However, it remains unknown whether these two methods can be used for plain, twill, and statistical-texture fabrics. In a preliminary evaluation, the BB and ID methods failed to recognize some defective samples. As shown in Figure 1, the BB and ID methods are weak at differentiating defects with directional features. These methods achieve good results on a certain texture, but it remains challenging to robustly and accurately handle the fabric defect image if it has a complicated patterns texture, low contrast between defect object and background, various colors, and a low signal-to-noise ratio.

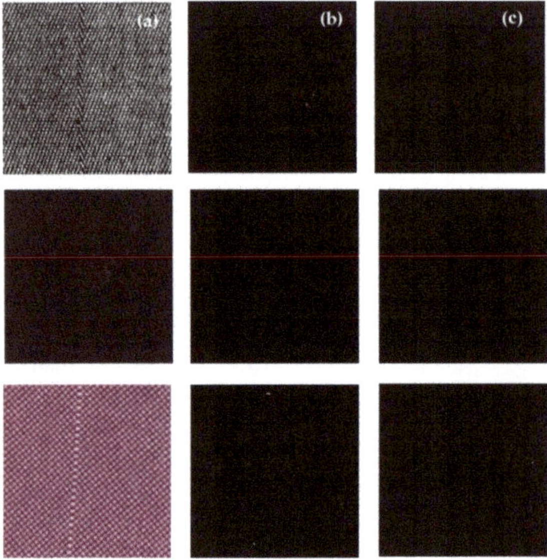

Figure 1. Failure of the BB and ID methods for detecting defects. (**a**) Defect sample, (**b**) detection result using the BB method, and (**c**) detection result using the ID method.

To address these problems, we present a novel method based on L0 gradient minimization (LGM) and fuzzy c-means (FCM), which provides a new perspective for the detection of fabric defects. Usually, a defect-free fabric image in industrial products has consistent texture, and the defect can be considered as the defective structure information and texture information. In our work, we first used the LGM method to filter the input image to eliminate the influence of texture information on fabric defects. Then, the filtered results with just defective information were segmented by applying the FCM. The proposed method can handle the defect with the plain fabric, twill fabric, star-patterned fabric, dot-patterned fabric, box-patterned fabric, striped fabric, and statistical-texture fabric (as shown in Figure 2).

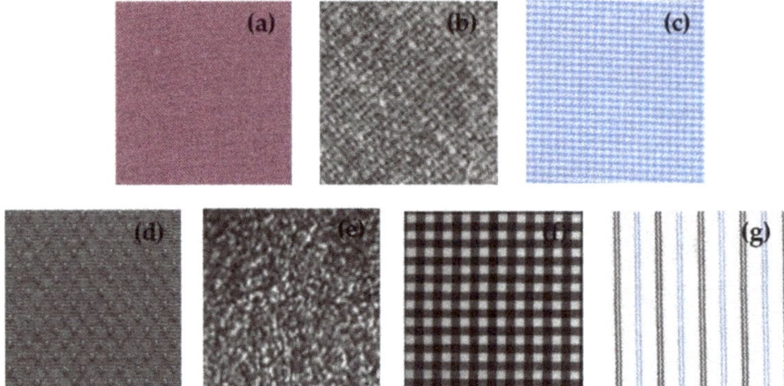

Figure 2. Various textures of fabric images. (**a**) Plain fabric, (**b**) twill fabric, (**c**) star-patterned fabric, (**d**) dot-patterned fabric, (**e**) statistical-texture fabrics, (**f**) box-patterned fabric, and (**g**) striped fabric.

The remainder of this paper is organized as follows. Section 2 briefly discusses some related works. Section 3 mainly focuses on presents the proposed method. The experiment results and discussions are given in Section 4. Conclusions and future works pertaining to this work are presented in Section 5.

2. Related Works

Plain and twill fabric detection methods can be classified into five aspects: Spectral [9,10], learning [11,12], statistical [13–15], model-based [16,17], and structural methods [18,19]. The spectral method based on the Wavelet transform [20] achieved 97.5% detection accuracy with five known defect types and a 93.3% detection accuracy (a slight drop) with three unknown defect types in an evaluation. The statistical method applied gray relational analysis with co-occurrence Matrix (CM) features [21] on Jacquard fabric images, reaching 94% detection accuracy for 50 defective samples. The learning method via three-layer back-propagation neural network and thresholding of the image analysis [22] was tested on the same kind of fabric; it achieved 94.38% accuracy, using 240 samples of the four defect classes. The limitation of their method includes a longer training time, because of the larger number of inner layers and the danger of over-training. In addition, model-based approach using the Gaussian mixture model [23] was successfully applied to Brodatz mosaic image segmentation and fabric defect detection. Structural approaches based on normalized cross-correlation algorithm [24] obtained a higher detection success rate of 95% on twelve defective plain and twill fabrics' images. In general, many pieces of research on plain and twill fabric inspection works have achieved fruitful results; however, these methods were not efficiently evaluated on complicated patterned fabrics and statistical-texture fabrics.

The defect detection of complicated patterned fabric has been increasing during the last decade. The Bollinger bands (BB) and regular bands (RB) [25] methods employed the regularity property in the patterned texture to carry out defect detection on dot-, box- and star-patterned fabrics. They also obtained accuracy rates of 98.59% (167 defect-free and 171 defective images) and 99.4% (80 defect-free and 86 defective images), respectively. The wavelet-preprocessing golden image subtraction (WGIS) method [20] achieved 96.7% accuracy on 30 defect-free and 30 defective patterned images by using a golden image to perform moving subtraction of each pixel along each row of every wavelet-pre-processed tested image. The ID [8] method obtained the detection accuracies range from 94.9–99.6% for dot- (110 defect-free and 120 defective samples), star- (25 defect-free and 25 defective samples) and box-patterned fabrics (30 defect-free and 26 defective samples), which decomposed a fabric image into structures of cartoon (defective objects) and texture (repeated patterns). A recent Elo rating method [26] achieved an overall 97.07% detection success rate based on databases of [8]. However, it remains unknown whether the patterned fabric defect detection method can also be applied to twill, plain fabric, and statistical-texture fabrics.

In our work, we take full consideration of the original fabric images, which can be seen as defective structure information and texture information. Texture information often affects the fabric defect detection. LGM is used to filter the images to remove the texture information. In this way, the filtered fabric images would quickly locate and segment the defects. The proposed method can detect the defect in the plain fabric, twill fabric, star-patterned fabric, dot-patterned fabric, box-patterned fabric, striped fabric, and statistical-texture fabric.

3. Methods

In this section, procedures of the proposed LCM and FCM algorithm are described in detail. Figure 3 shows the overview of the presented method. It consists of two steps: Firstly, the L0 gradient minimization is applied to eliminate the influence of the background texture of fabric defects. Then the fuzzy c-means clustering is used to determine whether each pixel is defective.

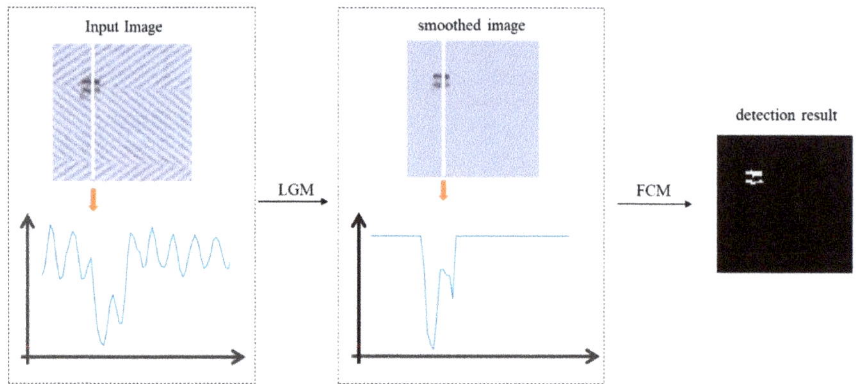

Figure 3. The framework of the proposed method.

3.1. Texture Removal by the L0 Gradient Minimization (LGM)

Due to the complexity of background texture information, it often increases the challenge of fabric defect detection. The L0 gradient minimization method [27] is widely used to smooth texture information. It is often adopted for filtering the image while preserving edge feature. The L0 gradient minimization method enhances the significant edge portion of the image by increasing the steepness of the transition portion of the image while removing the low-amplitude detail portions. Inspired by the L0 gradient minimization, we apply it to remove the background texture of fabric.

As shown in Figure 4a, the fabric defect sample has three different directions of texture, and it is a mesh diagram, as shown in Figure 5a. After smoothing via LGM, the unimportant background texture of the fabric is removed, as shown in Figure 4b. Notice that the high-contrast edges on the defect are preserved, and the defect feature is more prominent, as shown in Figure 5b.

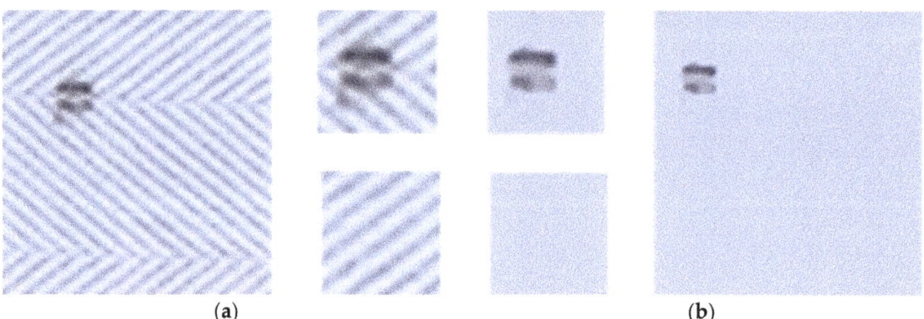

Figure 4. Smoothed fabric defect image using L0 gradient minimization (LGM). (**a**) Fabric defect image; (**b**) Smoothed fabric defect image.

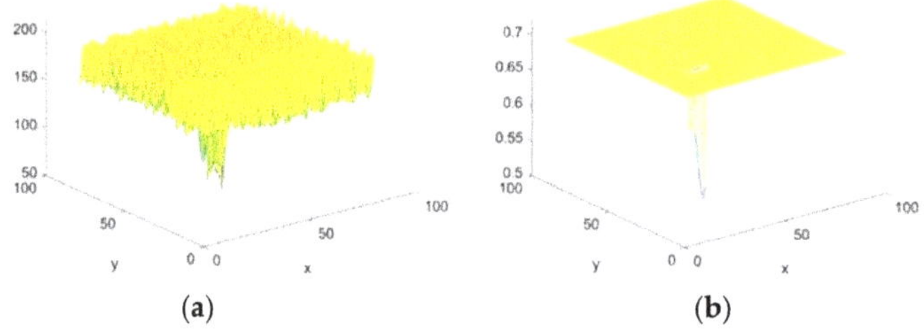

Figure 5. Detection effect of LGM. (**a**) Mesh diagram of fabric defect image; (**b**) Mesh diagram of smoothed fabric defect image.

In order to illustrate the method clearly, we briefly summarize the theory of the L0 gradient minimization model. Let I be the input fabric image, its smoothed output result is S, $\partial_x S_p$ and $\partial_y S_p$ are the partial derivative of the processed image in the x and y directions at p respectively, and the gradient of image S at pixel p is denoted by $\nabla S_p = \left(\partial_x S_p, \partial_y S_p\right)^T$. The image L0 gradient specific objective function is defined as:

$$min\left\{\sum_p \left|S_p - I_p\right|^2 + \beta|\nabla S_p - h|^2 + \lambda|h|_0\right\} \tag{1}$$

where λ is a non-negative parameter, which directly controls the weight of the smoothing term. β is an automatically adapting parameter. h is auxiliary variable. By alternatively computing h and S, it can obtain the output result. Figure 6 shows the result using the LGM method with their corresponding mesh diagrams. It is observed that the output result retains the defect information and removes the background texture information.

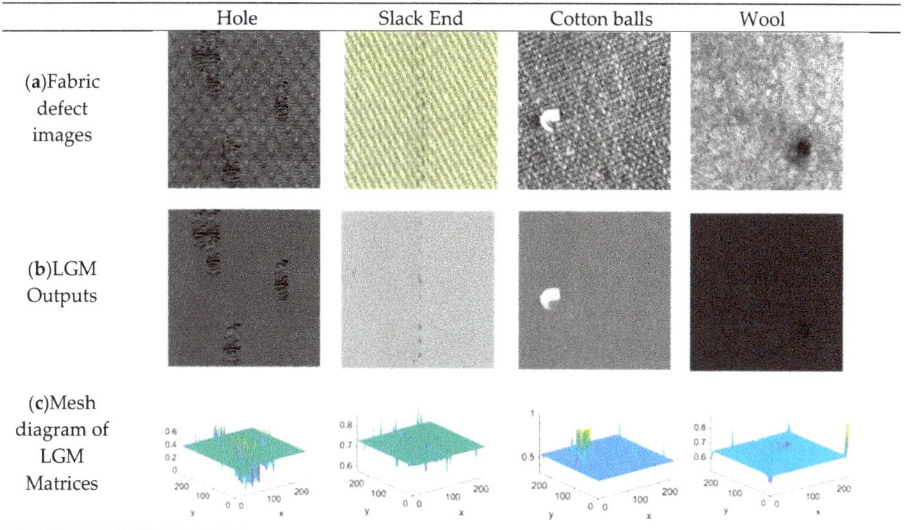

Figure 6. Results of fabric defect based on LGM.

3.2. Fuzzy C-Means Clustering Algorithm (FCM)

The output results via LGM are obtained in the previous section. We can easily process the defect information, which becomes more obvious. Then the FCM algorithm is applied to segment the defects.

The FCM algorithm [28,29] is a fuzzy and unsupervised clustering algorithm. Its classification capacity is flexible and simple to implement. The FCM algorithm defines the objective function represented the sum of squares of the weighted distances from each pixel in the target image to each cluster center, which is given by:

$$J(M, V) = \sum_{j=1}^{c}\sum_{i=1}^{n}\left[\mu_j(x_i)^b\right]\|x_i - m_j\|^2 \tag{2}$$

where c is the number of clusters, n is the number of pixels in the image, $\mu_j(x_i)$ is the membership degree of the i-th pixel belonging to the j-th class, normally $0 \leq \mu_j(x_i) \leq 1$ and $\sum_{j=1}^{c}\mu_j(x_i) = 1$, b is a preset fuzzy factor which controls the degree of blur, normally $b \geq 1$, and $\|x_i - m_j\|^2$ represents the Euclidean distance from the i-th pixel to the j-th class.

The physical meaning of the objective function J represents the sum of squares of the weighted distances from each pixel in the target image to each cluster center. When the Euclidean-distance-weighted value of each pixel-point in the target image to the cluster center is the minimum, and the Euclidean distance from other cluster centers is as large as possible. The basic principle of the FCM algorithm is to find a set of suitable clustering centers and the membership matrix, such that the objective function J takes the minimum value min (J).

When calculating the minimum value of the objective function, it is necessary to continuously update the membership matrix and the cluster center according to Equations (3) and (4) until the minimum value is obtained:

$$m_j = \frac{\sum_{i=1}^{n}\left[\mu_j(x_i)\right]^b x_i}{\sum_{i=1}^{n}\left[\mu_j(x_i)\right]^b} \tag{3}$$

$$\mu_j(x_i) = \frac{\left(1/\|x_i - m_j\|^2\right)^{1/(b-1)}}{\sum_{k=1}^{c}\left(1/\|x_i - m_k\|^2\right)^{1/(b-1)}} \tag{4}$$

When the objective function obtains the minimum value, the membership degree is retained, and each pixel of the target image is clustered. In the iterative process, the appropriate iteration termination condition should be selected correctly, otherwise, it cannot obtain the ideal segmentation result. When the FCM algorithm is convergent, it can achieve the different clustering centers which are the clustering center of defects information and the clustering center of the normal fabric information.

4. Experimental Results and Discussion

The testing code was implemented under the MATLAB version R2014B. The proposed method was carried out on a standard workstation equipped with an Intel Core i5-4460 3.2 GHz CPU with 8 GB of main memory, an NVIDIA GeForce GT 745 graphics card and Windows 8.1 OS.

In our work, we used fabric defect images from the automation laboratory sample database of Hong Kong University, TILDA Textile Texture Database and Guang Dong Esquel Textiles with a resolution of 600 dpi, scanned by Canon Scanner 9000F. The images have a size of 256 × 256 pixels and an 8-bit grey level. Various fabric images (including plain fabric, twill fabric, star-patterned fabric, dot-patterned fabric, box-patterned fabric, striped fabric, and statistical-texture fabrics) were used for evaluating our method.

4.1. Parameter Setting

The parameter λ plays a key role in detecting defects accurately in our method. The influence of λ that changed the defection results is shown in Figure 7. The best filter results using LGM are marked with a red block. From the Figure 7, it can be seen that if the parameter λ is set too small, the noise and background texture are almost unchanged. On the contrary, if the parameter λ is set too large, the defect is smoothed over. If the parameter λ is set moderate, the defect area will be easily distinct from the surrounding background in the following segmentation, and we can obtain a better detection result, as shown in Figure 7.

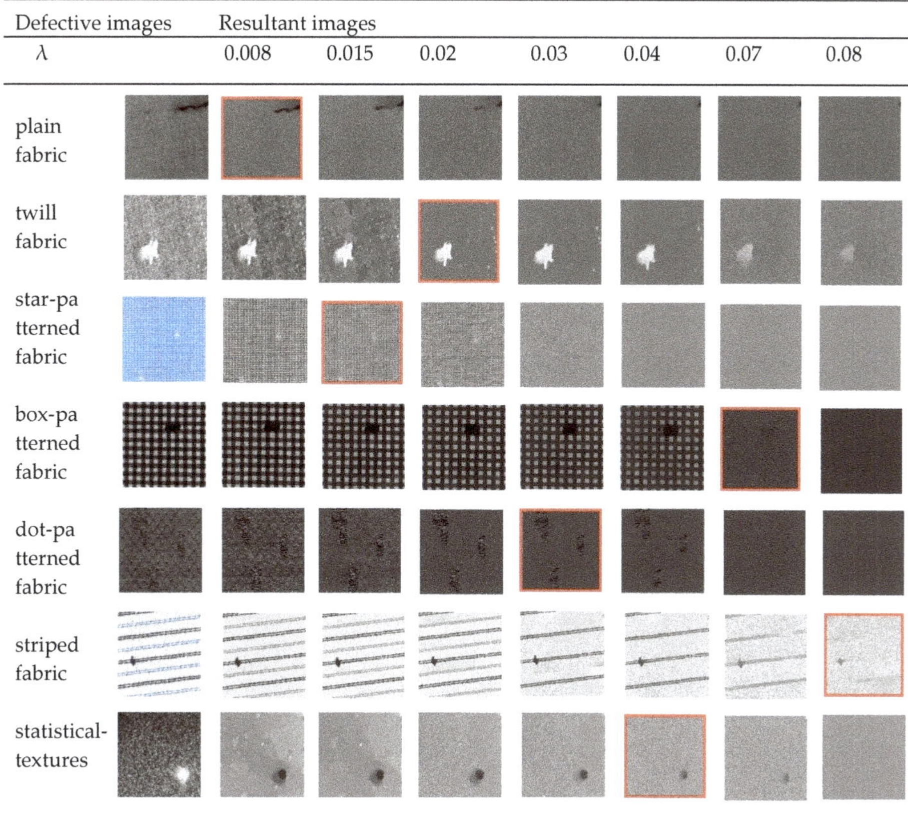

Figure 7. The optimum parameters about various fabric types.

To verify the impact of value of parameter λ on various fabric types, the parameter λ is 0.008, 0.015, 0.02, 0.03, 0.04, 0.07, and 0.08 respectively, and the seven fabric types' detection accuracies are presented by a bar chart as shown in Figure 8. The experiment proved that the number of λ selected should be between 0.008 and 0.08 to meet the requirements of various fabric types. As the Figure 8a,c shown, when λ is set between 0.07 and 0.08, the plain fabric and star-patterned fabric will be excessively smooth, so the accuracy rate cannot be detected.

4.2. Experimental Results

Figure 9 shows the results of plain fabric defect detection; it can be seen that the position and shape of fabric image defects have been successfully detected. Furthermore, nine twill fabric defects were tested, and the defect detection results are shown in Figure 10. The fabric defects were well detected; it reveals that the defection method can detect the twill fabric defects.

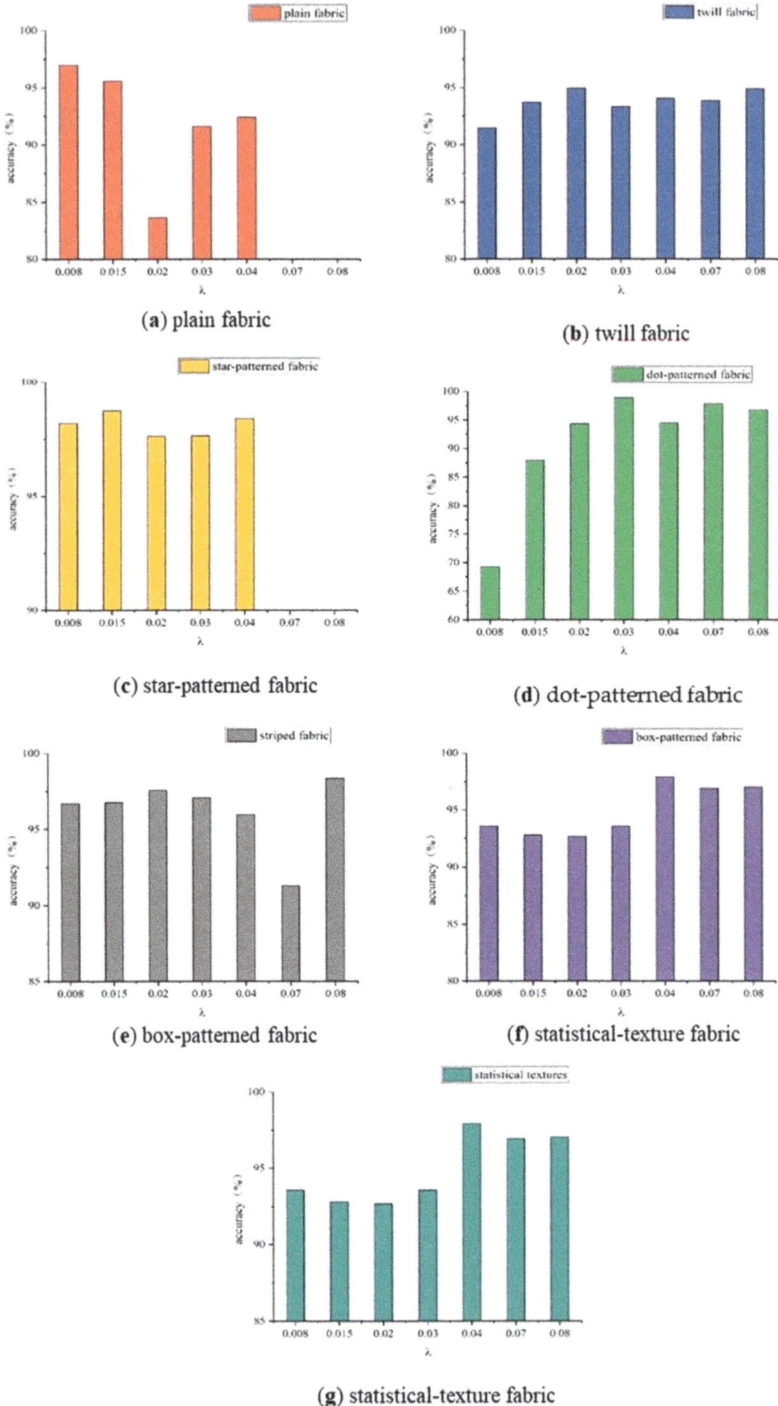

(**a**) plain fabric

(**b**) twill fabric

(**c**) star-patterned fabric

(**d**) dot-patterned fabric

(**e**) box-patterned fabric

(**f**) statistical-texture fabric

(**g**) statistical-texture fabric

Figure 8. The accuracy bar chart of various fabric types' detection under the number of different parameters λ.

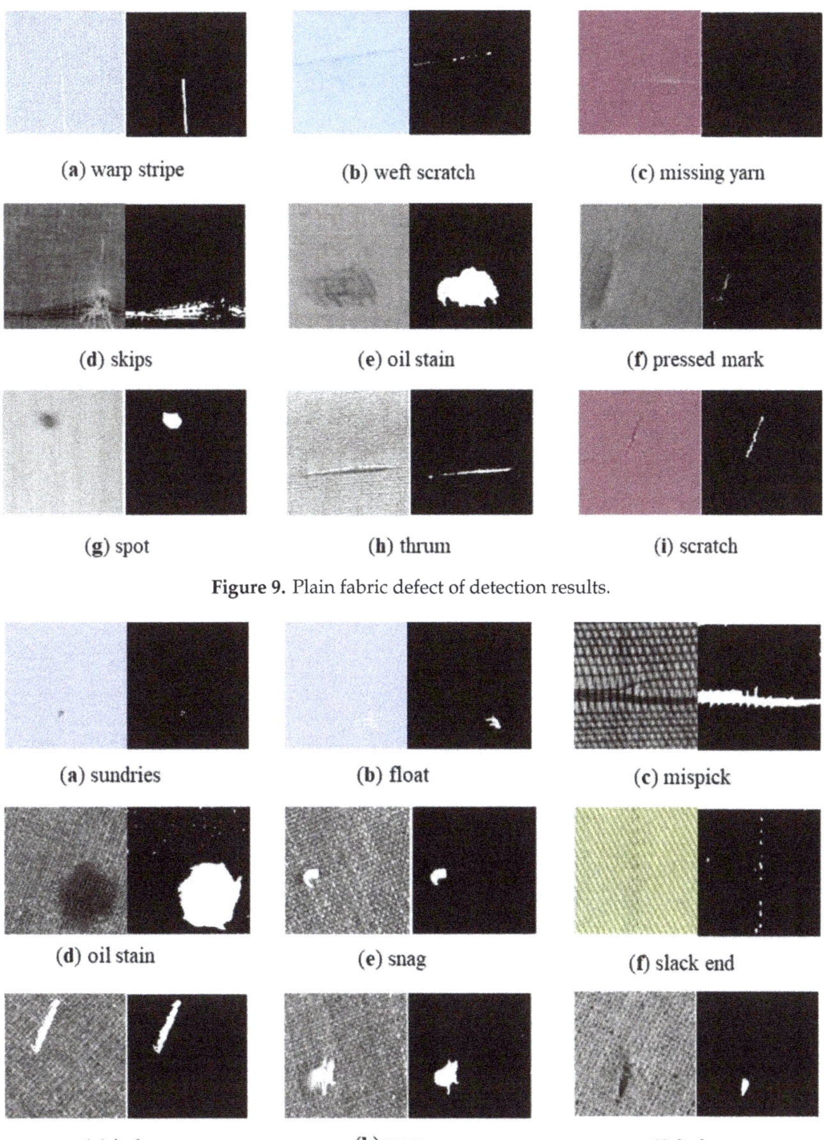

(a) warp stripe

(b) weft scratch

(c) missing yarn

(d) skips

(e) oil stain

(f) pressed mark

(g) spot

(h) thrum

(i) scratch

Figure 9. Plain fabric defect of detection results.

(a) sundries

(b) float

(c) mispick

(d) oil stain

(e) snag

(f) slack end

(g) hole

(h) snag

(i) hole

Figure 10. Twill fabric defect of detection results.

Star-patterned defects, such as linear defects and blob-shaped defects, are also successfully segmented using the proposed method, as shown in Figure 11. The results highlight the utility of our technique of accurate defect detection and segmentation. Our method can defect the box-patterned fabric defect, which includes broken ends, thick bars, and thin bars, as shown in Figure 12.

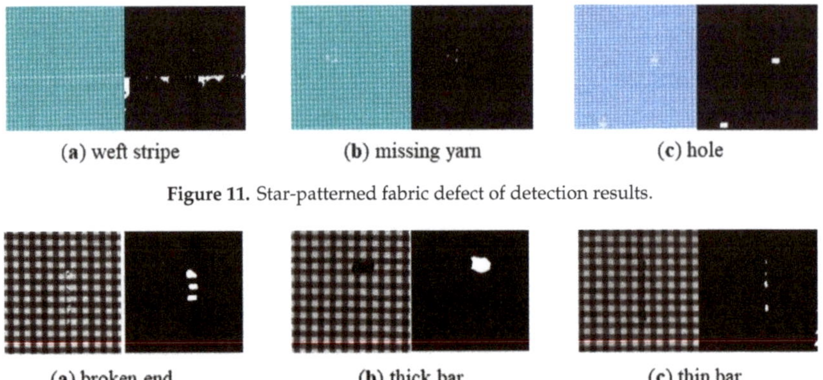

(a) weft stripe   (b) missing yarn   (c) hole

Figure 11. Star-patterned fabric defect of detection results.

(a) broken end   (b) thick bar   (c) thin bar

Figure 12. Box-patterned fabric defect of detection results.

The Figure 13 shows a representative set of different types of defects in dot-patterned fabric obtained from a group of images. In addition, even in a complicated background with a pattern, striped fabric, and statistical-textures, our method also can achieve an outperform the results of others, as shown in Figures 14 and 15. It can be seen that the proposed method can detect a variety of fabric samples with different defect types, shapes and textured backgrounds.

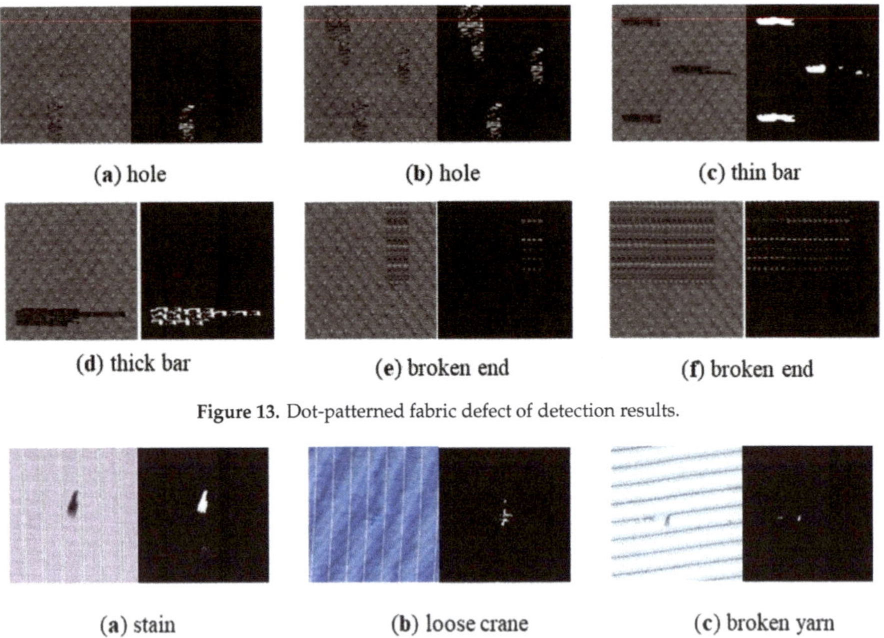

(a) hole   (b) hole   (c) thin bar

(d) thick bar   (e) broken end   (f) broken end

Figure 13. Dot-patterned fabric defect of detection results.

(a) stain   (b) loose crane   (c) broken yarn

Figure 14. Striped fabric defect of detection results.

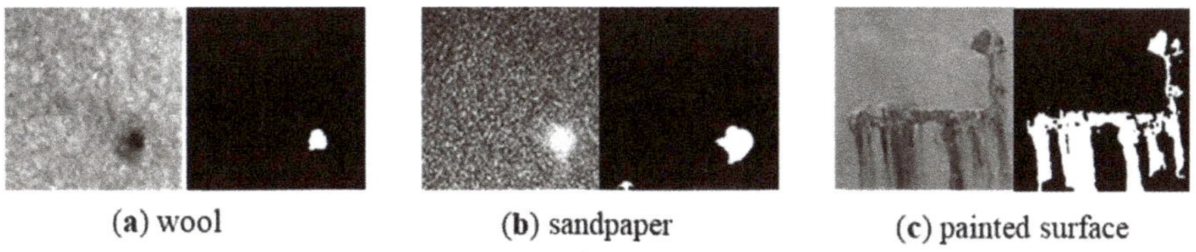

(a) wool   **(b) sandpaper**   **(c) painted surface**

Figure 15. Statistical-texture defect of detection results.

We further provide the computational time complexity of the proposed method, which is shown in Figure 16. We fit the curve. From the trend of the curve, the time complexity is $O(N^2)$, where N is the size of image.

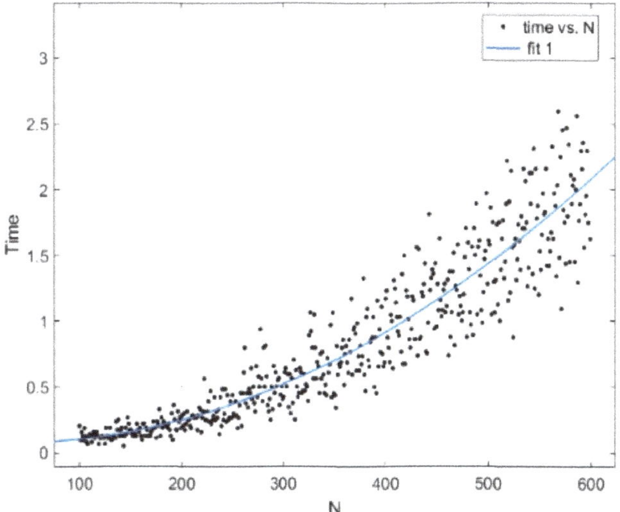

Figure 16. The computational time complexity curve line.

4.3. Qualitative Comparison

We compared our method with other state-of-the-art detection methods, including the EADL method [5] and the automatic band selection method [6], and detection carried out by human inspectors (the defects are marked by experienced factory workers). In each process, we used the parameters suggested in the original papers and followed the instructions provided in the authors' code distributions. The comparison results are shown in Figure 17, where the input original seven texture types' fabric defect images are given in Figure 17a. From top to bottom, they are plain fabric, twill fabric, star-patterned fabric, box-patterned fabric, dot-patterned fabric, statistical-texture fabric, and striped fabric respectively. The columns in Figure 17b–e are the defection results by our method, the ground-truth images from segmentation carried out by human inspectors, the Tsai [6], and the EADL method. It is found that the detection results using the EADL method [5] are located accuracy than using the automatic band selection method [6]. Figure 17b is the detection result using our method, it can be seen that the detection results are consistent with those carried out by human inspectors and outperform other methods.

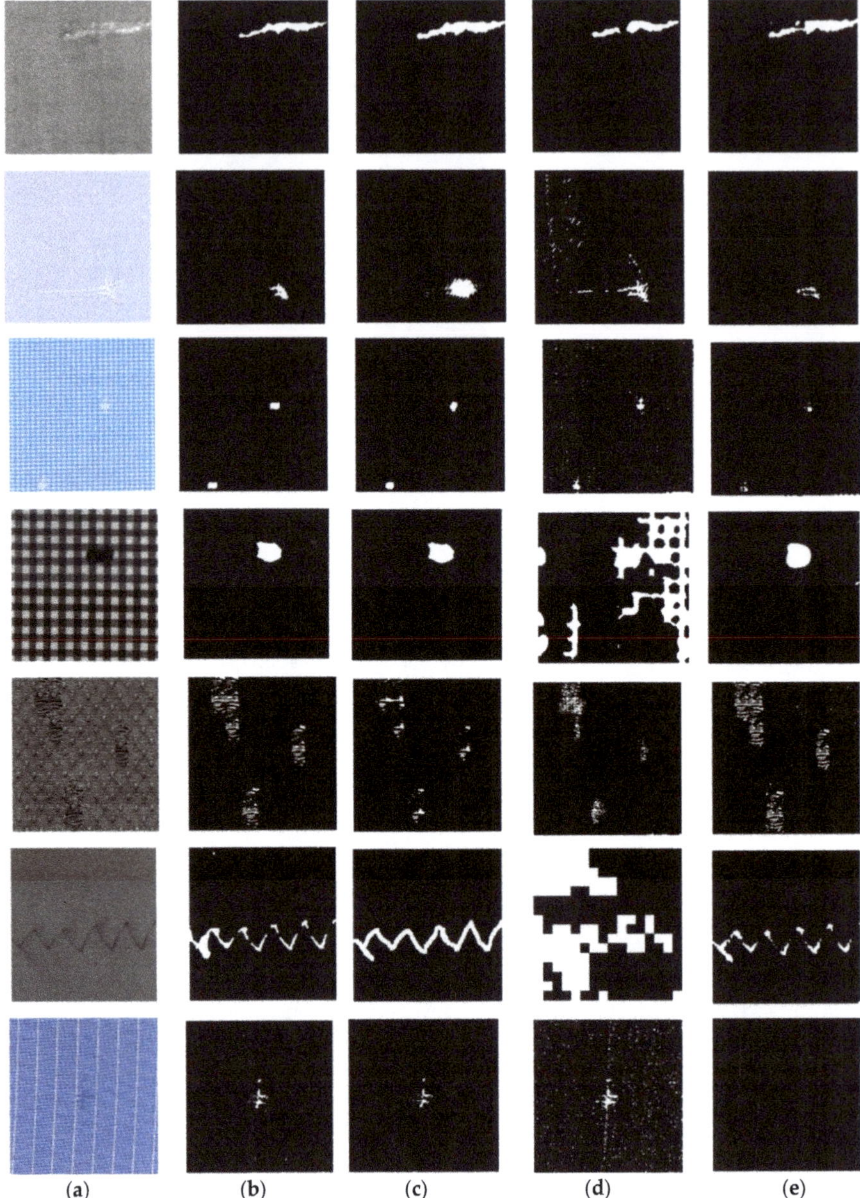

Figure 17. Various texture fabric defect detection results using different methods (**a**). Different fabric type defects (**b**). Our method (**c**) carried out by human inspectors. (**d**) The automatic band selection method [6]. (**e**) The EADL method [5].

4.4. Quantitative Comparison

Besides visual qualitative comparisons, we also did quantitative comparisons. A group of metrics, including accuracy (ACC), true positive rate (TPR), false positive rate (FPR), and positive predictive value (PPV), were employed to quantify the accuracies of the other methods. A number of measurement metrics; namely, true positive (TP), false positive (FP), true negative (TN), and false negative (FN), should be calculated. Based on those parameters, accuracy is calculated as: ACC = (TP + TN)/(TP + FN + FP + TN); true positive rate: TPR = TP/(TP + FN); false positive rate: FPR = FP/(FP + TN); positive predictive value: PPV = TP/(TP + FP). For these evaluation indicators, the proposed L0 gradient minimization and fuzzy c-means can obtain higher ACC and PPV values, and lower TPR, FPR values.

We also adopted the intersection-over-union (IOU) to quantitatively evaluate the performance of different methods. For the segmentation task, IOU is defined as: IOU = TP/(TP + FN + FP). For intersection-over-Union (IOU), the ideal case is a ratio of 100%, which usually stipulates that when the IOU value is greater than 50%, the detection is considered correct (which is also taken into consideration in the detection).

Figure 18 illustrates the ACC, TPR, FPR, PPV, and IOU results of the plain fabric, twill fabric, star-patterned fabric, dot-patterned fabric, box-patterned fabric, striped fabric, and statistical-texture fabric. Our method achieves most of the high scores in 35 testing items. The proposed method is better than other methods in Figure 18a,c–e. As shown in Figure 18a, the ACC by the proposed method, for the plain fabric dataset, twill fabric dataset, the dot-patterned fabric dataset, box-patterned fabric dataset, and statistical-textures defect dataset, are 96.99%, 98.40%, 98.87%, 96.13%, and 97.63%, respectively. As shown in Figure 18b, the proposed method can obtain the lowest TPR value for the star-patterned fabric defect detection. Figure 18c shows that the FPR values of our method are almost smaller than other approaches. Figure 18d shows that our method obtains the highest PPV value for the twill fabric, box-patterned fabric, striped fabric, and statistical-texture fabric. Figure 18e clearly indicates that our method provides the optimal IOU for all types of fabric. Furthermore, it can be observed that the proposed method obtains higher ACC, PPV, and IOU values, and lower TPR, and FPR values. These results verify the effectiveness of our proposed method, which performed better than the EADL method [5] and the automatic band selection method [6].

Considering the fact that the captured fabric images are often affected by noise, light intensity, and blurring, we analyzed the robustness of each method in different conditions. Table 1 shows the detection results of different methods (the proposed method, the EADL method [5], and the automatic band selection method [6]) in noisy, luminously intense, and blurry conditions. According to Table 1, when signal-noise ratio (SNR) decreases gradually, the ACC and IOU can remain a high level, especially when SNR = 10 dB; then, ACC can remain around 0.85. It was shown that the proposed method is robust when dealing with noise. In addition, we found that when luminous intensity decreases 20% or increases 20%, ACC can remain above 0.90. When increasing the blur with a radius of 20, the ACC and IOU can remain at a high level.

Table 1. Experimental results with different conditions.

Condition		Ours		Pedro [5]		Tsai [6]	
		ACC	IOU	ACC	IOU	ACC	IOU
Normal		0.9673	0.7575	0.9153	0.7707	0.9349	0.6599
SNR	20 dB	0.9489	0.6512	0.8346	0.7015	0.8773	0.5973
	15 dB	0.8976	0.6217	0.7793	0.6584	0.8315	0.5517
	10 dB	0.8532	0.5576	0.7544	0.5542	0.7966	0.5044
Luminous intensity	+20%	0.9245	0.6851	0.8972	0.6645	0.8645	0.5754
	−20%	0.9097	0.6544	0.8733	0.6497	0.8142	0.5945
Blur	Radius = 20	0.9456	0.6956	0.8546	0.6701	0.8599	0.5482

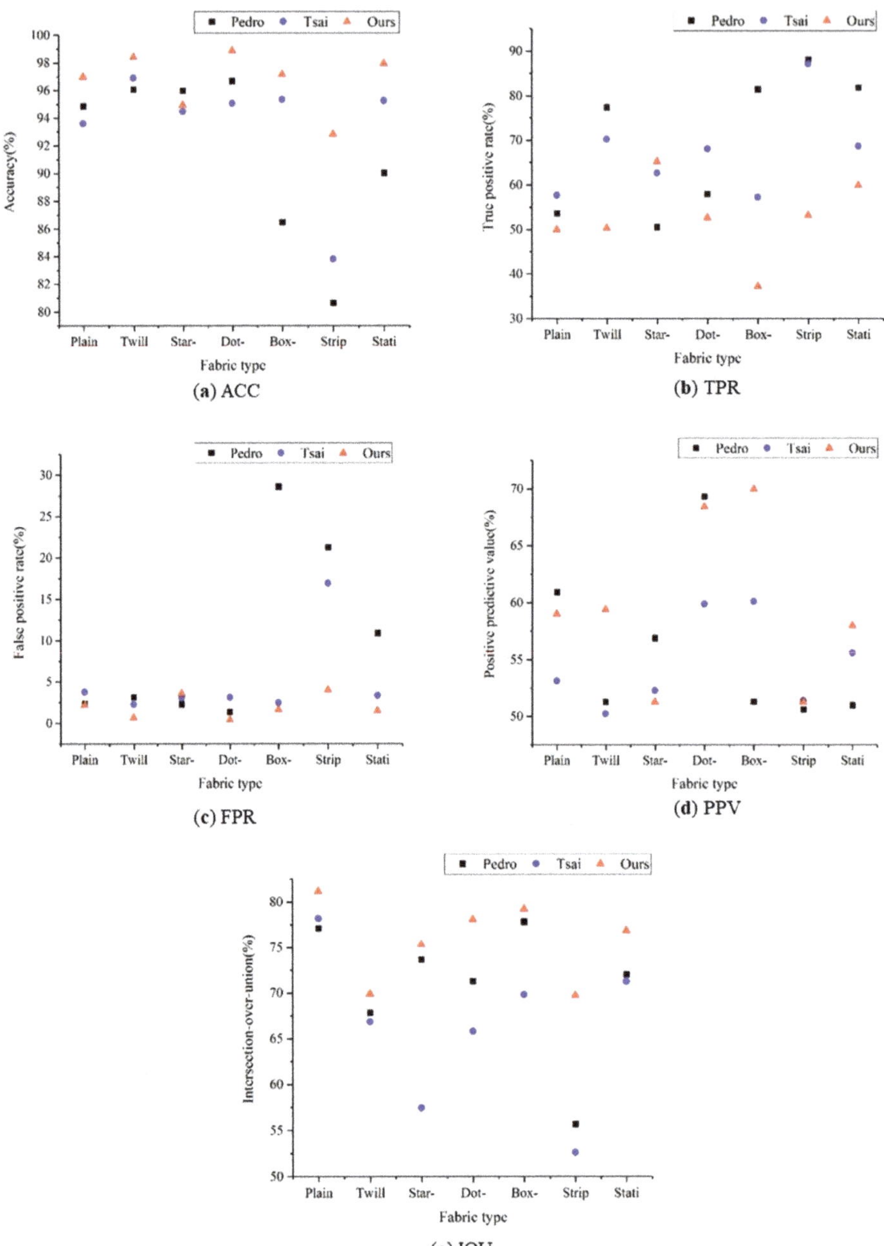

Figure 18. Inspection results of defect inspection methods with different methods on different types of fabric.

The computational comparison result is shown in the Table 2, which reports the average computational time (in seconds) of four methods while processing plain, twill, star-patterned, dot-patterned, box-patterned, striped, and statistical fabrics. As can be seen from Table 2, our method is faster than the automatic band selection method [6] and human inspectors. Even though the EADL method is faster than our method, it has fatal limitations. Their method cannot segment defects, as shown in Figure 17d, and cannot maintain high accuracy, as shown in Figure 18a. From the average calculation speed of the algorithm, the method proposed in this paper takes less time in comparison with the other methods when detecting various types of textured fabrics. In addition, our method performed better than Pedro [5] in terms of TPR, FPR, and IOU.

Table 2. Comparison of average computational time by four different methods.

	Plain	Twill	Star-	Dot-	Box-	Strip.	Stati.	Average Time/s
Pedro [5]	1.54	**0.98**	9.90	**1.85**	4.38	**4.99**	0.97	3.52
Tsai [6]	4.32	4.99	8.52	6.11	4.78	8.12	10.34	6.74
Human inspectors	8.19	10.11	16.78	8.53	9.78	15.96	39.85	15.6
Ours	**1.32**	1.50	**7.56**	2.54	**3.01**	5.12	**1.52**	**3.22**

5. Conclusions

We have proposed a novel method based on LGM and the FCM for fabric defect detection of a wide variety of textures. Extensive experimental results demonstrate that the proposed method could be applied to detect and segment fabric defects from a broad range of fabric defects datasets: Plain fabric, twill fabric, star-patterned fabric, dot-patterned fabric, box-patterned fabric, striped fabric, and statistical-texture fabric, with different defect types and shapes. It can achieve more accurate defect detection than other state-of-art competitors. Despite the effectiveness of the proposed method for fabric images with complicated patterns, it is still clumsy in computational time for detecting the defects. Our future work will be to improve our algorithm, to reduce the computational time in a real-time fabric defect detection system.

Author Contributions: H.Z. and J.M. conceived and designed the experiments; H.Z. performed the experiments; H.Z., J.J., and P.L. analyzed the data; H.Z. and J.M. wrote the paper.

Funding: This research is supported in part by National Natural Science Foundation of China under Grant No.61902302; in part by grants from the Shaanxi Provincial College of Science and Technology Youth Talent Support Project (Grant Number 20180115), the Science Research Program Funded by Shaanxi Provincial Education Department (Grant number 18JK0338), and the Youth Innovation Team of Shaanxi Universities.

Conflicts of Interest: The authors declare no conflict of interest.

References

1. Wang, M.J.J.; Huang, C.L. Evaluating the eye fatigue problem in wafer inspection. *IEEE Trans. Semicond. Manuf.* **2004**, *17*, 444–447. [CrossRef]
2. Ngan, H.Y.T.; Pang, G.K.H.; Yung, N.H.C. Automated fabric defect detection-A review. *Image Vis. Comput.* **2011**, *29*, 442–458. [CrossRef]
3. Susan, S.; Sharma, M. Automatic texture defect detection using Gaussian mixture entropy modeling. *Neurocomputing* **2017**, *239*, 232–237. [CrossRef]
4. Chen, L.C.; Papandreou, G.; Kokkinos, I.; Murphy, K.; Yuille, A.L. DeepLab: Semantic image segmentation with deep convolutional nets, atrous convolution, and fully connected crfs. *IEEE Trans. Pattern Anal. Mach. Intell.* **2018**, *40*, 834–848. [CrossRef] [PubMed]
5. Navarro, P.J.; Fernández-Isla, C.; Alcover, P.M.; Suardíaz, J. Defect detection in textures through the use of entropy as a means for automatically selecting the wavelet decomposition level. *Sensors* **2016**, *16*, 1178. [CrossRef] [PubMed]
6. Tsai, D.M.; Chiang, C.H. Automatic band selection for wavelet reconstruction in the application of defect detection. *Image Vis. Comput.* **2003**, *21*, 413–431. [CrossRef]
7. Ngan, H.Y.T.; Pang, G.K.H. Novel method for patterned fabric inspection using Bollinger bands. *Opt. Eng.* **2016**, *45*, 087202.
8. Ng, M.K.; Ngan, H.Y.T.; Yuan, X.; Zhang, W. Patterned fabric inspection and visualization by the method of image decomposition. *IEEE Trans. Autom. Sci. Eng.* **2014**, *11*, 943–947. [CrossRef]

9. Danesh, M.; Khalili, K. Determination of tool wear in turning process using undecimated wavelet transform and textural features. *Procedia Technol.* **2015**, *19*, 98–105. [CrossRef]

10. Li, P.F.; Zhang, H.H.; Jing, J.F.; Li, R.Z.; Zhao, J. Fabric defect detection based on multi-scale wavelet transform and Gaussian mixture model. *J. Text. Inst.* **2014**, *106*, 1–6. [CrossRef]

11. Qian, W. Research on Fabric Defect Detection Based on Neural Network. Master's Thesis, Donghua University, Shanghai, China, 2018.

12. Li, Y.; Zhao, W.; Pan, J. Deformable patterned fabric defect detection with fisher criterion-based deep learning. *IEEE Trans. Autom. Sci. Eng.* **2017**, *14*, 1256–1264. [CrossRef]

13. Huang, X.; Chen, L.; Tian, J.; Zhang, X. Blind image noise level estimation using texture-based eigenvalue analysis. *Multimed. Tools Appl.* **2016**, *75*, 2713–2724. [CrossRef]

14. Dutta, S.; Pal, S.K.; Sen, R. On-machine tool prediction of flank wear from machined surface images using texture analyses and support vector regression. *Precis. Eng.* **2015**, *43*, 34–42. [CrossRef]

15. Hoseini, E.; Farhadi, F.; Tajeripour, F. Fabric defect detection using auto-correlation function. *Int. J. Comput. Theory Eng.* **2013**, *5*, 114–117. [CrossRef]

16. Zhang, E.H.; Zhang, Y.; Duan, J.H. Color inverse half-toning method with the correlation of multi-color components based on extreme learning machine. *Appl. Sci.* **2019**, *9*, 841. [CrossRef]

17. Singh, S.; Kaur, M. Machine vision system for automated visual inspection of tile's surface quality. *IOSR J. Eng.* **2012**, *2*, 429–432. [CrossRef]

18. Chetverikov, D. Pattern regularity as a visual key. *Image Vis. Comput.* **2000**, *18*, 975–985. [CrossRef]

19. Narendra, V.G.; Hareesh, K.S. Study and comparison of various image edge detection techniques used in quality inspection and evaluation of agricultural and food products by computer vision. *Int. J. Agric. Biol. Eng.* **2011**, *4*, 83–90.

20. Yang, X.Z.; Pang, G.K.H.; Yung, N.H.C. Discriminative fabric defect detection using directional wavelets. *Opt. Eng.* **2002**, *41*, 3116–3126. [CrossRef]

21. Kuo, C.J.; Su, T.L. Gray relational analysis for recognizing fabric defects. *Text. Res. J.* **2003**, *73*, 461–465. [CrossRef]

22. Kuo, C.F.J.; Lee, C. A back-propagation neural network for recognizing fabric defects. *Text. Res. J.* **2003**, *73*, 147–151. [CrossRef]

23. Li, M.; Cui, S.; Xie, Z. Application of Gaussian mixture model on defect detection of print fabric. *J. Text. Res.* **2015**, *36*, 94–98.

24. Bodnarova, A.; Bennamoun, M.; Kubik, K.K. Defect detection in textile materials based on aspects of the HVS. In Proceedings of the 1998 IEEE International Conference on Systems, Man, and Cybernetics (Cat. No.98CH36218), San Diego, CA, USA, 14 October 1998.

25. Ngan, H.Y.H.; Pang, G.K.H. Regularity analysis for patterned texture inspection. *IEEE Trans. Autom. Sci. Eng.* **2009**, *6*, 131–144. [CrossRef]

26. Kang, X.J.; Zhang, E.H. A universal defect detection approach for various types of fabrics based on the Elo-rating algorithm of the integral image. *Text. Res. J.* **2019**. [CrossRef]

27. Xu, L.; Lu, C.; Xu, Y.; Jia, J. Image smoothing via L0 gradient minimization. *ACM Trans. Graph.* **2011**, *30*, 1–12.

28. Yang, M.S.; Nataliani, Y. Robust-learning fuzzy c-means clustering algorithm with unknown number of clusters. *Pattern Recognit.* **2017**, *71*, 45–59. [CrossRef]

29. Ding, Y.; Fu, X. Kernel-Based fuzzy c-means clustering algorithm based on genetic algorithm. *Neurocomputing* **2015**, *188*, 233–238. [CrossRef]

Article

The Self-Calibration Method for the Vertex Distance of the Elliptical Paraboloid Array

Zekui Lv [1,†], Zhikun Su [1,†], Dong Zhang [1], Lingyu Gao [1], Zhiming Yang [1], Fengzhou Fang [2,*], Haitao Zhang [3,*] and Xinghua Li [1,*]

[1] State Key Laboratory of Precision Measuring Technology and Instruments, School of Precision Instruments and Opto-electronics Engineering, Tianjin University, Tianjin 300072, China
[2] Centre of Micro/Nano Manufacturing Technology, University College Dublin, D04 V1W8 Dublin, Ireland
[3] Key Laboratory of Advanced Transducers and Intelligent Control System, Ministry of Education, Taiyuan University of Technology, Taiyuan 030024, China
* Correspondence: Fengzhou.Fang@ucd.ie (F.F.); zhanghaitao@tyut.edu.cn (H.Z.); lixinghua@tju.edu.cn (X.L.)
† These authors contributed equally to this work.

Received: 21 July 2019; Accepted: 21 August 2019; Published: 23 August 2019

Abstract: The elliptical paraboloid array plays an important role in precision measurement, astronomical telescopes, and communication systems. The calibration of the vertex distance of elliptical paraboloids is of great significance to precise 2D displacement measurement. However, there are some difficulties in determining the vertex position with contact measurement. In this study, an elliptical paraboloid array and an optical slope sensor for displacement measurement were designed and analyzed. Meanwhile, considering the geometrical relationship and relative angle between elliptical paraboloids, a non-contact self-calibration method for the vertex distance of the elliptical paraboloid array was proposed. The proposed self-calibration method was verified by a series of experiments with a high repeatability, within 3 μm in the X direction and within 1 μm in the Y direction. Through calibration, the displacement measurement system error was reduced from 100 μm to 3 μm. The self-calibration method of the elliptical paraboloid array has great potential in the displacement measurement field, with a simple principle and high precision.

Keywords: elliptical paraboloid array; self-calibration method; vertex distance; optical slope sensor; geometric relationship; relative angle

---

1. Introduction

Sensor arrays are widely used in modern scientific research and industrial production [1–3]. Depending on the application requirements, sensor arrays can be designed with different geometries, including those that are linear [4], circular [5], planar [6], L-shaped [7], and so on. Rui et al. [8] proposed a capacitance-sensor-array-based imaging system to detect water leakage inside insulating slabs with porous cells. Tan et al. [9] developed a giant magneto resistance sensor array which included various types of small gaps, curling wires, wide fractures, and abrasion to detect defects in various types of wire rope. Gao et al. [10] utilized a large area sinusoidal grid surface array as the measurement reference of a surface encoder for multi-axis position measurement. Wu et al. [11] presented a novel instrumentation system, including an infrared laser source and a photodiode sensor array, to provide an accurate measurement of the gas void fraction of two-phase CO_2 flow. Dario [12] demonstrated a novel sensor array based on SnO_2, CuO, and WO_2 nanowires, which is able to discriminate four typical compounds added to food products. Atte et al. [13] demonstrated that the bioimpedance sensor array has the ability to achieve long-term monitoring of intact skin and acute wound healing from beneath primary dressings. Forough et al. [14] developed a fluorometric sensor array for the detection of TNT (trinitrotoluene), DNT (dinitrotoluene), and TNP (trinitrophenyl), which demonstrated a promising capacity to detect structurally similar nitroaromatics in mixtures and the complex media of soil and groundwater samples. The advantage of using a sensor array instead of a single sensor is that the array adds new dimensions to the observations, helping to estimate more parameters and improve the estimation performance. Of course, the calibration of relevant parameters is a requirement that cannot be ignored before using the sensor array.

Zhang et al. [15] described a method for calibrating the distances between the balls of a 1-D ball array using a specially designed device with a laser interferometer, which gives the distances between the balls in two directions. Ouyang et al. [16] proposed a new alternative method for calibrating the ball array based on a CMM (coordinate measuring machine) and a gage block. To obtain a high accuracy in ball calibration, repeated measurements of the balls must be taken at least 10 times. Guenther et al. [17] introduced a self-calibration method for a ball plate, through which not only the pitch position of the balls, but also their radial and height position on the circular ball plate, were calibrated. Through computing the intersections of 2π-phase lines to detect the phase-shifting wedge grating arrays centers, Tao et al. [18] completed camera calibration. Xu et al. [19] presented a calibration method for a camera array and a rectification method for generating a light field image from the captured images. Zhang et al. [20] used a high-precision machine developed by the research of Halcon to measure the aperture diameter and aperture distance of a microstructure array. Solórzano et al. [21] presented a calibration model that can be extended to uncalibrated replicas of sensor arrays without acquiring new samples, favoring mass-production applications for gas sensor arrays. Sun et al. [22] purposed a new array geometry calibration method for underwater compact arrays to improve the robustness and accuracy of the calibration results. Zhai et al. [23] researched a calibration method for an array complementary metal–oxide–semiconductor photodetector using a black-box calibration device and an electrical analog delay method to address the disadvantages of traditional PMD (photonic mixer device) solid-state array lidar calibration methods. Through the above research, it can be concluded that the calibration of sensor arrays is helpful to improve their accuracy and practicability.

Elliptical paraboloid arrays have considerable application potential in the accurate detection of machine geometry errors. Nowadays, the elliptical paraboloid used in the field of precision measurement has a small surface area and high machining accuracy. To expand the measuring range of displacement, the elliptical paraboloid array has become a research focus. However, the calibration method for the vertex distance of the elliptical paraboloid array has seldom been investigated. There are two reasons accounting for this. On the one hand, contact measurement will destroy the surface properties of the elliptical paraboloid, and they are too small to be detected by probe. On the other hand, traditional visual inspection methods are often affected by specular images and ignore the relative angle between the elliptical paraboloids.

To overcome the above difficulties, a self-calibration method for the vertex distance of the elliptical paraboloid array using a non-contact optical slope sensor is proposed. A self-calibration method utilizes the existing displacement measurement system to achieve calibration with a low cost and convenient operation. The relative angle and geometric relationship between elliptical paraboloids was considered in the self-calibration method to improve the calibration accuracy. This self-calibration method has the characteristics of a high accuracy and high repeatability, is non-destructive, and has a long adaptive distance. After calibration, the displacement measurement system error was reduced from 100 μm to 3 μm, which satisfies the measuring requirement.

2. Materials and Methods

2.1. Elliptical Paraboloid Array

The elliptical paraboloid is a common type of quadric surface, which has wide application prospects in the field of precision measurement. As shown in Figure 1a, the three-dimensional model of the elliptical paraboloid with a superior optical reflection performance was machined by an ultra-precision single-point diamond lathe. The processing quality of the elliptical paraboloid was detected by an optical 3D surface profiler, and the results are shown in Figure 1b. In Figure 1b, different colors in the image represent different depths. According to Figure 1b, it can be seen that the machining height (80.679 μm) and contour of the elliptical paraboloid are consistent with the theoretical design.

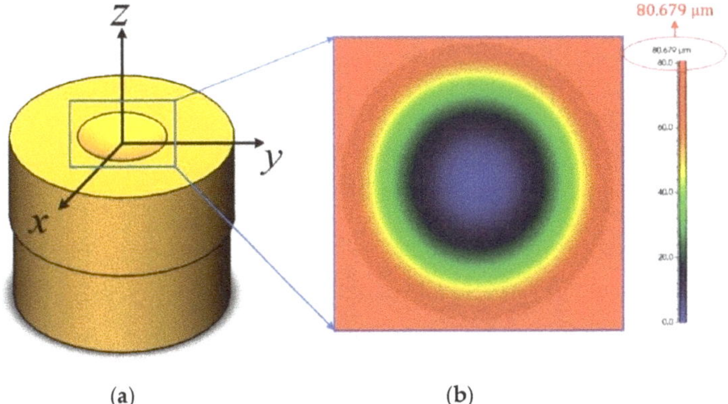

(a)   (b)

Figure 1. The model of the elliptical paraboloid and detection results of the 3D surface profile: (**a**) the model of the elliptical paraboloid; (**b**) the 3D surface profile for the model.

Lv et al. have described in detail the principle of 2D micro-displacement based on a single elliptical paraboloid [24]. However, the range of this measurement system is limited by the surface area of the elliptical paraboloid. A larger elliptical paraboloid will cause inconvenience in manufacturing and application. In this paper, a linear elliptical paraboloid array is proposed, in which the distance of the centers between adjacent elliptical paraboloids is only 50 mm in the X direction. As shown in Figure 2, due to the existence of machining errors and installation errors, the vertex distance of the actual elliptical paraboloid array may be biased in both the X and Y directions. Therefore, it is our responsibility to accurately calibrate the vertex distance between elliptical paraboloids in the array to reduce systematic errors in actual applications.

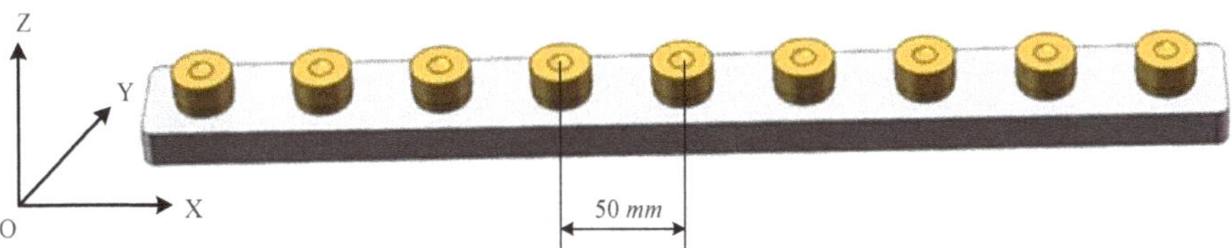

Figure 2. The linear elliptical paraboloid array.

2.2. Optical Slope Sensor

The optical slope sensor used for the self-calibration method consists of a laser, a reflector, a spatial filter, a beam splitter, two lenses, a CCD (charge coupled device) camera, and more, as shown in Figure 3. The laser emitted from a laser source (power 5 mW, wavelength $\lambda = 650$ nm, unpolarized light) is reflected by a reflector (offset 135° angle in a horizontal direction) and passes through a spatial filter (diameter 300 μm). After going through the spatial filter and focusing Lens 1 (focal length 30 mm), the laser beam is divided into two beams by the beam splitter (reflection factor of 50%). The reflected beam, as the measuring beam, reaches the surface of the elliptical paraboloid array and is reflected again. Finally, the measuring beam is received by a CCD camera (pixels number 2592×1944, pixel size 2.2×2.2 μm) located on the focal plane of object Lens 2 (focal length 100 mm) and is transmitted to the computer for analysis and processing. The actual spot image is displayed on the computer screen in Figure 3, whose central position will be determined by the centroid algorithm. Through the processing of the spot image, the angle and displacement can be calculated precisely. In short, the optical slope sensor is simple, portable, and accurate.

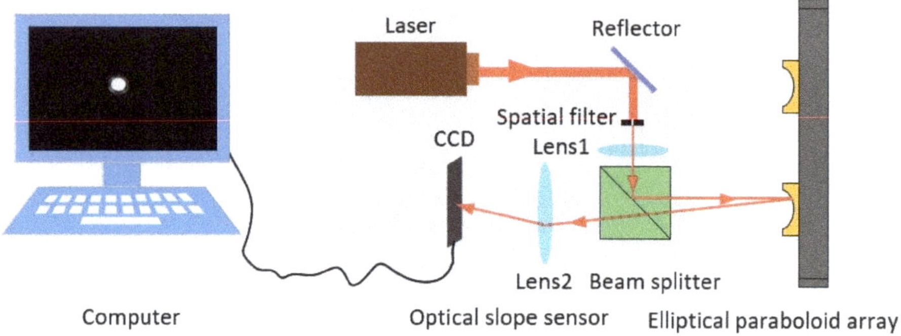

Figure 3. Schematic diagram of the optical slope sensor.

2.3. Experimental Apparatus

Experiments to certify the self-calibration method for the vertex distance of the elliptical paraboloid array were undertaken several times with the help of an XL80 laser interferometer with a resolution of 0.001 μm and system accuracy of ±0.5 ppm (Renishaw, Gloucestershire, UK), and a Brown & Sharpe Chameleon 9159 bridge-type coordinate measuring machine (CMM) with a technical target of $U12.3 + 2.8 \, L/1000$ (Hexagon, Stockholm, Sweden). The experimental apparatus of the self-calibration system is shown in Figure 4. The optical slope sensor mounted on the quill of the CMM moves along the direction of the elliptical paraboloid array, whose direction is parallel to the X direction of the CMM. The function of the laser interferometer is to monitor the actual displacement of the quill for accurate calibration. It should be noted that the apparatus of the calibration experiment is also suitable for the measurement experiment, which is the reason the method is called self-calibration.

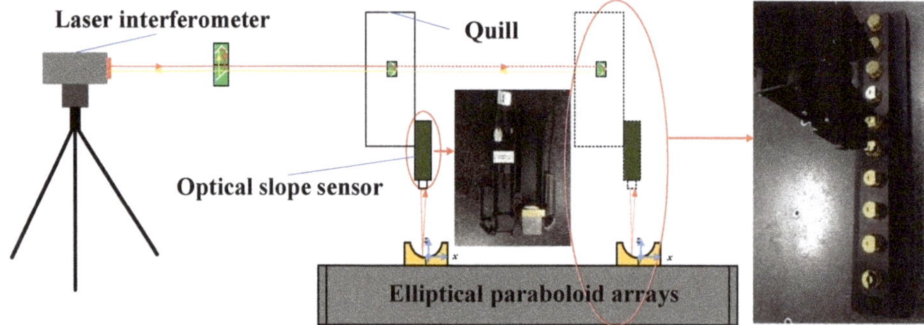

Figure 4. Experimental apparatus of the calibration system.

2.4. Mathematical Model

As shown in Figure 5, the vertex distance between elliptical paraboloid i and j is d_x and d_y in the X and Y direction, respectively. First, point A on the elliptical paraboloid i is measured by the optical slope sensor to obtain the light spot central position (x_{cA}, y_{cA}). Then, an optical slope sensor moves about 50 mm along the X direction to point B on the elliptical paraboloid j and the central position (x_{cB}, y_{cB}) is acquired. Point A′ is the mapping point of point A, which means that point A′ and point A will be in the same position if they are on an identical elliptical paraboloid. Therefore, the distance between point A′ and point A is the same as the vertex distance of two elliptical paraboloids.

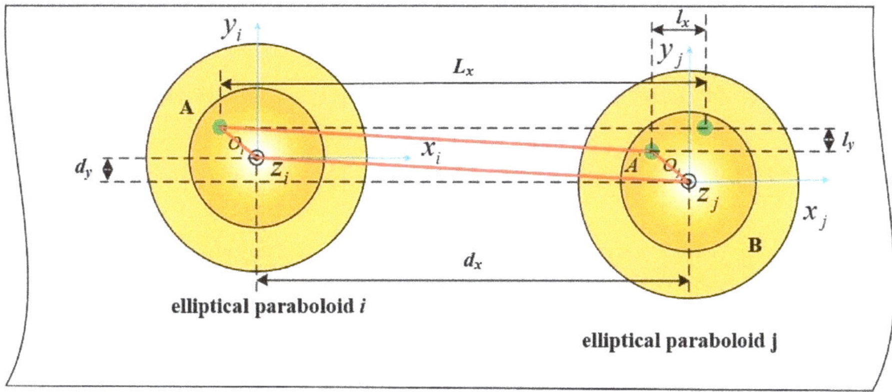

Figure 5. Calibration principle according to the geometric relationship.

According to the geometric relationship in Figure 5, the following equation can be determined:

$$\begin{cases} d_x = L_x - l_x \\ \quad d_y = l_y \end{cases} \tag{1}$$

where l_x and l_y are the distance between point A′ and B, and L_x is the length between point A and B along the X direction monitored by a laser interferometer. l_x and l_y are calculated by displacement of the spot, as follows:

$$\begin{cases} l_x = (x_{cB} - x_{cA})c_1 \\ l_y = (y_{cB} - y_{cA})c_1 \end{cases} \tag{2}$$

where c_1 represents the displacement coefficient that needs to be calibrated.

The above model used the calibration formula in the case of only translation occurring between the elliptical paraboloids. However, the relative angular (pitch and yaw) elliptical Y and X axis between them is neglected. The pitch between two elliptical paraboloids is shown in Figure 6, which will result in the influence on spot displacement. The relative pitch ($\Delta\theta$) and yaw ($\Delta\varphi$) between each elliptical paraboloid should be calibrated, and their influence on displacement can be calculated by the following formula:

$$\begin{cases} x_{cA'} - x_{cA} = \Delta\theta/c_2 \\ y_{cA'} - y_{cA} = \Delta\varphi/c_2 \end{cases} \tag{3}$$

where c_2 represents the angle coefficient that needs to be calibrated.

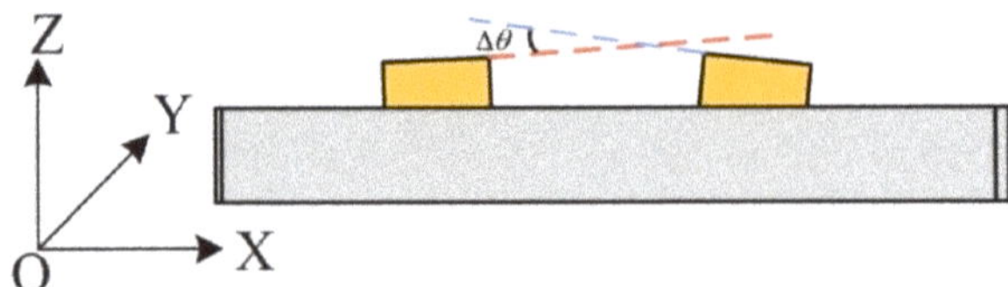

Figure 6. The pitch between two elliptical paraboloids.

Finally, the calibration formula for the vertex distance between paraboloids can be expressed as

$$\begin{cases} d_x = L_x - (x_{cB} - x_{cA} - \Delta\theta/c_2)c_1 \\ d_y = (y_{cB} - y_{cA} - \Delta\varphi/c_2)c_1 \end{cases} \tag{4}$$

2.5. Calibration Procedure

The procedure of calibration for the elliptical paraboloid array is shown in Figure 7.

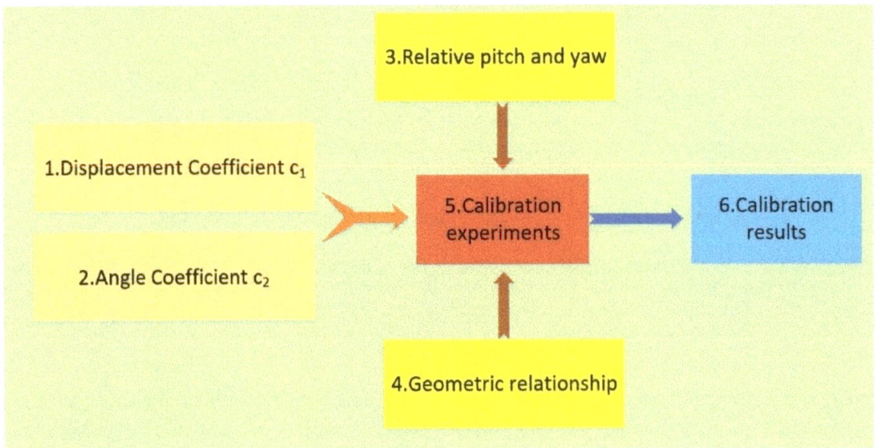

Figure 7. Calibration procedure.

Step one: The displacement coefficient c_1 is calibrated. The displacement coefficient c_1 refers to the actual displacement (μm) change corresponding to a change of a 1 μm length of the spot on the CCD camera. The calibration results for the displacement coefficient are shown in Figure 8. We verified that the displacement coefficient is about 3.5045 from the linear fitting equation, with a correlation coefficient of 0.9999.

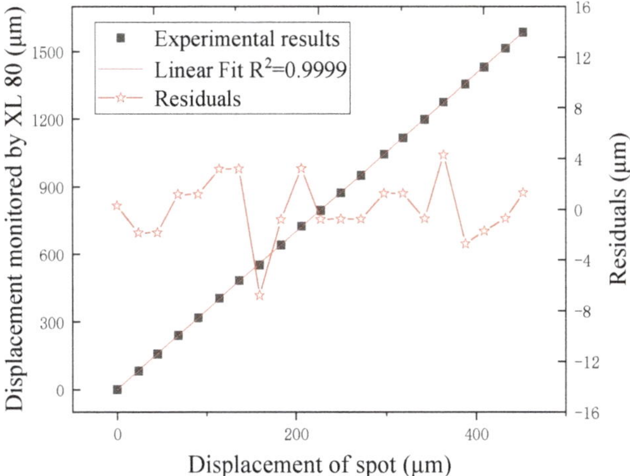

Figure 8. Calibration results in the X direction.

Step two: The angle coefficient c_2 is calibrated. The angle coefficient c_2 is described as the angular (arcsec) change associated with the 1 μm spot position on the CCD camera. The experimental results for angle coefficient calibration are shown in Figure 9. Figure 9 indicates that the result of the angle coefficient is approximately 4.975 from the linear fitting equation, with a correlation coefficient of 0.9999.

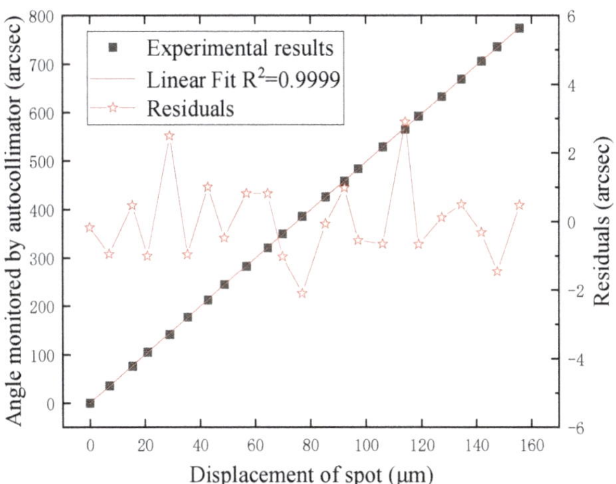

Figure 9. Calibration results in the Y direction.

Step three: Calculate the relative angles between elliptical paraboloids, including the pitch angle and yaw angle, using the optical slope sensor.

Step four: Consider the geometric relationship between the vertices and measuring points of the elliptical paraboloids, which is the basic principle of the whole calibration method.

Step five: Conduct calibration experiments. The quill of CMM moves 50 mm each time to reach the position of the next elliptical paraboloid; meanwhile, the light spot central position is recorded.

Step six: Analyze and compare calibration results. The calibration results require analysis and comparison to prove their correctness, especially in practical applications.

3. Experimental Results

3.1. Relative Pitch and Yaw

Before the experiment, it was obvious that the ring surrounding the elliptical paraboloid was machined into a mirror plane whose *pv* (peak to valley) was 0.038 µm. The reflection characteristics of the rings were similar to a K9 standard precision planar mirror. Therefore, the optical angle sensor based on the self-collimation principle and the ring could be utilized to measure the two-dimensional angle (pitch and yaw). Ultra-precision machining requires a guarantee that the optical axis of the elliptical paraboloid and the normal vector of the ring have good parallelism. Therefore, detection of the pitch and yaw of each elliptical paraboloid is replaced by each ring. The elliptical paraboloid array was placed along the X direction of the CMM, and the optical slope sensor was mounted on the quill of the CMM to perform multi-point measurement on each of the rings. Taking the first elliptical paraboloid as an angle reference, the relative pitch angle and yaw angle of other elliptical paraboloids could be obtained. The average result is shown in Table 1. According to Table 1, the relative pitch angle and yaw angle are clearly calibrated within 200″, which is due to installation and adjustment of the elliptical paraboloid.

Table 1. Calibration results of the relative pitch angle and yaw angle.

Elliptical Paraboloid	Relative Pitch Angle/Arcsec	Relative Yaw Angle/Arcsec
1	0	0
2	−149.8	103.9
3	−81.7	124.3
4	176.7	−71.5
5	186.5	155.8
6	82.7	−34.0
7	107.1	−92.2
8	−44.1	166.7
9	−175.4	185.7

3.2. Calibration Results

The calibration experiment was carried out three times, and the experimental results are shown in Figure 10. In order to eliminate the influence of errors caused by a single measurement, the average of three measurements was taken as the final calibration result, which is shown in Table 2. As can be seen from Figure 10, the calibration results have a good repeatability in both the X and Y directions. Through three repetitive experiments, we can conclude that the difference between them was within 3 µm in the X direction and within 1 µm in the Y direction. The experimental results shown in Table 2 are approaching the theoretical design values in two directions, which directly proves the correctness of the calibration method.

Table 2. Calibration results of the elliptical paraboloid array.

Type	1	2	3	4	5	6	7	8	9
X/mm	0	49.9681	100.0044	149.9427	199.9558	250.0217	300.0113	350.0063	400.0362
Y/mm	0	0.0306	0.0398	0.0359	0.0194	−0.0158	−0.0377	0.0488	−0.0344

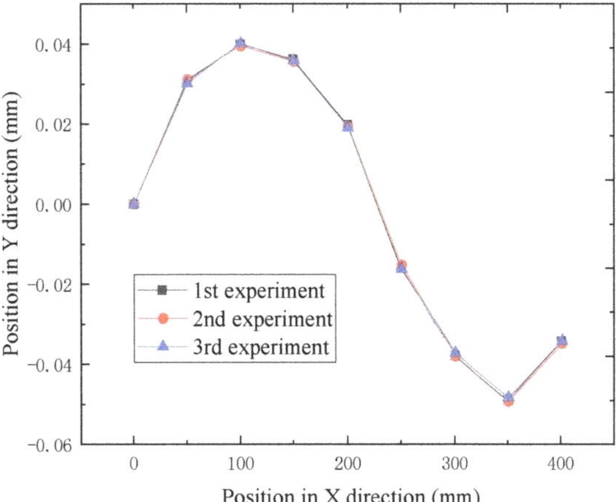

Figure 10. Repeatability of the calibration experiment.

3.3. Comparison Experiment

In order to verify the effect of system calibration, a comparison experiment were carried out in the same laboratory environment. The direction of the elliptical paraboloid array was adjusted to be parallel to the Y direction of the CMM. Correspondingly, the optical slope sensor was mounted on the quill to move in the Y direction. The measured data were processed with the calibration results and the original design values as reference values in turn. The comparison results in the moving direction are shown in Figure 11. As shown in Figure 11, the error of the measurement results before calibration is less than 100 μm due to the influence of installation and manufacture. However, the experimental results after system calibration show that the measurement error is controlled within 3 μm in the moving direction, which satisfies the requirement of the displacement measurement accuracy. Therefore, we can say that the necessity and correctness of system calibration have been proven by the comparison experiment. In fact, the calibration experiment results compensate for the displacement measurement error of the system in a sense.

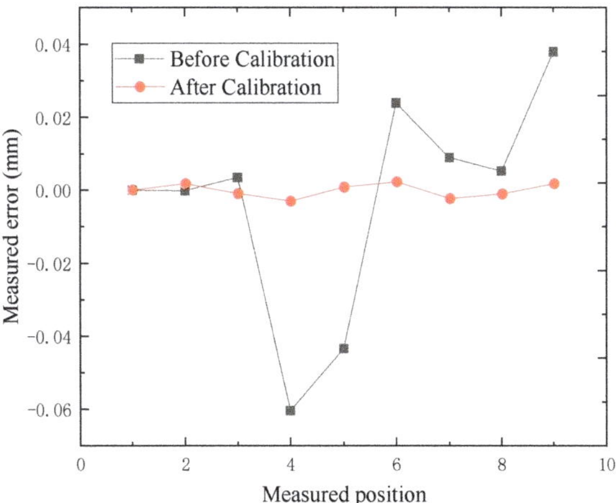

Figure 11. Comparison experiment.

4. Conclusions

The present study was designed to research a self-calibration method with respect to the vertex distance of the elliptical paraboloid array, which solves the benchmark problem in long-range displacement measurements. The self-calibration method, which was based on the geometric relationships between elliptical paraboloids, was verified by experiments using a Renishaw XL80, Hexagon CMM, and optical slope sensor designed by ourselves. The results of these experiments show that the calibration results are consistent with the design values, and the repeatability was within 3 μm in the X direction and within 1 μm in the Y direction. In addition, the comparison experiment proved that the displacement measurement system error was reduced from 100 μm to 3 μm after calibration. The results show that the self-calibration method can meet the calibration requirements of the elliptical paraboloid array, which lays a solid foundation for subsequent experiments related to displacement measurement. In order to improve the accuracy of the calibration results, it would be necessary to select high-precision motion guides and ensure that their motion direction is parallel to the elliptical paraboloid array. Further studies need to be carried out in order to validate the application value of the elliptical paraboloid array in the displacement-related measurement field, such as by focusing on the positioning error, straightness error, and perpendicularity error.

Author Contributions: X.L., F.F., and H.Z. proposed the method and modified the paper; Z.L. designed the experiments and wrote the paper; D.Z. and L.G. developed the system software and processed the data; Z.S. and Z.Y. designed the mechanical and optical structure.

Funding: This research was financially supported by the National Natural Science Foundation of China (NSFC) (No: 51775378); the Science Foundation Ireland (SFI) (No.15/RP/B3208); the National Key R&D Program of China (No.2017YFF0108102); and the Natural Science Foundation of Shanxi Province, China (Grant No. 201801D121180).

Conflicts of Interest: The authors declare no conflicts of interest.

References

1. GhazaI, N.; Ebrahim, G.-Z.; Antoine, L.; Mohamad, S. Smart Cell Culture Monitoring and Drug Test Platform Using CCD Capacitive Sensor Array. *IEEE Trans. Biomed. Eng.* **2019**, *66*, 1094–1104.
2. Jie, H.; Hanmin, P.; Ting, M.; Tingyu, L.; Mingsen, G.; Penghui, L.; Yalei, B.; Chunsheng, Z. An airflow sensor array based on polyvinylidene fluoride cantilevers for synchronously measuring airflow direction and velocity. *Flow Meas. Instrum.* **2019**, *67*, 166–175.
3. Kaisti, M.; Panula, T.; Leppänen, J.; Punkkinen, R.; Tadi, M.J.; Vasankari, T.; Meriheinä, U. Clinical assessment of a non-invasive wearable MEMS pressure sensor array for monitoring of arterial pulse waveform, heart rate and detection of atrial fibrillation. *NPJ Digit. Med.* **2019**, *39*, 1–10. [CrossRef] [PubMed]
4. Zhong, Y.; Xiang, J.; Chen, X.; Jiang, Y.; Pang, J. Multiple Signal Classification-Based Impact Localization in Composite Structures Using Optimized Ensemble Empirical Mode Decomposition. *Appl. Sci.* **2018**, *8*, 1447. [CrossRef]
5. Cui, X.; Yan, Y.; Guo, M.; Han, X.; Hu, Y. Localization of CO_2 Leakage from a Circular Hole on a Flat-Surface Structure Using a Circular Acoustic Emission Sensor Array. *Sensors* **2016**, *16*, 1951. [CrossRef] [PubMed]
6. Chung, S.; Park, T.; Park, S.; Kim, J.; Park, S.; Son, D.; Cho, S. Colorimetric Sensor Array for White Wine Tasting. *Sensors* **2015**, *15*, 18197–18208. [CrossRef]
7. Si, W.; Zhao, P.; Qu, Z. Two-Dimensional DOA and Polarization Estimation for a Mixture of Uncorrelated and Coherent Sources with Sparsely-Distributed Vector Sensor Array. *Sensors* **2016**, *16*, 789. [CrossRef]
8. Li, R.; Li, Y.; Peng, L. An Electrical Capacitance Array for Imaging of Water Leakage inside Insulating Slabs with Porous Cells. *Sensors* **2019**, *19*, 2514. [CrossRef]
9. Tan, X.; Zhang, J. Evaluation of Composite Wire Ropes Using Unsaturated Magnetic Excitation and Reconstruction Image with Super-Resolution. *Appl. Sci.* **2018**, *8*, 767. [CrossRef]
10. Gao, W.; Araki, T.; Kiyono, S.; Okazaki, Y.; Yamanaka, M. Precision nano-fabrication and evaluation of a large area sinusoidal grid surface for a surface encoder. *Precis. Eng.* **2003**, *27*, 289–298. [CrossRef]
11. Wu, H.; Duan, Q. Gas Void Fraction Measurement of Gas-Liquid Two-Phase CO_2 Flow Using Laser Attenuation Technique. *Sensors* **2019**, *19*, 3178. [CrossRef] [PubMed]

12. Zappa, D. Low-Power Detection of Food Preservatives by a Novel Nanowire-Based Sensor Array. *Foods* **2019**, *8*, 226. [CrossRef] [PubMed]

13. Kekonen, A.; Bergelin, M.; Johansson, M.; Kumar Joon, N.; Bobacka, J.; Viik, J. Bioimpedance Sensor Array for Long-Term Monitoring of Wound Healing from Beneath the Primary Dressings and Controlled Formation of H_2O_2 Using Low-Intensity Direct Current. *Sensors* **2019**, *19*, 2505. [CrossRef] [PubMed]

14. Ghasemi, F.; Hormozi-Nezhad, M.R. Determination and identification of nitroaromatic explosives by a double-emitter sensor array. *Talanta* **2019**, *201*, 230–236. [CrossRef] [PubMed]

15. Zhang, G.X.; Zang, Y.F. A method for machine geometry calibration using 1-D ball array. *CIRP Ann.* **1991**, *40*, 519–522. [CrossRef]

16. Ouyang, J.F.; Jawahir, I.S. Ball array calibration on a coordinate measuring machine using a gage block. *Measurement* **1995**, *16*, 219–229. [CrossRef]

17. Anke, G.; Dirk, S.; Gert, G. Self-Calibration Method for a Ball Plate Artefact on a CMM. *CIRP Ann.* **2016**, *65*, 503–506.

18. Tao, J.; Wang, Y.; Cai, B.; Wang, K. Camera Calibration with Phase-Shifting Wedge Grating Array. *Appl. Sci.* **2018**, *8*, 644. [CrossRef]

19. Xu, Y.; Maeno, K.; Nagahara, H.; Taniguchi, R.I. Camera array calibration for light field acquisition. *Front. Comput. Sci.* **2015**, *9*, 691–702. [CrossRef]

20. Zhang, X.; Liu, Q.; Yin, Z.; Zhao, R.; Lin, J. Research on In-situ Measurement System of Microstructure Array. *Modul. Mach. Tool Autom. Manuf. Tech.* **2018**, *6*, 93–97.

21. Solórzano, A.; Rodriguez-Perez, R.; Padilla, M.; Graunke, T.; Fernandez, L.; Marco, S.; Fonollosa, J. Multi-unit calibration rejects inherent device variability of chemical sensor arrays. *Sens. Actuators B Chem.* **2018**, *265*, 142–154. [CrossRef]

22. Sun, D.; Ding, J.; Zheng, C.; Huang, W. Array geometry calibration for underwater compact arrays. *Appl. Acoust.* **2019**, *145*, 374–384. [CrossRef]

23. Zhai, Y.; Song, P.; Chen, X. A Fast Calibration Method for Photonic Mixer Device Solid-State Array Lidars. *Sensors* **2019**, *19*, 822. [CrossRef] [PubMed]

24. Lv, Z.; Li, X.; Su, Z.; Zhang, D.; Yang, X.; Li, H.; Li, J.; Fang, F. A Novel 2D Micro-Displacement Measurement Method Based on the Elliptical Paraboloid. *Appl. Sci.* **2019**, *9*, 2517. [CrossRef]

Article

A Generic Automated Surface Defect Detection Based on a Bilinear Model

Fei Zhou, Guihua Liu *, Feng Xu and Hao Deng

School of Information Engineering, Southwest University of Science and Technology, Mianyang 621010, China
* Correspondence: liughua_swit@163.com

Received: 4 June 2019; Accepted: 31 July 2019; Published: 3 August 2019

Abstract: Aiming at the problems of complex texture, variable interference factors and large sample acquisition in surface defect detection, a generic method of automated surface defect detection based on a bilinear model was proposed. To realize the automatic classification and localization of surface defects, a new Double-Visual Geometry Group16 (D-VGG16) is firstly designed as feature functions of the bilinear model. The global and local features fully extracted from the bilinear model by D-VGG16 are output to the soft-max function to realize the automatic classification of surface defects. Then the heat map of the original image is obtained by applying Gradient-weighted Class Activation Mapping (Grad-CAM) to the output features of D-VGG16. Finally, the defects in the original input image can be located automatically after processing the heat map with a threshold segmentation method. The training process of the proposed method is characterized by a small sample, end-to-end, and is weakly-supervised. Furthermore, experiments are performed on two public and two industrial datasets, which have different defective features in texture, shape and color. The results show that the proposed method can simultaneously realize the classification and localization of defects with different defective features. The average precision of the proposed method is above 99% on the four datasets, and is higher than the known latest algorithms.

Keywords: automated surface inspection; D-VGG16; bilinear model; Grad-CAM; classification; localization

---

1. Introduction

Surface defect detection is an important part of industrial production, and has significant impact upon the quality of industrial products on the market. The traditional manual detection method is time-consuming, and its detection accuracy is easily affected by the subjectivity, energy and experience of the inspector. To overcome the shortcomings of manual inspection, automatic surface defect detection based on machine vision comes into being.

With the rapid development of computer technology, machine vision has been widely applied in industrial production, especially for defect detection in industrial products. Over the last decade, a large number of surface defect detection algorithms have emerged. These algorithms can be roughly classified into three categories: Traditional methods based on image structure features, methods combining statistical features with machine learning, and deep learning methods based on the Convolutional Neural Network (CNN). The traditional defect detection algorithm based upon image structure features mainly detects the surface defects by analyzing the texture, skeleton, edge and spectrum of the image. Shafarenko et al. [1] proposed a color similarity measurement for an automatic detection and segmentation of random texture surface defects, which was realized by using watershed transform for color images of random textures, and extracting the color and texture features of the images.

Ojala et al. [2] utilized histogram analysis to threshold the texture image and then map it into a special data structure of the skeleton representation, achieving the extraction of texture image defects.

Wen et al. [3] used the image edge intensity and the distribution of the gray values of pixels in the edge domain to model the surface defects. Zhou et al. [4] realized the defect of the metal surface by wavelet analysis. Although the detection and segmentation of defects can be realized by analyzing the structural features of the surface of the object, the parameters of the algorithm need to be set manually for most of these methods, making them easily affected by interference factors, such as illumination in the environment, thereby affecting the detection effect.

The methods of combining statistical features with machine learning are mainly to extract statistical features from the defect surface, and then use machine learning algorithms to learn these features in order to realize surface defect detection. Ghorai et al. [5] used a combination of discrete wavelet transforms and a Support Vector Machine (SVM) to detect surface defects in steel. Xiao et al. [6] realized the detection of the surface defects of steel strips by constructing a series of SVMs with a random subspace of the features, and an evolutionary separator with a Bayesian kernel to train the results from the sub-SVM to form an integrated classifier. The combination of statistical features and machine learning can obtain higher accuracy and robustness than traditional structure-based methods. However, in image feature modeling, the accuracy of detection may be altered due to the different selections of feature types, and is closely linked to the extracted features, so it is necessary to find a suitable feature descriptor for a specific detection object.

Recently, because of the rapid development of deep learning, especially in terms of its strong feature extraction ability, it has been widely used in image-related tasks, such as graphic analysis [7], semantic segmentation [8] and target tracking [9]. Many researchers have also applied deep learning to surface defect detection. Lin et al. [10] proposed a CNN-based LEDNet network for light-emitting diode (LED) defect detection, and used Class Activation Mapping (CAM) [11] to achieve an automatic location of defects. Tao et al. [12] used a novel cascade auto-encoder to segment and locate metal surface defects automatically. Di et al. [13] used a combination of the Convolutional Auto Encoder (CAE) and Semi-supervised Generative Adversarial Networks (SGAN) to detect surface defects in steel, where CAE was used to extract the fine-grained features of the steel surface, and SGAN was used to further improve the generalization ability of the network. The authors tested the steel defect dataset to verify the effectiveness of the proposed method. Compared with the traditional methods based on the image structure and statistical features, combined with machine learning, the advantage of using CNN-based deep learning for surface defect detection is that CNN can simultaneously realize the automatic extraction and recognition of features in a network, and get rid of the trouble of manually extracting features.

Defect localization can make the observer find and understand the location of surface defects more intuitively. In essence, defect localization belongs to the category of object detection. Therefore, some researchers regarded surface defect detection as the problem of defect detection. Lin et al. [14] used a Faster-Region Convolutional Neural Network (Faster-RCNN) [15] and a Single Shot MultiBox Detector (SSD) [16] object detection algorithm to detect steel surface defects, and achieved a higher accuracy and recall rate. Cha et al. [17] proposed a defect detection method based upon Faster-RCNN, and verified the effectiveness of the proposed defect detection method on concrete cracks, steel corrosion, bolt corrosion and steel delamination. The advantage of using an object detection algorithm to detect and locate surface defects, is that it can directly draw lessons from the successful and excellent algorithms in object detection tasks, but these algorithms require a large number of pixel-level labeled training samples, which is difficult to achieve in actual industrial production.

Aiming at the problem of sample labeling difficulty for defect detection in actual industrial production, Lin et al. [10] and Ren et al. [18] used Class Activation Mapping (CAM), which is a class-discriminative localization technique that generates visual explanations from the CNN-based network to automatically locate surface defects. The CAM replaced the last full connection layer of the CNN network with Global Average Pooling (GAP) [19] to calculate the spatial average of each feature mapping in the last convolution layer, serving as input features to the fully-connected layer.

In this way, the importance of the image region can be recognized by projecting the weights of the output layer back to the convolutional feature map. However, the network with CAM needs to change the original design structure of the network, resulting in the need to retrain the network, therefore its usage scenarios are limited. To overcome the shortcomings of CAM, Selvaraju et al. [20] proposed Gradient-based Class Activation Mapping (Grad-CAM), but calculated the weights by using the global average of the gradient, which is the generalization of CAM, and is suitable for any CNN-based network without modifying any architecture of the network or re-training.

Therefore, to solve the problems above, a generic method of automated surface defect detection based upon a bilinear model is presented in this paper. Firstly, the Double-VGG16 (D-VGG16) that consists of two completely symmetric sub-networks based on VGG16 [21] is proposed as the feature extraction network of the bilinear model [22]. The output of the bilinear model uses the soft-max function to predict the corresponding type of the input image, which is realized as the automatic detection of surface defects. Then the heat map of the original image is obtained by applying Grad-CAM to one of the output features of D-VGG16. Finally, the defects in the original input image can be located automatically after processing the heat map with a threshold segmentation method. For the problem of insufficient training samples in actual industrial production, the D-VGG16 is initialized by loading the VGG16 pre-training weights on ImageNet [23] with 1000 classes, and adopt the transfer learning [24] to train the whole network, attaining the target of small samples training. The training of the entire network only uses image-level annotation, and is carried out in an end-to-end manner. The main contributions of this paper are as follows:

(1) The bilinear model for the detect detection tasks was proposed. To the best of our knowledge, this is the first paper that uses the bilinear model for surface defect detection. Moreover, the proposed method has a generalization capability, and can be successfully applied to defective features with texture, shape and color.

(2) A D-VGG16 network based upon VGG16 for the feature function of the bilinear model was designed. The Experimental results show that such a network structure for defects detection applications has a higher average precision than that network using VGG16 as the feature function, and is also higher than the known latest methods.

(3) The training process of the whole network proposed in this paper has the characteristics of a small sample, end-to-end, and is weakly-supervised. In the training stage, only a few training images of image-level labeled are needed to locate the defects of input images in the prediction stage.

The rest of the paper is organized as follows: Section 2 describes in great detail the specific method of the paper, mainly about describing the overall structure of the proposed method. Section 3 presents the details and the results of performing experiments on the datasets, which is followed by the conclusions drawn in Section 4.

2. Methodology

There are two phases in the proposed method. The first phase is the automatic classification of defects, during which the features of the original input image are firstly handled by the bilinear model consisting of two fully symmetrical Double-Visual Geometry Group16 (D-VGG16) networks, and then the extracted features are sent to the soft-max function to achieve the automatic classification of these defects. The second phase is the automatic localization of the defects, during which Gradient-weighted Class Activation Mapping (Grad-CAM) is used to get the heat map of the original input image, and then the corresponding defects are located by employed threshold segmentation to the heat map. The overall structure of the automated surface defect detection, based on the bilinear model proposed in this paper, can be demonstrated in Figure 1. The whole network is a typical bilinear model structure.

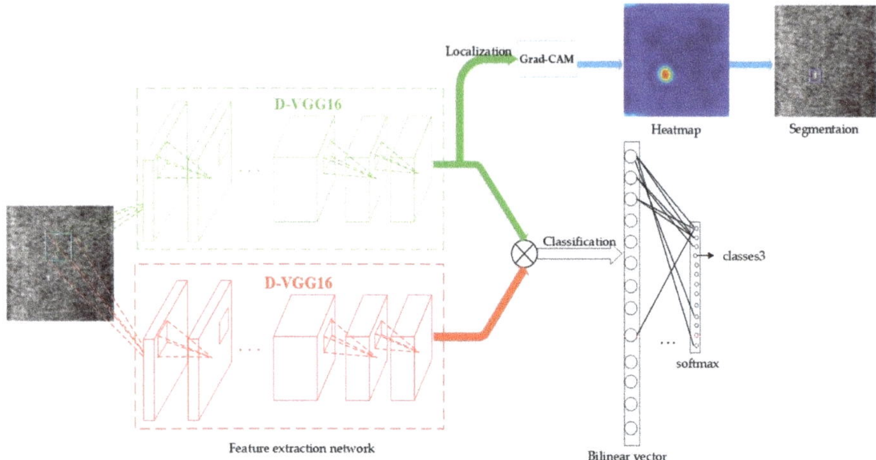

Figure 1. Network overall structure.

2.1. Defect Classification

The whole process of defect classification is as follows: Two features that function from D-VGG16 are concatenated to get the bilinear vector, which is fed into the soft-max function to obtain the probability of the corresponding defects in the input image and realize the defect classification. The whole process is a typical bilinear model structure, and its core is D-VGG16 that is used as a feature function.

2.1.1. D-VGG16

Feature function, as a function extraction network of a bilinear model, plays an important role whatever for locating and classifying in the whole network. In this paper we used two fully symmetrical D-VGG16 that were based on VGG16, as a feature extraction network of a bilinear model were used, where the structure of the network is shown in Figure 2.

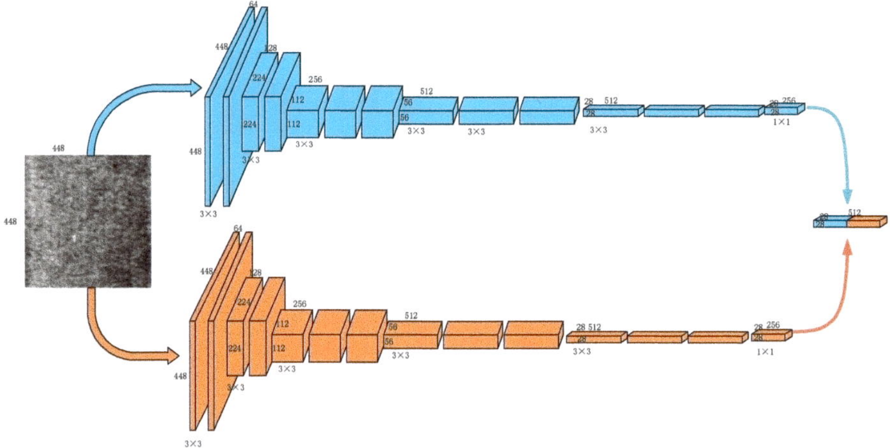

Figure 2. Double-Visual Geometry Group16 (D-VGG16) network structure. Feature maps with the same shape have the same width, height, number of channels and convolutional kernel.

For the classification task using the Convolutional Neural Network (CNN), the simplest way to improve the accuracy of small sample training and avoid over-fitting is to reduce the feature map of the last layer of CNN without decreasing the receptive field of the network. However, this will inevitably influence the output features of the network, thereby limiting the expressive capability of the network. Given the considerations above, D-VGG16 is designed, as shown in Figure 2. As a 1 × 1 convolutional kernel with 256 channels, and this is used after the last convolution layer of the VGG16 network, and then the outputs of two such networks are concatenated to form D-VGG16. On the one hand, it can not only reduce the risk of an over-fitting of complex CNN for small samples training, but also maintain the diversity of the network output features, and the output features of two sub-networks can be conditioned on each other. The feature extraction network consists of two symmetrical D-VGG16, i.e., the two D-VGG16 are identical in architectures, so the entire network is composed of four VGG16 with exactly the same structure. The advantage of this design is that the global and local features of the image can be adequately extracted, making the network more easily able to detect the subtle features in the image. In training, each sub-network loaded the pre-training weights of VGG16 on ImageNet directly, and used transfer learning, which achieves the goal of small samples training.

2.1.2. Bilinear Model

The bilinear model is composed of two-factors, and is mathematically separable, i.e., when one factor remains constant, its output is linear in any factor. A bilinear model B for defect classification consists of a quaternion function, as shown in Equation (1).

$$B = (f_A, f_B, P, F) \tag{1}$$

where f_A and f_B are feature functions, D-VGG16 is used in this paper, P represents the pooling function, and F represents the classification function, which here refers to the soft-max classifier.

The output of the feature function, f_A and f_B, are combined at each position of the image I using the matrix inner product, as shown in Equation (2).

$$bilinear(i, f_A, f_B) = f_A(i)^T f_B(i) \tag{2}$$

where $i \in I$. The feature dimensions of f_A and f_B must be equal, and the value should be greater than 1 to represent various descriptors that can be written as bilinear models.

To obtain the descriptor of the image, the pooling function P aggregates the bilinear features across all of the locations in the image. The pooling function can use the weighted sum of all bilinear features of the image, i.e., the sum of all bilinear features, which was calculated as follows.

$$\Phi(I) = \sum_{i \in I} bilinear(i, f_A, f_B) \tag{3}$$

If the feature sizes of the f_A and f_B output are $C \times M$ and $C \times N$ respectively, then the size of the bilinear vector $\Phi(I)$ is $M \times N$, and its corresponding class probability can be obtained by inputting the $\Phi(I)$ reshaped size $MN \times 1$ into the classification function F. The data stream of this bilinear model is shown in the Figure 3.

From an intuitive point of view, the structure of the bilinear can make the output features of the feature extraction function, f_A and f_B, to be fine tuned on each other by considering all of their pairwise interactions similar to quadratic kernel expansion. Because the entire network is a directed acyclic graph, and parameters of the network can be trained by the gradient of back-propagating loss.

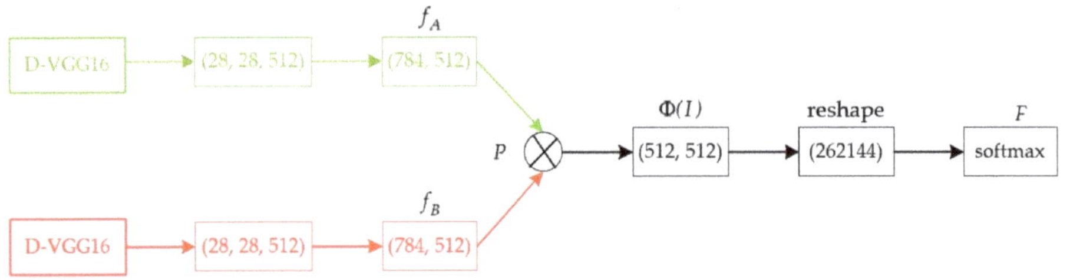

Figure 3. Data stream of the bilinear model.

2.2. Defect Localization

Defect localization of the input image enables the inspector to find and understand the specific location of the defect intuitively, and the implementation process is as follows: Firstly, the heat map of the original image is obtained by applying Gradient-weighted Class Activation Mapping (Grad-CAM) to one of the output features of D-VGG16, and then the corresponding defect location can be determined in the input image by a threshold segmentation to the heat map.

2.2.1. Grad-CAM

Although CNN has significant effects on image processing tasks for a long time, it has been a controversial method due to the poor interpretability of the CNN internal feature extraction, thus a new field, which is called the interpretable research of deep learning, appeared. Apart from that, Grad-CAM is a visualized method of the convolutional neural network, which can be used to visualize network category location results in the last level of the network's convolutional layer.

In order to obtain a class activation map $L^n_{Grad-CAM}$, the score gradient $\frac{\partial y^n}{\partial A^k}$ of the class n is firstly calculated, in which A^k represents the weight of the class n of the first k feature map, and y^k represents the score of the category before the soft-max. Then the gradient of the global average pooling layer is used to obtain the importance α^n_k of the first k feature map for the category n.

$$\alpha^n_k = \frac{1}{Z}\sum_i\sum_j \frac{\partial y^n}{\partial A^k_{ij}} \tag{4}$$

where Z represents the size of the feature map and A^k_{ij} represents the activation value of the position in (i, j) the k first feature map. Finally, the weighted sum of the forward activation features is performed according to Formula (4), and a Grad-CAM of a given class can be obtained using a rectified linear unit (ReLU).

$$L^n_{Grad-CAM} = ReLU(\sum_k \alpha^n_k A^k) = \max(0, \sum_k \alpha^n_k A^k) \tag{5}$$

Grad-CAM can explain the feature extraction results of the network and enhance the trust of the network performance, which is particularly important to the training network of small samples, because the insufficient number of training samples may lead to an inadequately trained network, thus causing a problem that the judgments of the network for a particular class may not be based on the real discriminant region in the image, and this results in serious over-fitting. In addition, Grad-CAM is used in the defect detection network, which can automatically locate the defects of input images in the prediction stage only by image-level annotation in the training stage.

In this paper, the Grad-CAM of defective images are generated. As shown in Figure 1, the Grad-CAM highlights the defect regions.

2.2.2. Segmentation

The threshold segmentation is performed after the heat map of the input image obtained from Grad-CAM to locate the defect regions. Let $f(m,n)$ represent the binarized image for the heat map, and $f(m,n)$ is as shown in

$$f(m,n) = \begin{cases} 255, & \text{if } f_{hm}(m,n) \geq \sigma \\ 0, & \text{otherwise} \end{cases} \tag{6}$$

where $f_{hm}(m,n)$ indicates the heat map after graying, and σ is the threshold, respectively. In $f(m,n)$, pixels whose gray value is 255 indicate the defect region, and pixels whose gray value is 0 present the non-defective area. In order to get better localization results, it is significantly important to choose the threshold segmentation method for σ. Experiments show that different types of defects and defects distribution in the entire image can result in different methods of threshold segmentation. For images with defects of limited distribution and signal type, a simple fixed threshold segmentation can be used to obtain a better result. For images with defects of scattered distributions and variable types, the adaptive Otsu [25] algorithm can obtain satisfactory results.

3. Experiments

This section evaluates the performance of the surface defect detection method proposed in this paper on two public and two collected defect datasets in real industrial scenes. Firstly, the experimental hardware environment and training details are briefly explained. Secondly, the datasets used will be expounded. Then, the number of images for training and testing is interpreted. Finally, the proposed method is compared with the latest experimental method of each data set in four datasets, which highlight the effectiveness and universality of the proposed method on the task of surface defect detection.

3.1. Hardware Platform and Training Details

Experiments in this paper are implemented on a workstation with 64 GB memory, and we also used TITAN XP for acceleration. Similar to most deep convolutional neural networks, the back-propagation algorithm was used as the training rule, and we then minimized the loss function with respect to the network parameters using Adam [26]. The training of the whole network is carried out in an end-to-end way. The training and testing images of each dataset are labeled only with the image-level. Input images were resized to 448×448, with no preprocessing of the images except for normalization.

The training process of the whole network was in the form of transfer learning. Specific implementation steps were as follows: Firstly, the pre-training weights of VGG16 on ImageNet was loaded to initialize two D-VGG16, and only parameters other than VGG16 were trained. At this time, the learning rate was 0.001, the momentum was 0.9, and the batch size was 64, and a model with relatively low loss was trained. Then we load the weights of the last step to continue training the entire network. At this time, the learning rate was 0.00001, the momentum was 0.9, and the batch size was 16. In this schedule, a model with lower loss will be obtained by several iterations.

3.2. Datasets Description

The open datasets are DAGM_2007 [27] and hot-rolled strip [28], respectively. The collected datasets are the diode glass bulb surface defect dataset and the fluorescent magnetic powder surface defect dataset. These datasets cover texture defects, shape defects and color defects on the actual industrial product.

3.2.1. DAGM_2007 Defect Dataset

The first open dataset is the DAGM_2007 surface defect dataset, which is manually generated and can be used for surface defect detection. The dataset contains six types of surface defects with different textures, where in each of these defects has 1000 defect-free and 150 defective grayscale images, the size of the image is 512 × 512 pixels and the pixel precision is 8 bits. The ground truths of all defective images are provided in the dataset. Examples of defect images are shown in Figure 4.

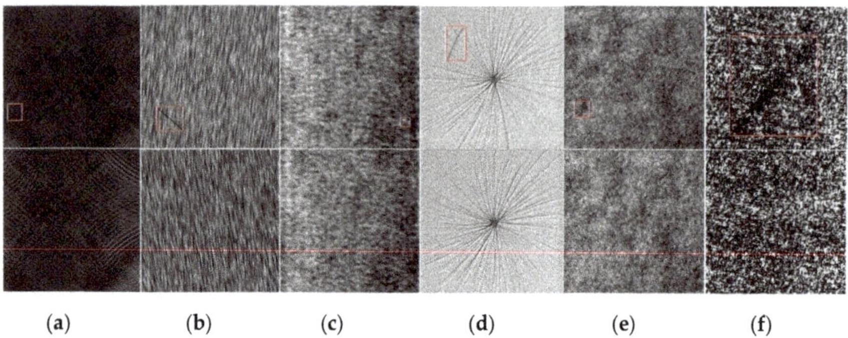

Figure 4. Examples of the DAGM_2007 defect dataset. Each column represents a type of defect, and the defect areas are labeled by the red bounding boxes. (**a**) classes1; (**b**) classes2; (**c**) classes3; (**d**) classes4; (**e**) classses5; (**f**) classes6.

3.2.2. NEU Defect Dataset

NEU [24] is a surface defect dataset of hot-rolled steel strips. There are six types of defects, including crazing, inclusion, patches, pitted-surface, rolled-in scale and scratches. Examples of the defect images are shown in Figure 5. Each class of defect has 300 grayscale images, and the size of the image is 200 × 200 pixels, and the pixel precision is 24 bits. The labels of all of the images are provided in the dataset, but the ground-truth of the defective images is not provided.

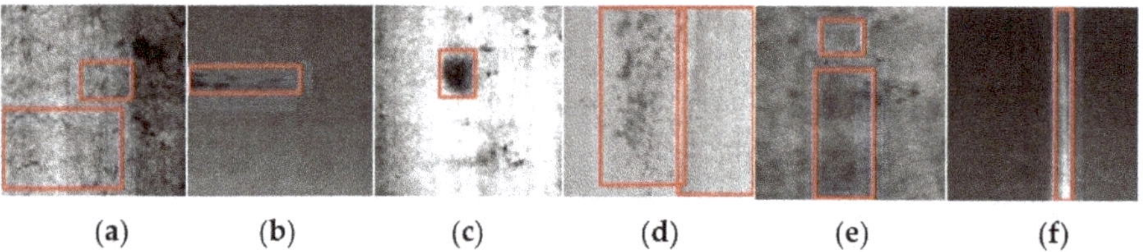

Figure 5. Examples of the NEU defect dataset. Each column represents a type of defect, and the defect areas are labeled by the red bounding boxes. (**a**) crazing; (**b**) inclusion; (**c**) patches; (**d**) pitted-surface; (**e**) rolled-in-scale; (**f**) scratches.

3.2.3. Diode Glass Bulb Surface Defect Dataset

The glass bulbs have been widely used as packaging material for diodes because of their heat resistance, damp-proof ability and high reliability. They play an important role in protecting the diode. In order to obtain surface images of the surface of the diode glass bulb, an image acquisition system consisting of an industrial camera with DAHENG IMAGING MER-131-75 GM/C-P, a telecentric lens with 1× magnification and a dark-field LED light source was used in the experiment. A total of 1730 color images were collected, of which 1020 were defective images, including 390 of shell wall damage, 360 breaks, 270 stains, and the size of the image is 661 × 601 pixels and the pixel precision is 8 bits. Examples of these defect images are shown in Figure 6.

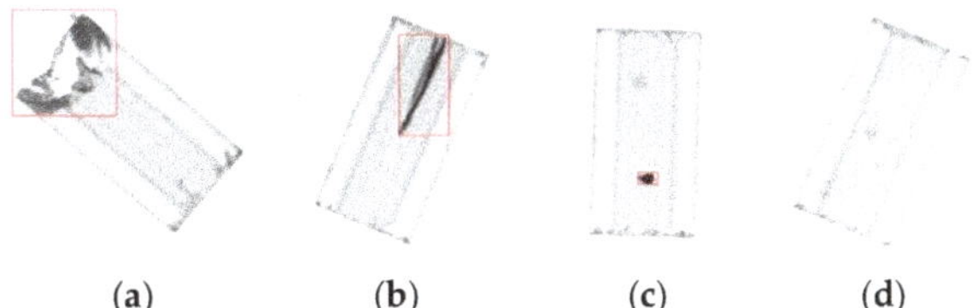

(a)   (b)   (c)   (d)

Figure 6. Examples of the diode glass bulb surface defect dataset, and the defect areas are labeled by the red bounding boxes. (**a**) break; (**b**) shell wall damage; (**c**) stain; (**d**) good.

3.2.4. Fluorescent Magnetic Powder Surface Defect Dataset

Fluorescent magnetic powder nondestructive testing is a common method for the detection of any surface and near-surface defects of ferromagnetic materials such as aero-turbines [29], turbines [30], and train bearings [31] in the aerospace, military and civil industry. Its working principle is that after the ferromagnetic material work-piece is magnetized, the magnetic force line will be locally distorted when there are defects on the surface and near-surface of the work-piece. It leads to magnetic leakage, absorbing fluorescent magnetic particles suspended on the surface of the work-piece, and the forming of visible magnetic marks under ultraviolet light. In the experiment, the image acquisition system consisted of an industrial camera with XIMEA MQ042CG-CM, a fixed focus lens with a focal length of 6 mm and an ultraviolet light. The system was used to detect surface cracks of the ferromagnetic cylindrical work-pieces with the height of 100 mm and diameter of 45 mm, in which the width and height of the crack range from 0.3 mm to 1.0 mm and from 7 mm to 90 mm, respectively. The experiment collected 800 defects and 1000 defects-free color images, and the size of the image is 468 × 1324 pixels, and the pixel precision is 24 bits. Examples of defect images are shown in Figure 7.

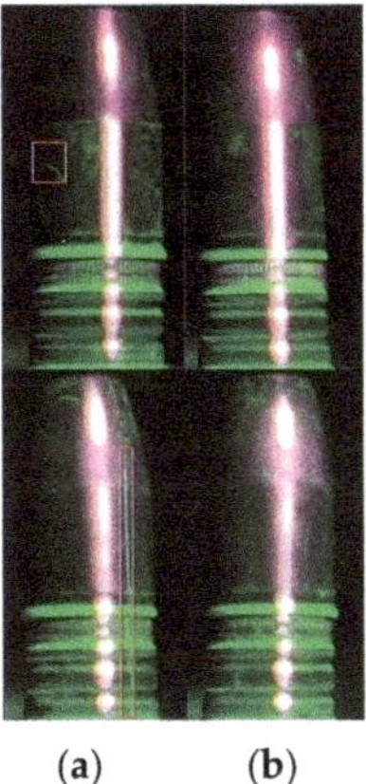

(a)   (b)

Figure 7. Examples of the fluorescent magnetic powder surface defects dataset. The defect areas are labeled by the red bounding boxes. (**a**) bad; (**b**) good.

3.3. Contrast Experiments

In order to test the performance of the proposed method in work-pieces surface defect detection, the proposed method is evaluated on two published and two collected work-pieces surface defect datasets. At present, most of the defect detection algorithms only aim at a specific category of defects; however, the surface defect detection method proposed in this paper is a kind of defect that can be applied to different types of work-pieces. It is unreasonable to apply a defect detection algorithm suitable for a specific category to other categories of defects and compare it with the method proposed in this paper.

Therefore, in each defect data set, not only GLCM + MLP [17], gcForest [32] and Bilinear Convolutional Neural Network (BCNN) are used to perform four kinds of generic surface defect detection algorithms, but also the open datasets will also be compared with the known latest experimental results on this dataset.

3.3.1. Open Datasets

Since the vast majority of the evaluations using the two datasets for performance evaluation had only the experimental results of average precision, and average precision is the main and most important performance indicator for the multi-category task, therefore only the average precision of the methods is compared on two open datasets.

(A) Localization and Classification Results of the DAGM_2007 Defect Dataset: For the DAGM_2007 dataset, the ratio of the training set to the test set is 1:1. Some experimental localization results of the proposed method running in the dataset are shown in Figure 8.

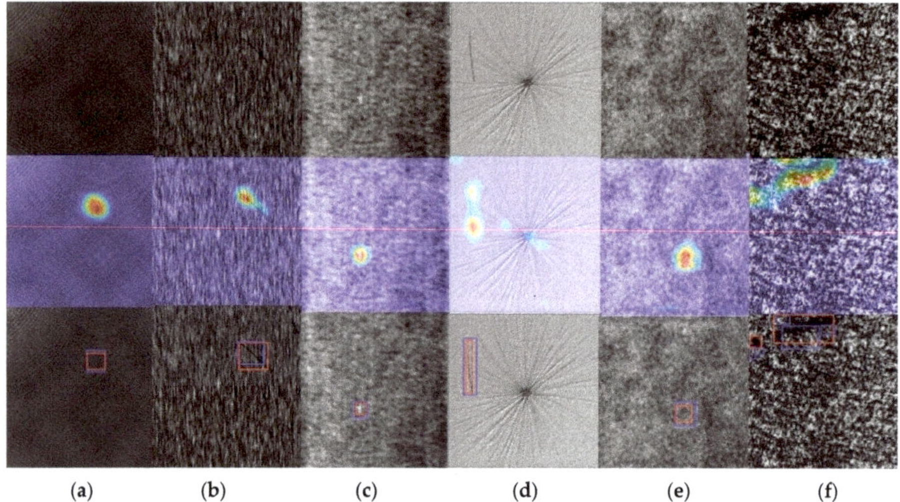

(a)   (b)   (c)   (d)   (e)   (f)

Figure 8. Examples of localization on DAGM_2007 defect dataset. From top to bottom are the original image, the combination of the original image and the heat map, and the location results of the defects. The Ground-Truth of the defect is marked with the red bounding boxes, while the localization results of the proposed method is marked with the blue bounding boxes. (**a**) classes1; (**b**) classes2; (**c**) classes3; (**d**) classes4; (**e**) classsses5; (**f**) classes6.

In the combination image of the original image and Grad-CAM, the red region represents the confidence level of the pixels that the network discriminates against. The deeper the color, the higher the confidence level of the pixels in the image. The dataset is compared with the results of surface defect detection algorithms proposed by Yu [33] and Zhao [34]. The experimental classification results are shown in Table 1.

Table 1. Comparison of results on DAGM_2007 surface defect dataset.

Method	Average Precision
GLCM + MLP	81.68%
gcForest	86.67%
BCNN	95.57%
FCN [33]	98.35%
Zhao [34]	98.53%
Ours	99.49%

As can be seen from Table 1, although high classification accuracy has been achieved on the DAGM_2007 surface defect data set at present, the proposed method can still further improve the classification accuracy on the data set and achieve the automatic location of defects at the same time.

(B) Localization and Classification Results of the NEU Defect Dataset: For the NEU surface defect dataset, a number of 150 images are randomly selected as the test set in each class of defects, and the remaining images are used as the training set. Some experimental localization results of the proposed method running in the dataset are shown in Figure 9.

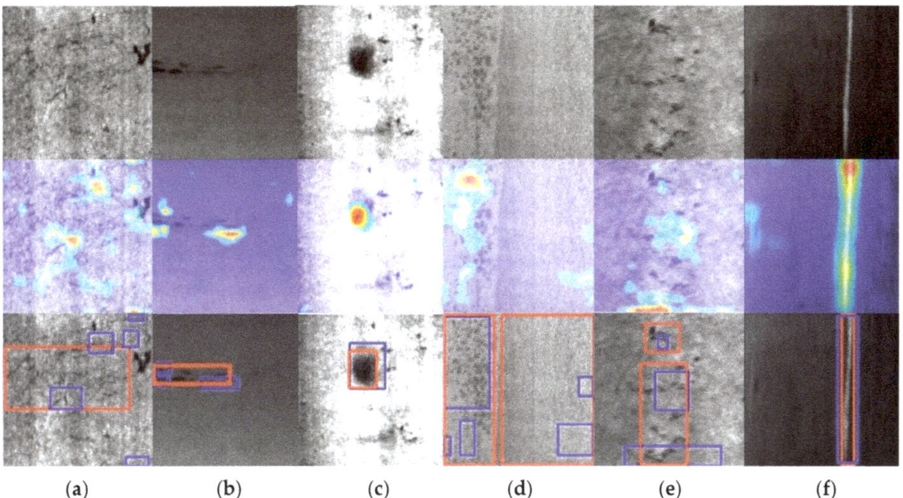

(a)   (b)   (c)   (d)   (e)   (f)

Figure 9. Localization results of the proposed method on the NEU defect dataset. From top to bottom are the original image, the combination of the original image and the heat map, and the location results of the defects. The Ground-Truth of the defect is marked with the red bounding boxes, and the localization results of the proposed method is marked with the blue bounding boxes. (**a**) crazing; (**b**) inclusion; (**c**) patches; (**d**) pitted-surface; (**e**) rolled-in-scale; (**f**) scratches.

Most images of the NEU defect dataset have multiple defects, and the texture of each type of defective image is different, which brings more challenges to automatic location. As shown in Figure 9, although the proposed method does not perform well in defect localization when applied to NEU datasets, it can extract specific pixel regions to identify a certain class of images. Using this dataset, the proposed method was compared with the algorithms proposed by BYEC [6], Song et al. [35] and Ren et al. [18]. To ensure the validity of the comparison results, the same training data generation method as the papers mentioned above is used. The experimental classification results are shown in Table 2.

Table 2. Comparison of results on NEU surface defect dataset.

Method	Average Precision
GLCM + MLP	98.61%
gcForest	61.56%
BCNN	98.56%
BYEC	96.30%
Song [35]	98.60%
Ren [18]	99.21%
Ours	99.44%

As can be seen from Table 2, compared with the latest methods proposed by Sun and Ren, the proposed method has a higher detection accuracy in the NEU defect detection dataset.

3.3.2. Real Collected Datasets

The two kinds of defect datasets collected contains defective and defect-free images, so they can be regarded as multi-classification or binary classification tasks.

(A) Localization and Classification Results of the Diode Glass Bulb Surface Defect Dataset: For the diode glass bulb surface defect dataset, the ratio of the training set and testing set images is 7:3. Some experimental localization results of defect detection on this dataset by the proposed method are shown in Figure 10.

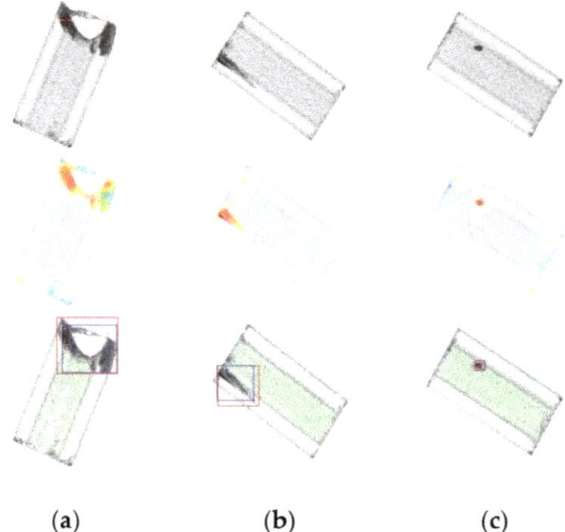

(a)   (b)   (c)

Figure 10. Examples of localization on the diode glass bulb surface defect dataset. From top to bottom are the original image, the combination of the original image and the heat map, and the location results of these defects. The Ground-Truth of the defect is marked with the red bounding boxes, while the localization results of the proposed method is marked with the blue bounding boxes. (a) break; (b) shell wall damage; (c) stain.

There is no significant texture difference around different defect types in the diode glass bulb surface defect dataset, and shell wall damage is a typical shape defect. However, it can be found that the proposed method can accurately extract the key pixel regions that discriminate each type of defect, which can not only explain the reason why it can achieve a higher precision than other methods, but also obtain the better effect of localization. The comparative experiments on this dataset are shown in Table 3.

Table 3. Comparison of results on the diode glass bulb surface defect dataset.

Method	Average Precision
GLCM + MLP	91.32%
gcForest	85.25%
BCNN	91.80%
Ours	99.87%

It can be seen from Table 3 that even in the work-piece surface defect detection task with few texture features, the proposed method has an advantage in detection accuracy compared with other algorithms.

(B) Localization and Classification Results of the Fluorescent Magnetic Powder Surface Defect Dataset: For this fluorescent magnetic powder surface defect dataset, the ratio of the training set and the testing set images is 7:3. The experimental localization results of the proposed method on this dataset are shown in Figure 11.

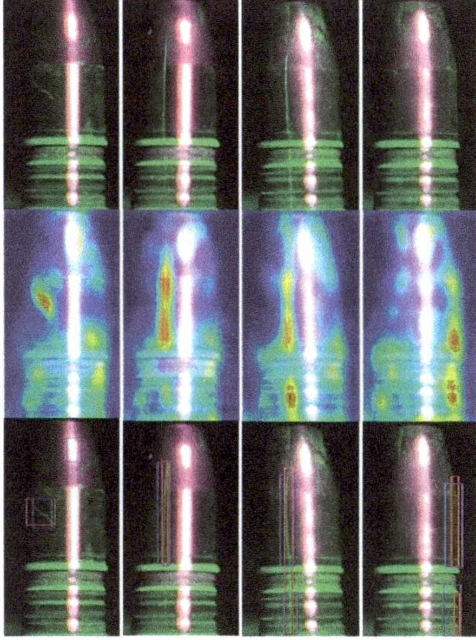

Figure 11. Localization results of the proposed method on the fluorescent magnetic powder surface defect dataset. From top to bottom are the original image, the combination of the original image and the heat map, and the location results of these defects. The Ground-Truth of the defect is marked with the red bounding boxes, and the localization result of the proposed method is marked with the blue bounding boxes.

When the ultraviolet light is irradiated on the smooth iron work-piece, the surface of the magnetized work-piece will reflect the violet light emitted by the ultraviolet light due to the principle of light reflection. This phenomenon is particularly prominent on the cylindrical work-piece. Therefore, the defect image of the fluorescent magnetic powder obtained in the experiment has a bright purple reflective area in the center of the work-piece, which will cause a great interference to the detection of any defects. In the experiment, the original image is zoomed into a size of 448 × 448, with no pre-processing having been performed on the images except for normalization, and then the image is sent to the network for training and testing. As shown in Figure 11, it can be seen that the network can effectively eliminate interference in the reflective area and extract the defective area. The classification results of the comparative experiments on this dataset are shown in Table 4.

Table 4. Comparison of results on fluorescent magnetic powder surface defect dataset.

Method	Average Precision
GLCM + MLP	90.56%
gcForest	92.59%
BCNN	93.33%
Ours	99.13%

It can be seen from Table 4 that even if there is a task of defect detection with strong interference factors, the detection accuracy of the proposed method is still nearly 6% higher than that of BCNN.

(C) Evaluation of Binary Classification Performance: The above experiments have shown that the average precision of the proposed method on four datasets is higher than that of other methods. However, the detection rate of defects and the precision of non-defects are often emphasized in defect detection, and at this time, only the dataset is divided into defects and non-defects. TP and TN denote the number of true positives and true negatives respectively, FP and FN denote the number of false positives and false negatives, respectively. Then the definitions of the Precision Rate (PR), True Positive Rate (TPR), False Positive Rate (FPR) and False Negative Rate (FNR) are as follows.

$$PR = \frac{TP}{TP + FP} \tag{7}$$

$$TPR = \frac{TP}{TP + FN} \tag{8}$$

$$FPR = \frac{FP}{FP + TN} \tag{9}$$

$$FNR = \frac{FN}{FN + TP} \tag{10}$$

Results of the four methods PR, TPR, FPR and FNR on the diode glass bulb and fluorescent magnetic powder surface defect dataset are shown in Table 5.

Table 5. Results of the four methods PR, TPR, FPR and FNR on the diode glass bulb and fluorescent magnetic powder surface defect datasets.

Dataset Method	Diode Glass Bulb Surface Defect Dataset				Fluorescent Magnetic Powder Surface Defect Dataset			
	PR	TPR	FPR	FNR	PR	TPR	FPR	FNR
GLCM + MLP	93.86%	88.21%	6.14%	11.79%	85.71%	89.55%	14.28%	10.45%
grForest	83.19%	79.84%	16.81%	20.16%	95%	91.94%	5%	8.06%
BCNN	81.25%	100%	18.75%	0%	99.65%	94%	0.35%	6%
Ours	100%	100%	0%	0%	98.36%	99.67%	1.64%	0.33%

Precision Rate and True Positive Rate are often a pair of contradiction measure, and generally speaking, when the Precision Rate is high, the True Positive Rate tends to be low, and the higher True Positive Rate, the lower the Precision Rate. Therefore, the Precision Rate and the True Positive Rate cannot accurately reflect the effectiveness of the detection method, but usually F_1 is used, which is defined as follows.

$$F_1 = \frac{2 \times PR \times TPR}{PR + TPR} \tag{11}$$

F_1 value of GLCM + MLP, gcForest, BCNN and the proposed method on the diode glass bulb surface defect dataset and fluorescent magnetic powder surface defect dataset are shown in Figure 12.

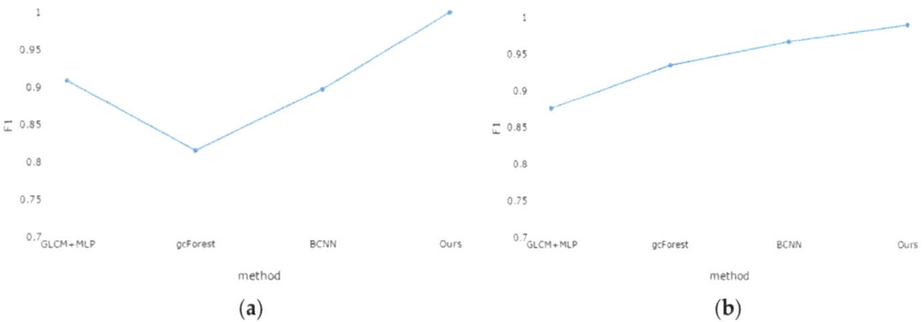

Figure 12. Comparison of F_1 curve obtained from four methods. (**a**) Diode glass bulb surface defect dataset; (**b**) Magnetic powder surface defect dataset.

The results are shown in Figure 12. The proposed surface defect detection method achieves a higher F_1 among all of the methods. It outperforms both methods combining statistical features with machine learning (GLCM + MLP) and the generic deep learning method based on a Convolutional Neural Network (BCNN).

There are many kinds of defects in actual industrial production, and one method which works well in a specific category is usually not applicable to the other types of defects. Experimental results show that the surface defect detection method proposed in this paper demonstrates excellent detection performance in surface defects with features of texture, shape and color. Furthermore, it can simultaneously realize an automatic localization and classification of defects. In the prediction phase, it takes an average of 0.292 s to a localization and classification of defects for an image at the same time.

4. Conclusions

The conclusions from the work are presented as follows.

- A generic method of automated surface defect detection based on a bilinear model is proposed. Firstly, as a feature extraction network of the bilinear model, D-VGG16, which consists of two completely symmetric VGG16, is designed, and the features extracted from the bilinear model are output to the soft-max function to realize the automatic classification of defects. Then the heat map of the original image is obtained through applying Grad-CAM to one of the output features in D-VGG16. Finally, the defects in the input image can be located automatically after processing the heat map with a threshold segmentation algorithm.
- The training of the proposed method is carried out in a small sample, end-to-end, and in a weakly-supervised way. Even though the number of training images used in the experiments were no more than 1300, over-fitting did not occur during the training process of all the datasets, and the surface defects can be automatically located using only training images labeled at image-level.
- The experiments has been performed on four datasets with different defective features. This shows that the proposed method can be effectively applied to surface defect detection scenarios with texture, color and shape features, even a diode glass bulb surface defect dataset with complex texture and the fluorescent magnetic powder surface defect dataset with strong interference factors. The overall performance of the proposed method is superior to other methods.

The proposed method has certain limitations for automatic localization in the datasets with complex textures. Since the whole network is composed of four VGG16, and the Grad-CAM used in automatic localization is time-consuming, it takes a long time to detect and locate defect in the testing stage. Future work will focus on solving the above effect of automatic location and real-time performance of the method in this paper.

Author Contributions: F.Z. designed the algorithm, performed the experiments and wrote the paper. G.L. modified the paper. F.X. and. H.D. supervised the research.

Funding: This work was supported by National Natural Science Foundation of China (Grant Nos. 11602292, 61701421, 61601381).

Conflicts of Interest: The authors declare no conflict of interest.

References

1. Shafarenko, L.; Petrou, M.; Kittler, J. Automatic watershed segmentation of randomly textured color images. *IEEE Trans. Image Process.* **1997**, *6*, 1530–1544. [CrossRef] [PubMed]
2. Ojala, T.; Pietikäinen, M.; Mäenpää, T. Multiresolution gray-scale and rotation invariant texture classification with local binary patterns. *IEEE Trans. Pattern Anal. Mach. Intell.* **2002**, *7*, 971–987. [CrossRef]
3. Wen, W.; Xia, A. Verifying edges for visual inspection purposes. *Pattern Recognit. Lett.* **1999**, *20*, 315–328. [CrossRef]
4. Zhou, P.; Xu, K.; Liu, S. Surface defect recognition for metals based on feature fusion of shearlets and wavelets. *Chin. J. Mech. Eng.* **2015**, *51*, 98–103. [CrossRef]
5. Ghorai, S.; Mukherjee, A.; Gangadaran, M.; Dutta, P.K. Automatic defect detection on hot-rolled flat steel products. *IEEE Trans. Instrum. Meas.* **2012**, *62*, 612–621. [CrossRef]
6. Xiao, M.; Jiang, M.; Li, G.; Xie, L.; Yi, L. An evolutionary classifier for steel surface defects with small sample set. *EURASIP J. Image Video Process.* **2017**, *2017*, 48. [CrossRef]
7. Santoyo, E.A.R.; Lopez, A.V.; Serrato, R.B.; Garcia, J.A.J.; Esquivias, M.T.; Fernandez, V.F. Reconocimiento de patrones y evaluación del daño generado en aceros de baja aleación a partir del procesamiento digital de imágenes e inteligencia artificial. *DYNA Ing. Ind.* **2019**, *94*, 357.
8. Li, Y.; Chen, X.; Zhu, Z.; Xie, L.; Huang, G.; Du, D.; Wang, X. Attention-guided unified network for panoptic segmentation. In Proceedings of the IEEE Conference on Computer Vision and Pattern Recognition (CVPR), Long Beach, CA, USA, 16–20 June 2019.
9. Feng, W.; Hu, Z.; Wu, W.; Yan, J.; Ouyang, W. Multi-Object Tracking with Multiple Cues and Switcher-Aware Classification. *arXiv* **2019**, arXiv:1901.06129.
10. Lin, H.; Li, B.; Wang, X.; Shu, Y.; Niu, S. Automated defect inspection of LED chip using deep convolutional neural network. *J. Intell. Manuf.* **2018**, *30*, 2525–2534. [CrossRef]
11. Zhou, B.; Khosla, A.; Lapedriza, A.; Oliva, A.; Torralba, A. Learning deep features for discriminative localization. In Proceedings of the IEEE conference on Computer Vision and Pattern Recognition, Las Vegas, NV, USA, 27–30 June 2016.
12. Tao, X.; Zhang, D.; Ma, W.; Liu, X.; Xu, D. Automatic metallic surface defect detection and recognition with convolutional neural networks. *Appl. Sci.-Basel* **2018**, *8*, 1575. [CrossRef]
13. Di, H.; Ke, X.; Peng, Z.; Dongdong, Z. Surface defect classification of steels with a new semi-supervised learning method. *Opt. Lasers Eng.* **2019**, *117*, 40–48. [CrossRef]
14. Lin, W.-Y.; Lin, C.-Y.; Chen, G.-S.; Hsu, C.-Y. Steel Surface Defects Detection Based on Deep Learning. In Proceedings of the International Conference on Applied Human Factors and Ergonomics (AHFE), Orlando, FL, USA, 22–26 July 2018.
15. Ren, S.; He, K.; Girshick, R.; Sun, J. Faster r-cnn: Towards real-time object detection with region proposal networks. In Proceedings of the Advances in Neural Information Processing Systems, Montreal, QC, Canada, 7–12 December 2015; pp. 91–99.
16. Liu, W.; Anguelov, D.; Erhan, D.; Szegedy, C.; Reed, S.; Fu, C.-Y.; Berg, A.C. Ssd: Single shot multibox detector. In Proceedings of the European Conference on Computer Vision (ECCV), Amsterdam, The Netherlands, 8–16 October 2016.
17. Cha, Y.J.; Choi, W.; Suh, G.; Mahmoudkhani, S.; Büyüköztürk, O. Autonomous structural visual inspection using region-based deep learning for detecting multiple damage types. *Comput. Aided Civ. Infrastruct. Eng.* **2018**, *33*, 731–747. [CrossRef]
18. Ren, R.; Hung, T.; Tan, K.C. A generic deep-learning-based approach for automated surface inspection. *IEEE Trans. Cybern.* **2017**, *48*, 929–940. [CrossRef] [PubMed]
19. Lin, M.; Chen, Q.; Yan, S. Network in network. In Proceedings of the International Conference on Learning Representations (ICLR), Scottsdale, AZ, USA, 2–4 May 2013.

20. Selvaraju, R.R.; Cogswell, M.; Das, A.; Vedantam, R.; Parikh, D.; Batra, D. Grad-cam: Visual explanations from deep networks via gradient-based localization. In Proceedings of the IEEE International Conference on Computer Vision (ICCV), Venice, Italy, 22–29 October 2017.

21. Simonyan, K.; Zisserman, A. Very deep convolutional networks for large-scale image recognition. In Proceedings of the International Conference on Learning Representations, Banff, AB, Canada, 14–16 April 2014.

22. Lin, T.-Y.; RoyChowdhury, A.; Maji, S. Bilinear cnn models for fine-grained visual recognition. In Proceedings of the IEEE International Conference on Computer Vision, Santiago, Chile, 7–13 December 2015.

23. Deng, J.; Dong, W.; Socher, R.; Li, L.-J.; Li, K.; Li, F.-F. Imagenet: A large-scale hierarchical image database. In Proceedings of the 2009 IEEE Conference on Computer Vision and Pattern Recognition, Miami, FL, USA, 20–25 June 2009.

24. Yosinski, J.; Clune, J.; Bengio, Y.; Lipson, H. How transferable are features in deep neural networks? In Proceedings of the Advances in Neural Information Processing Systems, Montreal, QC, Canada, 8–13 December 2014; pp. 3320–3328.

25. Ostu, N. A threshold selection method from gray-histogram. *IEEE Trans. Syst. Man Cybern.* **1975**, *9*, 62–66.

26. Kingma, D.P.; Ba, J. Adam: A method for stochastic optimization. In Proceedings of the International Conference on Learning Representations, Banff, AB, Canada, 14–16 April 2014.

27. DAGM 2007 Datasets. Available online: https://hci.iwr.uni-heidelberg.de/node/3616 (accessed on 29 July 2019).

28. Song, K.; Yan, Y. A noise robust method based on completed local binary patterns for hot-rolled steel strip surface defects. *Appl. Surf. Sci.* **2013**, *285*, 858–864. [CrossRef]

29. Coro, A.; Abasolo, M.; Aguirrebeitia, J.; López de Lacalle, L. Inspection scheduling based on reliability updating of gas turbine welded structures. *Adv. Mech. Eng.* **2019**, *11*, 1687814018819285. [CrossRef]

30. Artetxe, E.; Olvera, D.; de Lacalle, L.N.L.; Campa, F.J.; Olvera, D.; Lamikiz, A. Solid subtraction model for the surface topography prediction in flank milling of thin-walled integral blade rotors (IBRs). *Int. J. Adv. Manuf. Technol.* **2017**, *90*, 741–752. [CrossRef]

31. Zhao, M.; Lin, J.; Miao, Y.; Xu, X.J.M. Detection and recovery of fault impulses via improved harmonic product spectrum and its application in defect size estimation of train bearings. *Measurement* **2016**, *91*, 421–439. [CrossRef]

32. Zhou, Z.-H.; Feng, J. Deep forest: Towards an alternative to deep neural networks. In Proceedings of the International Joint Conference on Artificial Intelligence (IJCAI), Melbourne, Australia, 19–25 August 2017.

33. Yu, Z.; Wu, X.; Gu, X. Fully convolutional networks for surface defect inspection in industrial environment. In Proceedings of the International Conference on Computer Vision Systems (ICVS), Venice, Italy, 22–29 October 2017.

34. Zhao, Z.; Li, B.; Dong, R.; Zhao, P. A Surface Defect Detection Method Based on Positive Samples. In Proceedings of the Pacific Rim International Conference on Artificial Intelligence (PRICAI), Nanjing, China, 28–31 August 2018.

35. Song, K.; Hu, S.; Yan, Y. Automatic recognition of surface defects on hot-rolled steel strip using scattering convolution network. *J. Comput. Inf. Syst.* **2014**, *10*, 3049–3055.

Article

Blister Defect Detection Based on Convolutional Neural Network for Polymer Lithium-Ion Battery

Liyong Ma [1,*], Wei Xie [1,*] and Yong Zhang [2]

[1] School of Information Science and Engineering, Harbin Institute of Technology, Weihai 264209, China
[2] Department of Civil Engineering, Harbin Institute of Technology, Weihai 264209, China; yz@hit.edu.cn
* Correspondence: maly@hitwh.edu.cn (L.M.); xw1248@163.com (W.X.); Tel.: +86-631-2679199 (L.M.)

Received: 25 January 2019; Accepted: 11 March 2019; Published: 14 March 2019

Abstract: To ensure the quality and reliability of polymer lithium-ion battery (PLB), automatic blister defect detection instead of manual detection is developed in the production of PLB cell sheets. A convolutional neural network (CNN) based detection method is proposed to detect blister in cell sheets employing cell sheet images. An improved architecture for dense block and a learning method based on optimization of learning rate are discussed. The proposed method was superior to other machine learning based methods when the classification performance and confusion matrix were compared in experiments. The proposed CNN method had the best defect detection performance and real-time performance for industry field application.

Keywords: defect detection; polymer lithium-ion battery; convolutional neural network; deep learning; blister defect; flower pollination algorithm

---

1. Introduction

The application of lithium-ion batteries has changed consumer electronic products, which greatly reduce the weight and volume of mobile phones, notebooks, and other portable products. Lithium-ion batteries have been widely studied, including model, optimal design, and so on [1–6]. At present, the most commonly used lithium-ion battery is polymer lithium-ion battery (PLB). The cathode materials of PLB are commonly lithium cobalt acid, lithium manganese acid, ternary materials or lithium iron phosphate. The anode materials usually use carbon materials, such as artificial graphite, natural graphite, intermediate phase carbon microspheres, petcoke, carbon fiber, pyrolysis resin carbon and so on. C-rate is the measurement of the charge and discharge current with respect to its nominal capacity. At present, most PLB use polymer gel electrolyte instead of liquid electrolyte, which makes PLB have the advantages of thinning, arbitrary area and arbitrary shape. These characteristics improve the capacity of PLB, and PLB has the characteristics of miniaturization, thinning, and light quantification. PLB has been widely used in portable electronic equipment, and it is gradually being applied to more fields. With the application of PLB in more and more electronic products, the quality of PLB has more and more influence on the quality of electronic products. The quality of PLB is critical to the quality and reliability of electronic products.

Recently, an automatic line of PLB has been developed in Dongsheng Energy Corporation, Weihai, China. The PLB has a voltage of 48 V, a charging current of 2–3 A, and nominal capacity of 16 AH. The anode of PLB is a conductive high molecular polymer, the catrode is graphite, and a colloidal polymer electrolyte is used. During the production process of the PLB, several cathode pieces, anode pieces, and separator pieces are combined to produce a cell sheet. Great battery capacity can be provided with more cell sheets combined together. Thus, the quality inspection of cell sheets in automated production lines is essential to the quality of the final battery product. The cell sheet defect needs to be detected to ensure product quality. Blister is a kind of common defect in the grid net of cell sheet. There are two main reasons for blister. One is that the coating components do not mix properly with appropriate time to form a homogeneous slurry, and the other is that the feed rate is not appropriate [7]. Blister can damage the chemical properties of the battery and cause micro-short circuits. Blister seriously affects the safety, service life and other quality characteristics of PLB. The detection of blister in cell sheets is inefficient and laborsome when it is performed manually. Automated blister detection methods need to be developed.

Some automated defect detection methods for lithium-ion battery have been developed. For example, X-ray is used for electrode coating detection [8], computed tomography is employed to inspect defects and structural deformations [9], laser and thermography methods are used to detect battery electrodes [10], and thermography is used to detect defects [11,12]. Reconstruction of object surface is an important method for these detection methods [13]. Combining computer vision with senseless detection is also effective [14]. Currently, X-Ray, computed tomography, and thermography methods are mainly used for inner defect detection of lithium-ion battery and are not used for blister detection, which appears on the surface of PLB. Laser and vision inspection can be used for blister detection. This article uses vision inspection technology. Compared with laser, the main advantages of vision inspection are: First, the hardware cost of the device is low. Second, the speed is fast. Laser detection requires scanning the PLB surface line-by-line, while visual inspection can complete image capture at once.

Visual inspection is a fast, convenient and economical method for detecting surface defects. Compared with manual processing, vision inspection does not suffer from fatigue, emotion, repetitive work boredom and other factors leading to reduced detection efficiency. Due to its real-time and low-cost features, it is widely used in automated production lines and other fields [15–18]. Visual inspection has been considered for lithium-ion battery production recently employing industry camera and image processing technique [15–17]. These studies focus on applying traditional image processing method to the inspection of flaws, scratches, and defects in battery separator or electrode surface. Structured light is important for visual inspection, and a novel classifier subset selection for stacked generalization is reported in [19]. In these studies, feature extraction of defects or flaws is the key to successful detection.

Defect detection can be regarded as a classification problem of battery components. The battery components are classified into qualified and unqualified according to whether there are defects. Since machine learning methods have made great progress in classification problem and have produced many examples of successful applications, machine learning methods have also been used for defect detection for batteries. Some defect detection applications based on common machine learning algorithms have been developed. The neural network method is an early machine learning method. It is applied to Li-ion battery, defect diagnosis or evaluation of battery module state [20–23]. Support vector machine (SVM) is able to solve the nonlinear classification problem when the number of samples in the training dataset is small. SVM is applied to the classification of post-weld defects for battery [24]. Another improved tensor based SVM method is used for bubble detection in cell sheets [25,26]. In these machine learning methods, the selected features determine the accuracy of the classification [27,28]. These features are selected by hand, and how to select these features is a difficult problem.

With the development of machine learning recently, deep learning technology has shown impressive results in image classification applications [29–36]. As a machine learning technology, deep learning simulates the human brain. It can automatically complete feature extraction, and the features can be employed for image classification with superior performance. As a widely used deep learning model, convolutional neural network (CNN) performs well in a variety of visual recognition tasks, especially in the field of image classification. CNN can automatically extract typical and representative features from the input image. CNN uses a hierarchical structure to gradually obtain the required advanced features from the low-level features, and then it uses these advanced features to complete image classification and other tasks. CNN has been used in defect detection of industry field. For example, CNN is employed to detect whether solders, chips or circuit boards have defects [37–45].

In this paper, a novel blister detection method based on CNN is proposed employing images of PLB sheets. The contribution of this paper includes two aspects. On the one hand, an improved CNN architecture with optimization based learning strategy is proposed. Trainable weight parameters are added to each skip connection to improve dense block. Optimization of the learning rate is used to improve the efficiency of training process. Experimental results indicate that the proposed CNN method is superior to other machine learning based methods for blister detection. On the other hand, this paper shows that deep learning based method has potential for defect detection application of PLB.

The rest of this paper is organized as follows. Blister defect detection for PLB and the proposed CNN method is described in Section 2. The experiments and performance evaluations are discussed in Section 3. Finally, this paper is concluded in Section 4.

The main abbreviations used in this paper are listed in Table 1. The main symbols used in this paper are listed in Table 2.

Table 1. Abbreviations.

Abbreviation	Definition
CNN	Convolutional Neural Network
CPU	Central Processing Unit
FPA	Flower Pollination Algorithm
GPU	Graphic Processing Unit
NN	Neural Network
PLB	Polymer Lithium-ion Battery
RAM	Random Access Memory
RCNN	Region Convolutional Neural Network
ReLU	Rectified Linear Unit
SVM	Support Vector Machine
STM	Support Tucker Machine
VGG	Visual Geometry Group

Table 2. Alphabetic symbols.

Symbol	Meaning
d	layer number
F_i	nonlinear transformation function of ith layer
g	function
$k_{i,j}$	parameters which determinate weights of x_j in ith layer
l_i	ith learning rate
l_{best}	optimal learning rate solution
L	variable drawn from Levy distribution
m	total pollen number
P_C	probability of choosing cross-pollination
s	step
t	iteration number
x	input of CNN network
x_i	input of ith layer
y	output of CNN network
y_i	output of ith layer
w	network parameters
W_i	parameters of ith layer
γ	scaling factor
ε	variable drawn from uniform distribution
λ	variable of gamma function
Γ	standard gamma function

2. Blister Defect Detection for PLB Based on CNN

2.1. Data Capture

Visual inspection is employed to detect blister of sheet net of PLB. PLB sheet is controlled by the manipulator and images of both sides are captured. The field image acquisition in the automated production line is shown in Figure 1. The damages and scratches of PLB sheet are easy to be detected using traditional image processing methods. Due to the inconsistent background color, shape and size, blister defects cannot be well detected using the usual image processing method. The image of the PLB sheet is divided into multiple patch images, and each patch image is detected. Some images of blister defects are shown in Figure 2.

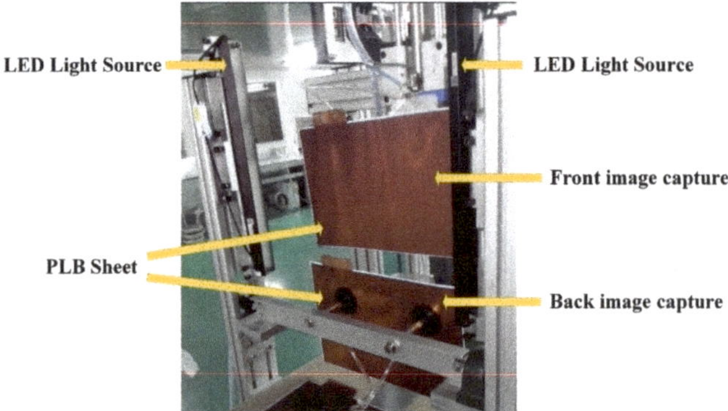

Figure 1. Image capture of PLB sheets.

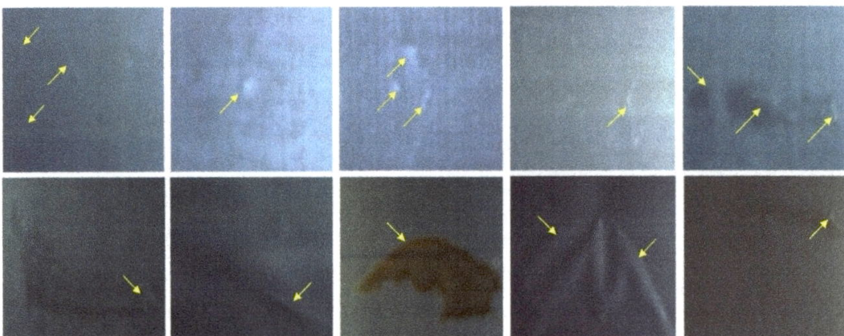

Figure 2. Blister defect samples of PLB sheet.

2.2. Detection Scheme Based on CNN

Blister defect detection problem of PLB is to judge whether PLB sheet has blister from images. It can be considered as an image classification problem. CNN method has achieved great success in the image classification applications. In this paper, an optimization based CNN method is proposed to detect blister defect of PLB sheet.

As an important deep learning model, CNN uses an architecture of multi-layer stack. Each layer in the stack can be considered an input-to-output conversion that is used to achieve selective extraction of image feature representations. CNN learns the mapping relationship between a large number of input samples and outputs with a combination of input layer, convolution layer, ReLU layer, pooling layer and fully connected layer. Input layer completes image preprocessing. The convolution layer implements the perception of local feature information of the image. These local information will be combined at a higher level to get global information. The convolution layer also greatly reduces the amount of computation through parameter sharing, and it extracts different features by employing multiple kernels. ReLU layer performs a nonlinear mapping of the output of convolutional layer. The pooling layer is used in the middle of a continuous convolutional layer to reduce overfitting and compress the amount of data and parameters. Fully connected layer is used to achieve the final classification using the advanced features extracted from the previous layers. The major feature of CNN is the shared convolution kernel, which works well for high-dimensional data processing. Another feature is that it is not necessary to manually select features and train weights.

Thus far, many successful architectures have been proposed for CNN, which include DenseNet [46], ResNet [47], VGG16, RCNN, etc., and some of them have been applied to defect detection [37–45]. VGG16 architecture is employed for pavement crack detection in [43]. Single shot multibox detector network is adopted to detect surface defects of container [44]. RCNN architecture is used for polymeric polarizer detection of liquid crystal display panel [45].

In this paper, an efficient CNN based detection method is provided for blister detection in PLB sheets. The experimental results indicate that the proposed method is superior to other machine learning based methods.

2.3. Improved Architecture for CNN

In the deep learning approach, deeper networks are used to accomplish complex tasks. The learning process of the neural network adopts the strategy of backpropagation, that is, the error calculated by the loss function is used to guide the update and optimization of the weights of the deep network through the backpropagation of the gradient. The deep neural network is composed of many nonlinear layers, and each nonlinear layer can be regarded as a nonlinear function. Therefore, the entire deep network can be regarded as a composite nonlinear function. The purpose of neural network learning is to make this nonlinear function perform a good mapping between input and output. To find the optimal solution of different input and output, the learning process is to find the appropriate depth network weights so that the loss function takes a minimum value. The gradient descent method is used to solve this minimum problem. Its idea is to take the negative direction of the current gradient as the search direction, and adjust the weights to make the loss function approach the local minimum, that is, let the loss function become smaller and smaller.

In backpropagation, the gradient is updated layer by layer. The gradient update can be seen as multiplying the output of the upper layer network nonlinear function by a factor. If the factor is less than 1, as the number of layers increases, the gradient update will decay exponentially and become smaller and smaller, gradually disappearing. This is called the gradient vanishment; it causes poor learning and training effect. To solve the gradient vanishment problem, short paths are often created from early layers to later layers in CNN architecture.

DenseNet is an efficient architecture of CNN for image classification. In DenseNet, all layers are connected directly to ensure maximum information transmission to solve the gradient vanishment problem. DenseNet uses dense block to create short paths from early layers to later layers. Dense block uses skip connections not only to connect the upper and lower layers, but also to achieve cross-layer connections. The gradients obtained from each layer are derived from the gradient concatenation of the preceding layers. Because the gradient transfers directly between layers, the effect of gradient vanishment is reduced. This kind of architecture also strengthens the transmission and usage of features.

Denote the input and output of the dth layer as x_d and y_d, respectively, then

$$y_d = F_d([x_0, x_1, ..., x_{d-1}], W_d), \tag{1}$$

where F_d is a nonlinear transformation function, the symbol [] indicates the concatenation operation, and W_d is the parameters of F_d in the dth layer. In the dense block of DenseNet, previous layer features are skipped and connected with concatenation operation. That is to say, the features of different layers are treated in an equivalent manner in this architecture.

However, this is not the case in actual classification systems. Not all features of the previous layers play a key role in image classification. Only some of the key features are important for classification. Inspired by this, a novel weight-based architecture is proposed to improve the network performance of dense block. An improved dense block architecture is proposed in this paper, as shown in Figure 3. Its detailed architecture is shown in Figure 4. In this architecture, the output is

$$y_d = F_d([x_0 k_{d,0}, x_1 k_{d,1}, ..., x_{d-1} k_{d,d-1}], W_d), \tag{2}$$

where $k_{d,0}, k_{d,1}, ..., k_{d,d-1}$ are the parameters that determine the weights of $x_0, x_1, ..., x_{d-1}$ to be concatenated together. These parameters are trained during the CNN training process. The whole CNN architecture proposed in this paper is illustrated in Figure 5.

In the above architecture, the weight parameters will be effectively trained in the training of CNN network. The weight parameters here have practical meanings for indicating how important the corresponding feature map is. The greater is the weight value, the more important is the role of the corresponding feature map in the classification task. That is, the corresponding features contain more useful information for classification. When the trainable weight parameters are introduced in our proposed architecture, the important features can be quickly found and efficiently represented for image classification.

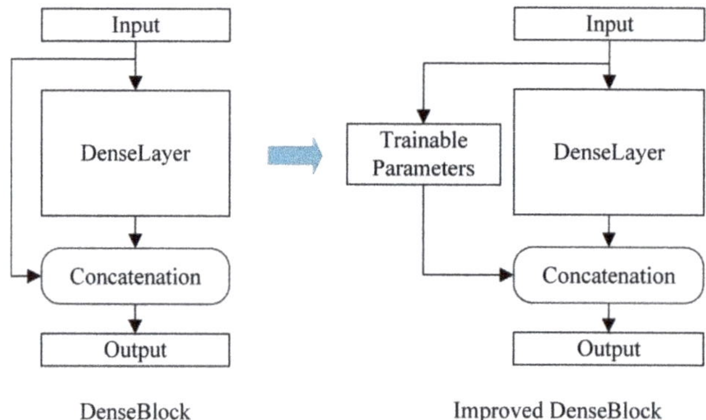

Figure 3. Improved dense block with trainable parameters for concatenation.

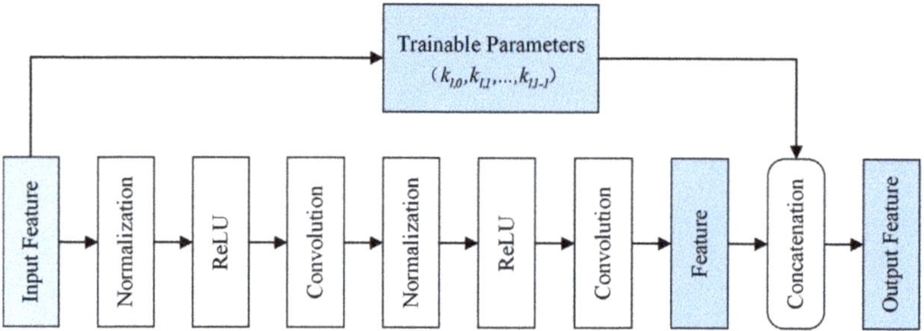

Figure 4. Improved dense block architecture.

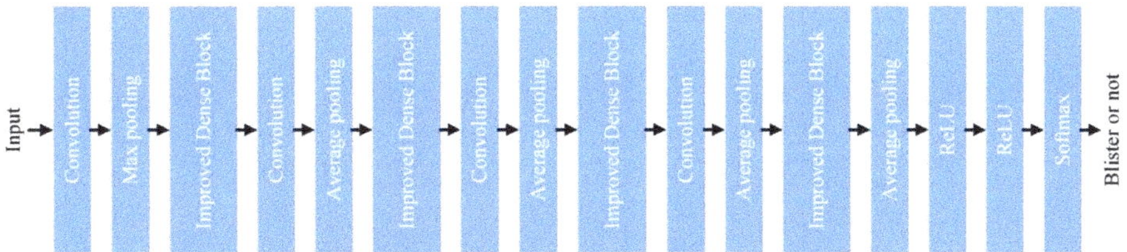

Figure 5. Proposed CNN architecture for blister detection.

2.4. Training Method Based on Optimization of Learning Rate

Learning rate is an important super-parameter in CNN. How to adjust the learning rate is one of the key elements for training a good CNN model. When the learning rate is too large, the learning process becomes unstable, and small learning rate leads to extremely long training time. By properly setting the learning rate, it is possible to reasonably improve the training speed and reduce the training time while stabilizing the training.

When the number of samples is large, the calculation of gradient descent processing on the entire sample is slow and inefficient. The method of dividing the samples into mini-batch is usually used to increase the speed. Let x be the input of CNN network in mini-batch processing, w be the network parameter, l be the learning rate, and the output is

$$y = g(x, w). \tag{3}$$

Using loss function can obtain the loss by comparing the output y with its label. The gradient is obtained with $\triangledown w = \partial C / \partial w$. w is updated as

$$w^{t+1} = w^t - l \cdot \triangledown w^t, \tag{4}$$

where t is the current iteration number.

In the process of above mini-batch-based learning, after the current mini-batch parameter w is updated, the processing and parameters of the next mini-batch data are continuously updated. However, the effect of parameter update for mini-batch learning has not been verified in this process. At the same time, the learning rate is usually manually selected based on experience, so it is likely that the calculation loss of the current mini-batch cannot be effectively reduced.

In this paper, a mechanism for optimizing the learning rate is provided. For each mini-batch, the optimal learning rate is found before the update formula (Equation (4)) is applied. Thus, the current mini-batch can reduce the loss function value. In other words, the original mini-batch does not guarantee that each update will be done in the direction of the correct gradient. The mechanism provided in this paper makes mini-batch update in the correct direction every time, which improves the efficiency of training and reduces the training time.

Flower pollination algorithm (FPA) is a new optimization algorithm of meta heuristic swarm intelligent. FPA is optimized by simulating the pollination process of flowering plants in nature. The pollination process includes two modes, self-pollination and cross-pollination, which represent local search and global search, respectively. Cross-pollination occurs between the pollens of different plants. Pollinators can fly for a long time and transmit pollen over a long distance. In contrast, self-pollination is the implantation of the same flower or different flowers of the same plant in the pollen, usually without pollinators. In the existing engineering applications, FPA shows the ability to search in the space with multiple local optima adaptively. FPA can avoid premature convergence, thus it has better performance.

FPA uses four rules as follows [48,49].

Rule 1. Biological biotic and cross-pollination is a process of global pollination by pollinators with pollen.

Rule 2. Biological abiotic and self-pollination is the process of local pollination.

Rule 3. Flower constancy. Plants and pollinators form a partnership to maximize the reproduction.

Rule 4. Switch probability controls the conversion between global pollination and local pollination.

FPA has achieved good results in solving multi-objective optimization problem and other application problems [50–53]. FPA has robust performance for applications. FPA has only few parameters. FPA is employed to find optimal learning rate in this paper.

FPA simulates two kinds of pollination, which are cross-pollination and self-pollination. Each flower in FPA is regarded as a solution to the destination function. Cross-pollination or self-pollination is selected by a flower to reproduce. This choice is selected by switch probability, the probability of choosing cross-pollination is P_C, and the probability of choosing self-pollination is $1 - P_C$. Cross-pollination operations draw on the cross-pollination method of different flowers by bees and butterflies at a long distance. The flight of pollinators is regarded as Levy flight, so the global pollination is modeled using a Levy distribution. Similarly, self-pollination models near-distance pollination in nature.

The optimal learning rate for CNN training is found in this paper employing FPA, and the main steps are summarized as below.

Step 1. Initialize parameters. The initial parameters include: maximum iteration number N, total pollen number m, and probability P_C. The learning rate is the pollen in FPA model. It conforms to the standard distribution and takes values in the range of $[l_{min}, l_{max}]$. m learning rates are created, which are listed as $l_1, l_2, ..., l_m$.

Step 2. FPA operation. Probability P is chosen randomly.

When $P \leq P_C$, the current learning rate l_i is updated as below to simulate cross-pollination

$$l_i^{t+1} \leftarrow l_i^t + \gamma \cdot L \cdot (l_{best}^t - l_i^t), \tag{5}$$

where l_{best} is the optimal learning rate solution in the global. γ is the scaling factor; its value is suggested to be in the range of (0,10) in previous studies [49]. It was found that the best result can be obtained when it is set to 0.1 in this application. Thus, it was set to 0.1 in this study. L can be drawn from Levy distribution as

$$L \backsim \frac{\lambda \Gamma(\lambda) sin(\pi \lambda / 2)}{\pi} \frac{1}{s^{1+\lambda}}, \tag{6}$$

where $\Gamma(\lambda)$ is a standard gamma function, s is step, and λ was set to 1.5 in this study as recommended [49].

When $P > P_C$, the current learning rate l_i is updated as below to simulate self-pollination operation

$$l_i^{t+1} \leftarrow l_i^t + \varepsilon \cdot (l_u^t - l_v^t), \tag{7}$$

where ε is drawn from the uniform distribution of $[0, 1]$, l_u and l_v are two randomly pollens, which represent learning rates, and $1 \leq u, v \leq m$. The implementation flow chart of FPA is illustrated in Figure 6. Previous studies have suggested that the range of P_c is $[0.1, 0.9]$, and the recommended value is 0.8 [54]. In the FPA implementation of this study, P_c was set to 0.8.

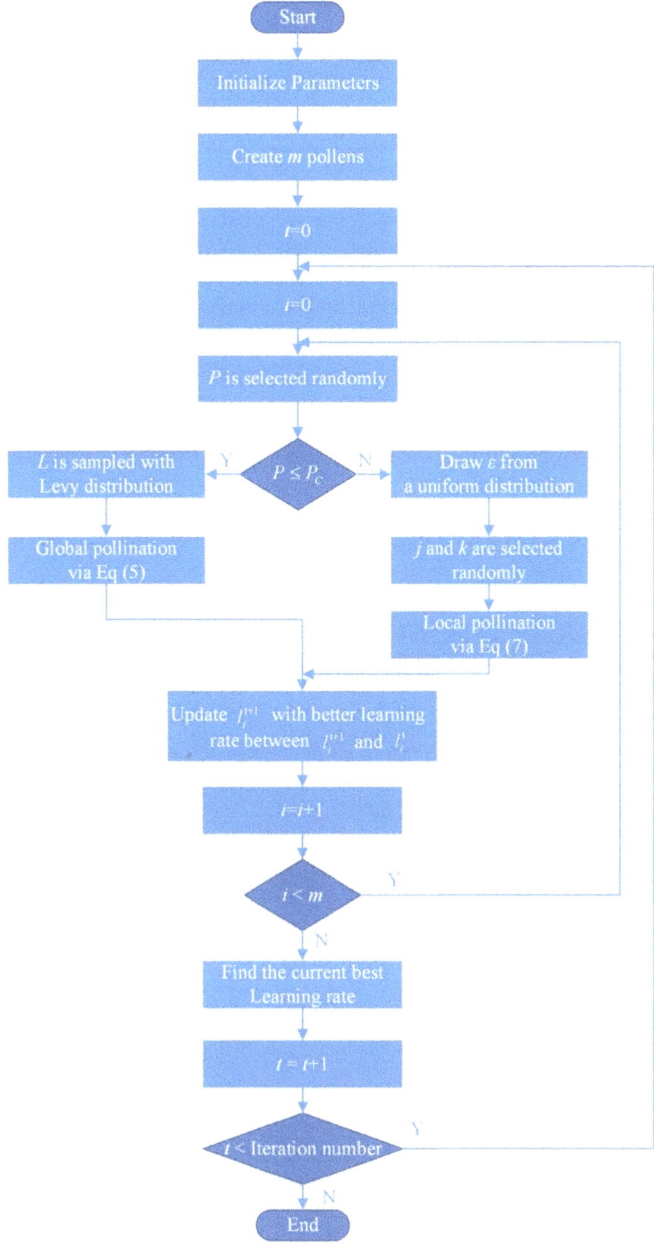

Figure 6. Flow chart of FPA implementation.

2.5. Dataset and Training

Some images of both sides of PLB sheets were captured in the automatic line. The size of the PLB image obtained from the camera was 2448 × 2048. The size of the input image in the CNN method was 219 × 219. We cut the polar area in each image of the PLB into 160 patch images with the size of 219 × 219. Because each PLB has two sides, 320 patch images of each PLB were taken. Professional engineers selected 600 qualified PLBs and 600 blister PLBs to obtain images. After cutting out patches, a typical patch was selected as sample image. Finally, a blister sample image dataset was created including 11,600 qualified images and 10,460 blister images. Then, 1800 qualified images and 1400 blister images were selected randomly for test in the CNN training, and the other images were employed for training.

It was obvious that the dataset was not large enough. To overcome the overfitting problem caused by small dataset training in CNN, transfer learning was employed. A set of pre-trained weights was transferred from ImageNet to the network proposed in this paper. After transferring network weights, the proposed network could be trained employing the blister image dataset. To enhance CNN image classification performance, batch normalization, dropout strategy and early stop scheme were used as in other image classification task.

3. Results and Discussions

To evaluate the proposed blister detection method based on CNN, some other machine learning based methods were employed for comparison: neural network (NN) [22], support vector machine (SVM) [24], support Tucker machine (STM) [25], and CNN methods with DenseNet model [46], ResNet model [47], VGG16 model [43], and fast RCNN model [45]. The comparisons included classification performance evaluation and confusion matrix.

3.1. Classification Performance Evaluation

According to whether the classification results are correct, TP, TN, FP, and FN can be determined. TP means that the classification result is true and positive, TN means true negative, FP means false positive, and FN means false negative.

Recall, precision, accuracy, specificity and F1-score were employed as classification performance indicators to evaluate different methods. They are defined as follows [55].

$$\text{Recall} = \frac{TP}{TP + FN}. \tag{8}$$

$$\text{Precision} = \frac{TP}{TP + FP}. \tag{9}$$

$$\text{Specificity} = \frac{TN}{FP + TN}. \tag{10}$$

$$\text{Accuracy} = \frac{TP + TN}{TP + TN + FP + FN}. \tag{11}$$

$$\text{F1-score} = \frac{2TP}{2TP + FP + FN}. \tag{12}$$

Recall measures the proportion of actual positives that are correctly identified as such. Specificity represents the proportion of actual negatives that are correctly identified as such. Accuracy is defined as the proportion of all samples that have been successfully classified. Precision is the ratio of samples correctly classified as positive to all the samples that are classified. F1-score is the harmonic mean of precision and sensitivity. When the above performance index is greater, the classification performance is better.

The above-mentioned five indicators of different blister classification methods are listed in Table 3. The method proposed in this paper had the greatest value for all performance indicators, meaning that the proposed CNN based method was superior to other classification methods for blister recognition of PLB sheets. The performance comparison is also shown in Figure 7, which indicates that the proposed method had the best classification performance. The improved CNN architecture with optimization based training method was efficient for blister detection when trainable weight parameters were added to skip connections in dense block.

Table 3. Classification performance of different methods for blister detection.

Method	NN	SVM	STM	DenseNet	ResNet	VGG	RCNN	Proposed
Recall	0.680	0.824	0.932	0.959	0.954	0.937	0.937	0.982
Precision	0.799	0.855	0.898	0.949	0.934	0.923	0.939	0.993
Accuracy	0.724	0.813	0.902	0.948	0.936	0.920	0.931	0.986
Specificity	0.781	0.798	0.864	0.934	0.914	0.899	0.922	0.991
F1-score	0.735	0.839	0.915	0.954	0.944	0.930	0.938	0.988

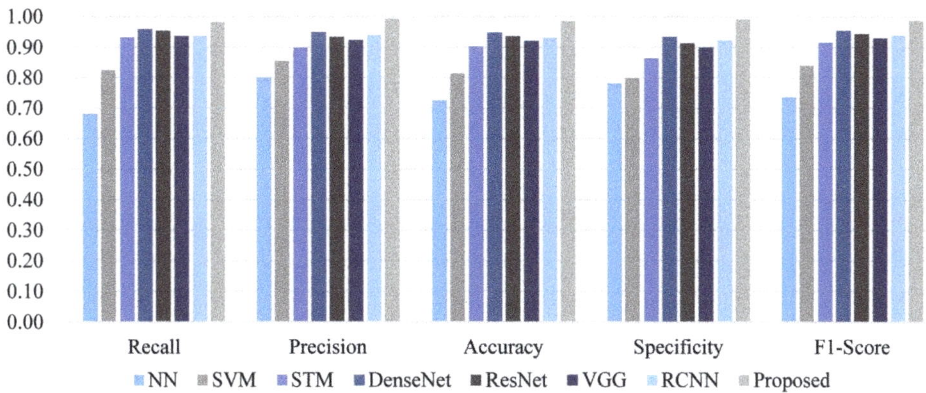

Figure 7. Classification performance comparison of different methods.

To observe the effectiveness and efficiency of the proposed FPA optimization in this paper, an ablation study was performed on the proposed FPA optimization. The FPA optimization based method was compared with the method without FPA optimization in the experiment. The results of the comparison are shown in Table 4, and the data are also illustrated in Figure 8.

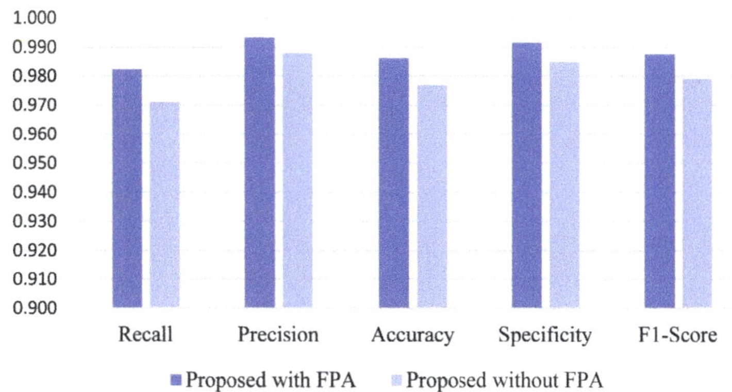

Figure 8. Classification performance comparison on FPA optimization.

Table 4. Classification performance comparison on the proposed FPA optimization.

Method	Proposed Method with FPA	Proposed Method without FPA
Recall	0.982	0.971
Precision	0.993	0.988
Accuracy	0.986	0.977
Specificity	0.991	0.985
F1-score	0.988	0.979

By observing the data in Table 4 and Figure 8, it can be seen that the performance was obviously improved when applying the proposed FPA optimization. Because FPA optimization could find the optimum learning rate, the training process was more efficient and the classification results were improved.

3.2. Confusion Matrix

Confusion matrix is often used to visualize the performance of supervised learning based classification. The matrix row represents samples in a predicted class while matrix column indicates the samples in an actual class [55]. Confusion matrices of the experiments in this paper are illustrated in Figure 9. The confusion matrix of the proposed method obtainEd the maximum value on the main diagonal and the minimum value in the secondary diagonal, showing that the proposed method had the best classification performance. This is consistent with the analysis results of the performance data presented in Section 3.1. This also indicates that our proposed method was the most efficient for blister detection.

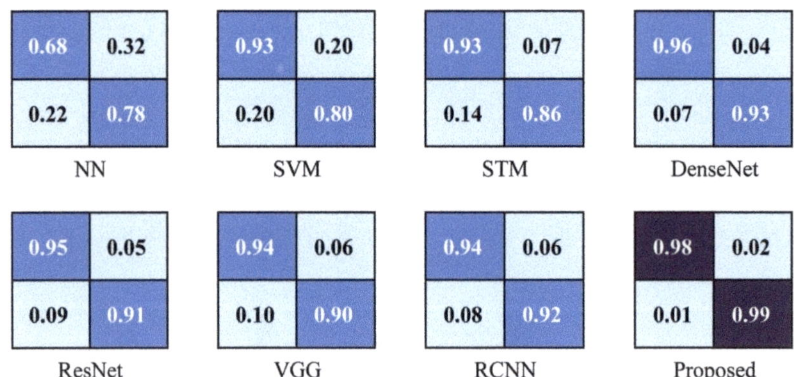

Figure 9. Confusion matrices of different methods.

3.3. Real-Time

All tests were performed on a computer with 32G RAM, Intel Xeon E5-2620 CPU, and NVIDIA GeForce 1080Ti GPU. Each sample test took no more than 0.3 s. The total processing time including capturing images of two sides and transportation on the production line for each PLB sheet was less than 10 s. If parallel processing were used, the processing speed could be improved. In the actual production line, parallel pipeline processing was adopted. The image acquisition and detection on both sides of the PLB were performed simultaneously, the time was shortened to half of the original, and the efficiency was increased to twice the manual processing. The proposed CNN based defect detection method was fast enough for real-time industry application.

4. Conclusions

Blister, a kind of common defect in the grid net of cell sheet, needs to be detected to ensure product quality of PLB. A blister detection method based on visual inspection was employed using images of PLB sheet. A deep learning detection method is proposed and it is fast enough for real-time industry application. The following conclusions can be drawn from this paper.

(1) Trainable weight parameters are added to each skip connection to improve dense block in CNN architecture. These parameters are trained during the CNN training process. This improvement efficiently finds important features for the image classification.

(2) CNN training is improved with the optimization of learning rate by FPA. Mini-batch is updated in the correct direction to improve the efficiency of training.

(3) A deep learning application for blister detection of PLB is developed. The proposed method had the best classification performance when it was compared with other methods.

(4) The proposed CNN based blister detection method is fast enough for real-time industry application.

Author Contributions: Conceptualization, L.M. and W.X.; Methodology, L.M. and Y.Z.; Writing—original draft, L.M.; and Writing—review and editing, W.X.

Funding: This research was funded by National Key R&D Program of China grant number 2018YFC0114800, Shandong Province Natural Science Foundation grant number ZR2018MF026, and University Co-construction Project at Weihai grant number ITDAZMZ001708.

Conflicts of Interest: The authors declare no conflict of interest. The funders had no role in the design of the study; in the collection, analyses, or interpretation of data; in the writing of the manuscript, or in the decision to publish the results.

References

1. Liu, L.; Guan, P.; Liu, C. Experimental and simulation investigations of porosity graded cathodes in mitigating battery degradation of high voltage lithium-ion batteries. *J. Electrochem. Soc.* **2017**, *164*, A3163–A3173. [CrossRef]
2. Liu, C.; Liu, L. Optimal design of Li-ion batteries through multi-physics modeling and multi-objective optimization. *J. Electrochem. Soc.* **2017**, *164*, E3254–E3264. [CrossRef]
3. Liu, C.; Liu, L. Optimizing battery design for fast charge through a genetic algorithm based multi-objective optimization framework. *ECS Trans.* **2017**, *77*, 257–271. [CrossRef]
4. Liu, L.; Park, J.; Lin, X.K.; Sastry, A.M.; Lu, W. A thermal-electrochemical model that gives spatial-dependent growth of solid electrolyte interphase in a Li-ion battery. *J. Power Sources* **2014**, *268*, 482–490. [CrossRef]
5. Lin, X.K.; Park, J.; Liu, L.; Lee, Y.; Sastry, A.M.; Lu, W. A Comprehensive capacity fade model and analysis for Li-ion batteries. *J. Electrochem. Soc.* **2013**, *160*, A1701–A1710. [CrossRef]
6. Guan, P.J.; Liu, L.; Lin, X.K. Simulation and experiment on solid electrolyte interphase (SEI) morphology evolution and Lithium-ion diffusion. *J. Electrochem. Soc.* **2015**, *162*, A1798–A1808. [CrossRef]
7. Mohanty, D.; Hockaday, E.; Li, J.; Hensley, D.K.; Daniel, C.; Wood, D.L. Effect of electrode manufacturing defects on electrochemical performance of lithium-ion batteries: Cognizance of the battery failure sources. *J. Power Sources* **2016**, *312*, 70–79. [CrossRef]
8. Etiemble, A.; Besnard, N.; Adrien, J.; Tran-Van, P.; Gautier, L.; Lestriez, B.; Maire, E. Quality control tool of electrode coating for lithium-ion batteries based on X-ray radiography. *J. Power Sources* **2015**, *298*, 285–291. [CrossRef]
9. Wu, Y.; Saxena, S.; Xing, Y.J.; Wang, Y.R.; Li, C.; Yung, W.K.C.; Pecht, M. Analysis of manufacturing-induced defects and structural deformations in Lithium-ion batteries using computed tomography. *Energies* **2018**, *11*, 925. [CrossRef]
10. Mohanty, D.; Li, J.L.; Born, R.; Maxey, L.C.; Dinwiddie, R.B.; Daniel, C.; Wood, D.L. Non-destructive evaluation of slot-die-coated lithium secondary battery electrodes by in-line laser caliper and IR thermography methods. *Anal. Methods* **2014**, *6*, 674–683. [CrossRef]

11. Robinson, J.B.; Engebretsen, E.; Finegan, D.P.; Darr, J.; Hinds, G.; Shearing, P.R.; Brett, D.J.L. Detection of internal defects in lithium-ion batteries using lock-in thermography. *ECS Electrochem. Lett.* **2015**, *4*, A106–A109. [CrossRef]

12. Sharp, N.; Oregan, P.; Adams, D.; Caruthers, J.; David, A.; Suchomel, M. Lithium-ion battery electrode inspection using pulse thermography. *NDT E Int.* **2014**, *64*, 41–51. [CrossRef]

13. Qian, X.Y.; Huang, X.G. Reconstruction of surfaces of revolution with partial sampling. *J. Comput. Appl. Math.* **2004**, *163*, 211–217. [CrossRef]

14. Pena, B.; Aramendi, G.; Rivero, A.; de Lacalle, L. Monitoring of drilling for buff detection using spindle torque. *Int. J. Mach. Tools Manuf.* **2005**, *45*, 1614–1621. [CrossRef]

15. Huber, J.; Tammer, C.; Kempter, A.; Seidel, C.; Reinhart, G. Optical quality inspection of battery separators. *Tm-Tech. Messen* **2015**, *82*, 495–505.

16. Huber, J.; Tammer, C.; Krotil, S.; Waidmann, S.; Hao, X.; Seidel, C.; Reinhart, G. Method for classification of battery separator defects using optical inspection. *Procedia CIRP* **2016**, *57*, 585–590. [CrossRef]

17. Li, K.; Dan, T. Research and design of inspection of LR6 battery negative surface scratches online defects based on computer vision. In Proceedings of the International Conference on Communications, Circuits and Systems, Chengdu, China, 15–17 November 2013; pp. 120–123.

18. Alberdi, A.; Rivero, A.; de Lacalle, L.; Etxeberria, I.; Suárez, A. Effect of process parameter on the Kerf geometry in abrasive water jet milling. *Int. J. Adv. Manuf. Technol.* **2010**, *51*, 467–480. [CrossRef]

19. Álvarez, A.; Sierra, B.; Arruti, A.; López-Gil, J.M.; Garay-Vitoria, N. Classifier subset selection for the stacked generalization method applied to emotion recognition in speech. *Sensors* **2016**, *16*, 21. [CrossRef]

20. Panchal, S.; Dincer, I.; Agelin-Chaab, M.; Fraser, R.; Fowler, M. Experimental and theoretical investigation of temperature distributions in a prismatic lithium-ion battery. *Int. J. Therm. Sci.* **2016**, *99*, 204–212. [CrossRef]

21. Panchal, S.; Dincer, I.; Agelin-Chaab, M.; Fraser, R.; Fowler, M. Thermal modeling and validation of temperature distributions in a prismatic lithium-ion battery at different discharge rates and varying boundary conditions. *Appl. Therm. Eng.* **2016**, *96*, 190–199. [CrossRef]

22. Zhao, Y.; Liu, P.; Wang, Z.; Zhang, L.; Hong, J. Fault and defect diagnosis of battery for electric vehicles based on big data analysis methods. *Appl. Energy* **2017**, *207*, 354–362. [CrossRef]

23. Liang, X.; Bao, N.; Zhang, J.; Garg, A.; Wang, S. Evaluation of battery modules state for electric vehicle using artificial neural network and experimental validation. *Energy Sci. Eng.* **2018**, *6*, 397–407. [CrossRef]

24. Wu, B.; Qin, L.; Zhang, Q.; Ma, Q. Research on vision-based post-welding quality inspection of power battery. *Hanjie Xuebao/Trans. China Weld. Inst.* **2018**, *39*, 122–128.

25. Ma, L.Y.; Hu, Y.H.; Zhang, Y. Support Tucker machines based bubble defect detection of lithium-ion polymer cell sheets. *Eng. Lett.* **2017**, *25*, 46–51.

26. Ma, L.Y. Support Tucker machines based marine oil spill detection using SAR images. *Indian J. Geo-Mar. Sci.* **2016**, *45*, 1445–1449.

27. Murua, M.; Suarez, A.; de Lacalle, L.; Santana, R.; Wretland, A. Feature extraction-based prediction of tool wear of Inconel 718 in face turning. *Insight* **2018**, *60*, 443–450. [CrossRef]

28. Coro, A.; Abasolo, M.; Aguirrebeitia, J.; de Lacalle, L. Inspection scheduling based on reliability updating of gas turbine welded structures. *Adv. Mech. Eng.* **2019**, *11*. [CrossRef]

29. LeCun, Y.; Bengio, Y.; Hinton, G. Deep learning. *Nature* **2015**, *521*, 436–444. [CrossRef]

30. Pouyanfar, S.; Sadiq, S.; Yan, Y.; Tian, H.; Tao, Y.; Reyes, M.P.; Shyu, M.L.; Chen, S.C.; Iyengar, S.S. A survey on deep learning: Algorithms, techniques, and applications. *ACM Comput. Surv.* **2019**, *51*, 92. [CrossRef]

31. Rawat, W.; Wang, Z. Deep convolutional neural networks for image classification: A comprehensive review. *Neural Comput.* **2017**, *29*, 2352–2449. [CrossRef]

32. Voulodimos, A.; Doulamis, N.; Doulamis, A.; Protopapadakis, E. Deep learning for computer vision: A brief review. *Comput. Intell. Neurosci.* **2018**, *4*, 7068349. [CrossRef] [PubMed]

33. Shin, H.C.; Roth, H.R.; Gao, M.C.; Lu, L.; Xu, Z.Y.; Nogues, I.; Yao, J.H.; Mollura, D.; Summers, R.M. Deep convolutional neural networks for computer-aided detection: CNN architectures, dataset characteristics and transfer learning. *IEEE Trans. Med. Imaging* **2016**, *35*, 1285–1298. [CrossRef] [PubMed]

34. Ma, L.Y.; Ma, C.K.; Liu, Y.J.; Wang, X.G.; Xie, W. Diagnosis of thyroid diseases using SPECT images based on convolutional neural network. *J. Med. Imaging Health Inform.* **2018**, *8*, 1684–1689. [CrossRef]

35. Ma, L.Y.; Ma, C.K.; Liu, Y.J.; Wang, X.G. Thyroid diagnosis from SPECT images using convolutional neural network with optimization. *Comput. Intell. Neurosci.* **2019**, *5*, 6212759. [CrossRef] [PubMed]

36. Sun, M.J.; Zhang, X.; Qu, G.; Zou, M.S.; Du, H.; Ma, L.Y.; Qu, Y.W. Automatic polyp detection in colonoscopy images: Convolutional neural network, dataset and transfer learning. *J. Med. Imaging Health Informat.* **2019**, *9*, 126–133.

37. Chen, F.C.; Jahanshahi, M.R. NB-CNN: Deep learning-based crack detection using convolutional neural network and Naive Bayes data fusion. *IEEE Trans. Ind. Electron.* **2018**, *65*, 4392–4400. [CrossRef]

38. Kim, J.; Kim, S.; Kwon, N.; Kang, H.; Kim, Y.; Lee, C. Deep learning based automatic defect classification in through-silicon Via process: FA: Factory automation. In Proceedings of the 29th Annual SEMI Advanced Semiconductor Manufacturing Conference, Saratoga Springs, NY, USA, 30 April–3 May 2018; pp. 35–39.

39. Park, J.K.; Kwon, B.K.; Park, J.H.; Kang, D.J. Machine learning-based imaging system for surface defect inspection. *Int. J. Precis. Eng. Manuf. Green Technol.* **2016**, *3*, 303–310. [CrossRef]

40. van Veenhuizen, M. Void detection in solder bumps with deep learning. *Microelectron. Reliab.* **2018**, *88*, 315–320. [CrossRef]

41. Cai, N.; Cen, G.; Wu, J.; Li, F.; Wang, H.; Chen, X. SMT solder joint inspection via a novel cascaded convolutional neural network. *IEEE Trans. Compon. Packag. Manuf. Technol.* **2018**, *8*, 670–677. [CrossRef]

42. Tao, X.; Zhang, D.P.; Ma, W.Z.; Liu, X.L.; Xu, D. Automatic metallic surface defect detection and recognition with convolutional neural networks. *Appl. Sci.* **2018**, *8*, 1575. [CrossRef]

43. Gopalakrishnan, K.; Khaitan, S.K.; Choudhary, A.; Agrawal, A. Deep convolutional neural networks with transfer learning for computer vision-based data-driven pavement distress detection. *Constr. Build. Mater.* **2017**, *157*, 322–330. [CrossRef]

44. Li, Y.T.; Huang, H.S.; Xie, Q.S.; Yao, L.G.; Chen, Q.P. Research on a surface defect detection algorithm based on MobileNet-SSD. *Appl. Sci.* **2018**, *8*, 1678. [CrossRef]

45. Lei, H.W.; Wang, B.; Wu, H.H.; Wang, A.H. Defect detection for polymeric polarizer based on faster R-CNN. *J. Inf. Hid. Multimed. Sign. Process.* **2018**, *9*, 1414–1420.

46. Huang, G.; Liu, Z.; Van Der Maaten, L.; Weinberger, K.Q. Densely connected convolutional networks. In Proceedings of the 30th IEEE Conference on Computer Vision and Pattern Recognition, Honolulu, HI, USA, 21–26 July 2017; pp. 2261–2269.

47. He, K.; Zhang, X.; Ren, S.; Sun, J. Deep residual learning for image recognition. In Proceedings of the IEEE Conference on Computer Vision and Pattern Recognition, Las Vegas, NV, USA, 27–30 June 2016; pp. 770–778.

48. Yang, X.S.; Karamanoglu, M.; He, X.S. Flower pollination algorithm: A novel approach for multiobjective optimization. *Eng. Optim.* **2014**, *46*, 1222–1237. [CrossRef]

49. Abdel-Basset, M.; Shawky, L.A. Flower pollination algorithm: a comprehensive review. *Artif. Intell. Rev.* **2018**, *50*, 1–25. [CrossRef]

50. Galvez, J.; Cuevas, E.; Avalos, O. Flower Pollination Algorithm for multimodal optimization. *Int. J. Comput. Intell. Syst.* **2017**, *10*, 627–646. [CrossRef]

51. Nabil, E. A modified flower pollination algorithm for global optimization. *Expert Syst. Appl.* **2016**, *57*, 192–203. [CrossRef]

52. Shen, L.; Fan, C.; Huang, X. Multi-level image thresholding using modified flower pollination algorithm. *IEEE Access* **2018**, *6*, 30508–30519. [CrossRef]

53. Xu, S.H.; Wang, Y.; Liu, X. Parameter estimation for chaotic systems via a hybrid flower pollination algorithm. *Neural Comput. Appl.* **2018**, *30*, 2607–2623. [CrossRef]

54. Zhou, Y.; Zhang, S.; Luo, Q.; Wen, C. Using flower pollination algorithm and atomic potential function for shape matching. *Neural Comput. Appl.* **2018**, *29*, 21–40. [CrossRef]

55. Power, D. Evaluation: from precision, recall and F-factor to ROC, informedness, markedness & correlation. *J. Mach. Learn. Technol.* **2011**, *2*, 37–63.

Article

Contour Detection for Fibre of Preserved Szechuan Pickle Based on Dilated Convolution

Hongyang Li [1,2], Lizhuang Liu [1,*], Zhenqi Han [1] and Dan Zhao [1]

1   Shanghai Advanced Research Institute, Chinese Academy of Sciences, Shanghai 201210, China
2   University of Chinese Academy of Sciences, Beijing 100049, China
*   Correspondence: liulz@sari.ac.cn

Received: 26 April 2019; Accepted: 27 June 2019; Published: 1 July 2019

Featured Application: Our method is applied to the automatic detection of the fibre of preserved Szechuan pickle. This method can be extended to the detection of non-salient contours and provide ideas for the detection of special contours.

Abstract: Peeling fibre is an indispensable process in the production of preserved Szechuan pickle, the accuracy of which can significantly influence the quality of the products, and thus the contour method of fibre detection, as a core algorithm of the automatic peeling device, is studied. The fibre contour is a kind of non-salient contour, characterized by big intra-class differences and small inter-class differences, meaning that the feature of the contour is not discriminative. The method called dilated-holistically-nested edge detection (Dilated-HED) is proposed to detect the fibre contour, which is built based on the HED network and dilated convolution. The experimental results for our dataset show that the Pixel Accuracy (PA) is 99.52% and the Mean Intersection over Union (MIoU) is 49.99%, achieving state-of-the-art performance.

Keywords: fibre of preserved Szechuan pickle; contour detection; dilated convolutions; HED

---

1. Introduction

The production of preserved Szechuan pickle includes peeling fibre, cleaning, cutting and so on. The cleaning and cutting processes can be easily and efficiently implemented with automatic devices, but peeling is always done manually, which limits the yield and quality of preserved Szechuan pickle. An automatic peeling device is needed to effectively solve this problem. Contour detection, as an important part of the device, is a necessary process to identify the location of fibre and guide the cutting tool.

In this paper, we address the problem of detecting contour in the fibre of preserved Szechuan pickle. The preserved Szechuan pickles consist of fibre, flesh and peel, as shown in Figure 1. The fibre of preserved Szechuan pickle is inedible, while the flesh can be eaten as a dish. In the productive process, the existence of fibre is not allowed and it is necessary to peel off the fibre.

By analyzing the section of the preserved Szechuan pickle stem, the contour of the preserved Szechuan pickle can be seen as one kind of non-salient contour. It is unsmoothed, irregular and the gray distinction between the fibre and flesh is small. The characteristics are not obvious enough and there is a large difference between individuals. The contour we studied in this paper is non-salient, which means that it is difficult to distinguish the contour from the background in texture, color and other aspects. This special contour also puts forward higher requirements for detection.

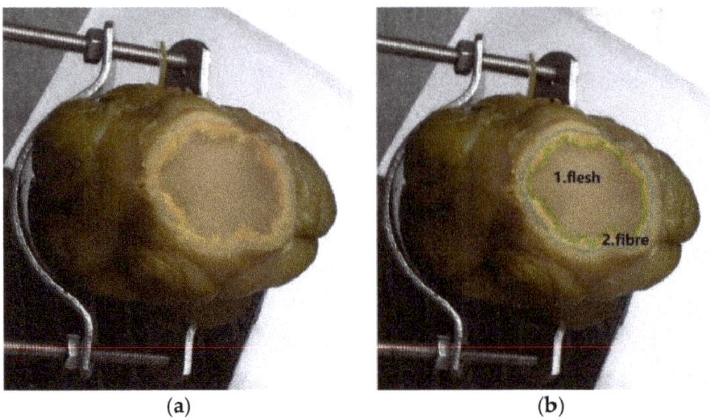

(a)   (b)

Figure 1. (**a**) Picture of a section of preserved Szechuan pickle; (**b**)the region within the green line is flesh, and the region surrounded by the blue and green line is fibre.

Contour detection has broad application prospects and is an important part of object segmentation, target detection, recognition and tracking [1]. There is a trend of using convolutional networks to detect contour since convolution neural networks show a strong capability to learn high-level representations of images. The effective integration of these methods and machines has promoted great developments in detection, such as the fully convolutional network (FCN) [2], holistically-nested edge detection (HED) [3], richer convolutional features for edge detection (RCF) [4] and so on. The features of the contour have the characteristics of large intra-class differences and small differences between classes. The general contour detection method has limited ability for these features. At present, there is little research into non-salient contour, but it is also one of the necessary technical means for production.

For this contour, we improve the structure of the HED network and use the output of each stage effectively. Our method automatically learns rich hierarchical representations and is able to make multi-scale predictions. We use dilated convolutions [5] to increase the receptive field. It is useful to reduce the loss of information and enhance the acquisition of spatial hierarchical information to improve the detection effect of the non-salient contours. In a complex background, our method can detect the region of contour. Finally, the Pixel Accuracy (PA) of our method [2] is 99.52% and the mean intersection over union (MIoU) [2] is 49.99%. Compared with the HED and RCF, the PA of our results increased by 3.80% and 9.61%, and the MIoU increased by 1.91% and 4.83% respectively.

2. Related Work

Contour detection has broad application prospects in computer vision, medical image and industrial production. At present, the main detection methods include the shallow feature-based method and deep feature-based method.

Shallow feature-based methods can be divided into edge-based, pixel-based and local region-based methods. Edge-based approaches are based on contour related edges or curves provided by edge detectors or human prior experience, aiming to determine whether they are contained in a certain contour [1]. Traditional operators in edge detection are widely used for high efficiency and strong applicability, such as Sobel, Laplace and Canny. Sobel [6] is a typical edge detection operator based on first derivative. Because it introduces a local average operation and has a smooth effect on noise and can eliminate the influence of noise very well. Laplace [7] is an isotropic operator and a second-order differential operator, which responds more strongly to isolated pixels than to edges or lines, so it is only suitable for noiseless images. Canny [8,9] is a multi-stage optimization operator with filtering, enhancement and detection. Its function is better than the previous examples. In pixel-based approaches, features are constructed and then employed to determine whether each pixel of the image belongs to a contour [1]. The following three methods are pixel-based contour detection methods. Pb [10] is a probabilistic detector that combines discontinuity features with other gradient features, including color and texture gradients. Sparse code gradient (SCG) features can automatically learn from image data through sparse coding, thus minimizing human involvement [1,11]. In view of this point, an improved normalized cuts algorithm called multiscale combinatorial grouping (MCG) is proposed, providing a 20-fold increase in speed to the eigenvector computation [1,12]. Regarding contours as boundaries of interesting regions, region-based approaches take advantage of internal information of the regions to enhance their effectiveness and robustness [1]. The oriented watershed transform (OWT) was proposed by Jones et al. [13] and can form initial regions for the construction of an ultrametric contour map (UCM) [1,14,15], which also belongs to the region-based approaches.

Convolutional neural networks can extract deep features, which refer to the high-level representation of images. We will detail the method based on a convolutional neural network that is the most effective and suitable for a large number of samples. Deep convolutional neural networks (DCNNs) have recently shown impressive performance in various tasks such as classification, image and video detection, and segmentation [1]. AlexNet [16] is designed by Hinton and Alex Krizhevsky and it uses a GPU to speed up operations. ReLU, Dropout and LRN have been successfully applied in CNN for the first time. VGG net (Visual Geometry Group Network) [17] is a deep convolution neural network, which has 16 layers or 19 layers. In the structure, 3 x 3 filters which can effectively extract image features were used completely. The filters of GoogLeNet [18] have multiple scales, which solve the limitation of depth and width by stacking modules together. Jonathan Long et al. proposed FCN [2] for image segmentation. It converts the fully-connected layers of VGG into convolutional ones and attempts to harness information from multiple layers to better estimate the object boundaries [1]. The conditional random field (CRF) proposed by Lafferty et al. combines the characteristics of the maximum entropy model and hidden Markov model. It is also an undirected graph model [19]. In the same FCN framework, the dilated convolution is used to obtain more information about features, and the fully connected CRF is used to refine the label maps. The net is called DeepLab [20] and it can produce high-resolution segmentation. HED is an end-to-end approach based on FCN and VGG [4]. The edges of different scales are output through multiple side outputs, and the final edge output is obtained through a training weighted-fusion layer. HED improves the accuracy of edge detection through feature fusion. Inspired by the HED network, RCF was proposed in 2017 and achieved state-of-the-art performance on several available datasets [4]. It is very helpful to use this rich hierarchical information at each stage, so the model increases the number of output layers on the basis of HED.

With the development of contour detection technology, it is no longer difficult to recognize the general contour. Because of the non-saliency of the contour, many technical means cannot achieve the expected results. Under this condition, the detection of non-salient contour can be studied in depth on the basis of the existing contour detection technology and look for a technical means to distinguish this non-salient feature to form a complete and accurate contour.

3. Contour Detection for Fibre of Preserved Szechuan Pickle

3.1. HED Architecture

HED is improved and adjusted based on VGG16 net, and it shows a good performance in various tasks, containing 13 convolution layers and five side-outputs. The side-outputs contain multi-scale features extracted by the network. Supposing we have M side-output layers in the network, the classifier corresponding to each side-output can be defined as

$$L_{side}(W, w) = \sum_{m=1}^{M} \alpha_m l_{side}^{(m)}\big(W, w^{(m)}\big), \tag{1}$$

where L_{side} denotes the image-level loss function of the side-output, W denotes the set of all standard network layer parameters, the corresponding weight can be denoted as w and M is the number of layers of the side-output [3].

According to the position of side-outputs, the structure of HED can be divided into five stages. Multiple stages with different strides can capture the inherent scales of contour maps [3]. After inputting pictures, we obtain the contour map predictions from both the side-output layers and the weighted-fusion layer:

$$\left(\hat{Y}_{HED}\right) = \text{Average}(\text{CNN}(X, (W, w, h)^*)),\tag{2}$$

where $\text{CNN}(\cdot)$ denotes the map produced by the network, $\hat{Y}_i$ denotes the predictions through i stage, X denotes the raw input image and h is the fusion weight.

This multi-scale and multi-level feature information is conducive to the transmission of holistic information and helps the network to obtain better prediction results. This structure is beneficial to our method. In the process of testing, the output results also prove that our choice is effective.

3.2. Dilated Convolutions

Dilated convolution increases the reception field by injecting holes into standard convolution maps. Compared with the original normal convolution, dilated convolution has a hyper-parameter called the dilation rate, which refers to the number of kernels intervals (e.g., the normal convolution is dilatation rate 1). The discrete convolution operator $*_l$ can be defined as

$$(F *_l k)(p) = \sum_{s+lt=p} F(s)k(t),\tag{3}$$

where F denotes a discrete function, and k is a discrete filter of $\text{size}(2r+1)^2$. We will refer to $*_l$ as a dilated convolution [5].

Dilated convolution can solve problems such as loss of data structure, loss of spatial hierarchical information, and an inability to reconstruct small object information. For the contour of preserved Szechuan pickle, dilated convolution has a better effect on information extraction when the difference of feature information is small. We hope to give full rein to its advantages and it will play an active role in our research.

3.3. Dilated-HED

Inspired by many proposed contour detection models, we designed our own model for the contour detection of preserved Szechuan pickle on the basis of HED. In our repeated tests, this structure has the best result, as can be seen in Figure 2.

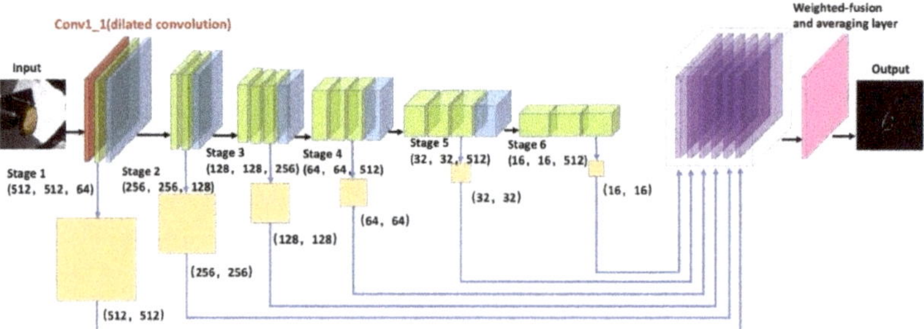

Figure 2. The structure of our model. The dark red cube represents the dilated convolution layer, the green cubes represent the convolution layer, the blue cubes represent the pooling layer, and the side-output layers are yellow. The purple cubes in the transparent cube represent the fusion of the output of each layer, and finally, a weighted-fusion layer is formed, represented by pink.

The input is an image and our network outputs a contour possibility map of the same size. Our model is improved on the basis of HED, adding a stage and dilated convolution, which is more sensitive to features. There are 16 convolution layers and five max-pooling layers. Its convolution layers are divided into six stages, in which a side-output layer is inserted after the last convolution of each stage. There are two convolution layers and one max-pooling layer in the first and the second stage. From the third to the fifth stage, each stage contains three convolution layers and one max-pooling layer. The last stage contains only three convolution layers. The outputs of dilated-HED are multi-scale and are finally integrated into a weighted-fusion layer by deconvolution. The weighted-fusion layer can automatically learn how to combine and average outputs from multiple scales, which are scaled according to 1, 2, 4, 8, 16 and 32. Our network architecture has different receptive field sizes and it will become larger with the deepening of the network. The convolution kernel of 3×3 is used in the HED network, and its receptive field is small. Increasing receptive field can capture more regional features. We use dilated convolution, which is located on the first layer of the whole structure. Our experimental results for the test data will be presented in the next section.

4. Experiments

4.1. Dataset of the Contour of Fibre

Due to the lack of data, we built a fibre contour dataset of preserved Szechuan pickle which was collected from production workshop. The object is the section of the fibre of preserved Szechuan pickle. We used a 5-million-pixel industrial camera with a 16 mm fixed-focus lens. The distance between the section of preserved Szechuan pickle and the camera is more than 30 cm to simulate the actual production process.

The data collected are in two forms: pictures and videos. Pictures serve as the main body of the dataset and videos serve as data supplements. Labels of each image are quality-controlled and human-annotated. The collected data are labeled manually by LabelMe [21], which is an open source annotation tool. We labeled images with two colors: black and white. The region of contour is white and the background is black. The dataset contains 2120 pairs of pictures. Each pair of pictures has an original picture and a label. It contains about 600 different individuals, with an average of approximately 3 images per stem. Some examples of the dataset are shown in Figure 3.

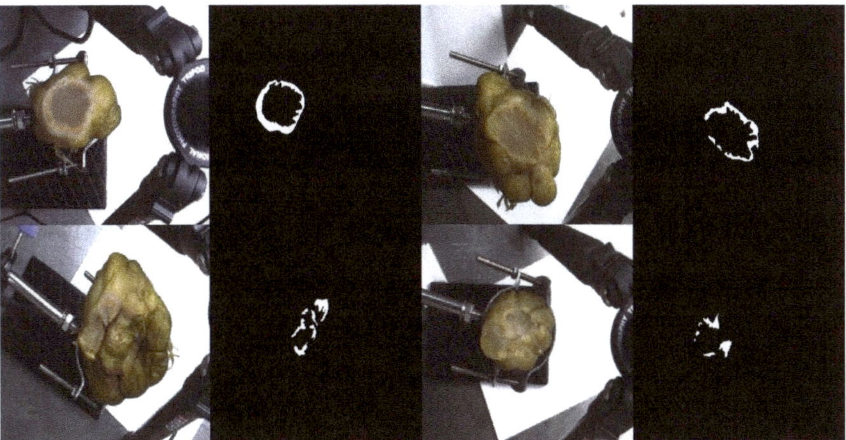

Figure 3. Some examples of the dataset. The two figures are a pair of data, with the image on the left and label on the right.

The method of marking a region can reduce the inaccuracy of contour lines and creates more possibilities to extract information from the original picture. We divide the dataset into the train set and test set in the ratio of 8:2, which comprise 1696 pairs and 424 pairs, respectively. The fibres of preserved Szechuan pickle are irregular. Some of them have contours that are continuous, while others are discrete regions. A continuous contour means that the contour consists of only one connected domain, while a discrete contour consists of two or more connected domains. The shape of the region of the continuous contour is similar to a ring, except that the inner and outer lines of the ring are unsmooth and irregular. The shape of the region of a discrete contour is usually a part of the ring and is mixed with many irregular shapes. The number of cases is shown in Table 1.

Table 1. Introduction to our dataset.

Categories	Continuous contour	Discrete contour	Total
Train set	910	786	1696
Test set	256	168	424

The fibre contour dataset of preserved Szechuan pickle is quite different from the public datasets, such as BSD500, PASCAL VOC2012 which are mainly for people, roads, vehicles and so on. The dataset of preserved Szechuan pickle is an object with small inter-class differences and large intra-class differences, and public datasets are the opposite. There are fewer common characteristics of objects between the public datasets and ours. It is necessary to set up a non-salient contour dataset and our method is also designed for this kind of dataset.

4.2. Comparison of RCF, HED and Dilated-HED

We evaluate dilated-HED, HED and RCF on the test set, which is composed of 424 pairs of pictures. The detection accuracy is evaluated using two measures: Pixel Accuracy (PA) and Mean Intersection over Union (MIoU). Both are proposed in [2] and are standard measures for semantic segmentation. Pixel accuracy (PA) is the simplest indicator used to calculate the ratio between the number of pixels correctly classified and the total number of pixels. It can be defined as

$$\text{PA} = \frac{\sum_{i=0}^{k} p_{ii}}{\sum_{i=0}^{k} \sum_{j=0}^{k} p_{ij}} , \tag{4}$$

$k+1$ represents the number of categories of contours, including background. p_{ij} represents the number of pixels that belong to class i but are predicted to be class j. MIoU is a standard metric to calculate the ratio of intersection and union between sets. Its definition is as follows:

$$MIoU = \frac{1}{k+1} \sum_{i=0}^{k} \frac{p_{ii}}{\sum_{j=0}^{k} p_{ij} + \sum_{j=0}^{k} p_{ji} - p_{ii}} , \tag{5}$$

Before we loaded the data into the model for training, we first adjusted the size of the picture to 512×512. The main reason for this is memory constraints and control parameters. The output images on the dataset are grayscale, with gray values ranging from 0 to 255. The value of 0 represents the background, and the value of pixel is closer to 255, the more likely it is to be a contour. By setting the threshold, we can filter out the possible points to get the predicted contours.

Compared with HED and RCF networks, our method achieves better processing results. For a specific comparison, see Figure 4.

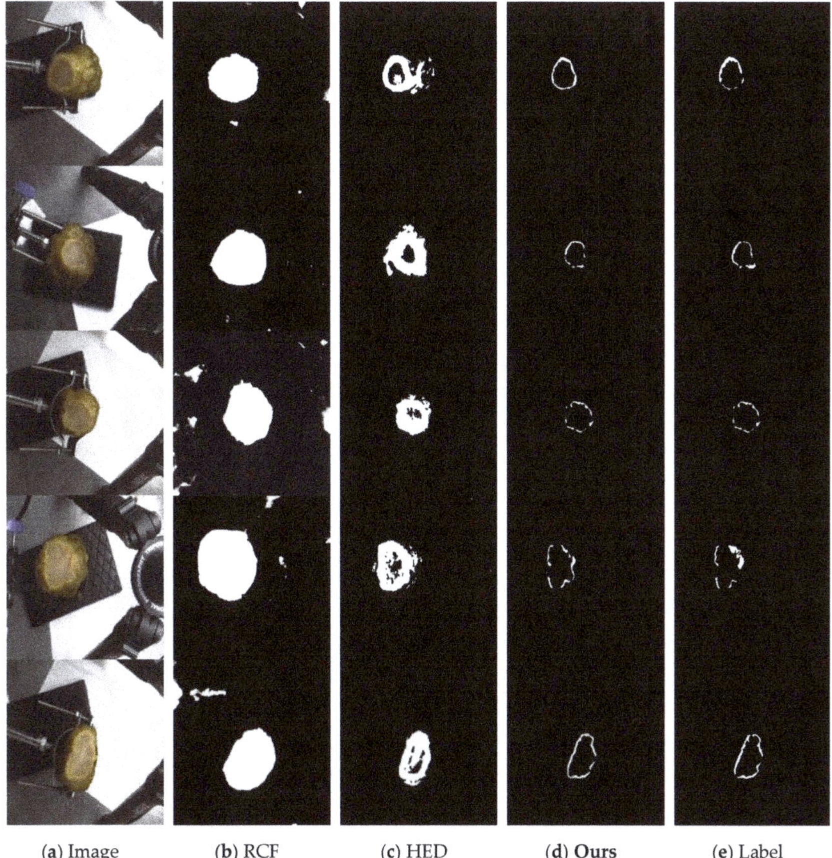

(a) Image   (b) RCF   (c) HED   (d) **Ours**   (e) Label

Figure 4. Comparison of richer convolutional features for edge detection (RCF), holistically-nested edge detection (HED) and dilated-HED. From left to right: (**a**) original images; (**b**) the results of RCF; (**c**) the results of HED; (**d**) the results of our method; (**e**) labels, also known as ground-truth.

From the comparison of the above results, we can see that RCF outputs more false regions, which are distributed on both sides of the right contour, and the background area of non-preserved Szechuan pickle is also recognized as a contour. HED can detect the area where the section of preserved Szechuan pickle is located, but there are also many false regions. RCF and HED have low accuracy and high misjudgment rates, which means they are not suitable for contour detection in our project. Our model can detect the contour of fibre accurately, which is distributed irregularly. In addition, the PA and MIoU of the three methods are measured as shown in Table 2.

Table 2. Pixel Accuracy (PA) and Mean Intersection over Union (MIoU) by three methods (Unit: %).

Categories	RCF	HED	Ours	Continuous Contour (Ours)	Discrete Contour (Ours)
PA	89.91	95.72	**99.52**	**99.52**	**99.53**
MIoU	45.16	48.08	**49.99**	**49.99**	**49.98**

In the evaluation index, the background is also a category, and the background is much larger than the target in the image, so the value of PA is large no matter which method is used. The result shows that PA of our method is 99.52% and MIoU is 49.99%. We use dilated convolution at conv1-1 and the dilated rate is [(2,2), (1,1)]. From the data, we can see that the values of PA and MIoU obtained by

RCF are the lowest, followed by HED, and our model gets the best results. This is consistent with the conclusion obtained from Figure 3. Compared with the HED and RCF, the PA of our results increased by 3.80% and 9.61%, and the MIoU increased by 1.91% and 4.83% respectively. Dilated-HED is more effective in the detection of fibre contours as shown in the experiment.

We divided the test set into two sets according to the contour types, and then use our proposed model to test on two sets of data. The results in Table 2 show that the PA of the discrete contour is higher than the continuous contour by about 0.01%, but the MIoU of the discrete contour is lower than the continuous contour by about 0.01%. By analyzing the data and predicted images, we believe that our model will smoothen regions with large variation, which has a greater impact on the continuous contour. So, the PA of the two sets will be different. The model is more effective for large contours and the smaller connected areas in discrete contours are more easily omitted, so the MIoU of continuous contours is higher.

The network structure of HED improves the performance of the model by incorporating multi-scale information. We retain this structure and add the number of layers of the model, so that the model can get more useful information. When we obtain the feature map, it is beneficial to use dilated convolution first in our model. Dilated convolution enlarges the receptive field and is more sensitive to non-salient features. This extraction method can capture the key features of contour, so it can make effective use of non-salient features. From the results, we can see that these changes have a promotion effect.

4.3. Training Details

In this part we discuss our detailed implementation. We experimented with our method and HED in the Tensorflow [22] deep learning framework and trained it on a Nvidia GTX 1080-TI GPU. RCF was implemented in the PyTorch [23] deep learning framework and trained on the same GPU. Like HED, our model has filters with size 3×3. The pool size of the pooling layer is 2×2 and the strides is 2×2. We initialized the network with the weight of VGG16 training on ImageNet ILSVRC—2014 submission [17]. To get a better optimization, we changed the initial values of the parameters in the convolution layer. The initial values of the parameters are different at different layers. An Adam optimizer with a learning rate of 0.0001 is used in our method. The batch size is 2, the number of decay_steps is 10000 and decay_rate is 0.1. We set pos_weights at 0.7, weight_decay_ratio at 0.0002 and sides_weights at 1.0 in every output-layer. It takes about 5 h to train our model.

5. Discussion and Future Work

In this paper, we proposed a dilated convolutional network for non-salient contour detection. Our method shows promising results in performing image-to-image learning by combining dilated convolution and multi-scale visual responses. Dilated-HED takes advantage of HED in contour recognition, combines dilated convolution and makes the network more sensitive to the fibre of preserved Szechuan pickle by enhancing the ability to acquire non-salient features. In order to achieve better results, we will try to extract the section of preserved Szechuan pickle and then detect the contour. In the case of large background and small target, this method can reduce the interference of background to accuracy. We believe that the work is a step towards non-salient contour detection which is a crucial step in industrial production. Non-salient contour detection will greatly improve the accuracy of production and make products more sophisticated. We will also add the categories of contours and make our method applicable to more fields in the future.

Author Contributions: All the authors contributed to this work. Conceptualization, H.L. and Z.H.; Data curation, H.L.; Formal analysis, H.L. and Z.H.; Investigation, H.L. and D.Z.; Methodology, H.L. and L.L.; Project administration, L.L.; Validation and visualization, H.L.; Writing—original draft, H.L.; Writing—review & editing, L.L. and Z.H.

Funding: This research was funded in part by grants from the Science and Technology Commission of Shanghai Municipality (17511106400, 17511106402).

Conflicts of Interest: The authors declare no conflict of interest.

References

1. Gong, X.Y.; Su, H.; Xu, D.; Zhang, Z.T.; Shen, F.; Yang, H.B. An Overview of Contour Detection Approaches. *Int. J. Autom. Comput.* **2018**, *15*, 656–672. [CrossRef]
2. Long, J.; Shelhamer, E.; Darrell, T. Fully convolutional networks for semantic segmentation. *IEEE Trans. Pattern Anal. Mach. Intell.* **2017**, *39*, 640–651. [CrossRef]
3. Xie, S.; Tu, Z. Holistically-Nested Edge Detection. *Int. J. Comput. Vis.* **2017**, *125*, 3–18. [CrossRef]
4. Liu, Y.; Cheng, M.M.; Hu, X.; Wang, K.; Bai, X. Richer Convolutional Features for Edge Detection. *IEEE Trans. Pattern Anal. Mach. Intell.* **2019**. Available online: http://mmcheng.net/rcfEdge/ (accessed on 29 January 2019).
5. Yu, F.; Koltun, V. Multi-scale context aggregation by dilated convolutions. *arXiv* **2016**, arXiv:1511.07122.
6. Kanopoulos, N.; Vasanthavada, N.; Baker, R.L. Design of an image edge detection filter using the Sobel operator. *IEEE J. Solid-State Circuits* **2002**, *23*, 358–367. [CrossRef]
7. Van Vliet, L.J.; Young, I.T.; Beckers, G.L. A nonlinear laplace operator as edge detector in noisy images. *Comput. Vis. Gr. Image Process.* **1989**, *45*, 167–195. [CrossRef]
8. Ding, L.; Goshtasby, A. On the Canny edge detector. *Pattern Recognit.* **2001**, *34*, 721–725. [CrossRef]
9. Bao, P.; Zhang, L.; Wu, X. Canny Edge Detection Enhancement by Scale Multiplication. *IEEE Trans. Pattern Anal. Mach. Intell.* **2005**, *27*, 1485–1490. [CrossRef] [PubMed]
10. Martin, D.R.; Fowlkes, C.C.; Malik, J. Learning to detect natural image boundaries using local brightness, color, and texture cues. *IEEE Trans. Pattern Anal. Mach. Intell.* **2004**, *26*, 530–549. [CrossRef] [PubMed]
11. Ren, X.F.; Bo, L. Discriminatively trained sparse code gradients for contour detection. In Proceedings of the 25th International Conference on Neural Information Processing Systems, Lake Tahoe, NV, USA, 3–6 December 2012; pp. 584–592.
12. PArbeláez, P.; Pont-Tuset, J.; Barron, J.T.; Marques, F.; Malik, J. Multiscale combinatorial grouping. In Proceedings of the 2014 IEEE Conference on Computer Vision and Pattern Recognition, Columbus, OH, USA, 23–28 June 2014; pp. 328–335. [CrossRef]
13. Jones, H.E.; Grieve, K.L.; Wang, W.; Sillito, A.M. Sillito. Surround suppression in primate V1. *J. Neurophysiol.* **2001**, *86*, 2011–2028. [CrossRef] [PubMed]
14. Arbelaez, P. Boundary extraction in natural images using ultrametric contour maps. In Proceedings of the 2006 Conference on Computer Vision and Pattern Recognition Workshop (CVPRW'06), New York, NY, USA, 17–22 June 2006; pp. 182–182. [CrossRef]
15. Arbelaez, P.; Maire, M.; Fowlkes, C.; Malik, J. From contours to regions: An empirical evaluation. In Proceedings of the 2009 IEEE Conference on Computer Vision and Pattern Recognition, Miami, FL, USA, 20–25 June 2009; pp. 2294–2301. [CrossRef]
16. Krizhevsky, A.; Sutskever, I.; Hinton, G.E. ImageNet classification with deep convolutional neural networks. In Proceedings of the 25th International Conference on Neural Information Processing Systems-Volume 1, Lake Tahoe, NV, USA, 3–6 December 2012; pp. 1097–1105.
17. Simonyan, K.; Zisserman, A. Very Deep Convolutional Networks for Large-Scale Image Recognition. *arXiv* **2014**, arXiv:1409.1556.
18. Szegedy, C.; Liu, W.; Jia, Y.; Sermanet, P.; Reed, S.; Anguelov, D.; Erhan, D.; Vanhoucke, V.; Rabinovich, A. Going Deeper with Convolutions. In Proceedings of the 2015 IEEE Conference on Computer Vision and Pattern Recognition (CVPR), Boston, MA, USA, 7–12 June 2014; Available online: https://arxiv.org/abs/1409.4842 (accessed on 15 March 2018).
19. Lafferty, J.; McCallum, A.; Pereira, F.C.N. Conditional Random Fields: Probabilistic Models for Segmenting and Labeling Sequence Data. In Proceedings of the 18th International Conference on Machine Learning 2001 (ICML 2001), San Francisco, CA, USA, 28 June–1 July 2001; pp. 282–289.
20. Chen, L.C.; Papandreou, G.; Kokkinos, I.; Murphy, K.; Yuille, A.L. DeepLab: Semantic image segmentation with deep convolutional nets, atrous convolution, and fully connected CRFs. *IEEE Trans. Pattern Anal. Mach. Intell.* **2018**, *40*, 834–848. [CrossRef] [PubMed]
21. Russell, B.C.; Torralba, A.; Murphy, K.P.; Freeman, W.T. LabelMe: A Database and Web-Based Tool for Image Annotation. *Int. J. Comput. Vis.* **2008**, *77*, 157–173. [CrossRef]

22. Abadi, M.; Barham, P.; Chen, J.; Chen, Z.; Davis, A.; Dean, J.; Devin, M.; Ghemawat, S.; Irving, G.; Isard, M.; et al. TensorFlow: A system for large-scale machine learning. In Proceedings of the 12th USENIX conference on Operating Systems Design and Implementation, Savannah, GA, USA, 2–4 November 2016; Volume 16, pp. 265–283. Available online: https://arxiv.org/abs/1605.08695v2 (accessed on 10 August 2018).
23. Ketkar, N. Introduction to PyTorch. In *Deep Learning with Python: A Hands-on Introduction*; Apress: Berkeley, CA, US, 2017; pp. 195–208.

Article

An Intelligent Vision System for Detecting Defects in Micro-Armatures for Smartphones

Jiange Liu, Tao Feng, Xia Fang, Sisi Huang and Jie Wang *

School of Manufacturing Science and Engineering, Sichuan University, Chengdu 610065, Sichuan, China;
liujiange666@163.com (J.L.); 15528248895@163.com (T.F.); FangXia1991@163.com (X.F.); hssylyn@163.com (S.H.)
* Correspondence: wangjie@scu.edu.cn; Tel.: +86-138-0801-5321

Received: 18 April 2019; Accepted: 25 May 2019; Published: 28 May 2019

Abstract: Automatic vision inspection technology shows a high potential for quality inspection, and has drawn great interest in micro-armature manufacturing. Given that the inspection process is highly influenced by the lack of real standardization and efficiency performed with the human eye, thus, it is necessary to develop an automatic defect detection process. In this work, an elaborated vision system for the defect inspection of micro-armatures used in smartphones was developed. It consists of two parts, the front-end module and the deep convolution neural networks (DCNNs) module, which are responsible for different areas. The front-end module runs first and the DCNNs module will not run if the output of the front-end module is negative. To verify the application of this system, an apparatus consisting of an objective table, control panel, and a camera connected to a Personal Computer (PC) was used to simulate an industrial position of production. The results indicate that the developed vision system is capable of defect detection of micro-armatures.

Keywords: micro-armature; defect detection; convolutional neural networks; computer vision

1. Introduction

With the rapid expansion of network applications, computer vision technology has been successfully applied to the quality inspection of industrial production [1–6], including glass products [1], fabrics [2,3], steel surfaces [4], bearing rollers [5], and casting surfaces [6]. The inspection of these mentioned examples needs a matching algorithm to extract image features based on the actual defect situation. For glass products for packaging and domestic use, an intelligent system based on the classical computer vision technology was proposed for the automatic inspection of two types of defects [1]. For fabric quality control, an unsupervised learning-based automated approach to verify and localize fabric defects was proposed [2]. This approach was realized by using a convolutional denoising autoencoder network at multiple Gaussian pyramid levels to reconstruct image patches and synthesize detection results from the corresponding resolution channels. A two-fold procedure was proposed to extract powerful features [3]. First, the sample class was determined based on its background texture information, then the image was divided into 49 blocks to figure out which images contain defective regions. For the defects of steel surfaces, an inspection system with a dual lighting structure was proposed to distinguish uneven defects and color changes by surface noise [4]. In a previous study [5], a multi-task convolutional neural network applied to recognize defects was raised. Although there are many detection systems based on computer vision technology to solve product defects, few studies have focused on the defect inspection of micro-armatures. For the inspection of surface aluminum, a vision based approach and neural network techniques in surface defects inspection and categorization are proposed. The new vision inspection system, image processing algorithm, and learning system based on artificial neural networks (ANNs) were successfully implemented to inspect surface aluminum die casting defects [6].

Recently, deep convolutional neural networks (DCNNs) have been proved to be important methods in visual detection. However, some classical methods should still be considered. Literature [6] mentioned above is an example. For instance, in order to get better parameters, a suitable smart manufacturing strategy for real industrial conditions was proposed. The results of this dataset showed that the Adaboost ensembles provided the highest accuracy and were more easily optimized than ANNs [7]. Obviously, for now, there are some limitations with using classical methods alone or using DCNNs directly. For classical methods, for example, they usually come with high complexity of programming and less tolerance to data variability. With regard to the DCNNs, a large number of samples are needed for network training and the iterative optimization of parameters is partly a black-box operation [8].

In the industrial applications, part defects in the samples can be easily identified with a picture by the classical computer vision method. Therefore, we do not need to feed the whole picture into the network for discrimination, which can reduce the number of features to be identified. As a result, the network is easier to converge. Therefore, by combining the classical image recognition method and DCNNs, the landing speed of deep learning technology in industrial applications can be effectively improved.

In this work, an intelligent detection system that combined the classical computer vision method and DCNNs was designed to automatically detect the quality defects of micro-motor armatures. Firstly, the quality, excluding the region of copper wire crossing (ROC), is decided based on the classical computer vision technology. If the first result is positive, the ROC will be extracted and sent to the DCNNs for identification. If the result is still positive, the image is a defect-free sample, otherwise the whole image is labeled defective. In the experiments, this system works very fast and presents a high hit rate, which can bring practical benefits for industrial applications.

2. Related Works and Foundations

In motor armatures prepared for smartphones, the key component in the process of inter-conversion between mechanical energy and electrical energy is usually very small. During the service of the product, the poor performances of low-quality armatures will significantly affect the comfort of users. Furthermore, the magnetic field may change due to the poor quality of the armature, which can deteriorate the mechanical properties of the product.

Currently, many micro-motor armatures are manually placed under the microscope by the operator to adjust the armature position through the observation of the staff, according to the experience to achieve defect detection, which shows various disadvantages, such as time-consuming processes [9–11] and the lack of real standardization. Therefore, there is an urgent need to bring a related defect inspection system into the production process of micro-armatures.

As previously stated, currently the defect detection method for micro-motor armatures is achieved by transferring the armature to the inspection area after soldering, and then the armature defect is inspected by staff with microscopes. This detection method is not only expensive, but also inefficient and fluctuates with the flow of employees. To address these problems, the main aim of this work is, therefore, to design a new set of armature positioning and imaging devices, as well as the matching discriminating procedures. The armature and apparatus are shown in Figures 1 and 2, respectively.

Although the armature has three commutator terminals, a stepper motor was used to turn the armature. Thus, only one camera is needed to get the picture of each commutator terminal. We used the plane-array camera with 1.3 million pixels and the telecentric lenses with 0.66 mm depth of field and 110 mm working distance to capture the sample images. Fiber-optic sensor model E32DC200B4 and fiber optic amplifier model FX-101 were used to get the position of the armature.

Proper illumination can ensure the high quality of the image. We set up two area light sources, and a ring light source. The ring light source was arranged in front of the armature, and two area light sources were arranged on the left and right of the armature.

The software system was programmed with Python. The detection algorithm was developed by OpenCV and Tensorflow deep learning platform.

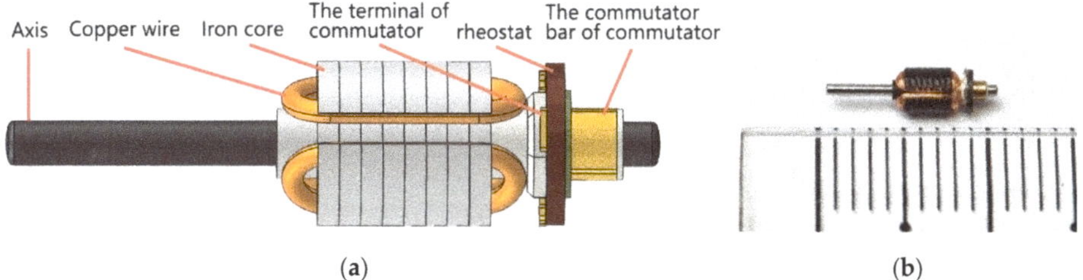

Figure 1. (**a**) The model of armature and (**b**) the real sample.

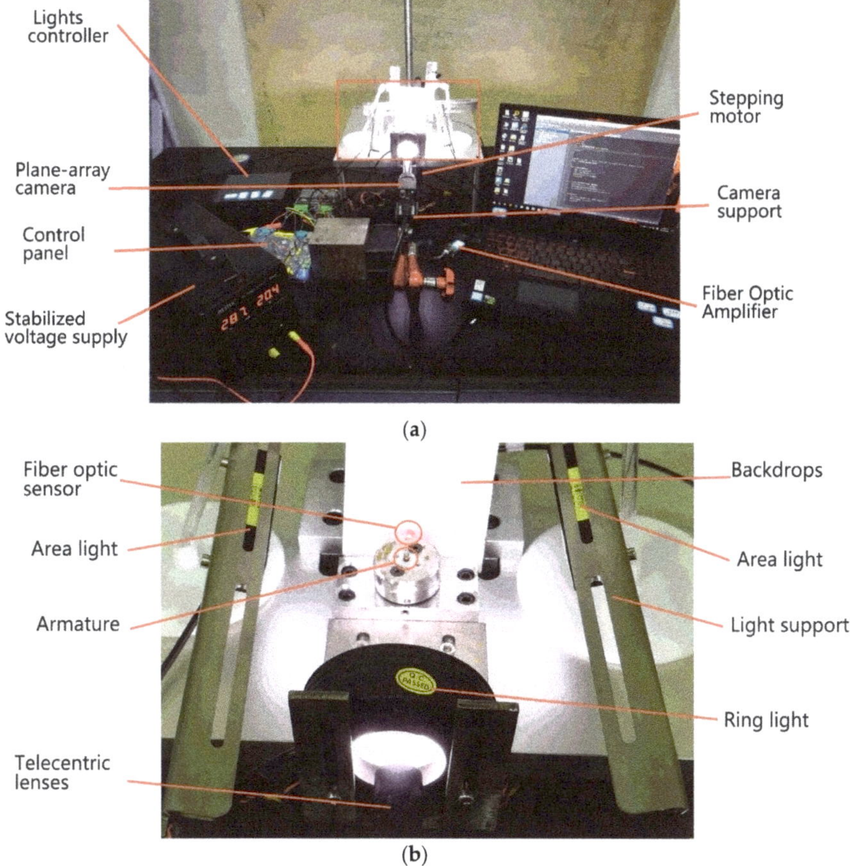

Figure 2. (**a**) Apparatus for the proposed inspection system. (**b**) shows the details in the red box of (**a**).

3. Methodology

Figure 3 provides an overview of the processing workflow, which displays the component of our proposed vision system. The system can be roughly divided into two parts, i.e., the classical computer vision method and DCNNs. Initially, the theoretical background for the model is depicted in Sections 3.1 and 3.2. Then, the database processed in this work is briefly described in Section 3.3, while the details of the algorithm are discussed in Section 3.4. Finally, the experimental details are shown in Section 3.5.

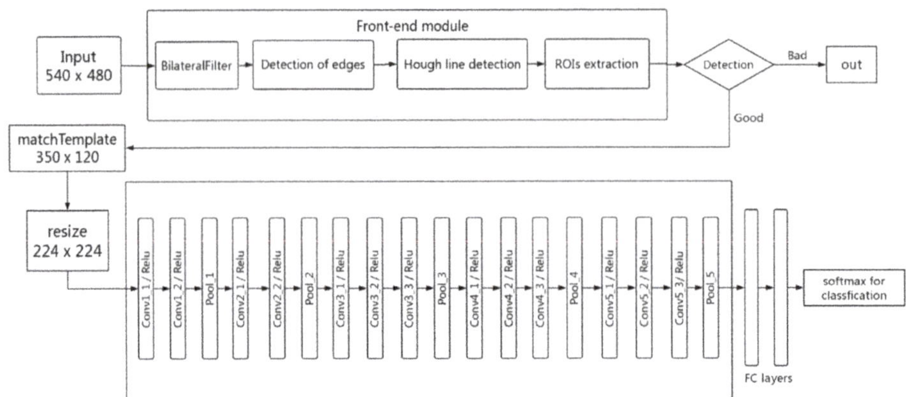

Figure 3. Architecture of our proposed intelligent vision system. The front-end module part takes the original images as inputs and the DCNNs part takes the extracted ROC as input if needed.

3.1. Front-End Module

3.1.1. Filters

Bilateral filtering is a kind of non-linear filter [12,13], which represents the intensity of a pixel by weighted average of the brightness values of the surrounding pixels. The parameters of bilateral filtering take the Euclidean distance and color distance of the pixels in the range into account. Thus, the boundary is preserved, while the noise is reduced. Equation (1) gives the specific operation of bilateral filtering:

$$I(m) = \frac{\sum_{n \in \Omega} K_n H_n V_n}{\alpha} \tag{1}$$

where I is the denoising image, m is a pixel in the image, n is an adjacent pixel with value V_n in the m neighborhood range Ω, and K_n and H_n are, respectively, the space and gray weighting coefficient; α is the normalized constant, which represents the product of the spatial weighting coefficient and the gray weighting coefficient.

3.1.2. Edge Detection

The purpose of edge detection is to find the pixels with sharp changes in brightness for an image, that is, a discontinuity in the intensity of grayscale. If the edges in the image can be accurately measured and positioned, it means that the actual object can be positioned and measured [14,15]. Many edge detection operations are based on the gradient and direction of brightness. The gradient and direction of an image g at point (x, y) can be separately defined by Equation (2) and Equation (3):

$$\nabla g(x, y) = \left[\frac{\partial g}{\partial x}, \frac{\partial g}{\partial y} \right] \tag{2}$$

$$\|\nabla g(x, y)\| = \sqrt{\left(\frac{\partial g(x, y)}{\partial x}\right)^2 + \left(\frac{\partial g(x, y)}{\partial y}\right)^2} \tag{3}$$

where $\partial g / \partial x$ and $\partial g / \partial y$ are the discrete partial derivatives in the x and y directions, respectively. With the appropriate threshold t, the edges can be detected.

3.1.3. Detecting of Lines

The Hough transform, proposed by Hough [16], is a method widely used in image processing and computer vision for detecting parameterized shapes. The simplest application is straight line detection, the main principle of which is firstly to convert the line detection problem in the image space to the point detection problem in the parameter space, then complete the line detection task by finding the peak in the parameter space. That is, if using Equation (4) to represent a line in the image space, the line is equal to the point (θ, r) in Hough space.

$$r = x * \cos\theta + y * \sin\theta \tag{4}$$

where r is the distance between the point (x, y) and the origin in image space, and θ is the angle between r and the positive direction of x in image space. Note that some authors, such as Ballard [17] and Davies [18], have proposed enhancements to the method.

3.2. Deep Convolutional Neural Networks (DCNNs)

Recently, DCNNs has shown great detection power in computer vision, which has been widely used in various applications, e.g., classification [19], image segmentation [20], object tracking [21], and so on. A classic DCNNs architecture consists of several layers of convolutional, activation, and pooling layers, followed by fully connected layers at the end. A simplified version of DCNNs can be composed of the following five parts:

1. Input: The input of DCNNs is usually a batch of 3-channel color image matrices with fixed size, which depends on the network structure you are using.
2. Conv: The convolutional layers perform feature extraction and feature mapping by employing a set of fixed-size filters sliding on local receptive fields after receiving feature maps. The filter sizes are usually odd, such as 3×3 or 5×5. The weight-sharing scheme is applied in the convolution operations.
3. Activation: Since convolution is a linear operation, it is necessary to use the activation layer to nonlinearly map the output of the convolutional layer, thereby increasing the expression ability of the model. Nowadays, rectified linear units (ReLU) have become the most widely used activation function because they can effectively prevent the gradient from disappearing and accelerate the convergence speed during the training process [16]. The mathematical transformation between each input value x and its output y can be formulated as

$$y = \begin{cases} x, & x > 0 \\ 0, & x \leq 0 \end{cases} \tag{5}$$

4. Pool: The pooling layer performs a form of non-linear down-sampling to compress the input feature map, which makes the feature map smaller in spatial size and reduces the complexity of the network computation. The most common pooling operation is max pooling, which outputs the max value from the neighborhood of the input feature map.
5. FC: Each node of the fully connected layer is connected to all nodes of the previous layer to combine the features extracted from the front. The last fully connected layer generates the output of the overall network, the dimensions of which are the same as the dimensions of the input label, then achieving the classification with the transformation of the softmax classifier:

$$p(x_i) = \frac{exp(x_i)}{\sum_{i=1}^{k} exp(x_i)} \tag{6}$$

where x_i indicates the value of the *ith* dimension computed by the last layer and $p(x_i)$ indicates the probability of the corresponding label.

3.3. The Dataset

In this paper, our dataset is collected by the device showed in Figure 2. We evaluate our method on this dataset, shown in Figure 4, which contains image samples of six representative defect types with a size 540 × 480 pixel. Further, we roughly divided the picture into two parts, respectively the ROC and the remaining region.

Figure 4. Original image samples taken by the device showed is Figure 2. The defective regions are marked out by a surrounding red ellipse; the ROC is marked by a blue rectangle.

For the ROC required to train the neural network, our dataset contains 5106 positive samples and 3322 negative samples; some of the samples are shown in Figure 5. As some pictures are prone to controversy in classification, we first assigned several groups of skilled employees to manually classify them, and then synthesized the classification results of several groups to finally determine the classification label.

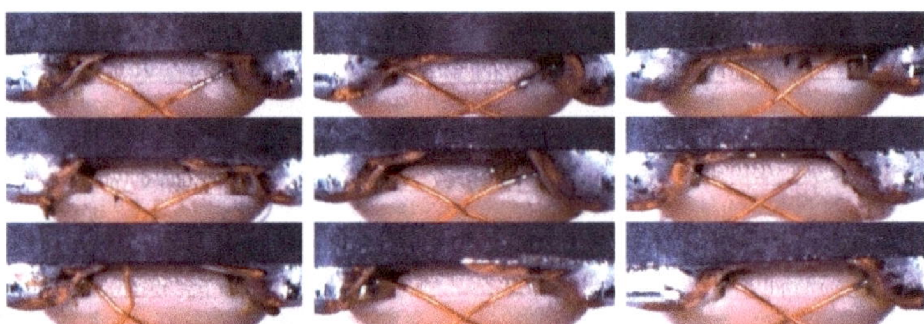

Figure 5. The ROC clipped from the original picture makes the proportion of defective parts larger, thus making the network easier to train and optimize.

3.4. Model Implementation

Considering the distribution of defective areas, as shown in Figure 4, the most complex defects are concentrated in the ROC (the region marked by a blue rectangle), and only this block is difficult to identify by the front-end module. Therefore, the implementation of our model is two-staged. We design a joint detection architecture, which contains two major parts: the front-end module and the DCNNs part. The front-end module was designed to detect these defects, including the resistance, the tin package, and the cilia. The DCNNs detection part was developed to only identify whether the ROC contains defects or not. If the front-end module has determined that the armature contains defects, the DCNNs part will not be executed.

As shown in Figure 3, the front-end module can be roughly divided into five steps: filter, detection of edges, Hough transformation, regions of interests (ROIs) extraction, and identification. When the armature is placed on the workpiece table, the optical fiber sensor will conduct the armature to the initial position to make the sample face the camera. The picture at this point is shown in Figure 4. After the initial position correction is completed, the discrimination system starts to work.

The input of the system is a color image of the armature. In the front-end module, the RGB color image was firstly converted to a grayscale one by averaging the RGB channel. At this time, the effect of a canny edge detector used directly is shown in Figure 6. Note that Figure 6 shows a positive sample.

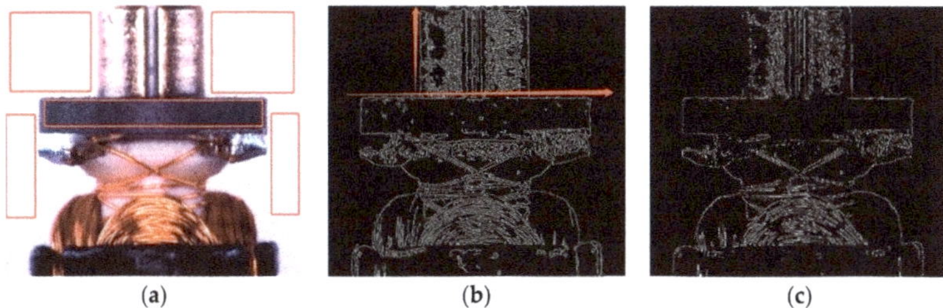

Figure 6. The original image and direct edge detection results. The ROIs have been marked by red rectangles (**a**). The two ideal baselines are marked by red lines (**b**). The noise is not reduced dramatically with the increase of the threshold, while the baseline disappears rapidly (**c**).

According to the above description, the areas we want to detect are the ones in the red checkboxes shown in Figure 6 in the front-end module. Combined with the edge detection results described by Figure 6, it is easily found that there are two obvious lines in the figure. A large number of pictures have proved that the two lines are stable, while both the horizontal line and vertical line are expected to be used as the reference line to locate other areas. However, the experimental result is different from the intuitive prediction, which may result from the inaccurate position of the lines detected by Hough transform. To improve the accuracy of detection, we increased the value of the double threshold to keep the edge and reduce the noise. However, the noise is not reduced dramatically with the increase of the threshold, while the baseline disappears rapidly. Thus, in our system, the bilateral filter is applied firstly to image noise reduction before edge detection. After the experimental verification of a large number of pictures, our canny operator achieved the best edge detection effect after the bilateral filter used the 60 mixVal and the 140 maxVal. The edge detection results are shown in Figure 7.

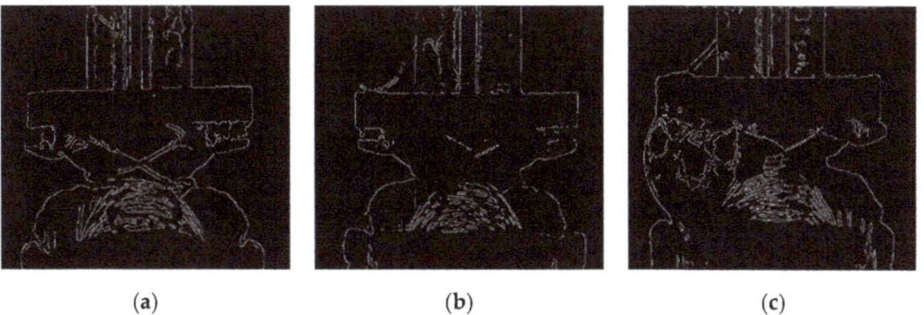

Figure 7. The edge detection results after using a bilateral filter. (**a**) shows that the picture is free of cilium and tin bags are too large defects; (**b**) reveals cilium defects. (**c**) represents the tin package is oversized.

In our dataset, the desired area can be located easily if the horizontal and vertical lines can be located stably. The experiment proved that the bilateral filter has no effect on the edge detection of the cilia and the tin package. The defects of the cilia and the package can be found directly by the edge detection result. For the three edge detection results shown in Figure 7, it can be found that the workpiece corresponding to (b) has cilium defects and the (c) tin package is oversized. For the rheostat region, we need to go back to the original image and cut it out, then use the new filter and threshold to carry out edge detection, and finally judge whether the rheostat is bad or not.

Although the poor resistance can be determined by edge detection according to the jump of color on the damaged location, it is easy to cause a lot of misjudgments if we still directly use the existence of the edge as the criterion, even with the new filter and threshold value. This is why firstly we convert the pixel edge to the pixel blocks by morphological transformation, secondly accumulate the area of the pixel blocks that exceed the threshold (although noise usually can be eliminated, some still converted into small pixel blocks), then set a threshold to output the result. The edge detection results of the resistance are shown in Figure 8.

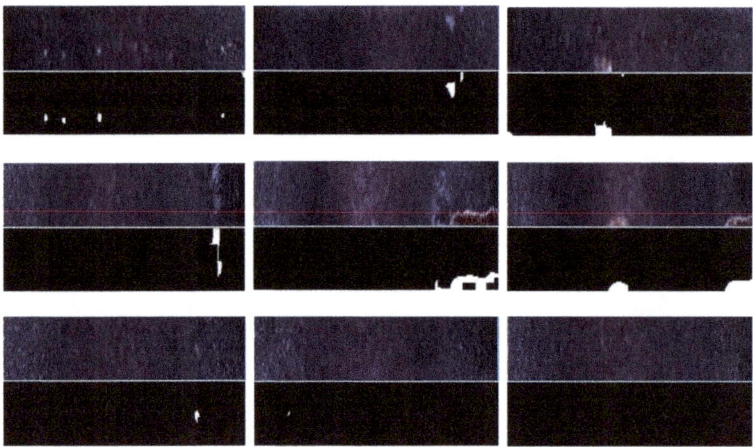

Figure 8. A total of nine images are shown. Each image is composed of two parts: the original image of resistance on the top and the edge detection image of resistance on the bottom. We will determine whether the resistance is damaged according to the white area in the edge detection diagram.

At this point, the inspection task of the front-end module has been completed. The system will directly judge whether the product is bad if any defective area is identified. Otherwise the template matching algorithm will be used to cut out the most difficult areas for discrimination and feed them into the trained neural network for detection. Thanks to the elimination of most interference, our DCNNs module did not need to be specially designed, while satisfactory results can be achieved directly using the classical classification network VGG_16 shown in Figure 3.

The front-end classical computer vision part works in conjunction with the DCNNs detection part for the purpose of defect detection. Given an image, the first part decides whether the external part is bad or not. If the external part is already defective, it is no longer necessary to enable the DCNNs part. If DCNNs is needed, it only needs to focus on a small part of the original picture.

3.5. Experiments Details

Our model is trained on one NVIDIA GTX1080 GPU with 8GB memory for roughly 10 h. Experiments are implemented based on the deep learning framework Tensorflow. The operating system is Windows 10.

The DCNNs network has proven its powerful ability in image detection, while the structure of DCNNs is very complex and deep, which is a kind of black box operation for us. At the beginning, we directly fed the original image into to the VGG_16 network for classification [22]. However, overfitting occurred in the process of training the network. The highest accuracy of the test set was only 0.784, while the training accuracy is 0.998. Similar results also happened to Alex-net [16] and Resnet_50 [23]. We speculated that the reason for the failure is that with the multiple feature extraction, the network extracts more representative information and naturally loses details. However, in some cases, the differences between bad products and good products in our data lie in the small change; for some pictures, even people may be wrong. The original samples of data and ROI are shown in Figure 9. Therefore, we used the idea of SSD: Single Shot MultiBox Detector [24] for reference to integrate multiple scale features to make predictions based on VGG_16. To be specific, we concatenate conv5_3, conv4_3, and conv3_3 features, followed by a 1×1 convolution layer to form the final convolution feature representation. The method dose work and the accuracy was improved to 0.812, but it was still too low for our demand. In this case, we analyzed the data again and found that the three defects of armature winding cilia, resistance cracking, and tin package were relatively easy to be realized by classical computer vision technology, while only the small defect area where the two copper wires crossed was difficult to be distinguished by classical methods. We suspect that it may have achieved a higher accuracy rate if only such a small area was fed into the neural network. The experimental results show that our method is feasible. We obtained the difficult block with a size of 350×120 through template matching from each image to form the new training set [25]. Obviously, when a picture is switched left and right, the network needs to obtain the same discriminant result. Our workpiece does not switch up and down, nor does it tilt at an excessive angle. Therefore, in order to prevent the network from overfitting, we only adopt mirroring to augment data. In this way, the training set for the DCNNs now has 10,212 defect-free blocks and 6644 defective blocks.

Figure 9. In some cases, the differences between bad products and good products in our data lie in the small changes, where even people may be wrong. However, at the same time, there are some areas in the picture that we do not care about and areas that are easy to identify, and the location of this area is relatively stable. If the whole picture is directly fed into the neural network, obviously we hope that the network can focus on the ROC region and give less consideration to other parts. In other words, we hope that the network can acquire stronger feature selection ability. However, according to our experimental results, the network does not have good feature selection ability. Therefore, we cut out the ROC from the original picture and only fed the ROC into the network, i.e., to help the network complete the process of feature selection through the traditional computer vision method.

Inspired by transfer learning [26–28], we used VGG16's convolution layers parameters trained on ImageNet as our initial convolution parameters. For FC layers, the weight parameters are initialized from a truncated random normal distribution subject to $N \sim (0, \frac{2}{n})$, where n denotes the number of connections between two layers. We selected the cross-entropy function as the loss function of our model. During the training process, the stochastic gradient descent with mini-batches of 16 samples was applied to update the weight parameters. We set the base learning rate to 0.001, momentum to 0.9, and weight decay to 0.005.

4. Results

In our experiments, we used accuracy to evaluate the performance. The formula for calculating the accuracy is defined by Equation (7). To evaluate the performance of our method, we validated our module on our own dataset and achieved the final detection accuracy of 92.1%. The misclassified cases are listed in Table 1, which represents the confusion matrix, and the comparisons to directly feed the total image into network are listed in Table 2. It can be seen from Table 2 that the defective inspection can be achieved by directly feeding the total image into the network with accuracy of about 80%, whereas our two-stage module acquired a final score of over 92% in the classification task. We believe that this comparison result verifies our previous conjecture; that is, the network does not acquire good feature selection ability when the original image is directly fed into the neural network, resulting in low accuracy. The experiment shows that our model has the ability to distinguish between defect-free and defective images in our dataset and achieve a higher accuracy than others, which proves the effectiveness of our two-stage module.

$$Accuracy = \frac{TP + TN}{TP + TN + FP + FN} \tag{7}$$

where TP represents the number of positive sample that are judged to be positive samples, FP represents the number of negative samples that are judged to be positive samples, FN represents the number of positive samples that are judged to be negative samples, and TN represents the number of negative samples that are judged to be negative samples.

Table 1. Confusion matrix.

	Predicted Positive	Predicted Negative
True Positive	2785 (*TP*)	207 (*FN*)
True Negative	187 (*FP*)	1809 (*TN*)

Table 2. The detection results of our model and the comparison to only DCNNs. "B345" represents the combination of "conv3_3", "conv4_3", and "conv5_3", "FM" represents our front-end module, "VGG-16-ROC" means that the neural network input is only ROC.

Method	Accuracy (%)
VGG-16	78.4
AlexNet	78.2
ResNet-50	78.3
VGG-16+B345	81.2
FM	95.8
VGG-16-ROC	93.6
VGG-16+FM	92.1

5. Conclusions

In industrial applications, workpiece images in many cases are similar to our dataset due to automation and standardization. ANNs and complex algorithms are in use today in artificial vision, and the use is in some cases very useful. For example, some defects of a sample can be easily detected by classical computer vision technology, while the combination of deep learning techniques, and traditional computer vision techniques should be considered. Our experimental results provide a reference and demonstration for the cooperation of DCNNs and complex algorithms.

In this study, a system based on classical computer vision and deep learning was proposed to detect the micro-armature defects. In our dataset, the rough location of defects is relatively stable, and partial defect detection can be easily achieved by classical computer vision technology. In order to improve the accuracy of our system, for the defective parts that are easy to be identified, we will achieve the defection work by classical computer vision technology, and then only feed the complex parts into the neural network if needed. This is similar to L1 regularization, which can reduce part parameters of the network to 0 and achieve the effect of parameter sparsity, for which we use the front-end module to make the neural network have similar ability. We conducted many experiments, the results of which proved that our method was superior in accuracy, and met the requirements of industrial manufacturing.

In general, the main contributions of this work are as follows:

1. According to the workpiece, the supporting fixture, optical positioning system, and lighting system are designed, and the defect detection algorithm matching our hardware is designed.
2. Through experiments, it is proven that the combination of DCNNs and complex algorithms is very useful in some cases, providing a reference and demonstration for the application of computer vision technology in industrial detection.
3. In our work, we combine traditional computer vision technology and DCNNs to achieve the task, so as to improve the detection accuracy. This two-stage idea could also be considered for use in deep learning techniques, i.e., a two-stage approach similar to Faster R-CNN [29]. This two-stage idea is worth considering as we focus more on accuracy.

The difficult in our experiments is that some of the ROCs are very prone to controversy in classification. More data processing methods will be explored to solve this problem. Note that we are not talking about data augmentation. In the future, we will continue to optimize the front-end module and network structure to improve the accuracy.

Author Contributions: The authors J.L. and T.F. contributed equally to this work; project administration, J.W.; resources, X.F. and S.H.

Funding: This presented work was supported by Sichuan Science and Technology Program (No. 2019YFG0359).

Conflicts of Interest: The authors declare no conflict of interest.

References

1. Truong, M.T.N.; Kim, S. Automatic image thresholding using Otsu's method and entropy weighting scheme for surface defect detection. *Soft Comput.* **2018**, *22*, 4197–4203. [CrossRef]
2. Keser, T.; Hocenski, Z.; Hocenski, V. Intelligent Machine Vision System for Automated Quality Control in Ceramic Tiles Industry. *Strojarstvo* **2010**, *52*, 105–114.
3. Cabral, J.D.D.; de Araujo, S.A. An intelligent vision system for detecting defects in glass products for packaging and domestic use. *Int. J. Adv. Manuf. Technol.* **2015**, *77*, 485–494. [CrossRef]
4. Mei, S.; Wang, Y.D.; Wen, G.J. Automatic Fabric Defect Detection with a Multi-Scale Convolutional Denoising Autoencoder Network Model. *Sensors* **2018**, *18*, 1064. [CrossRef]
5. Wang, T.; Chen, Y.; Qiao, M.N.; Snoussi, H. A fast and robust convolutional neural network-based defect detection model in product quality control. *Int. J. Adv. Manuf. Technol.* **2018**, *94*, 3465–3471. [CrossRef]
6. Świłło, S.J.; Perzyk, M. Surface Casting Defects Inspection Using Vision System and Neural Network Techniques. *Arch. Foundry Eng.* **2013**, *13*, 103–106. [CrossRef]
7. Bustillo, A.; Urbikain, G.; Perez, J.M.; Pereira, O.M.; Lopez de Lacalle, L.N. Smart optimization of a friction-drilling process based on boosting ensembles. *J. Manuf. Syst.* **2018**, *48*, 108–121. [CrossRef]
8. Li, J.; Tao, F.; Cheng, Y.; Zhao, L. Big Data in product lifecycle management. *Int. J. Adv. Manuf. Technol.* **2015**, *81*, 667–684. [CrossRef]
9. Yun, J.P.; Kim, D.; Kim, K.; Lee, S.J.; Park, C.H.; Kim, S.W. Vision-based surface defect inspection for thick steel plates. *Opt. Eng.* **2017**, *56*, 053108. [CrossRef]
10. Wen, S.; Chen, Z.; Li, C. Vision-Based Surface Inspection System for Bearing Rollers Using Convolutional Neural Networks. *Appl. Sci.* **2018**, *8*, 2565. [CrossRef]

11. Cao, P.; Shu, H.; Yang, B.; Dong, J.; Fang, Y.; Yu, T. Speeded-up robust features based single-ended travelling wave fault location: A practical case study in Yunnan power grid of China. *IET Gener. Transm. Distrib.* **2018**, *12*, 886–894. [CrossRef]

12. Nain, A.K.; Gupta, S.; Bhushan, B. An extension to switching bilateral filter for mixed noise removal from colour image. *Int. J. Signal Imaging Syst. Eng.* **2016**, *9*, 1–19. [CrossRef]

13. Tomasi, C.; Manduchi, R. Bilateral filtering for gray and color images. In Proceedings of the Sixth International Conference on Computer Vision, Bombay, India, 7 January 1998; pp. 839–846.

14. Perona, P.; Malik, J. Scale-Space and Edge Detection Using Anisotropic Diffusion. *IEEE Trans. Pattern Anal. Mach. Intell.* **1990**, *12*, 629–639. [CrossRef]

15. Canny, J. A Computational Approach to Edge Detection. *IEEE Trans. Pattern Anal. Mach. Intell.* **1986**, *PAMI-8*, 679–698. [CrossRef]

16. Duda, R.O.; Hart, P.E. Use of the Hough Transform to Detect Lines and Curves in Pictures. *Commun. ACM* **1975**, *15*, 11–15. [CrossRef]

17. Ballard, D.H. Generalizing the Hough Transform to Detect Arbitrary Shapes. *Pattern Recognit.* **1987**, *13*, 111–122. [CrossRef]

18. Davies, E. Image space transform for detecting straight edges in industrial images. *Pattern Recognit. Lett.* **1986**, *4*, 185–192. [CrossRef]

19. Krizhevsky, A.; Sutskever, I.; Hinton, G.E. ImageNet Classification with Deep Convolutional Neural Networks. *Adv. Neural Inf. Process. Syst.* **2012**, *25*, 1097–1105. [CrossRef]

20. Long, J.; Shelhamer, E.; Darrell, T. Fully Convolutional Networks for Semantic Segmentation. In Proceedings of the IEEE Conference on Computer Vision and Pattern Recognition, Boston, MA, USA, 7–12 June 2015.

21. Redmon, J.; Divvala, S.; Girshick, R.; Farhadi, A. You Only Look Once: Unified, Real-Time Object Detection. In Proceedings of the IEEE Conference on Computer Vision and Pattern Recognition, Seattle, WA, USA, 27–30 June 2016.

22. Simonyan, K.; Zisserman, A. Very Deep Convolutional Networks for Large-Scale Image Recognition. *arXiv* **2014**, arXiv:1409.1556.

23. He, K.M.; Zhang, X.Y.; Ren, S.Q.; Sun, J. Deep Residual Learning for Image Recognition. In Proceedings of the 2016 IEEE Conference on Computer Vision and Pattern Recognition (CVPR), Las Vegas, NV, USA, 27–30 June 2016; pp. 770–778.

24. Liu, W.; Anguelov, D.; Erhan, D.; Szegedy, C.; Reed, S.; Fu, C.Y.; Berg, A.C. SSD: Single Shot MultiBox Detector. *Lect. Notes Comput. Sci.* **2016**, *9905*, 21–37.

25. Tsai, D.M.; Huang, C.K. Defect Detection in Electronic Surfaces Using Template-Based Fourier Image Reconstruction. *IEEE Trans. Compon. Packag. Manuf. Technol.* **2019**, *9*, 163–172. [CrossRef]

26. Lim, J.J. *Transfer Learning by Borrowing Examples for Multiclass Object Detection*; Curran Associates Inc.: Vancouver, BC, Canada, 2011; pp. 118–126.

27. Huh, M.; Agrawal, P.; Efros, A. What makes ImageNet good for transfer learning? *arXiv* **2016**, arXiv:1608.08614.

28. Zhuang, F.Z.; Luo, P.; He, Q.; Shi, Z. Survey on transfer learning research. *J. Softw.* **2015**, *26*, 26–39.

29. Ren, S.; He, K.; Girshick, R.; Sun, J. Faster R-CNN: Towards real-time object detection with region proposal networks. In Proceedings of the International Conference on Neural Information Processing Systems, Montreal, Canada, 7–12 December 2015; MIT Press: Cambridge, MA, USA, 2015; Volume 39, pp. 91–99.

 *applied sciences*

Article

Hybrid Edge–Cloud-Based Smart System for Chatter Suppression in Train Wheel Repair

Ruben Merino *, Iñigo Bediaga, Alexander Iglesias and Jokin Munoa

IK4-Ideko, Basque Country, 20870 Elgoibar, Spain; ibediaga@ideko.es (I.B.); aiglesias@ideko.es (A.I.); jmunoa@ideko.es (J.M.)
* Correspondence: rmerino@ideko.es; Tel.: +34-943748000

Received: 2 July 2019; Accepted: 4 October 2019; Published: 12 October 2019

Abstract: The contact profile of a train wheel has a key role in its operation performance. Rolling smoothly and with reduced resistance results in an increase in the efficiency and safety of rail transport. The original shape and dimensions of the profile of the wheel are altered under operation of the train, especially due to braking events and the presence of external objects between the wheel and the railway. With the purpose of recovering the optimum contact profile, train wheels are periodically machined using special lathes. This repair operation is particularly critical in freight trains, which are only reshaped a few times throughout their service life and, therefore, high depths of cut are required to recover the wheel in a productive way. As the presence of chatter vibrations limits the productivity of these operations, a hybrid edge–cloud computing approach has been developed for chatter vibration suppression. An expert system based on automatic chatter detection and suppression has been developed in the edge. The expert system is based on continuous real-time vibration monitoring and combines continuous spindle speed variation (CSSV) and cutting speed reduction to suppress chatter. Cloud computing is used to extract wheel profile machining fingerprints and obtain insights from multiple aggregated machined wheels. An industrial implementation of the system is described in the present work.

Keywords: chatter; train wheel; smart system; turning; edge computing

---

1. Introduction

The maintenance and periodic replacement of railway wheelsets represents a significant cost faced by train operating companies. Railway vehicle wheels wear away relatively slowly and could be expected to last for many years (sometimes more than 20 years) based only on wear considerations. However, they are also subject to tread damage caused by wheel slide events, rolling contact fatigue, flange wear, and tread roll over. These events are more likely in the case of poorly maintained railways, where debris is accumulated on highly eroded tracks. Wheels, therefore, require regular re-profiling by machining on a wheel lathe. Massive maintenance service on a wheelset is done one or two times in its entire service life. The reprofiling of the worn wheels and the machining of the brake-disc are the usual main corrective actions.

This reprofiling operation is carried out by reducing the diameter of the wheel in accordance with local standards [1,2]. For train operating companies, an optimum strategy for wheel maintenance and lathe operation is required to achieve two main goals: maintaining wheels within operational safety and efficiency limits, and minimizing the costs. Thus, performing the repair operation as fast as possible while retaining the tolerances of the wheel profile is essential. The problem arises when trying to reshape severely damaged freight train wheelsets, where the high deterioration of the tread profile requires a high depth of cut turning operation. This fact, linked to the usual forged steel material of the wheel and its hardened outer layer, makes machining of the parts very demanding. For this reason, chatter vibrations that compromise wheel surface finish and machine-tool life expectancy and reliability may arise during turning. It is usual to reprofile a freight wheelset in several cutting passes instead of one to avoid chattering, leading to a significant loss of productivity.

The first step in designing a smart system for chatter suppression is the automatic identification of chatter vibrations. Most of the available in-process methods for chatter identification rely on extracting certain features from the acoustic or vibration signals and comparing them against some predefined chatter benchmarks. They can be broadly divided into two groups: categorization of feature matrix extraction (signal spectrum and decompositions) and classifiers (supervised and non-supervised) [3,4]. In the first group, different methods have been proposed based on analysis of the signal spectrum [5], and decomposition of the signal based on wavelet packet transform [6] or empirical mode decomposition [7]. The general detection of chatter vibrations is a complex problem, and alternative types of classifier have been applied, including the hidden Markov model [8] and neural network classification [9].

Chatter vibrations in turning operations have been extensively studied, and many solutions to overcome the problem have been presented [10]. It is well known that the stability lobe diagram can be used to optimize the cutting capacity by selecting the correct process parameters [11–14]. The selection of an adequate tool–tool holder geometry based on stability diagrams can also increase the cutting capacity [15,16]; however, this solution is rarely applicable to structural chattering cases in turning. The process damping effect raises the effective damping by increasing the friction between tool and workpiece [17], and can be applied to stabilize the cutting [18]. The increase of stiffness of the vibrating structure could also be a feasible way to avoid chatter, especially if there is a dominant mode, although massive structural changes are sometimes required to achieve a substantial effect [10,19]. Passive and active damping solutions, on the other hand, can be used to add damping to the critical mode/s by means of external subassembly devices, leading to a higher stability limit [20–24]. Finally, continous spindle speed variation (CSSV) techniques increase stability by disrupting the regenerative effect [25–28].

By applying the previous structured engineering thinking and techniques to the industrial use of AI or machine learning, a transformation from collected raw data to enriched information or smart data [29] will be generated. The smart data are responsible for detecting relevant signals and patterns through intelligent digital signal-processing algorithms. Smart data makes sense out of big data, providing actionable information and improving decision making.

The present work addressed the implementation of a hybrid edge–cloud smart system in a portal lathe for optimal wheelset reprofiling. The edge computing expert system consisted of a chatter detection and suppression tool, which generated and evaluated the smart data. The process was continuously monitored through embedded sensors which decided when chatter was occurring and, depending on the severity, applied different chatter suppression strategies. Additionally, a cloud computing system obtained the fingerprint of wheel profile machining.

Finally, the correct behavior of the system was experimentally validated through industrial real wheel reprofiling tests where the vibration of the process was removed through the smart system and the turning process was optimally performed. Moreover, the data provided from the system were collected, creating a fingerprint for each machining operation in the cloud. The data were remotely accessed and analyzed allowing a forensic examination.

2. Description and Diagnosis: Engineering Knowledge for Smart Data Collection

The productivity of the machine in the case of freight train wheel reprofiling is limited by the presence of strong marks in the surface of the reshaped wheels and the premature breakage of mechanical elements due to the presence of strong vibrations. Proper diagnosis of the origin of the problem by domain experts is fundamental for successful design of the system. In this case, several cutting and impact tests were performed to identify the problem and select the proper sensors, broadbands, and admissible vibration limits. These tests were the key to defining the hardware of the smart system and the different vibration levels that were used to classify the status of the cutting process.

2.1. Portal Lathe Machine

This work was carried out in a portal lathe, which is a machine specifically conceived for wheelset machining (see Figure 1). The machine has a machining unit composed of two pairs of crossed carriages (X, radial and Z, axial), one for each wheel, as well as a measuring unit to determine the amount corrective depth of cut required.

Figure 1. General view of the portal lathe.

The wheelset rotation is driven by electric motors, which in turn drive rollers attached at the end of the shaft. These rollers are hydraulically preloaded against the wheel, thus rotating the wheelset by friction.

2.2. Cutting Tests

Cutting tests in different diameter wheelsets (33″, 36″, and 38″) were performed while varying the depth of cut until the stability limit was found. The material was a standard wheelset steel approved by the Association of American Railroads (AAR). A temporary set up of the sensors based on external shielded cables permitted a fast change of the position, being the type of sensor used suitable for an industrial environment. Finally, vibrations were successfully recorded by means of a triaxial piezoelectric accelerometer (IMI 604B31) placed in the left turret and a uniaxial (IMI 627A01) accelerometer in the left upper roller (see Figure 2). Sound pressure was also measured with a microphone (MG MK250). All the signals were processed in a vibration analyzer.

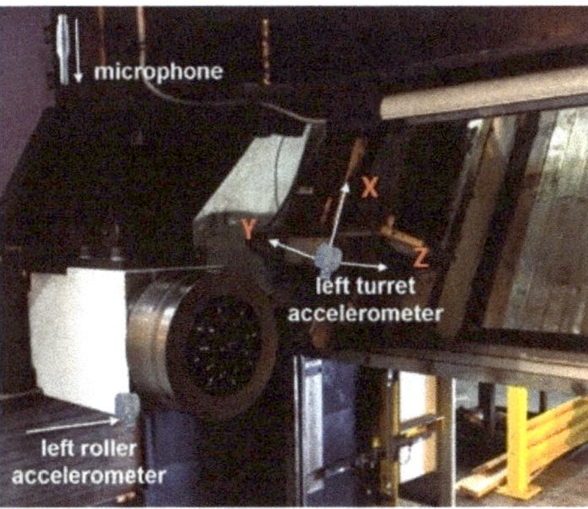

Figure 2. Sensor location during the cutting tests and frequency response function (FRF) determination. The left turret accelerometer was integrated into the hybrid edge–cloud-based smart system.

In these tests, the presence of the self-excited vibrations popularly known as chatter was detected and related to the marks in the wheel surface. A summary of the tests is shown below in Table 1. The green ticks stand for the stable tests, whereas the red crosses represent the unstable ones (chatter tests). Four different wheelsets were studied. The depth of cut (mm in radius) was increased until the stability limit was reached at different cutting speeds.

Table 1. Cutting test results.

| | Cutting Speed/V_c (m/min) | | | | | | | | |
| | Wheelset 33″ | | | Wheelset 36″ | | | Wheelset 38″ | | |
a_p (mm)	40	50	60	54	60	66	40	50	60
7.5		✓			✓				
11		✓	✓		✓	✓		✓	
15	✓	✗	✗	✓	✓	✗			
18	✓								
21									

The recorded chatter cases featured a high vibration amplitude (see Figure 3a), which turned out to be very harmful for the machine and the machined wheelset, resulting in scrapped parts due to severe surface marks (see Figure 3b). Vibration amplitudes up to 16 mm/s rms in the left turret and up to 40 mm/s rms in the upper roller were measured. The chatter frequency lay between 60–62 Hz at every unstable test. Both accelerometers showed similar vibration patterns and were more robust at detecting chatter than the microphone. Therefore, only one accelerometer, the left turret triaxial accelerometer, was selected to be included in the smart system. Additionally, based on the vibration levels of the left turret accelerometer, different machining stability conditions were defined in this phase.

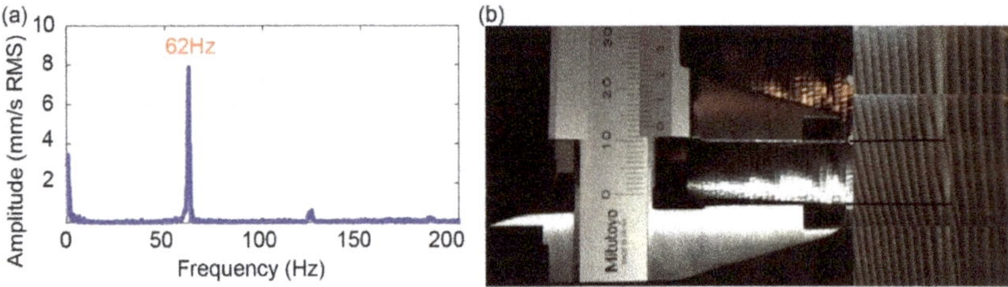

Figure 3. Machining of 36″ wheelset (Vc = 66 m/min; a_p = 15 mm): (**a**) Vibration spectrum from the left turret accelerometer (*X* axis). (**b**) Corresponding chatter marks.

The measured distance between chatter marks in the wheel was around 13 mm, which in a 900 mm diameter wheel and at a rotating speed of 18 rpm approximately matched the measured chatter frequency (62 Hz).

The lobe number *k* or the number of waves per revolution was calculated through Equation (1):

$$k = \frac{f_c}{f_N} = \frac{60 f_c}{ZN} = \frac{60 \cdot 62}{18} = 206.7, \tag{1}$$

where: f_c: chatter frequency (Hz), Z: number of inserts, N: spindle speed (rpm).

Therefore, the part was marked with around 207 marks in the peripheral surface of the machined train wheel, with an approximate distance of 13 mm.

2.3. Frequency Response Functions

The frequency response functions (FRF) of the machine were obtained by means of impact tests. The objective was to identify the vibration modes responsible for the chatter vibrations during machining. The FRFs shown in Figure 4 were obtained by shaking the structure in both turrets close to the tip (see Figure 2) using an impact hammer (PCB086D20/C41) and measuring the response at the same point using a high sensitivity triaxial accelerometer (PCB356A17). The averaged FRFs were obtained by repeating the process four times in the machining position with the wheelset clamped.

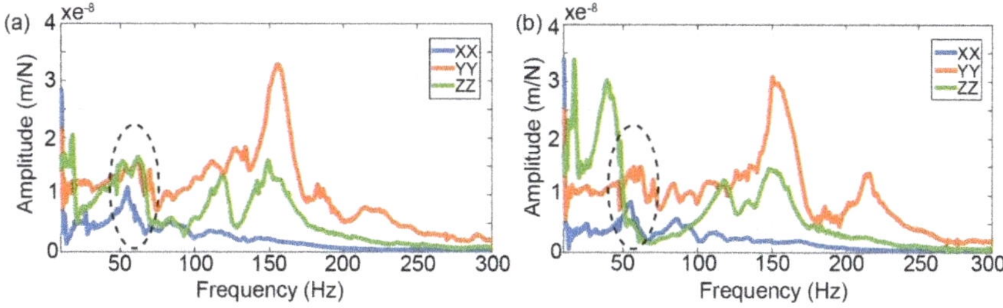

Figure 4. FRF on the (**a**) left turret and (**b**) right turret.

The dynamic properties of the main modes are summarized in Table 2:

Table 2. Dynamic properties of the main modes.

	Left Turret			Right Turret		
	Frequency (Hz)	Modal Stiffness (N/μm)	Damping (%)	Frequency (Hz)	Modal Stiffness (N/μm)	Damping (%)
X	55	94.5	5.8	55	116.8	5.8
Y	154	31.3	5	151	33.7	5.5
Z	149	62.2	5.4	39.3	33.2	-

Considering the high vibration amplitude at around 60 Hz that was recorded during the chatter tests, the mode at 55 Hz in X direction was determined to be causing the instability. The presence of this mode at 55 Hz definitely confirmed the presence of chatter. Although the dynamic stiffness (94–117 N/μm) and the damping value (5.8%) of this mode were high, the demanding operation over a heavily worn out wheel jeopardized the cutting stability. In addition, the high damping of the critical mode at 55 Hz ruled out the inclusion of the active damping technique in the chatter suppression smart system [10,23,24].

Three important conclusions were obtained from the tests carried out in Section 2 to inform the design of a hybrid edge–cloud-based smart system for train wheel repair. First, the cutting tests and the FRF measurement showed that regenerative chatter related to a structural mode around 55 Hz was the responsible for the marks that are endangering the productivity of the portal lathe. Therefore, the smart system focused on the detection, control, and suppression of chatter. Secondly, the industrial accelerometer located close to the tool tip in the left turret was selected as the main sensor to determine the status of the process, and the critical vibration levels in this position for the sensors were defined. Finally, the self-excited mode had considerable damping and the ratio between the chatter frequency and the cutting frequency k was very high ($k > 200$). These two characteristics were considered to select the most proper techniques to suppress chatter.

3. Smart Chatter Suppression in Wheel Reprofiling

3.1. Selection of the Adaptative Chatter Suppression Techniques for the Smart System

In Section 1, several chatter suppression techniques were mentioned. However, not all solutions are applicable in every scenario. Munoa et al. [10] studied the different chatter suppression techniques and discussed the most suitable application for each chatter problem. The number of waves per revolution or lobe order (k) was defined as one of the key factors in selecting the optimal solution. In the present case of structural vibrations arising in wheel reprofiling turning operations, the ratio between the chatter frequency and the cutting frequency k was very high ($k > 200$), due to the low rotating frequency of the machined wheel. This located the process in the A zone of the stability diagram (Figure 5), or the so-called process damping zone [10].

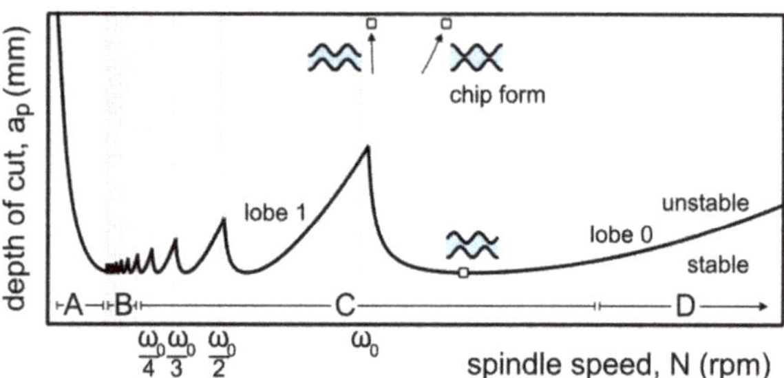

Figure 5. Relative location on the stability lobe diagram.

In this zone, according to the domain experts [10], two techniques are especially effective, the above-mentioned process damping maximization, and CSSV [10]. Although active damping could be also a suitable solution for chatter removal in this zone, the high damping of the critical mode at 55 Hz discouraged its use for the chatter suppression smart system [10,23,24].

3.2. Process Damping Maximization

For the case described in Figure 3, $k = 207$. This implies that the working conditions were in Zone A, or the process damping zone in the stability lobe diagram (Figure 5). The effectiveness of process damping grows as the ratio between the tooth passing frequency and the chatter frequency decreases, therefore, the lower the spindle speed is set, the higher the stability boundary is. The physical effect for this stability increase is the rubbing phenomenon between the flank face and the wavy surface (see Figure 6).

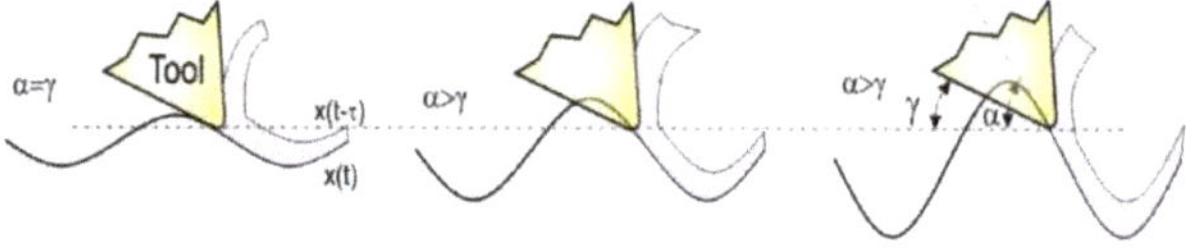

Figure 6. Process damping due to rubbing of the flank face with the wavy surface left [10,18].

In order to take advantage of the process damping effect, two practical alternatives are feasible. On one hand, the spindle speed can be reduced to increase the number of vibration waves per revolution, and thus increase process damping. However, the application of this technique can reduce productivity and can be limited by machinability constraints. On the other hand, worn tools can be used to avoid chatter, as the process damping increases with flank wear. However, worn tools increase the static cutting loads, and excessive wear may lead to tool breakage and poor surface finish. Special edge geometries can be also used to avoid sharp edges and reproduce the performance of a worn tool. Several authors [18,30] have studied the effect of cutting edge geometry on process damping.

Process damping was proven to be a valid solution for chatter avoidance in the studied portal lathe case. The applied solution consisted of reducing the cutting speed and thus the revolutions per minute of the wheelset. Figure 7 shows how reducing the cutting speed from 60 m/min to 40 m/min made the chatter vibration disappear, leading to a stable cutting condition.

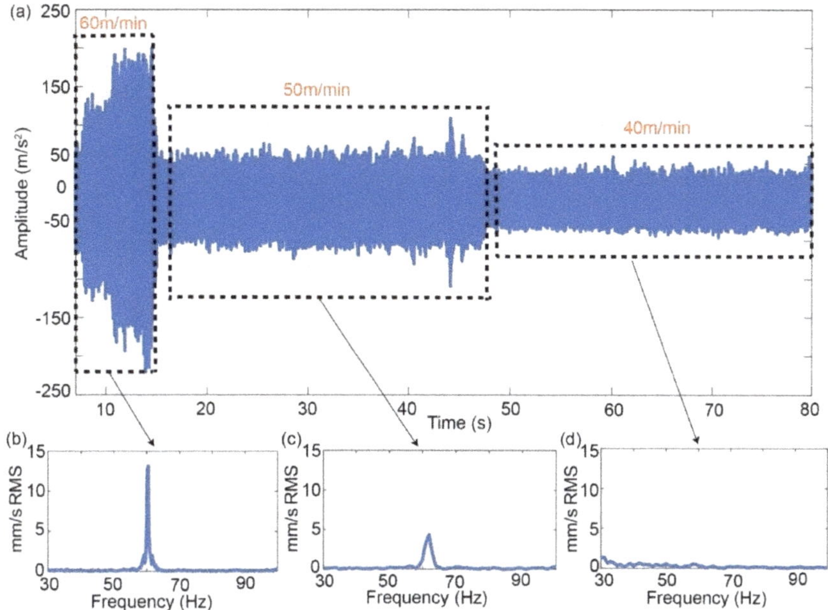

Figure 7. (**a**) Time signal of the test and vibration spectrum at different spindle speeds: (**b**) 60 m/min; (**c**) 50 m/min; (**d**) 40 m/min.

This technique was effective and easy to implement, but it reduced the productivity as the spindle speed was also decreased. Feed per tooth could be increased at the same time to preserve productivity, but there was a risk of side effects such as early tool wear or breakage.

3.3. Continuous Spindle Speed Variation (CSSV)

Some authors [25–28,31] have proposed the distortion of the regenerative effect by CSSV around a nominal speed. CSSV creates a time dependency in the delay that can improve stability. The effectiveness of this technique has been demonstrated in several studies in the literature [32–34].

This technique is based on the introduction of a perturbation in the spindle speed command. There are different methods for varying the spindle speed, including sinusoidal [35–38], triangular [39], rectangular [31], random [40], or continuous linear perturbation [41]. The sinusoidal signal has emerged as the most efficient technique compared to random, rectangular, or triangular shapes. Therefore, the sinusoidal spindle speed variation (SSSV) was selected. This consisted of a simple harmonic variation applied on the spindle speed around the nominal speed value.

When the machining process takes place at low spindle speeds (A & B zones in Figure 5), small variations in the spindle speed bring about relatively large variations in the delay between successive waves. Therefore, low spindle speed amplitude variations can increase the stability in these zones without any relevant side effects. In these areas, the application of CSSV supposes a small increase in the surface roughness and does not jeopardize the life of the main drive if the correct variation is applied [42]. Out of these zones, CSSV is not effective because the variation requirements are not physically reachable [10]. The selection of the amplitude and frequency for sinusoidal CSSV is already an open issue. Al Regib et al. [37] proposed simple formulae to define the parameters of the variation. These formulae are only approximate and do not consider the effect of process damping, which was dominant in the present case. Nevertheless, some facts related to CSSV are well-known, such as the fact that the variation of the amplitude has a bigger effect than the frequency variation [36,43].

The portal lathe chattering problem in this work was placed in Zone A, which made the application of CSSV appropriate. In the case, the amplitude and frequency were selected experimentally, introducing the minimum possible variation in terms of amplitude and frequency. A soft variation based on a frequency of 0.5 Hz and an amplitude variation of 5% of the original speed were enough to stabilize all the cutting tests shown in Table 1.

Figure 8 shows the variation of the vibration spectra in a cutting test of 15 mm depth of cut, 60 m/min cutting speed, and 1.3 mm/rev feed rate with and without CSSV.

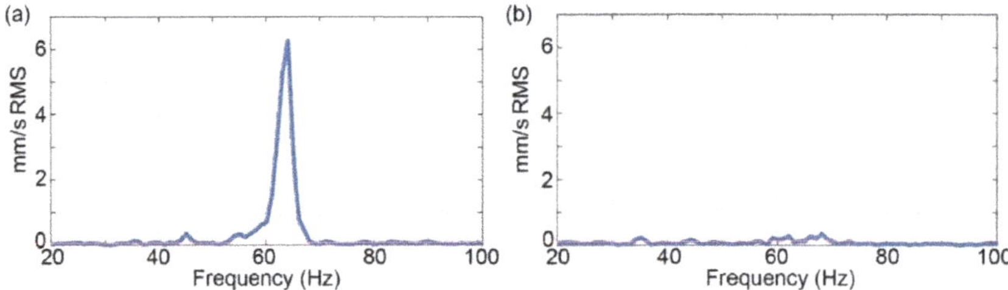

Figure 8. Vibration spectra with (**a**) continuous spindle speed variation (CSSV) off and (**b**) CSSV on at amplitude = 5%, frequency = 0.5 Hz, on a 36″ wheelset machining.

It can be concluded that CSSV technique for chatter suppression worked effectively in the present case study and, in addition, it did not alter productivity. However, sinusoidal variation of the rotating speed may lead to slipping between the wheels and the rollers due to the continuous acceleration/deceleration. On the other hand, the application of CSSV in stable conditions can reduce tool life [44], but its application in unstable condition drastically increases the tool life [45]. Therefore, this solution should be considered as a secondary option to be used only when the presence of chatter is confirmed.

4. Industrial Internet of Things (IIoT) Platform for Hybrid Edge and Cloud Computing

An Industrial Internet of Things (IIoT) platform (see Figure 9) is a form of middleware that sits between the layers of Internet of Things (IoT) devices (hardware, sensors) and IoT gateways (and thus, data) on the one hand, and applications (software) on the other hand. The reality of IIoT platforms is complex because applications and solutions come with different architectures, ways of connecting and managing devices, possibilities for managing and analyzing data, capabilities to build applications, and options to leverage IIoT in a meaningful way for any given use of IIoT in any given context.

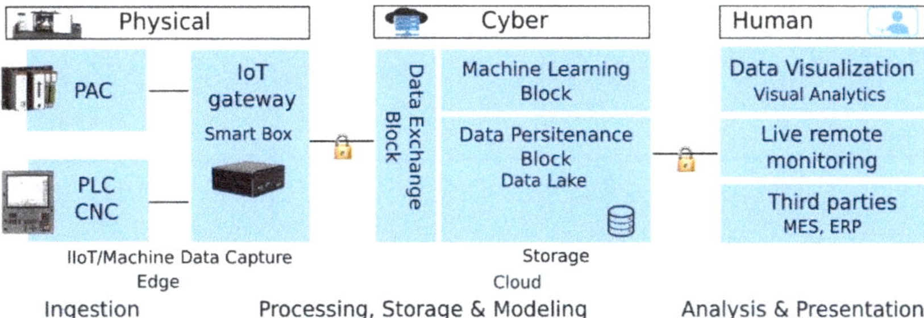

Figure 9. Block diagram of the Industrial Internet of Things (IIoT) platform for edge and cloud computing.

An IIoT platform needs hardware, such as sensors or devices (the "physical" part). These sensors and devices collect data from the machine/components, machining process, and performed actions. Three requirements can be established for the design of an IIoT platform:

An IIoT platform needs connectivity. The hardware needs a way to transmit all the collected data to the cloud, or needs a way to receive commands from the cloud. There is usually an intermediate step between hardware and the connection to the cloud, such as an IIoT gateway.

An IIoT platform needs software (the "cyber" part). This software is hosted in the cloud and is responsible for storing and analyzing the vast amount of data collected from the sensors, and for making decisions based on learned patterns due to the high computing power of the cloud. Nevertheless, the analysis, processing, and transmission of some types of data can be critical in contexts where fast actions need analytics and intelligence at the so-called edge (where the devices/assets and IIoT gateways are). This is what edge computing and fog computing are partially about and where edge platforms come into play.

Finally, *a complete IIoT platform needs a user interface.* To make all this useful, there must be a way for users to interact with the IIoT platform (for instance, a web-based app with dashboards for visual data analytics or live remote data monitoring).

4.1. The Edge

At the edge, the data come in from the physical world via sensors, and actions are taken to change physical state via various forms of output and actuators. The data processing, analysis, and evaluation are performed at the edge. The communications have real-time and deterministic behavior, while the quantity of the analyzed data is reduced.

A triaxial accelerometer was embedded in the machine close to the cutting point and its signal was sent to a real-time controller located in the electrical cabinet of the machine. An expert algorithm which ran in this real-time programmable automation controller (PAC) processed this signal and when the machining becomes unstable, it commanded an action from the computer numerical control (CNC) to solve the vibration problem. This controller was surveyed by a higher-level cloud-based monitoring system, which was used as a follow-up and analytics tool. This surveillance was done through an IoT monitoring hub or gateway which gathered the data from the controller, managed them, and uploaded them onto the cloud. The physical platform used is shown in Figure 10.

Figure 10. Installed hardware: (**a**) real-time controller, IoT monitoring hub, and (**b**) triaxial accelerometer.

4.2. The Cloud

The cloud is a huge, interconnected network of powerful servers that performs services for business. The Savvy Data Systems cloud, used in this system, can be divided in three major parts: communication to the cloud, cloud processing, storage and modeling, and cloud data analysis and presentation.

Communications: The cloud provider, Savvy Data Systems, uses transport layer security (TLS) to offer secure data transmission to the cloud. TLS is a widely adopted security protocol designed to facilitate privacy and data security for communication over the Internet. A primary use of TLS is encrypting the communication between web applications and servers, such as web browsers loading a website.

Cloud processing, storage and modeling: There are three main parts. The first one, named the data exchange block, allows connectivity to the smart box, the IoT hub, through optimal security actions. Companies gain secure and real-time access to new data and lightweight, instant interactions with other blocks; the second block is formed by the data persistence block that deals with persisting (storing and retrieving) data from a data lake. The machine CNC data (power consumption, depth of cut, feed rates, spindle speeds, program name, alarms ...) and PAC data (vibration data) are stored in an unstructured way in the Savvy Data Systems data lake. Finally, the machine learning (ML) block is a managed service that allows developers and data scientists to create superior models of machine learning and take them to production. The Cloud ML Engine offers preparation and prediction services that can be used together or separately. In the proposed use, the gathered data from the CNC and the PAC for the many train wheels machined were aggregated using the Cloud ML Engine, which helped to find weak points during the machining operation for different workpiece references.

Cloud data analysis and presentation: The data visualization block is responsible for the graphical display of the acquired information (historical data) for two purposes: sense-making (also called data analysis) and communication. The data can be effectively visualized following design principles that are derived from an understanding of human perception. In the train wheel repair manufacturing process, machining data fingerprint visualization was proposed (see Figure 11). The unstructured data of the cloud were transformed for human understanding, mixing process data, manufacturing orders, and quality results. On the other hand, the live remote monitoring block was used for the visualization of machine operation data in the moment. Live machine data could be visualized remotely in real-time. Finally, interoperability with third party software, manufacturing execution system (MES) and enterprise resource planning (ERP), was also possible using the full-rest API. Interoperability with human devices is also possible; the cloud could send an e-mail in case of failure or damage detection during the machining operation.

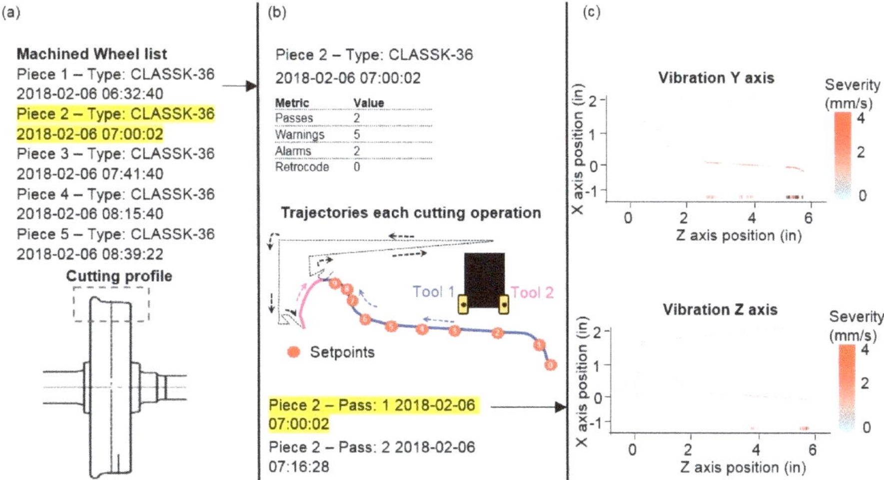

Figure 11. Data visualization for wheel cutting process fingerprint of the workpiece: (**a**) machined wheel list; (**b**) tool trajectory and reset points; (**c**) vibration severity graphs.

5. Hybrid Edge–Cloud-Based Smart System for Chatter Detection and Suppression

The complete solution required processing algorithms on both edge and cloud. Therefore, a hybrid edge and cloud platform was required to develop the smart system for chatter detection and suppression.

Considering the tests carried out in Section 3, process damping and CSSV techniques were determined to be, in principle, the most appropriate techniques to remove chatter vibrations, since the ratio of the chatter frequency and the tooth passing frequency lay in Zone A (Figure 5; $k = 207$ for the portal lathe case) of the stability diagram, which is called the process damping zone [10,18]. Both techniques were tested for the selected case to verify their effectiveness and learn about the vibration signature of the cutting process. The hybrid edge–cloud-based smart system for chatter detection and suppression was based on the computation of signals obtained by the accelerometers for chatter detection, and the smart combination of the aforementioned two techniques to suppress it.

5.1. Edge-Computing-Based Smart System for Chatter Detection and Suppression

The applied expert algorithm first processes the time domain signal captured by the accelerometer and converts it to the frequency domain through a fast Fourier transform (FFT). A frequency domain integration is then performed based on the acceleration spectrum, and a vibration severity is calculated. The severity calculation is a simple quadratic sum of the calculated vibration velocity values over the selected range, in accordance with the ISO-10816 standard [46], as in Equation (2):

$$v_{rms} = \sqrt{v_1^2 + v_2^2 + \ldots + v_n^2} \tag{2}$$

It was determined from the experimental testing carried out in Section 2 that if the cutting process was stable, the severity in this range would lie below 0.5 mm/s rms. When light chatter appeared, this value rapidly rose over 1 mm/s rms, up to around 2.4 mm/s rms. Strong chatter cases clearly exceeded 4 mm/s rms. With this experimental learning in mind, the implemented algorithm is roughly described in Table 3.

Table 3. Algorithm logic.

Status	Severity [mm/s rms]	Action
STABLE	0–2.4	Normal operation
WARNING	2.5–4 for 4 s	Process damping: 80% of spindle speed
ALARM	>4 for 4 s	Process damping: 80% of spindle speed + CSSV
DAMAGE	>4 for 10 or more seconds	Cutting process interruption
RESET	0–2.4	Resetting conditions to normal operation.

Depending on the severity value detected, different actions were taken. First, a decrease in cutting speed was programmed, which was observed to be the most effective solution. If the severity value did not go below 4 mm/s rms, CSSV was added to the cutting speed decrease. In case that both solutions applied simultaneously were not capable of reducing the vibration values after 10 s, the system prompted an alarm and moved the tool out of the workpiece to a safety position, preserving the machine mechanics and avoiding having to scrap a wheelset. The system then proposed to finish the cutting operation by doing an additional roughing pass at half the original depth of cut.

If the chatter is suppressed, various resetting points were programmed all over the wheel profile, in order not to carry out the machining entirely either at low spindle speed or with SSV active, since this could have side effects in terms of productivity decrease or early tool wear. Based on the experience obtained in the cutting tests carried out in Section 2, the process was divided in 10 zones connected by 9 resetting points (see Figure 11).

5.2. Cloud Computing for Machining Fingerprint Analytics

The cutting conditions for each workpiece were recorded. The recorded data consisted of the instantaneous position, feed rate, and power consumption for each axis, the rotational speed and power consumption of the spindle head, and the vibration severity values in X, Y, and Z directions. Thus, each workpiece had its own fingerprint, which was stored in the data persistence block in the cloud. Traceability of the cutting process was therefore made possible thanks to the cloud's massive storage capacity.

Using the data visualization block, the user was able to access the desired workpiece, for instance. Piece2-Type Class-36 2018-02-06 07:00:02 (see Figure 11). It was then possible to check the number of cutting passes required for machining the wheel profile and to observe relevant data such as required machining time, number of alarms, and vibration severity. If one of the cutting passes was selected, then all the recorded data were shown, including a vibration map indicating the maximum vibration levels on the wheel profile, a power consumption map showing the load required at each position for the axes and the spindle head, and finally a feed rate map showing the different speeds during the trajectory of the tool (see Figure 11). Machine analytics was used to describe the chatter problem, crossing the data coming from the accelerometers with the cutting position obtained from the machine sensors. Only descriptive analytics were used in this study.

It is important to remember that this process is a maintenance operation where the part is not homogenous, and the depth of cut and hardness of the surface can vary considerably. The aggregation of the same type of workpiece fingerprints can highlight hidden insights that would be very difficult to detect any other way. For instance, Figure 12 shows the probability of a certain machining position having a vibration higher than the alarm level when the smart system is working. The ratio between the number of occurrences (cases where the heuristically extracted vibration limit was exceeded) and the total number of possible cases (the total number of machined wheels) was analyzed for each position and reference.

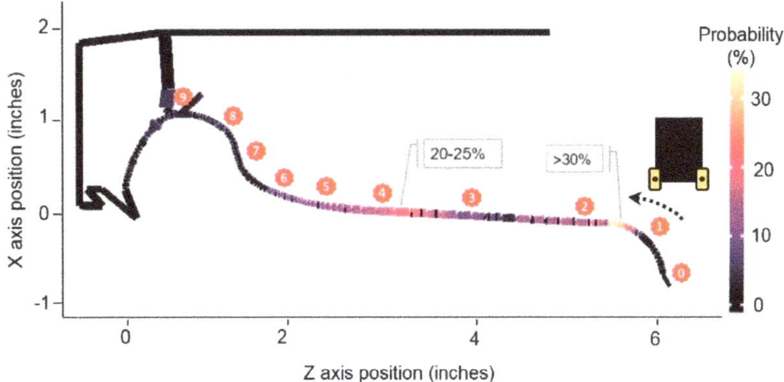

Figure 12. Resultant machining behavior extracted from data analytics for a large quantity of same type of rail wheel with the smart system activated. The red dots mark the reset points.

It was noticed that the points with the highest probability of having the maximum vibration level were always the same (Figure 12). Two points concentrated the risk of having high vibration from the same references and were not close to reset points.

The trajectory of the first insert crossed the wheel in the negative Z direction and found the first critical zone between reset point 1 and 2 in the initial radius. The wheel withstood a plastic deformation during its operative life, and a considerable difference in terms of depth of cut between the central part and this radius were found. Therefore, the depth of cut was bigger during the reshaping process and the risk of chatter increased.

The second critical point was located before the fourth reset point. In this area, the depth of cut did not have any important increase, but the material was hardened due to the deformation created during operation. The deformation process of the wheel profile was comparable with a rolling process in the positive Z direction [47].

Once a great number of fingerprints had been recorded, the smart system had a machine learning block prepared to house any machine learning technique. The idea was to use a machine learning technique that will enable the automatic customization of section length and reset point locations for each railway wheel reference. Therefore, the actual cutting-test-based section and reset point definition will be optimized using machine learning techniques and deployed in the Cloud ML engine. This way, the CSSV and the reduction of spindle speed will be managed automatically for each train wheel reference.

6. Experimental Validation

After the full integration of the chatter detection and suppression system, it was tested under normal production conditions. The process was performed using Kennametal KRR658675 SP97CV rectangular inserts with TiAlN coating and 3/16 inch corner radius. A theoretical depth of cut of 15 mm, a feed per revolution of 1.2 mm/rev, and a cutting speed of 50 m/min were used in the validation tests (Figures 13 and 14).

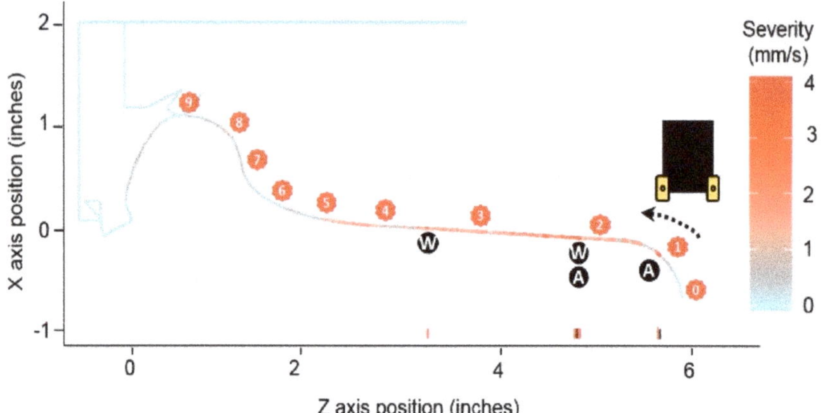

Figure 13. Vibration severity values over the wheel profile with the smart system activated. The red dots mark the reset points and the black circles show where the warning (w) and the alarm (a) were activated.

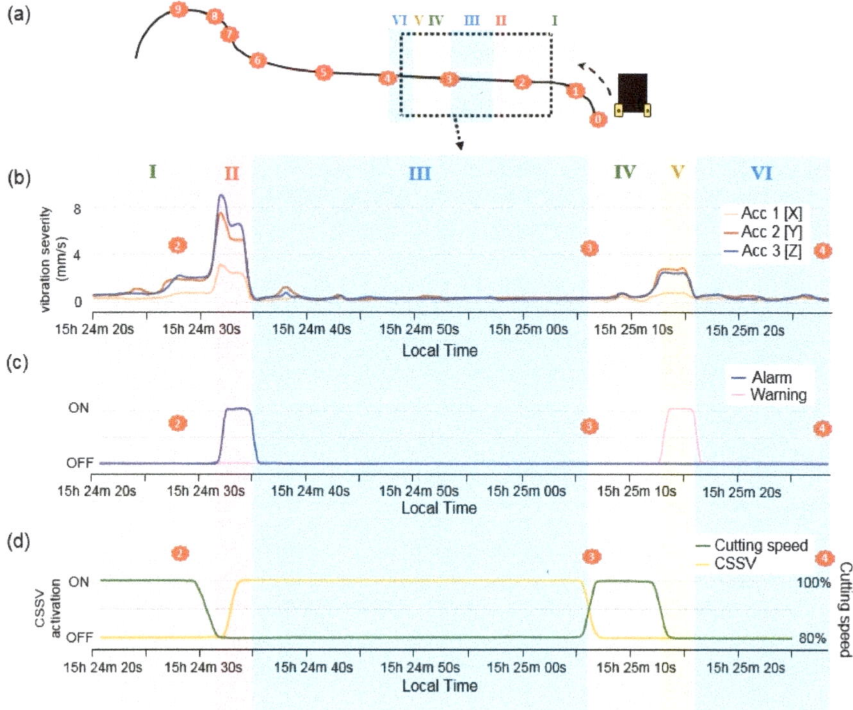

Figure 14. Chatter detection and suppression system performance, analyzed at the data visualization block: (a) wheel profile and tool trajectory; (b) vibration severity; (c) warnings and alarms; (d) activation of CSSV and the spindle speed (override percentage).

The cloud-based monitoring system was used to analyze the cutting process and valuable information was obtained, such as the graph shown in Figure 13, where the wheel profile was drawn from the tool trajectory. The severity values recorded at each instant were overlapped on this graph, thus obtaining a graphic representation of the location of the main critical machining areas.

As highlighted in Figure 12, the critical zones were concentrated in the segment between the first and the fourth reset points. In the beginning of the initial radius, chatter grew until the alarm level was reached. The spindle speed was reduced and CSSV was activated, stabilizing the process. The vibration gradually increased below the warning level until the second reset point.

Table 4 shows in detail the performance of the chatter detection and suppression system in a chronological sequence, starting before the second reset point and continuing the fourth reset point. All data were gathered and analyzed using the cloud-based monitoring system.

Table 4. Algorithm action sequence.

Sequence	Status	Severity (mm/s rms)	Action
I	STABLE	Lower than 2.5	Normal operation
II	ALARM	Higher than 4	Spindle speed 80% + CSSV are activated
III	STABLE	Lower than 2.5	Spindle speed 80% + CSSV remain on
IV	RESET	Lower than 2.5	Resetting point. Normal operation.
V	WARNING	Between 2.5–4	Spindle speed 80% is activated
VI	STABLE	Lower than 2.5	Spindle speed 80% remains on

At the beginning, the machining was stable (I) and the situation was maintained over the second reset point. At a certain point, a high vibration amplitude was measured (II). The smart system prompted a warning first and subsequently an alarm. Therefore, the real-time controller sent the command to the CNC to apply the spindle speed reduction and CSSV. The cutting became stable again (III) with the help of the chatter suppression techniques. After a while, the third resetting point was reached (IV) and cutting conditions returned to the original values. Machining remained stable until light chatter started (V) and a warning was displayed. In this case, only the spindle speed reduction technique was applied, and the cutting became stable again (VI). Figure 14 shows the actuation of the smart system in a graphical way.

7. Conclusions

A hybrid edge–cloud-based smart system for chatter detection and suppression in a specific operation of freight train wheelset reprofiling was developed. The system relies on an embedded accelerometer to detect chatter and uses a smart algorithm to select the most appropriate action in accordance with the measured signal. Thus, the process stability is assured. The data recorded by the sensor as well as the actions taken for every wheelset can be surveyed using the cloud-based monitoring system, which is connected to the machine PLC and to the process controller. This provides very useful information for follow-up and the traceability of the production system.

The system was validated using a portal lathe in an industrial environment. The results were very satisfactory, achieving a proper automatic machining performance optimization.

The developed system is an ad hoc utility for a specific industrial problem which has been previously characterized. The future challenge is to extend its application to different machining processes in which the actuation criteria can be determined automatically by the smart system itself, based on a thorough historical data analysis of the gathered information from previous events in a machine learning approach.

Author Contributions: Conceptualization and Supervision, J.M.; Investigation and Validation R.M. and I.B.; Review and editing, A.I.

Funding: This work was partially supported by the ASTRACOMP Project (EXP-00102217) from the Innoglobal program of the Spanish Ministry of Economy, Industry and Competitiveness.

Conflicts of Interest: The authors declare no conflict of interest.

References

1. EN, B. *13715: 2006+ A1: 2010–Railway Applications–Wheelsets and Bogies–Wheels–Tread Profile*; Technical Committee CEN/TC 256 held by German Association for Normalization (DIN): Berlin, Germany, 2010.
2. The American Public Transportation Association (APTA). *Standard for Wheel Flange Angle for Passenger Equipment*; APTA Press Task: Washington, DC, USA, 2007.
3. Sims, N.D. Dynamics diagnostics: Methods, equipment and analysis tools. In *Machining Dynamics*; Springer Series in Advanced Manufacturing; Springer: London, UK, 2009; pp. 85–115.
4. Yesilli, M.C.; Khasawneh, F.A.; Otto, A. On Transfer Learning for Chatter Detection in Turning Using Wavelet Packet Transform and Empirical Mode Decomposition. *arXiv* **2019**, arXiv:190501982.
5. Delio, T.; Smith, S.; Tlusty, J. Use of Audio Signals for Chatter Detection and Control. *J. Manuf. Sci. Eng.* **1992**, *114*, 146–157. [CrossRef]
6. Yao, Z.; Mei, D.; Chen, Z. On-line chatter detection and identification based on wavelet and support vector machine. *J. Mater. Process. Technol.* **2010**, *210*, 713–719. [CrossRef]
7. Ji, Y.; Wang, X.; Liu, Z.; Wang, H.; Jiao, L.; Wang, D.; Leng, S. Early milling chatter identification by improved empirical mode decomposition and multi-indicator synthetic evaluation. *J. Sound Vib.* **2018**, *433*, 138–159. [CrossRef]
8. Xie, F.-Y.; Hu, Y.-M.; Wu, B.; Wang, Y. A generalized hidden Markov model and its applications in recognition of cutting states. *Int. J. Precis. Eng. Manuf.* **2016**, *17*, 1471–1482. [CrossRef]
9. Lamraoui, M.; Barakat, M.; Thomas, M.; El Badaoui, M. Chatter detection in milling machines by neural network classification and feature selection. *J. Vib. Control* **2015**, *21*, 1251–1266. [CrossRef]
10. Munoa, J.; Beudaert, X.; Dombovari, Z.; Altintas, Y.; Budak, E.; Brecher, C.; Stepan, G. Chatter suppression techniques in metal cutting. *CIRP Ann.* **2016**, *65*, 785–808. [CrossRef]
11. Altintaş, Y.; Budak, E. Analytical Prediction of Stability Lobes in Milling. *CIRP Ann.* **1995**, *44*, 357–362. [CrossRef]
12. Iglesias, A.; Munoa, J.; Ciurana, J.; Dombovari, Z.; Stepan, G. Analytical expressions for chatter analysis in milling operations with one dominant mode. *J. Sound Vib.* **2016**, *375*, 403–421. [CrossRef]
13. Urbikain, G.; Olvera, D.; De Lacalle, L.N.L.; Elías-Zúñiga, A.; Pelayo, G.U. Stability and vibrational behaviour in turning processes with low rotational speeds. *Int. J. Adv. Manuf. Technol.* **2015**, *80*, 871–885. [CrossRef]
14. Urbikain, G.; Campa, F.J.; Zulaika, J.-J.; De Lacalle, L.-N.L.; Alonso, M.-A.; Collado, V. Preventing chatter vibrations in heavy-duty turning operations in large horizontal lathes. *J. Sound Vib.* **2015**, *340*, 317–330. [CrossRef]
15. Schmitz, T.; Donalson, R. Predicting High-Speed Machining Dynamics by Substructure Analysis. *CIRP Ann.* **2000**, *49*, 303–308. [CrossRef]
16. Mancisidor, I.; Zatarain, M.; Munoa, J.; Dombovari, Z. Fixed Boundaries Receptance Coupling Substructure Analysis for Tool Point Dynamics Prediction. *Adv. Mater. Res.* **2011**, *223*, 622–631. [CrossRef]
17. Eynian, M.; Altintas, Y. Chatter Stability of General Turning Operations with Process Damping. *J. Manuf. Sci. Eng.* **2009**, *131*, 041005. [CrossRef]
18. Tunc, L.T.; Budak, E. Effect of Cutting Conditions & Tool Geometry on Process Damping in Machining. *Int. J. Mach. Tool Manuf.* **2012**, *57*, 10–19.
19. Weck, M.; Miessen, W.; Muller, W. Visual Representation of the Dynamic Behaviour of Machine Tool Structures. *CIRP Ann.* **1976**, *25*, 263–266.
20. Hahn, R.S. Design of Lanchester Damper for Elimination of Metal- Cutting Chatter. *J. Eng. Ind.* **1981**, *73*, 331–335.
21. Munoa, J.; Iglesias, A.; Olarra, A.; Dombovari, Z.; Zatarain, M.; Stepan, G. Design of self-tuneable mass damper for modular fixturing systems. *CIRP Ann.* **2012**, *65*, 389–392. [CrossRef]
22. Cowley, A.; Boyle, A. Active Dampers for Machine Tools. *CIRP Ann.* **1969**, *18*, 213–222.

23. Bilbao-Guillerna, A.; Barrios, A.; Mancisidor, I.; Loix, N.; Munoa, J. Control laws for chatter suppression in milling using an inertial actuator. In Proceedings of the ISMA2010 International Conference on Noise and Vibration Engineering, Leuven, Belgium, 20–22 September 2010.

24. Mancisidor, I.; Munoa, J.; Barcena, R. Optimal control laws for chatter suppression using inertial actuator in milling processes. In Proceedings of the 11th International Conference on High Speed Machining (HSM2014), Prague, Czech Republic, 11–12 September 2014.

25. Inamura, T.; Sata, T. Stability Analysis of Cutting Under Varying Spindle Speed. *CIRP Ann.* **1974**, *23*, 119–120.

26. Hoshi, T.; Sato, M.; Sakisaka, N.; Moriyama, I. Study of Practical Application of Fluctuating Speed Cutting for Regenerative Chatter Control. *CIRP Ann.* **1977**, *25*, 175–179.

27. Takemura, T.; Kitamura, T.; Hoshi, T.; Okushimo, K. Active Suppression of Chatter by Programmed Variation of Spindle Speed. *CIRP Ann.* **1974**, *23*, 121–122.

28. Jemielniak, K.; Widota, A. Suppression of self-excited vibration by the spindle speed variation method. *Int. J. Mach. Tool Des. Res.* **1984**, *24*, 207–214. [CrossRef]

29. Lafrate, F. From Big Data to Smart Data. In *Advances in Information Systems Set Information Systems, Web and Pervasive Computing Series*; John Wiley & Sons: Hoboken, NJ, USA, 2015; Volume 1.

30. Yusoff, A.R.; Turner, S.; Taylor, C.M.; Sims, N.D. The role of tool geometry in process damped milling. *Int. J. Adv. Manuf. Technol.* **2010**, *50*, 883–895. [CrossRef]

31. Sexton, J.S.; Stone, B.J. The Stability of Machining with Continuously Varying Spindle Speed. *CIRP Ann.* **1978**, *27*, 317–326.

32. Altintas, Y.; Chan, P.K. In-process detection and suppression of chatter in milling. *Int. J. Mach. Tools Manuf.* **1992**, *32*, 329–347. [CrossRef]

33. Jayaram, S.; Kapoor, S.G.; DeVor, R.E. Analytical Stability Analysis of Variable Spindle Speed Machining. *J. Manuf. Sci. Eng.* **2000**, *122*, 391–397. [CrossRef]

34. Bediaga, I.; Egaña, I.; Munoa, J. Reducción de la inestabilidad en cortes interrumpidos en fresado a alta velocidad mediante variación de la velocidad del husillo. In Proceedings of the 16th Congreso de Máquinas-Herramienta y Tecnologías de Fabricación, San Sebastian, Spain, 18–20 October 2006.

35. Zatarain, M.; Bediaga, I.; Munoa, J.; Lizarralde, R. Stability of Milling Processes with Continuous Spindle Speed Variation: Analysis in the Frequency & Time Domains & Experimental Correlation. *CIRP Ann.* **2008**, *57*, 379–384.

36. Bediaga, I.; Zatarain, M.; Muñoa, J.; Lizarralde, R. Application of continuous spindle speed variation for chatter avoidance in roughing milling. *Proc. Inst. Mech. Eng. Part B J. Eng. Manuf.* **2011**, *225*, 631–640.

37. Al-Regib, E.; Ni, J.; Lee, S.-H. Programming spindle speed variation for machine tool chatter suppression. *Int. J. Mach. Tools Manuf.* **2003**, *43*, 1229–1240. [CrossRef]

38. Sastry, S.; Kapoor, S.G.; DeVor, R.E. Floquet Theory Based Approach for Stability Analysis of the Variable Speed Face-Milling Process. *J. Manuf. Sci. Eng.* **2002**, *124*, 10–17. [CrossRef]

39. Seguy, S.; Insperger, T.; Arnaud, L.; Dessein, G.; Peigne, G. On the Stability of High-Speed Milling with Spindle Speed Variation. *Int. J. Adv. Manuf. Technol.* **2010**, *48*, 883–895. [CrossRef]

40. Yilmaz, A.; Al-Regib, E.; Ni, J. Machine Tool Chatter Suppression by Multi-Level Random Spindle Speed Variation. *J. Manuf. Sci. Eng.* **2002**, *124*, 208–216. [CrossRef]

41. Alvarez, J.; Barrenetxea, D.; Marquinez, J.; Bediaga, I.; Gallego, I. Effectiveness of continuous workpiece speed variation (CWSV) for chatter avoidance in throughfeed centerless grinding. *Int. J. Mach. Tools Manuf.* **2011**, *51*, 911–917. [CrossRef]

42. Albertelli, P.; Musletti, S.; Leonesio, M.; Bianchi, G.; Monno, M. Spindle speed variation in turning: Technological effectiveness and applicability to real industrial cases. *Int. J. Adv. Manuf. Technol.* **2012**, *62*, 59–67. [CrossRef]

43. Otto, A.; Radons, G. Application of spindle speed variation for chatter suppression in turning. *CIRP J. Manuf. Sci. Technol.* **2013**, *6*, 102–109. [CrossRef]

44. Albertelli, P.; Mussi, V.; Monno, M. The analysis of tool life and wear mechanisms in spindle speed variation machining. *Int. J. Adv. Manuf. Technol.* **2014**, *72*, 1051–1061. [CrossRef]

45. Kayhan, M.; Budak, E. An Experimental Investigation of Chatter Effects in Tool Life. Proceedings of Institution of Mechanical Engineers. *Part B J. Eng. Manuf.* **2009**, *223*, 1455–1463. [CrossRef]

46. International Organization for Standarization (ISO). *Mechanical Vibration—Evaluation of Machine Vibration by Measurements on Non-Rotating Parts—Part 1—General Guidelines*; ISO-10816-1:1995(E); ISO: Geneva, Switzerland, 1995.
47. Farago, D.; Merino, R.; Dombovari, Z. On Basic Modeling of the Dynamics of Axles Rolling Process. *IFAC-PapersOnLine* **2018**, *51*, 282–287. [CrossRef]

Article

A Low-Cost Add-On Sensor and Algorithm to Help Small- and Medium-Sized Enterprises Monitor Machinery and Schedule Processes

Yi-Chung Chen [1], Kuo-Cheng Ting [2], Yo-Ming Chen [2], Don-Lin Yang [2], Hsi-Min Chen [2] and Josh Jia-Ching Ying [3,*]

[1] Department of Industrial Engineering and Management, National Yunlin University of Science and Technology, Yunlin 640, Taiwan; mitsukoshi901@gmail.com
[2] Department of Information Engineering and Computer Science, Feng Chia University, Taichung 407, Taiwan; zoncer15888@gmail.com (K.-C.T.); g2503886@gmail.com (Y.-M.C.); dlyang.tw@gmail.com (D.-L.Y.); seeme.goo@gmail.com (H.-M.C.)
[3] Department of Management Information Systems, National Chung Hsing University, Taichung 402, Taiwan
*   Correspondence: jashying@gmail.com

Received: 22 February 2019; Accepted: 8 April 2019; Published: 14 April 2019

Abstract: Since the concept of Industry 4.0 emerged, an increasing number of major manufacturers have incorporated relevant technologies to monitor machinery and schedule processes so as to increase yield and optimize production. However, most machinery monitoring technologies are far too expensive for small- and medium-sized enterprises. Furthermore, the production processes at small- and medium-sized enterprises are simpler and can thus be optimized without excessively complex scheduling systems. This study therefore proposed the use of cheaper add-on sensors for monitoring machinery and integrated them with an algorithm that can more swiftly produce results that meet multiple objectives. The proposed algorithm is meant to extend the capabilities of small- and medium-sized enterprises in monitoring machinery and scheduling processes, thereby enabling them to contend on an equal footing with larger competitors. Finally, we performed an experiment at an actual spring enterprise to demonstrate the validity of the proposed algorithm.

Keywords: Industry 4.0; anomaly detection; scheduling; neural network; skyline queries

---

1. Introduction

Since the term Industry 4.0 was coined, an increasing number of major manufacturers have incorporated relevant technologies, such as using well-designed and embedded sensors to collect data from machinery or the test results of various products. Then, a server receives the signal from multiple sensors, and unifies the format of these signals, for monitoring, scheduling, and analysis [1,2]. This makes it possible for enterprises to monitor the operational status of machinery on an ongoing basis, and in so doing enhance production efficiency, reduce maintenance costs, and optimize quality. Industry 4.0 protocols can indeed help enterprises to improve performance and reduce costs; however, the capital outlay required for the adoption of these technologies (e.g., sensors, cloud computing, big data analysis) can be prohibitively high. Furthermore, the base infrastructure differs greatly from that of conventional enterprises, thereby necessitating considerable upgrades prior to implementation. Most large-scale enterprises are able to make these changes without difficulty; however, most small- and medium-sized enterprises (SMEs) are unable to absorb the costs.

Many researchers have sought to lower the adoption threshold for Industry 4.0. Dassisti et al. [3] introduced a core-meta model to help SMEs implement Industry 4.0 in their enterprises. Uriarte et al. [4] used the technique of mechanistic modeling to predict the micromilling cutting forces. Rivelo et al. [5]

used the internal signals of machinery to do the tool wear detection. Plapper and Weck [6] proposed a new approach on using digital drive signals for monitoring the conditions of machine tool. Barrio et al. [7] discussed how to use the concept of Industry 4.0 to do modeling and process monitoring on machines. Finally, Birkel et al. [8] developed a risk framework for enterprises considering switching over to Industry 4.0.

It should be noted that most of the methods used to monitor machinery are impractical for SMEs, due to budgetary restrictions on the purchase of updated machinery equipped with the latest sensors. Moreover, it is doubtful that they will purchase large numbers of the same machines at the same time. To save money, the machines that they buy may even come from different manufacturers and countries. Consequently, these are reasons why mutual communication among machines is difficult for SMEs. Even if their machines had sensors that could send signals to the server, the formats of these signals vary and are almost impossible to analyze. In other words, it is extremely difficult for SMEs to introduce Industry 4.0 technologies [9,10]. Without these monitoring technologies, the gap between their yield and production costs and those of large manufacturers will gradually widen, and ultimately they will be unable to escape the fate of elimination. Some new methods are needed to address this issue. Such methods must be inexpensive so that they are affordable by SMEs with little capital.

Optimizing process schedules plays a significant role in Industry 4.0 technologies. For this reason, a number of researchers have proposed scheduling-related algorithms [11,12]. For instance, Kyparisis and Koulamas [12] proposed a multi-stage scheduling algorithm that can process multiple machineries running in parallel. Furthermore, they demonstrated that this problem is an NP-hard problem. Next, Tahar et al. [13] presented an algorithm that can schedule a set of independent jobs; the algorithm enables users to set the time, break up jobs, and complete them in the shortest time. In recent years, Ivanov et al. [11] developed a scheduling algorithm that can optimize job schedules and consider machinery statuses at the same time. Another feature of their algorithm is that it can assist enterprises in finding multi-objective results rather than the single-objective results in previous papers. As a result, their algorithm can provide enterprises with a variety of options and satisfy options under different conditions. Nevertheless, it is important to understand that although the method above solves the problem of scheduling optimization in Industry 4.0, it is too complex for SMEs. This is because most of these enterprises, which may produce products such as screws, springs, molds, or plastic injections, need only one step to complete the manufacturing of their products. A multi-stage algorithm would be too time-consuming, and these SMEs rarely have the budget to purchase relevant software and hardware to perform these calculations. Thus, some simple approaches are needed to solve this problem.

In response to these machinery monitoring and scheduling issues, this study proposed two solutions. First, in terms of monitoring, we present arduino-based add-on sensors to monitor machinery status. Then a feature extraction algorithm and a dimensionality reduction algorithm are utilized on these collected signals. Finally, a neural network is used to determine whether machinery statuses are normal. The advantage of this method is that the arduino-based add-on sensors are extremely inexpensive, so even if an enterprise requires a large quantity of these sensors, their cost will not be a great burden. Of course, the precision of the data collected by the arduino-based add-on sensors is not as high as that of data collected by expensive sensors. However, the products of SMEs, such as screws, springs, molds, or plastic products, do not require very high precision, so these add-on sensors are adequate for monitoring.

Second, in terms of scheduling, multi-objective scheduling is the current trend [11]. Therefore, we developed an algorithm for scheduling with multiple objectives. To achieve this goal, we incorporated the recently well-known skyline query [14–16], which can consider situations with different conditions and send back the optimal solutions for various combinations of conditions. With Table 1 and Figure 1 as an example, Table 1 presents the total work time and costs of different schedules. For instance, the total work time of Schedule 1 is 560 min, and its total cost is USD 2,500. Figure 1 displays the total work time and cost of each schedule in Table 1 using coordinates; each point represents a schedule. The points in the very lower left corner of Figure 1 (i.e., Schedules 1,3, and 4) are the skyline schedule results. Schedule 1 has the shortest total work time, Schedule 4 has the lowest total cost, and the total work time and cost of Schedule 3 falls between those of Schedules 1 and 4. Schedule 2 is not a skyline schedule result because its total work time and cost are both higher than those of Schedule 1. Enterprise workers will certainly choose Schedule 1 over Schedule 2. With the concept of skyline schedules, we can help enterprises identify schedules that are superior in all aspects. However, skyline queries are known for requiring longer computation time; to resolve this issue, we developed an innovative algorithm.

Table 1. The example of skyline schedules.

Schedule	Cost	Total Work Time
1	USD 2500	450 min
2	USD 3000	1400 min
3	USD 2000	670 min
4	USD 1000	800 min
5	USD 4500	500 min
6	USD 7300	650 min
7	USD 12,000	1400 min

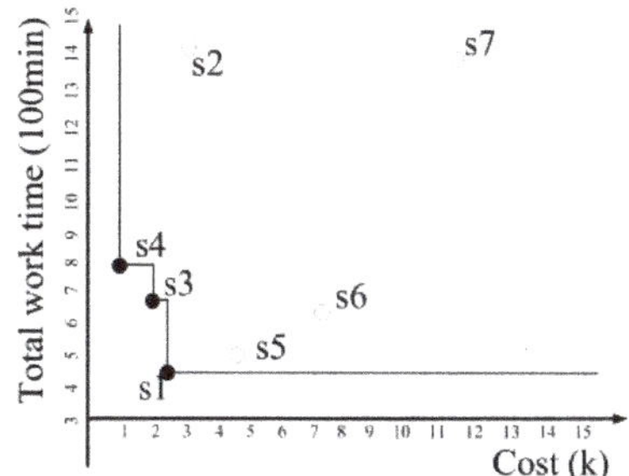

Figure 1. The example of skyline schedules.

This paper is arranged as follows. We first introduce some related work in Section 2. Section 3 explains the approaches for monitoring the machinery. Section 4 studies the methods for scheduling. Section 5 presents the experiment and discussion, and Section 6 contains the conclusion.

2. Related Works

2.1. Industry 4.0 in Small- and Medium-Sized Enterprises

Industry 4.0 initially enjoyed an enthusiastic response; however, many small- and medium-sized enterprises (SMEs) have been unable to participate in this movement due to prohibitively high adoption costs. This has led many researchers to look for ways to lower the entry threshold for SMEs. In the following, we examine a number of the aspects pertaining to the adoption of Industry 4.0 standards by SMEs.

One of the major issues is industrial performance objectives, which deal with flexibility, productivity, and delivery times [9]. Many enterprises seek to synchronize various flows through the supply chain; however, that requires considerable flexibility in responding to rapid market fluctuations. Peng et al. [17] proposed a real-time production flow scheme by which to modify production plans in accordance with changes in demand or disruptions in flow. Chalal et al. [18], introduced a model with one subsystem for modeling demand and another subsystem for modeling production, making it easier for firms to react to client demands.

Another major issue in the adoption of Industry 4.0 standards is productivity. Givehchi [19] sought to improve efficiency by sequencing machining tasks in a more efficient manner. Dombrowski and

Ernst [20] simulated various growth scenarios at the enterprise-level with the aim of reconfiguring production lines to improve production flows. Other researchers focused on reducing delivery times by synchronizing production processes using cloud computing platforms. Denkena et al. [21] used the Internet of Things (IoT) to improve production flows by focusing on waiting times and bottlenecks as a guide for system reconfiguration.

A third major issue in the adoption of Industry 4.0 standards is industrial managerial capacity; that is, monitoring, control, and optimization. Monitoring system status is meant to facilitate proactive responses. For example, Denkena et al. [21] used the IoT to measure the real time flow of parts, whereas Velandia et al. [22] used Radio Frequency Identification (RFID) to record data throughout the entire production process. Control refers to the interaction between employees and the system using historical data and predetermined thresholds [9]. For example, MacKerron et al. [23] proposed an RFID-based system to monitor supply processes and automatically alert managers of impending inventory shortfalls. Optimization refers to efforts aimed at improving systems and processes. However, despite the numerous methods that could be employed to reach this goal, most previous work related to SMEs focused on simulating current industrial practices [18,24].

2.2. Feature Extraction and Dimensionality Reduction

Feature extraction and dimensionality reduction [25] are the preprocessing methods most commonly used in the analysis of signals such as those related to speech and activity recognition. Feature extraction refers to the identification of specific characteristics or patterns within a given signal. For example, the average value of a speech signal could be used to indicate the loudness of speech. The features used in the analysis of signals include mean values, correlations among axes, mean absolute deviation, root mean square, variance, standard deviation, signal magnitude area, and average energy. It should be noted however that the choice of feature set depends on the specifics of the case; that is, not all features are relevant, and irrelevant features can increase the execution time and/or undermine the precision of analysis [25]. Following feature extraction, a number of methods can be used to identify the features best suited to the analysis of a given case.

Dimensionality reduction is the method most commonly used in the selection of features suitable for the analysis of a particular signal. For example, dimensionality reduction can be used to sort through a large number of features identified in the signals collected by automated sensors, such as those found on various forms of machinery. This makes it possible to identify the features best able to verify that a machine is working smoothly or identify cases of imminent failure. Dimensionality reduction is commonly implemented using principal component analysis (PCA) [26] or linear discriminant analysis (LDA) [25]. PCA is an unsupervised method, whereas LDA is a supervised method. PCA is used to identify the projections best suited to "representing data", whereas LDA is used for "discriminating among data" [27,28]. This makes LDA more suitable than PCA for the monitoring of machinery status.

2.3. Classifier for Recognizing the Status of Signal

Classifiers are among the most useful tools for the recognition of signal status. A classifier is first trained using well-defined features, the corresponding status of which is stipulated by experts. This makes it possible to use a trained classifier to determine the status of a signal based solely on its features. The most common classifiers include decision trees [29], K-nearest neighbors (KNN) [30], and neural networks [31]. Bao et al. [29] formulated the C4.5 decision tree to recognize 20 daily activities performed by humans. Karantonis et al. [32] developed an hierarchical binary classifier to enhance the precision of classification and recognition rates. Unfortunately, the computational and memory requirements of decision trees make them inapplicable in most real-world situations. Kose et al. [30] and Kaghyan and Sarukhanyan [33] achieved good results using KNN to deal with signals collected associated with human activities; however, the effectiveness of KNN can be compromised by a small training dataset and computational overhead limits its applicability.

Neural networks are currently the most popular method used in the recognition of signal status [34–36], due to their excellent learning capability in the discrimination of nonlinearly separable classes. Watanabe [36] used a feedforward neural network with triaxial accelerometer to recognize the movements of hemiplegic patients. Jatoba et al. [34] utilized an adaptive neuro-fuzzy inference system based on signals collected from a triaxial acceleration sensor. Yang et al. [25] recently merged a fuzzy system with a neural network to identify human activities based on signals collected using only one accelerometer. The results of these studies demonstrate the efficacy of using neural-network-based techniques for signal recognition.

2.4. Scheduling

Scheduling can be used in enterprises to optimize job flow by arranging tasks in a suitable order. Blazewicz et al. [37] and Lauff and Werner [38] discussed the problem of scheduling in multi-stage systems. Kyparisis and Koulamas [12] considered the same problem, while taking into account the effect of machinery processing speed on job flow. Tahar et al. [13] considered a scheduling problem in which jobs can be split up and assigned sequence-dependent setup times. These methods have proven effective in many situations; however, they are applicable only to single-objective scheduling problems. Ivanov et al. [11] proposed a multi-objective, multi-stage flexible scheduling problem using in each stage alternative machineries with different time-dependent processing speeds.

2.5. Skyline Query

Skyline query algorithms are quickly becoming the mainstay of decision support systems. They function as an alternative to top-k algorithms, which are unable to retain the characteristic features of the various dimensions. The skyline approach makes it possible to identify the best result for every dimension, and tailor the results according to criteria designated by the user. Consider a situation in which the user seeks accommodation close to a particular venue. As shown in Table 2, various forms of information pertaining to nearby hotels must be taken into account in the selection of accommodation. If cost and location were the only considerations, then the information could be arranged within a two-dimensional scatter diagram, as shown in Figure 2. The black line is the skyline; it indicates that the cost and distance of C are better than those of D. In other words, C dominates D on both criteria. Unfortunately, a skyline algorithm is unable to choose between A and C, due to the fact that C is closer to the desired venue but A has a lower price. In this situation, the skyline query would return both hotels, leaving the ultimate decision up to the user. This problem becomes increasingly complex when the number of criteria grows. For example, hotels A and C could all be considered skyline hotels, due to the fact that they are not dominated by other hotels.

Three skyline algorithms are particularly well known: block nested loops (BNL) [39], divide and conquer (DAC) [39], and sort limit skyline (SaLSa) [14]. The BNL algorithm compares a pair of data points and eliminates the point that is dominated. All of the remaining data points are then compared to one another iteratively in order to identify a single dominant candidate. This is the simplest and most intuitive skyline method; however, computational complexity grows exponentially with an increase in the quantity of data.

The DAC algorithm [39] breaks down data into groups, conducts a separate skyline query in each group, and then combines the group results into a final skyline query to produce a final decision. The elimination of many data points during the initial grouping process enhances the execution speed.

The SaLSa algorithm [14] uses summation values obtained from the raw data as threshold values to enable the filtering out of unnecessary data points. This algorithm arranges data sequentially according to the summation values and checks whether a given data point falls within the skyline threshold before testing for dominance; that is, only those below the threshold value are tested for dominance. The SaLSa algorithm has proven particularly effective in time and cost reduction.

Table 2. Information pertaining to hotels in the vicinity of the beach.

Hotel	Cost	Distance to the Beach	Internet Ratings
A	$100	5 km	3
B	$750	3.5 km	4
C	$230	1 km	3
D	$450	2 km	1
E	$320	8 km	3

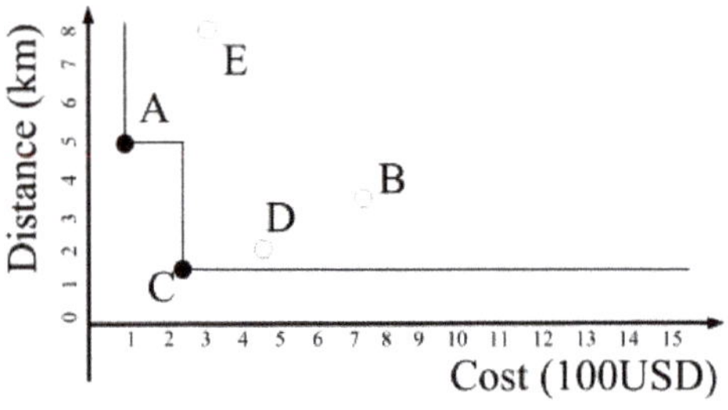

Figure 2. The example of skyline hotels.

3. Machinery Monitoring

In this chapter, we introduce a novel method for monitoring the status of machinery. The monitoring process involves the following steps: (1) collection of signals pertaining to machinery status using add-on sensors; (2) filtering and normalization of collected signals; (3) extraction of feature values and the export of each piece of data as a data point using windows with a designated range; (4) dimensionality reduction using LDA; and (5) input of results into neural network for training and generation of neural network parameters to enable the recognition of data pertaining to machinery status.

3.1. Add-On Sensors for the Collection of Data Related to Machinery Status

A number of researchers [40,41] have proposed the analysis of vibration data as a means by which to monitor the status of machines. In this study, we developed a monitoring system based on an inexpensive single chip controller, Arduino, in conjunction with triaxial accelerometers, infrared sensors, and temperature sensors. The proposed system is inexpensive and extensible, making it ideal for SMEs.

3.2. Filtering and Normalization of Collected Signals

The signals collected from accelerometers are filtered to remove noise prior to feature value calculation. To deal with the various forms of noise found in different machinery, we use the main frequency of the carrier signal for the selection of filter parameters. For example, if the frequency of the carrier were around 10 Hz, then a low-pass filter would be employed for the removal of noise.

3.3. Feature Extraction

The calculation of feature values is achieved using a window with data length determined by the user, such as an enterprise employee. Generally, the length should not exceed half the wavelength of the signal. In accordance with the recommendations in a previous study [25], 50% of each window is overlapped by the following window to enhance the precision of recognition. Figure 3 presents an example in which the signal was collected from a machine in a spring manufacturing facility. Based on a signal wavelength of approximately 200 Hz, the length of the window is set to 80 points with a 40-point overlap of the preceding and succeeding windows.

Window values are represented as $[X_1, X_2, \ldots, X_{|w|}]$, where $|W|$ refers to the length of a window, and extracted features are presented as follows:

1. Mean: Mean value of signal data in each window
2. Energy: Sum of squares of data in each window divided by the window length:

$$\text{Energy} = \frac{\sum_{i=1}^{|W|} |X_i|^2}{|W|}, \tag{1}$$

3. Root mean square:

$$\text{RMS} = \sqrt{\frac{1}{|W|} \sum_{i=1}^{|W|} x_i^2} \tag{2}$$

4. Variance:

$$\text{variance} = \frac{1}{|W|-1} \sum_{i=1}^{|W|} (x_i - m)^2, \tag{3}$$

where m is the mean value of Xi.

5. Average absolute deviation:

$$\text{MAD} = \frac{1}{|W|} \sum_{i=1}^{|W|} |x_i - m|. \tag{4}$$

6. Standard deviation: Root mean square of variance
7. Maximum value: Maximum value of each window.

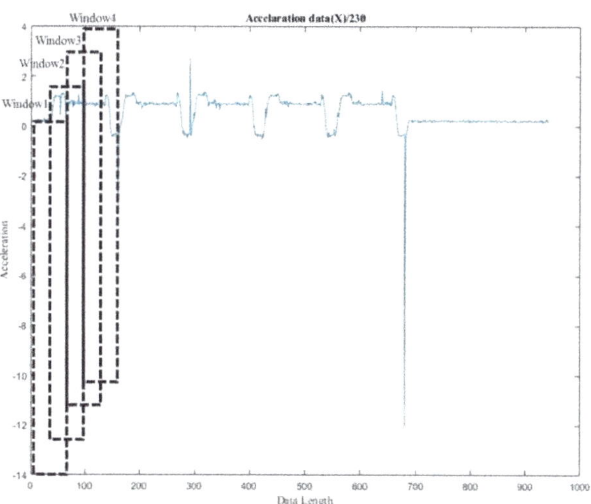

Figure 3. Example of feature extraction using windows.

3.4. Dimensionality Reduction Using LDA

In a system using triaxial accelerometers, infrared sensors, and temperature sensors, the following 5 signals would be received at one time: triaxial accelerometer (3 signals), infrared sensor (1 signal), and temperature sensor (1 signal). Extracting 7 features from each signal would result in a total of 35 features. However, this may also include irrelevant data, which could compromise recognition performance. Thus, we employ dimensionality reduction using LDA, which maps high-dimensional data onto a calculated vector, separates data into various categories, and tightens up data in each category. Dimensions with high feature value are then output as results. We adopted the two dispersion matrices outlined in [25], including covariance matrix S_B for between categories and covariance matrix S_W within categories. S_B is the predicted discrete vector values near the mixed averages, which is obtained as follows:

$$S_B = \sum_{\alpha=1}^{N} n_\alpha (m^\alpha - m)(m^\alpha - m)^T \tag{5}$$

and S_W is the predicted vector dispersion sample of each category, which is obtained as follows:

$$S_W = \sum_{\alpha=1}^{N} n_\alpha \sum_{i=1}^{n_\alpha} (x_i^\alpha - m^\alpha)(x_i^\alpha - m^\alpha)^T \tag{6}$$

where N denotes the number of categories; n_α is the number of samples in category α; x_i^α represents sample i of category α; m^α is the sample mean vector of category α, and m is the mean vector of all data. LDA produces mapping vector w to store the category integrity under low dimensionality, and w is used to maximize the formula below; that is, it maximizes the covariance between different categories and minimizes the covariance within a single category. In other words, the objective is to maximize $J(w)$ in Equation (7). Once w has been calculated, the data can be mapped onto coordinate systems and new coordinates can be identified to reduce dimensionality

$$J(w) = \frac{w^T S_B w}{w^T S_w w} \tag{7}$$

Using Figure 3 as an example of dimensionality reduction, we can obtain the results in Figure 4, in which the points in different colors represent different lengths of springs. The blue points are the springs with their lengths greater than 194 mm. The points in red are the springs with their lengths locate between 189 mm to 193 mm. Finally, the black points are the springs with their lengths smaller than 188 mm.

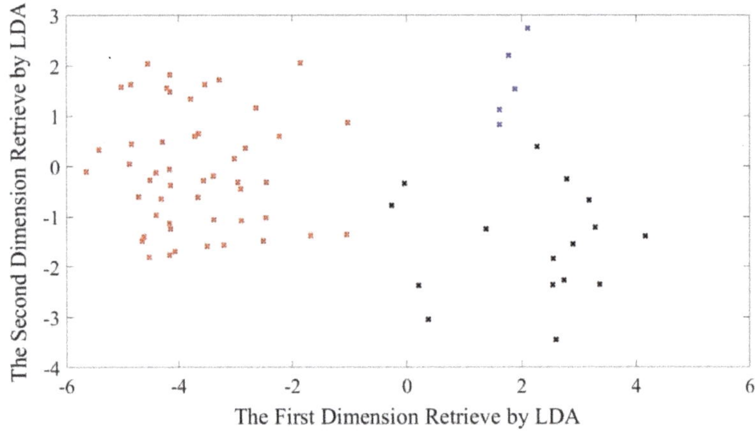

Figure 4. Example of dimensionality reduction.

3.5. Using the Neural Networks to Identify the Status of Machineries

In this section, we discuss the use of neural networks to identify the operating status of machinery. We introduce the structure of the proposed neural network, outline the training equations, and illustrate the process of identifying the operating status. Figure 5 presents the structure of the proposed neural network, which includes three layers: an input layer, a hidden layer, and an output layer. The input layer receives the features selected by the LDA, which are then relayed to the hidden layer. Thus, the number of nodes in this layer must equal the number of selected features. The hidden layer merges the features and then uses a set of activation functions to formulate the relationship between the features and the status of the machinery. Note that in this study, we selected the hyperbolic tangent sigmoid function as an activation function, due to the fact that it features dual polarity signals, which have proven highly effective in neural networks [42]. We opted for two nodes in this layer, based on the recommendation of Chen and Lee [15]. The output layer has only one node, which is responsible for determining the status of the machinery. A value is closer to 1 indicates that the machinery is working well, whereas a value closer to −1 is an indication of abnormal operating status.

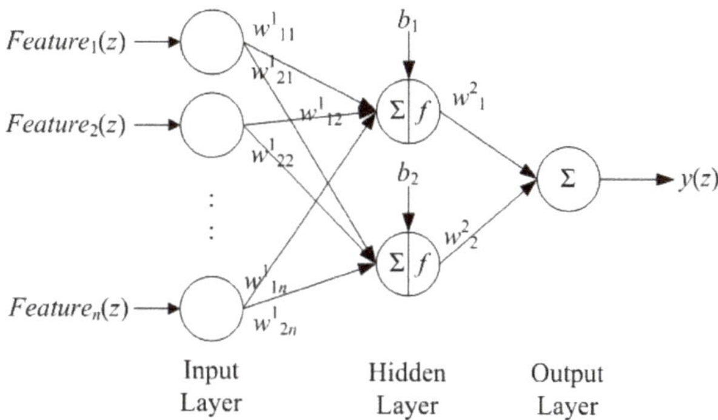

Figure 5. The structure of the proposed neural network.

In the following, we introduce the equations used in the proposed neural network [15]. Equations (8)–(10) are used for the input, hidden, and output layers, respectively. To facilitate this discussion, have also included z to represent the zth set of features.

$$out_i^{(\text{input layer})}(z) = input_i^{(\text{input layer})}(z) \tag{8}$$

$$out_i^{(\text{hidden layer})}(z) = \frac{\left(\exp\left(\sum_{j=1}^{r} w^1{}_{ij} out_j^{(\text{input layer})}(z) + b_i\right) - \exp\left(-\sum_{j=1}^{r} w^1{}_{ij} out_j^{(\text{input layer})}(z) + b_i\right) \right)}{\left(\exp\left(\sum_{j=1}^{r} w^1{}_{ij} out_j^{(\text{input layer})}(z) + b_i\right) + \exp\left(-\sum_{j=1}^{r} w^1{}_{ij} out_j^{(\text{input layer})}(z) + b_i\right) \right)} \tag{9}$$

$$y(z) = \sum_{j=1}^{2} w^2{}_j out_j^{(\text{hidden layer})}(z) \tag{10}$$

From the above equations, we derived the equations for training the neural network based on the concept of back propagation [15]. We first assume that the target function intended for the training network should be as follows:

$$Error(\mathbf{w}, z) = {}^{1\!}/{}_{2}(y_d(z) - y(z))^2 = {}^{1\!}/{}_{2}error(z)^2 \tag{11}$$

where $error(z)$ is the error value between the ideal output $y_d(z)$ and actual network output $y(z)$. Based on the theorem of back propagation, all parameter in the neural network can be adjusted using the following function:

$$w(z) = w(z-1) + \xi\left(-\frac{\partial^+ Error}{\partial w}\right) \tag{12}$$

where w is used as w^1, b, and w^2 in Equations (8)–(10), respectively. Thus, we obtain the training functions of this neural network, as follows:

$$w^1(z) = w^1(z-1) + \xi \left[\begin{array}{c} error(z)\mathbf{w}^2 \times \mathbf{input}^{(\text{input layer})}(z) \times \\ 4\Big/ \Big(\dfrac{\exp(\mathbf{out}^{(\text{input layer})}(z))}{+\exp(-\mathbf{out}^{(\text{input layer})}(z))} \Big)^2 \end{array} \right] \tag{13}$$

$$b(z) = b(z-1) + \xi \left[\begin{array}{c} error(z)\mathbf{w}^2 \times \\ 4\Big/ \Big(\dfrac{\exp(\mathbf{out}^{(\text{input layer})}(z))}{+\exp(-\mathbf{out}^{(\text{input layer})}(z))} \Big)^2 \end{array} \right] \tag{14}$$

$$w^2(z) = w^2(z-1) + \xi(e(z)\mathbf{out}^{(\text{hidden layer})}(z)), \tag{15}$$

After training the neural network using Equations (13)–(15), the feature set in input to enable the identification of machinery status. For example, if a signal collected from the machinery can be divided into 100 windows, then these windows are transformed into a set of 100 features. Situations in which the neural network recognizes more than half of these features are identified as abnormal; otherwise, the operating status of the machinery is regarded as normal.

4. Scheduling Algorithms

In the following chapter, we introduce a novel multi-objective scheduling algorithm based on the skyline query for SMEs. To the best of our knowledge, this is the first study to formulate a scheduling algorithm using skyline queries. It should be noted that the proposed scheduling algorithm was developed under the assumption that most of the jobs in smaller enterprises are implemented by single machinery, and that the operation of the machinery is seldom interrupted. Furthermore, for the sake of simplicity, we assumed that each machine in the enterprise is equal with regard to operating efficiency.

Figure 6 presents a flow chart of the proposed scheduling algorithm. For the sake of explanation, let us assume that there are n jobs and m machines ($n > m$) in this problem, and the maximum delay and the maximum flow time are our only concerns. The algorithm included a heap-based combination pool or the storage of all temporary schedules in the process and a skyline schedule list to store the results of the algorithm. The first step involves obtaining all possible schedules that each machine will be assigned the first job, to be evaluated based on the maximum delay time and maximum flow time. The example in Table 3 includes four jobs (a, b, c, and d) and three machines (A, B, and C).

Figure 6. Chart of the proposed scheduling algorithm.

Table 3. Example results obtained from the first step of the proposed scheduling algorithm.

Possible Schedules	Maximum Delay Time	Maximum Flow Time
{A(a), B(b), C(c)}	1 h	3 h
{A(a), B(b), C(d)}	2.5 h	6 h
{A(a), B(c), C(d)}	3 h	5 h
{A(b), B(c), C(d)}	4 h	5.5 h

The second step involves the creation of a heap-based combination pool, wherein all of the schedules obtained in step 2 are sorted in ascending order based on the summation of the maximum delay time and the maximum flow time. This sorting process is used to accelerate the execution speed of the algorithm, based on the fact that objects with a smaller summation value are more likely to be selected as the final result in a skyline query, and should therefore be examined first [14]. Using Table 3 as an example, we obtained the results in Table 4 with all of the schedules arranged according to the summation values of the maximum delay time and the maximum flow time.

Table 4. Example results obtained in the second step of the proposed scheduling algorithm.

Possible Schedules	Maximum Delay Time	Maximum Flow Time	Summation
{A(a), B(b), C(c)}	1 h	3 h	4 h
{A(a), B(c), C(d)}	3 h	5 h	8 h
{A(a), B(b), C(d)}	2.5 h	6 h	8.5 h
{A(b), B(c), C(d)}	4 h	5.5 h	9.5 h

The third step of this algorithm involves a comparison of the first schedule in the combination pool with the schedules in the skyline schedule list. Each comparison can lead to five possible results.

(1) The schedule in question includes all of the jobs and is not dominated by any schedule in the skyline schedule list. In this case, the schedule is deemed a skyline schedule, as it is dominated by no other schedule. It is therefore added to the skyline schedule list, whereupon we return to the third step for further comparisons.

(2) The schedule in question includes all of the jobs and is dominated by a schedule in the skyline schedule list. In this case, the schedule is deemed not to be a skyline schedule, as it is dominated by other schedules. It is therefore immediately deleted from the combination pool, whereupon we return to the third step for further comparisons.

(3) The schedule in question does not include all of the jobs and is not dominated by any schedule in the skyline schedule list. In this case, an extension of the schedule still could still be a skyline schedule. We therefore extend this schedule by adding new jobs and reinsert the extended schedule back into the combination pool, whereupon we return to the third step for further comparisons.

Table 5 presents an example to illustrate this case. If the first schedule {A(a), B(b), C(c)} in the heap is not dominated by any schedule in the skyline list, then we should extend it as {A(a, d), B(b), C(c)}, {A(a), B(b, d), C(c)}, and {A(a), B(b), C(c, d)}. We then evaluate the summation of the maximum delay time and maximum flow time of each schedule and reinsert them back into the heap. Table 4 presents one possible result.

Table 5. Example results obtained in the third step of the proposed scheduling algorithm.

Possible Schedules	Maximum Delay Time	Maximum Flow Time	Summation
{A(a, d), B(b), C(c)}	1 h	3 h	4 h
{A(a), B(c), C(d)}	3 h	5 h	8 h
{A(a), B(b, d), C(c)}	2.1 h	6 h	8.1 h
{A(a), B(b), C(d)}	2.5 h	6 h	8.5 h
{A(a), B(b), C(c, d)}	4.2 h	5 h	9.2 h
{A(b), B(c), C(d)}	4 h	5.5 h	9.5 h

(4) The schedule in question does not include all jobs and is dominated by a schedule in the skyline schedule list. In this case, no extension of the examined schedule could be a skyline schedule. It is immediately deleted from the heap-based combination pool, whereupon we return to the third step for further comparisons.

(5) No schedules remain in the heap-based combination pool. In this situation, all possible schedules have been examined; that is, the scheduling algorithm is completed.

Following the completion of the scheduling algorithm, the schedules remaining in the skyline scheduling list represent the final results.

5. Simulations

In this chapter, we adopt actual data obtained from a small-scale enterprise involved in the manufacture of metal springs to demonstrate the efficacy of the proposed algorithm.

5.1. Experiment Settings

Data were collected from a small Taiwanese enterprise engaged in the manufacture of springs. Most of the springs were meant for export to enterprises in Japan, where the tolerances are particularly low. As a result, the enterprise was eager to obtain assistance in monitoring machinery in order to maintain/improve product quality. Furthermore, this enterprise produces items for more than 50 companies, with new orders coming in every month (many of which are urgent). This greatly complicates scheduling, which often has to be updated several times each day. Clearly, this enterprise requires a systematic method by which to monitor and schedule the use of machinery.

To enable the monitoring of machinery, we employed a three-axis accelerometer (KSM001 ADXL345) with Arduino microcontroller (Uno r3) to collect vibration data from the machinery. This system was applied to a well-worn machine (approximately 10 years old), as shown in Figure 7. This machine was designed based on a cam mechanism mixed with an electrical 3-axis station. Also, this machine offers a Computer Numerical Control (CNC) with an AC servo motor. Metal wire is output from the small hole in the middle of the machine to be shaped into a spring by the surrounding tools. The accelerometer was attached along main axis of the machine, as shown in the middle in Figure 7. All data collect from the Arduino unit was stored in a computer as a Comma-Separated Values (CSV) file to facilitate subsequent processing.

Scheduling was based on data pertaining to actual orders, which was held in an enterprise database (Table 6). Note that all of these tasks were implemented on a single machine; that is, they were not split up among multiple machines. For the sake of simplicity, we established a set time for scheduling (31 May 2016; 22:00:00). Based on a suggestion from the enterprise workers, we focused only on the maximum delay time and maximum flow time.

Table 6. Jobs for the spring enterprise.

Job	Working Time	Deadline
A	350 min	17 June 2016
B	600 min	23 May 2016
C	300 min	3 June 2016
D	500 min	10 June 2016
E	240 min	6 June 2016
F	230 min	28 May 2016
G	300 min	29 June 2016
H	400 min	30 May 2016
I	240 min	27 May 2016
J	130 min	26 June2016

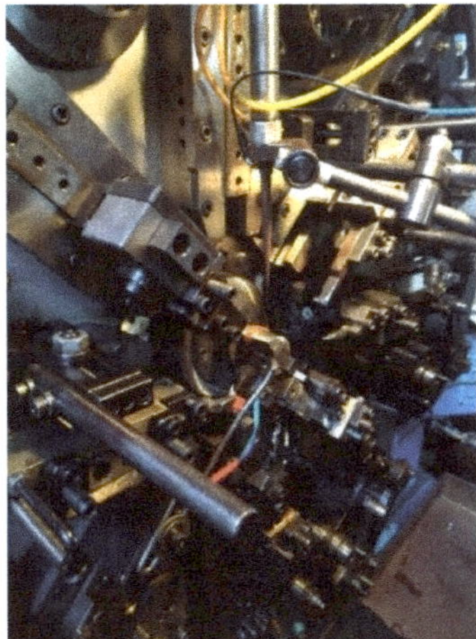

Figure 7. Spring fabrication machine used in experiment.

5.2. Identification of Machinery Status

Table 7 lists the dataset used for training, including 7 spring lengths indicative of normal status and 6 spring lengths indicative of abnormal status. According to an employee in the enterprise, a standard spring should be 191 mm in length, with variations of no more than 2 mm (i.e., between 189 mm and 193 mm). Springs outside this range of error are deemed anomalous. For each category of data, we collected ten signals over a period of 20 s, which is the time required for the fabrication of five springs. We then used the method outlined in Section 3 to identify the status of the machinery, the results of which are listed in Table 8. In this table, we can see that the recognition rate for normal springs is $(5 + 6 + 5 + 4)/(10 \times 4) = 50\%$, whereas the recognition rate for normal springsis $(7 + 7 + 8 + 6 + 8 + 6 + 7 + 7)/(10 \times 8) = 70\%$. The disappointing recognition rate can be attributed to two problems: (1) the differences between the signals associated with different lengths were too small to be recognized using the sensor; particularly in the range of 187 mm to 195 mm; and (2) the machinery used in this experiment was approximately 10 year old, and therefore produced more vibration-related noise than could be effectively filtered out.

Table 7. Training dataset (Unit of measure: millimeter).

Abnormal	171	181	188	187	195	210	230	194
Normal	189	190	191	192	193			

Table 8. Recognition results for springs of all lengths.

Dataset	Identity to Normal	Identity to Abnormal
171	3	7
181	3	7
187	2	8
188	4	6
189	5	5
190	6	4
192	5	5
193	4	6
194	2	8
195	4	6
210	3	7
230	3	7

5.3. Scheduling System

Table 9 lists the time costs obtained using the proposed scheduling algorithm in which ten jobs are divided up between 3 and 5 machines. As shown in the table, the time costs assigned by the algorithm are reduced with the number of machines. This is because an increase in the number of available machines reduces the number of possible schedules, such that fewer schedules need to be checked by the algorithm. Furthermore, reducing the number of possible schedules reduces the maximum delay time and maximum flow time, thereby reducing the number of schedules that eventually become skyline schedules. In other words, the algorithm no longer has to spend an excessive amount of time examining the dominance relationship between the first term of heap and the schedules in the skyline schedule list.

Table 9. Time cost of the proposed scheduling algorithm with 10 jobs.

Number of Machineries	Time Cost
3 machines	18.45 s
4 machines	11.39 s
5 machines	6.22 s

Table 10 lists the results of a comparison between the proposed algorithm and two existing scheduling algorithms: "shortest processing time first plus machinery's load" (SOT) and "earliest due date first plus machinery's load" (DDate). The two existing algorithms produced only one answer, rather than the multiple results returned by the proposed algorithm. We considered the following nine cases: 3 machines with 8, 9 or 10 jobs, 4 machines with 8, 9 or 10 jobs, and 5 machines with 8, 9 or 10 jobs. We then used the average results of these nine cases for comparisons. Table 9 clearly illustrates the superiority of the proposed algorithm with regard to maximum flow time as well as maximum delay time. The one exception is the maximum delay time of the DDate, which was designed to minimize the maximum delay time in scheduling problems. No existing algorithm is able to outperform DDate on this point; however, the proposed algorithm was able to match the performance of DDate. This is a clear demonstration that the proposed algorithm is always able to obtain the optimal solution for maximum delay time.

Table 10. Difference between the proposed scheduling algorithm and two existing algorithms.

Other Algorithms	Difference between SOT/DDATE and the Proposed Algorithm: Maximum Flow Time	Percentage Improvement Achieved by the Proposed Algorithm: Maximum Flow Time	Difference between SOT/DDATE and the Proposed Algorithm: Maximum Delay Time	Percentage Improvement Achieved by the Proposed Algorithm: Maximum Delay
SOT	+3.36 h	19.3%	+29.44 h	9.6%
DDATE	+0.81 h	5.55%	0 h	0

5.4. Information Integration

All of the information was uploaded to a web server to provide web page access to the working status of the machinery and scheduled order of all jobs in the queue. Figure 8 presents the interface of the web page. The green bar next to the machines indicates the current working status; that is, green for normal and red for abnormal. The table in the middle lists the job on which the machine is currently engaged, as well as the product quantity, time estimates, and other important data. The table at the bottom lists the schedule including the next job in the queue.

Machine 1		Machine 2		Machine 3	
Production status		Production status		Production status	
Order Name	A	Order Name	B	Order Name	F
The total amount	250	The total amount	300	The total amount	300
Number Completed	95	Number Completed	250	Number Completed	120
Remaining amount	155	Remaining amount	50	Remaining amount	180
Single-piece production time	1 sec	Single-piece production time	1.5 sec	Single-piece production time	1.9 sec
Time left	155 sec	Time left	75 sec	Time left	342 sec
Remaining raw materials	Full	Remaining raw materials	Full	Remaining raw materials	100

Scheduling results		Scheduling results		Scheduling results	
1:	A	1:	B	1:	F
2:	C	2:	E	2:	H
3:	D	3:	G		

Figure 8. Machine operating data integrated on web page.

6. Conclusions

Since the advent of Industry 4.0, a growing number of manufacturers have been adopting technologies for the monitoring and scheduling of machinery with the aim of increasing yields and optimizing production efficiency. Unfortunately, many of these technological innovations are too expensive for SMEs. In this study, we developed an inexpensive alternative to these two technologies, including add-on sensors by which to monitor the status of machines and a simple algorithm to solve multiple objective scheduling problems. The efficacy of the proposed algorithm was demonstrated in an actual enterprise involved in the manufacture of metal springs. However, our simulation results revealed two limitations of the proposed method: (1) when there is only a small difference between normal and abnormal springs, the use of a low-cost sensor provides disappointing recognition results, due to the low sampling rate and limited accuracy of the sensor; and (2) noise from the monitored machine can also lower the recognition rate, particularly when using a low-cost sensor. In the future, we will investigate the use of multiple low-cost sensors to overcome these problems. It is expected that this will allow the collection of more data that could be used by the algorithm for error recognition. It may also be possible to adopt another algorithm to perform wavelet analysis (i.e., to dismantling the sensor signal) prior to the recognition process. It is very likely that other types of neural network could also be used to enhance recognition performance.

Author Contributions: Y.-C.C. initialed the idea, addressed whole issues in the manuscript and wrote the manuscript. K.-C.T. and Y.-M.C. implemented algorithms. D.-L.Y. was responsible for communicating with the factory and collecting the data. H.-M.C. helped the construction of the scheduling system. Finally, J.J.-C.Y. revised and polished the final edition manuscript.

Funding: This research was funded by the Ministry of Science and Technology of Taiwan, R.O.C., grant number MOST 107-2119-M-224-003-MY3 and MOST 107-2625-M-224-003.

Conflicts of Interest: The authors declare no conflicts of interest.

References

1. Lee, J.; Kao, H.A.; Yang, S. Service Innovation and Smart Analytics for Industry 4.0 and Big Data Environment. *Procedia CIRP* **2014**, *16*, 3–8. [CrossRef]
2. Lee, J.; Bagheri, B.; Kao, H.A. A Cyber-Physical Systems architecture for Industry 4.0-based manufacturing systems. *Manuf. Lett.* **2015**, *3*, 18–23. [CrossRef]
3. Dassisti, M.; Giovannini, A.; Merla, P.; Chimienti, M.; Panetto, H. An approach to support Industry 4.0 adoption in SME s using a core-meta model. *Annu. Rev. Control* **2018**. [CrossRef]
4. Uriarte, L.; Azcárate, S.; Herrero, A.; Lopez de Lacalle, L.N.; Lamikiz, A. Mechanistic modeling of the micro end milling operation. *J. Eng. Manuf.* **2008**, *222*, 23–33. [CrossRef]

5.	Rivero, A.; Lopez de Lacalle, L.N.; Penalva, M.L. Tool wear detection in dry high-speed milling based upon the analysis of machine internal signals. *Mechatronics* **2008**, *18*, 627–633. [CrossRef]

6.	Plapper, V.; Weck, M. Sensor less machine tool condition monitoring based on open NCs. In Proceedings of the IEEE International Conference on Robotics and Automation, Seoul, Korea, 21–26 May 2001; pp. 3104–3108.

7.	Barrio, H.G.; Moran, I.C.; Ealo, J.A.; Barrena, F.S.; Beldarrain, T.O.; Zabaljauregui, M.C.; Zabala, A.M.; Arriola, P.J.A.; Marcaide, L.N.L.L.C. A reliable machining process by means of intensive use of modeling and process monitoring: Approach 2025. *Eng. J. DYNA* **2018**, *93*, 689–696.

8.	Birkel, H.S.; Veile, J.W.; Muller, J.M.; Hartmann, E.; Voigt, K.I. Development of a Risk Framework for Industry 4.0 in the Context of Sustainability for Established Manufacturers. *Sustainability* **2019**, *11*, 384. [CrossRef]

9.	Moeuf, A.; Pellerin, R.; Lamouri, S.; Tamayo-Giraldo, S.; Barbaray, R. The industrial management of SMEs in the era of Industry 4.0. *Int. J. Prod. Res.* **2018**, *56*, 1118–1136. [CrossRef]

10.	Muller, J.M.; Voigt, K.I. Sustainable Industrial Value Creationin SMEs: A Comparison between Industry 4.0 and Made in China 2025. *Int. J. Precis. Eng. Manuf.-Green Technol.* **2018**, *5*, 659–670. [CrossRef]

11.	Ivanov, D.; Dolgui, A.; Sokolov, B.; Werner, F.; Ivanova, M. A dynamic model and an algorithm for short-term supply chain scheduling in the smart factory industry 4.0. *Int. J. Prod. Res.* **2016**, *54*, 386–402. [CrossRef]

12.	Kyparisis, G.J.; Koulamas, C.P. Flexible Flow Shop Scheduling with Uniform Parallel Machineries. *Eur. J. Oper. Res.* **2006**, *168*, 985–997. [CrossRef]

13.	Tahar, D.N.; Yalaoui, F.; Chu, C.; Amodeo, L. A Linear Programming Approach for Identical Parallel Machinery Scheduling with Job Splitting and Sequence-dependent Setup Times. *Int. J. Prod. Econ.* **2006**, *99*, 63–73. [CrossRef]

14.	Bartolini, I.; Ciaccia, P.; Patella, M. SaLSa: Computing the skyline without scanning the whole sky. In Proceedings of the 15th ACM International Conference on Information and Knowledge Management, Arlington, VA, USA, 6–11 November 2006; pp. 405–414.

15.	Chen, Y.C.; Lee, C. The σ-Neighborhood Skyline Queries. *Inf. Sci.* **2015**, *322*, 92–114. [CrossRef]

16.	Papadias, D.; Tao, Y.; Fu, G.; Seeger, B. An optimal and progressive algorithm for skyline queries. In Proceedings of the 2003 ACM SIGMOD International Conferenceon Management of Data, SanDiego, CA, USA, 9–12 June 2003.

17.	Peng, G.; Jiang, Y.; Xu, J.; Li, X. A Collaborative Manufacturing Execution Platform for Space Product Development. *Int. J. Adv. Manuf. Technol.* **2012**, *62*, 443–455. [CrossRef]

18.	Chalal, M.; Boucher, X.; Marques, G. Decision Support System for Servitization of Industrial SMEs: A Modelling and Simulation Approach. *J. Decis. Syst.* **2015**, *24*, 355–382. [CrossRef]

19.	Givehchi, M.; Haghighi, A.; Wang, L. Generic Machining Process Sequencing Through a Revised Enriched Machining Feature Concept. *J. Manuf. Syst.* **2015**, *37*, 564–575. [CrossRef]

20.	Dombrowski, U.; Ernst, S. Scenario-Based Simulation Approach for Layout Planning. *Procedia CIRP* **2013**, *12*, 354–359. [CrossRef]

21.	Denkena, B.; Dengler, B.; Doreth, K.; Krull, C.; Horton, G. Interpretation and Optimization of Material Flow Via System Behavior Reconstruction. *Prod. Eng.* **2014**, *8*, 659–668. [CrossRef]

22.	Velandia, D.M.S.; Kaur, N.; Whittow, W.G.; Conway, P.P.; West, A.A. Towards Industrial Internet of Things: Crankshaft Monitoring, Traceability and Tracking Using RFID. *Robot. Comput.-Integr. Manuf.* **2016**, *41*, 66–77. [CrossRef]

23.	MacKerron, G.; Kumar, M.; Kumar, V.; Esain, A. Supplier Replenishment Policy Using E-Kanban: A Framework for Successful Implementation. *Prod. Plan. Control* **2014**, *25*, 161–175. [CrossRef]

24.	Barenji, A.V.; Barenji, R.V.; Roudi, D.; Hashemipour, M. A Dynamic Multi-Agent-Based Scheduling Approach for SMEs. *Int. J. Adv. Manuf. Technol.* **2017**, *89*, 3123–3137.

25.	Yang, J.Y.; Chen, Y.P.; Lee, G.Y.; Liou, S.N.; Wang, J.S. Activity Recognition Using One Triaxial Accelerometer: A Neuro-Fuzzy Classifier with Feature Reduction. *Lect. Notes Comput. Sci.* **2007**, *4740*, 395–400.

26.	Abidine, M.; Fergani, B. Evaluating a new classification method using PCA to human activity recognition. In Proceedings of the 2013 International Conference on Computer Medical Applications (ICCMA), Sousse, Tunisia, 20–22 January 2013.

27.	Uray, M.; Skocaj, D.; Roth, P.M.; Bischof, H.; Leonardis, A. Incremental LDA Learning by Combining Reconstructive and Discriminative Approaches. In Proceedings of the British Machinery Vision Conference (BMVC2007), Warwick, UK, 10–13 September 2007.

28. Ye, J.; Li, Q.; Xiong, H.; Park, H.; Janardan, R.; Kumar, V. IDR/QR: An incremental dimension reduction algorithm via QR decomposition. *IEEE Trans. Knowl. Data Eng.* **2005**, *17*, 1208–1222.

29. Bao, L.; Intille, S.S. Activity recognition from user- annotated acceleration data. In Proceedings of the International Conference on Pervasive Computing, Linzand Vienna, Austria, 21–23 April 2004; pp. 1–17.

30. Kose, M.; Incel, O.D.; Ersoy, C. Online Human Activity Recognition on Smart Phones. In Proceedings of the Workshop on Mobile Sensing: From Smart phones and Wearables to Big Data, Beijing, China, 16 April 2012.

31. Chen, Y.C.; Lee, C. A Neural Skyline Filter for Accelerating the Skyline Search Algorithms. *Expert Syst.* **2015**, *32*, 108–131.

32. Karantonis, D.M.; Narayanan, M.R.; Mathie, M.; Lovell, N.H.; Celler, B.G. Implementation of a real-time human movement classifier using a triaxial accelerometer for ambulatory monitoring. *IEEE Trans. Inf. Technol. Biomed.* **2006**, *10*, 156–167. [CrossRef] [PubMed]

33. Kaghyan, S.; Sarukhanyan, H. Activity recognition using K-nearestneighbor algorithm on smart phone with Tri-axialaccelerometer. *Inf. Models Anal.* **2012**, *1*, 146–156.

34. Jatoba, L.; Grossmann, U.; Ottenbacher, J.; Stork, W.; Muller-Glaser, K. Development of a self-constructing neuro-fuzzy inference system for online classification of physical movements. In Proceedings of the 2007 9th International Conference on e-Health Networking, Application and Services, Taipei, Taiwan, 19–22 June 2007; pp. 332–335.

35. Wang, S.; Yang, J.; Chen, N.; Chen, X.; Zhang, Q. Human activity recognition with user-free accelerometers in the sensor networks. In Proceedings of the IEEE International Conference on Neural Networks and Brain, Beijing, China, 13–15 October 2005; Volume 2, pp. 1212–1217.

36. Watanabe, T.; Yamagishi, S.; Murakami, H.; Furuse, N.; Hoshimiya, N. Recognition of lower limb movements by artificial neural network for restoring gait of Hemiplegic patients by functional electrical stimulation. In Proceedings of the IEEE International Conferenceon Engineering in Medicine and Biology Society, Istanbul, Turkey, 25–28 October 2001; pp. 1348–1351.

37. Blazewicz, J.; Dror, M.; Weglarz, J. Mathematical Programming Formulations for Machinery Scheduling: A Survey. *Eur. J. Oper. Res.* **1991**, *51*, 283–300.

38. Lauff, V.; Werner, F. On the Complexity and Some Properties of Multi-stage Scheduling Problems with Earliness and Tardiness Penalties. *Comput. Oper. Res.* **2004**, *31*, 317–345.

39. Borzsonyi, S.; Kossmann, D.; Stocker, K. The skyline operator. In Proceedings of the 17th International Conference on Data Engineering, Heidelberg, Germany, 2–6 April 2001; pp. 235–254.

40. Nawaz, J.M.; Arshad, M.Z.; Hong, S.J. Time Series Fault Prediction in Semiconductor Equipment Using Recurrent Neural Network. In Proceedings of the International Symposium on Neural Networks, Dalian, China, 4–6 July 2013; pp. 463–472.

41. Shao, Y.; Li, X.; Mechefske, C.K.; Zuo, M.J. Use of Neural Networks to Predict Rear Axle Gear Damage. In Proceedings of the International Conference on Reliability, Maintainability and Safety, Chengdu, China, 20–24 July 2009; pp. 986–990.

42. Wang, J.S.; Chen, Y.C. A Hammerstein- Wiener recurrent neural network with universal approximation capability. In Proceedings of the IEEE International Conference on Systems, Man and Cybernetics, Singapore, 12–15 October 2008; pp. 1832–1837.

 *applied* *sciences*

Article

Cutting Insert and Parameter Optimization for Turning Based on Artificial Neural Networks and a Genetic Algorithm

Bolivar Solarte-Pardo [1], Diego Hidalgo [2] and Syh-Shiuh Yeh [3,*

[1] Department of Mechanical and Electrical Engineering, National Taipei University of Technology, Taipei 10608, Taiwan; enrique.solarte.pardo@gmail.com
[2] Department of Mechanical and Automation Engineering, National Taipei University of Technology, Taipei 10608, Taiwan; hidalgocdiego@gmail.com
[3] Department of Mechanical Engineering, National Taipei University of Technology, Taipei 10608, Taiwan
* Correspondence: ssyeh@ntut.edu.tw; Tel.: +886-2-27712171

Received: 2 January 2019; Accepted: 28 January 2019; Published: 30 January 2019

Featured Application: People and companies involved on the manufacturing industry will be able to save plenty of time on the selection of cutting inserts and parameters by implementing and using the optimization system developed in this research.

Abstract: The objective of this present study is to develop a system to optimize cutting insert selection and cutting parameters. The proposed approach addresses turning processes that use technical information from a tool supplier. The proposed system is based on artificial neural networks and a genetic algorithm, which define the modeling and optimization stages, respectively. For the modeling stage, two artificial neural networks are implemented to evaluate the feed rate and cutting velocity parameters. These models are defined as functions of insert features and working conditions. For the optimization problem, a genetic algorithm is implemented to search an optimal tool insert. This heuristic algorithm is evaluated using a custom objective function, which assesses the machining performance based on the given working specifications, such as the lowest power consumption, the shortest machining time or an acceptable surface roughness.

Keywords: cutting insert selection; cutting parameter optimization; artificial neural networks; genetic algorithm

---

1. Introduction

Nowadays, there is great demand in the manufacturing industry for technologies that can deal with dynamic environments and customized products. Industry 4.0 and the notion of challenging trade by globalization have pushed companies to be more competitive in both large and small batches. This progress also shows a new way for computer numerical control (CNC) manufacturing industries to profit from large productions [1]. During the last few decades, there has been significant progress in improving the efficacy of CNC machining to meet world challenges. These breakthroughs come from the implementation of automation approaches, such as adaptive control and active control. They allow companies to achieve higher operation performances [1]. Manufacturing processes, such as computer-aided process planning (CAPP), expert processes planning systems (PP), computer-aided design (CAD) and computer-aided manufacturing (CAM) are now based on intelligent machining [2,3], which allow for the simulation and evaluation of variable environments. Complex cutting models can now predict fundamental variables related to machining operations performed in the industry [4]. These approaches mainly aim to obtain suitable cutting parameters and control them within certain working conditions. Thus, this increases the efficiency during the machining process and reduces the implementation time.

Cutting parameters for machining processes have a high impact on performance and they are usually the variables that need to be tuned for optimizing models. Generally, cutting parameters refer to cutting velocity, feed rate, depth of cut, cutting forces, torque, spindle speed, etc. On the other hand, the parameters for evaluating the machining results normally include surface roughness, power consumption, machining time, production cost, tool life, production rate, etc. [5–7]. For machining processes, accurate performance can be defined only within a working optimal range. This optimal range is evaluated by models, which generally relate the working conditions to cutting parameters and tool features. These models can be numeric, analytic, empiric, hybrid or AI-based models [4]. Nowadays, the trend is to use AI-based models, which clearly show adaptability and high performance in machining operations. Furthermore, the advances in computer science have allowed for the wide application of these models in the manufacturing industry [8–10].

In general, the implementation stage for machining processes takes a considerable amount of time and sometimes requires previous machining tests to reach admissible results. Because of this, some approaches seek to embed knowledge and technical data in machining processes. This results in the development of expert systems that are capable of optimally dealing with the changing conditions in shorter setting times. Many such expert systems use information from CAD models, databases, statement rules, tool preferences, suppliers, etc. [2–4,11,12]. Zarkti et al. [3] presented an automatic-optimized tool selector model based on CAD information to infer milling process stages. This approach uses a database from a tool supplier to build an expert system, which is capable of choosing suitable tools and suggesting optimal milling operation planning. Benkedjouh et al. [13] developed a model, which was based on a support vector machine, to predict the life of a cutting tool. This approach uses experimental testing to obtain a nonlinear regression model to estimate and predict the level of wear in a cutting tool. Several sensors, which are installed around the machine, gather information during the machining process and create a database to infer the model. Some significant remarks were taken from Özel et al. [14]. In this present study, the effects of cutting edge geometry on surface hardness are detailed. Moreover, this paper presents the relation between the cutting conditions and the surface roughness for turning processes. In addition, Arrazola et al. [4] detailed several models for chip formation, which can be used to predict cutting forces, temperatures, stress and strain. These models are based on insert geometries and cutting parameters. The approach [12] proposes a system software to optimize cutting parameters based on genetic algorithms. This model defines an objective function, which is based on theoretical models that relate fundamental variables in the machining operation. Ganesh et al. [5] showed an optimization of cutting parameters for the turning process. This research defines the surface roughness as an objective function for turning machining of EN 8 steel. A genetic algorithm is also applied in this approach. This model can be considered to be a hybrid model. Although the study by Li et al. [9] predicts annual power load consumption, the approach also shows an interesting hybrid model based on regression neural networks and the fruit fly optimization algorithm. This research presents a significant improvement in accuracy compared with previous approaches without neural network algorithms. Other models also show important achievements after applying training-based models that use the neural network algorithm. For instance, Xiong et al. [15] used the weld bead geometry prediction; Babu et al. [16] predicted the tensile behavior of tailor weld blanks; and Özel and Karpat [17] presented a model for surface roughness and tool wear for turning operations. Additionally, Malinov et al. [18] created an artificial neural network to predict the mechanical properties of titanium alloys as a function of the alloy composition. The research by Kuo et al. [19] poses a singular model for intelligent stock trading decision support systems. This model captures the stock expert's knowledge by a genetic-algorithm-based fuzzy neural network model.

Other important achievements for the optimization of cutting parameters have been proposed using only genetic algorithms. Although these approaches show dependency on the working conditions and the workpiece and tools, they can acquire models that can reach the optimal result with high efficiency. In the paper by Quiza Sardinas et al. [20], a multi-objective optimization of cutting parameters in turning operations is presented. This approach entails the use of an objective function based on power consumption, cutting forces and surface roughness. It also presents a qualification of chromosome population, which is inherent to genetic algorithms, based on feasible individuals. Cus and Balic [21] presents an approach for cutting parameters in turning operations. This paper also shows an experimental test for validating the model. Suresh et al. [22] proposes a model to predict surface roughness. This model uses the response surface methodology and a genetic algorithm to converge to an optimal solution. Yang and Tarng [23] showed an approach for optimization of cutting parameters using the Taguchi method and the analysis of variance (ANOVA) for a database obtained by testing surveys. Thamizhmanii et al. [24] presents a similar approach using the Taguchi method and the ANOVA analysis but for optimizing surface roughness.

Unlike other research approaches, our research considers the insert information from a tool supplier to obtain neural network models of insert before determining the optimal cutting parameters. This approach allows the selection of a tool-insert and the inference of the corresponding cutting parameters by simultaneously considering tool specifications and working conditions. To do so, the research is based on a considerable amount of information defined by the tool supplier. The proposed model is an integrated optimization system, which selects a suitable tool-insert and suggests optimal cutting parameters based on certain working conditions and a fitness function optimization, respectively. This approach models the relationships between the geometrical and mechanical features of a tool-insert and the working conditions, thus introducing a novel approach to the modeling of cutting parameters for a turning tool. The output of the proposed approach is a set of recommended cutting parameters for an optimally selected cutting tool. This approach considers the entire information defined by the tool supplier and its intrinsic relationship with the working material in a turning operation. Thus, this makes more assertive recommendations for the cutting parameters.

The objective of this research is to obtain a model for cutting insert selections and cutting parameter optimization. This model must be constrained by working conditions and evaluated by an objective function. The objective function of this approach is defined as a combination of the lowest power consumption, the shortest machining time and surface roughness within a certain range. However, this function must be customizable under external conditions. Furthermore, the cutting insert selection must be based on commercially available tools. Since this research proposes a model for cutting insert selections based on commercially available tools, it requires a database from a tool supplier. The chosen tool supplier was Sandvik Coromant and the selected tool-insert model for building the datasets was CoroTurn®107. The information about the recommended cutting parameters and the insert feature description was referenced from the official Sandvik Coromant website [25]. The models used on this approach are two artificial neural networks. The first neural network model defines the cutting parameter feed rate as a function of macro-geometrical features and recommended cutting depths. The second model defines the cutting speed as a function of material cutting specifications, working conditions and the feed rate cutting parameter. To find optimal cutting parameters and a suitable cutting insert, a genetic algorithm optimization is proposed based on working conditions. This algorithm is defined by a heuristic search of insert features and cutting parameters, which are evaluated by the neural network models. This heuristic search is set up under a defined objective function, which is a combination of the lowest power consumption, the shortest machining time and an acceptable surface roughness. Due to the heuristic search of the genetic algorithm, the result might be a non-existent tool-insert. Thus, the last stage of this approach is to evaluate a Euclidean distance to find the closest existent tool-insert in the commercial database based on a predefined threshold.

The structure of this paper is as follows. Section 2 aims to introduce the main features that define a tool-insert and the relations of the cutting parameters with geometrical features and the working conditions. This section also defines the database based on commercial data from a tool supplier. The section ends with the proposed dataset for this research. Section 3 explains the procedure used to obtain neural network models for this research. Furthermore, this section introduces data preparation and error validation. Section 4 details the mechanism behind the genetic algorithm implementation.

Some concepts, such as individual chromosomes, encoding and decoding procedures and fitness function, are introduced in this section. Section 5 aims to explain how to use this model for practical applications in more detail. Examples of applications shown in this section include light roughing machining, heavy roughing machining and finishing operations. Section 6 summarizes this paper.

2. Datasets Description and Preparation

Turning is a process to remove material, which is mainly oriented to metal machining. Within this field, several criteria are used to establish the conditions for an acceptable turning process, including surface roughness, measurement tolerance, power consumption, machining time, forces, loads, etc. These measurement criteria are highly related to the selected cutting insert, cutting parameters, working material and working conditions [25]. A cutting insert varies according to its geometry, shape and material composition. Thus, it is possible to apply a specific group of cutting inserts to a specific turning application, working material and working conditions. In general, a tool-insert can be defined by its geometrical features and material composition, which are hereinafter referred as tool-insert features. The most notable tool-insert features are the inscribed circle (ICmm), clearance angle (AN), cutting length (LEmm), corner radius (REmm), thickness (Smm) and shape angle (SC), as shown in Figure 1.

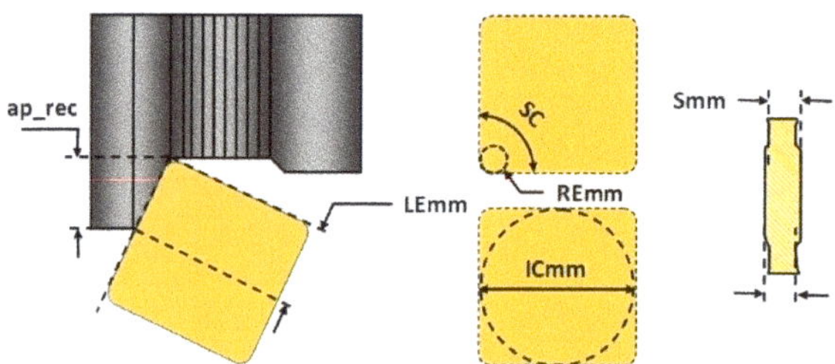

Figure 1. Tool-insert features considered in this study [25]; cutting length (LEmm); corner radius (REmm); thickness (Smm); and shape angle (SC).

The ICmm is a tool-insert feature that is highly related to the insert size. A large insert size increases the stability performance but oversizing can lead to high production costs. This feature must be linked with the depth of the cut, the entering angle and the cutting length to be used. The AN is the angle between the front face of the insert and the vertical axis of the workpiece [25]. This feature also defines the negative or positive quality of an insert [25]. On the other hand, the REmm feature is also related to the chip-breaking and cutting forces but it is mainly linked to finished surfaces. For a set feed rate cutting parameter, the machining yields a certain surface roughness. Furthermore, the radial forces that push away the insert in the turning machining become the axial forces as the depth of the cut increases with respect to the radius nose. As a suggestion from the supplier, the depth of cut should be greater or equal to 2/3 of the nose radius of the tool-insert [25]. The insert-shape feature is the most tangible feature, which is mainly selected by accessibility requirements. This feature defines a tool-holder and the depth of cut range in the process. The SC feature is defined by the nose angle of the tool-insert, which is related to strength and reliability issues. Large nose angles are stronger but require more machining power. They also increase the vibration tendency of the machine. On the contrary, small nose angles are weaker and have small cutting edges but are linked to better surface roughness [25].

The CTPT feature, which describes the cutting operation, is defined by three possible configurations: Finishing, Medium and Roughing operations. The WEP feature is a logical feature, which defines if the tool insert has a wiper radius. The GRADE feature describes the material composition of a tool-insert. Three features are linked to different grade features. These features describe the distribution of each grade in the range of area applications for ISO group materials.

The introduced features are MC_L (machine condition low), MC_H (machine condition high) and MC_Suitability (machine condition suitability). The MC_L and MC_H features represent the low and high borders of each grade in the range of area application. The MC_Suitability feature is defined as a percentage of range for each grade that belongs to wear resistance region, which is in the upper region of the total field of application. For instance, the grade 4325 is defined as MC_L = 40, MC_H = 10 and MC_Suitability = 50%; the grade 5015 is defined as MC_L = 0, MC_H = 20 and MC_Suitability = 100%; and the grade GC30 is defined as MC_L = 35, MC_H = 45 and MC_Suitability = 0% [25]. Figure 2 graphically shows the relation between the working material (hardness feature), cutting parameters (feed rate) and cutting velocity. It is important to note that the relation varies as a function of the insert grade feature.

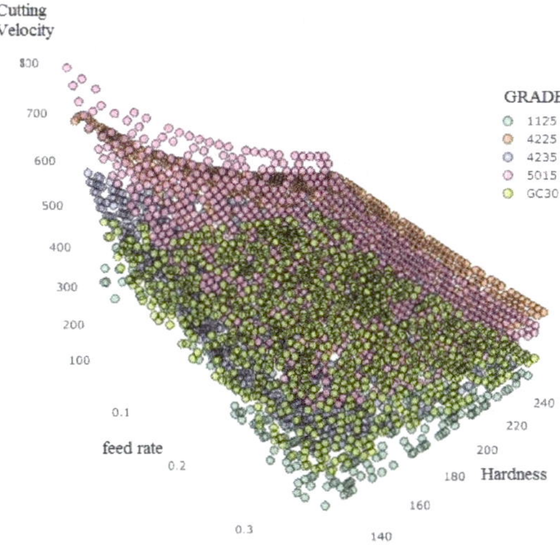

Figure 2. Cutting velocity, hardness and feed rate scatter.

There are two proposed datasets for this study. The first one is a dataset for insert features with cutting parameters. The second one is for working conditions with cutting parameters. The first dataset will be used to train a neural network model that infers the feed rate while the second one will be used to model the cutting velocity. The description for each dataset is shown in Tables 1 and 2. The main reason for using two neural network models is the nature of the dataset. The dataset for this study is based on two main sources, which are namely the data description for the tool-insert and the working conditions for a certain working material. The data description for a tool-insert relates the insert features of a tool-insert (i.e., the nose angle, thickness, cutting length and so on) and the suggested cutting parameters (depth of the cut and feed rate). The dataset, which ensues from working conditions, does not relate the insert features to the cutting parameter. It is defined by different working material specifications and their impact on the cutting parameters.

Table 1. Feed rate dataset–feature descriptions; CTPT: finishing, medium and roughing operations.

Feature Name	Data Type	Description
ICmm	Numeric	Inscribed circle
AN	Logical	Clarence angle
LEmm	Numeric	Cutting length
REmm	Numeric	Nose radius
Smm	Numeric	Thickness
SC	Numeric	Insert shape angle
Finishing	Logical	CTPT finishing operation
Medium	Logical	CTPT medium operation
Roughing	Logical	CTPT roughing operation
WEP	Logical	Wiper property
MC_L	Numeric	Low machine condition
MC_H	Numeric	High machine condition
MC_Suitability	Numeric	Suitability machine condition
ap_rec	Numeric	Recommended depth of cut

Table 2. Cutting velocity dataset–feature descriptions.

Feature Name	Data Type	Description
MC_L	Numeric	Low machine condition
MC_H	Numeric	High machine condition
MC_Suitability	Numeric	Suitability machine condition
HB	Numeric	Hardness material
fn_rec	Numeric	Recommended feed rate

3. Artificial Neural Network Models

An artificial neural network model is a powerful method to deal with nonlinear functions or to model systems with unknown input–output relations [26–28]. In fact, the purpose of the algorithm in this research is to find two models that relate the tool-insert features and working conditions to the cutting parameters. Some important issues in the implementation of a neural network model come from the nature of a neural network model. Since a neural network model considers nonlinear combinations and often uses a gradient descent algorithm to update its parameters, the main issues revolve around the need to reach a global solution instead of a local one. Furthermore, given the fact that a neural network model is built on an available dataset, any irrelevant, non-numerical and poor representations of input features can adversely affect the final model [29]. As a machine learning algorithm, neural network models mainly focus on the validation stage. For this reason, the error validation represents proof that the embedded information in the database is well represented by the resulting model. There are some concepts regarding neural network models, which need to be understood. These include overfitting, homogeneous representation and accuracy [30].

For a neural network model, the dataset plays a crucial role. In fact, an effective preparation of the data can lead to an increase in accuracy, reduction in computing time and prevention of overfitting in the model [31]. For the research's dataset, two scaling methods were considered to prepare the data for the training stage. The independent variables, which define the model inputs, have a scaling method called the z-score. The dependent variables, which describe the output for the neural network model, have a scaling method based on its minimum and maximum values, which has a range from 0 to 1 [31]. Equations (1) and (2) show the z-score and range scaling for inputs and outputs, respectively. It is also important to note that the output for a neural network model is highly dependent on the activation function, which is defined in the last layer (output layer). Generally, the activation function for the last layer is a hyperbolic tangent, softmax or a sigmoid function, which has an output range of 0 to 1.

Other activation functions are widely used for neural network models although this research uses a linear function because it can be easily implemented with less computational burden.

$$x_{i-scaled} = \frac{x_i - mean(x)}{\sigma(x)} \tag{1}$$

$$y_{i-scaled} = \frac{y_i - min(y)}{max(y) - min(y)} \tag{2}$$

The architecture and error validation of a neural network model are closely linked with each other. In fact, the performance of a neural network model is defined by its architecture, but this last one is selected by error validation. The architecture refers to the set of parameters that govern the complexity of the model, including the number of neurons, layers, updating and regularization algorithms among others. The error validation refers to an evaluation procedure for a certain architecture to find a balance between accuracy and error distribution without causing overfitting [32]. In practical applications, the training and testing datasets are used to achieve an architecture model that represents almost all information in the dataset. The training data are used to teach the model on the input–output relations. The testing data validate the model so it represents most of the total spectrum of possibilities in the dataset. For the present research, the databases for the feed rate and cutting velocity model, which are described in Tables 1 and 2, are divided into a ratio of 0.5 for training and testing validation. The algorithm for converging the weights in a neural network model also plays an important role in the architecture. For this implementation, a globally convergent training scheme, based on the resilient propagation, is used. A crucial advantage of this algorithm compared to the traditional back-propagation or normal resilient propagation is the computing time. This approach shows better accuracy performance with similar datasets to be used for this research (datasets that are compounded by factors and numerical mapped values [29]).

The most important part in error validation is the definition of a good performance, which qualifies the architecture of a neural network model. For this approach, the traditional root mean square error comparison was not the only one considered to validate the models [32]. Instead, the error validation for this research is defined by the comparison of error density functions for training and testing evaluations. Error density functions are continuous functions that represent the attained errors for training and testing evaluations by a known function. The Gaussian function is used to represent these density functions. In this way, the training and testing evaluations can be represented by a Gaussian error density function. It is important to note that a Gaussian function is defined by two parameters: mean or expectation and standard deviation. For this approach, the expectation value is zero and the entire range of errors in the validation is defined by $+/-3$ standard deviations by its definition. Under this premise, a certain trained model will present a certain Gaussian density function. For such an attained model, a good performance should be defined by a testing Gaussian density function that is similar to the reached model in the training stage. For instance, Figure 3 shows two models (model A and model B) that evaluate different architectures for the feed rate model as an example. Model A shows a testing evaluation with a particular density function, which is different to the one reached in the training evaluation. In contrast, model B shows a testing error density performance that is quite similar to the one reached in the training evaluations. Table 3 numerically shows the same concept shown in Figure 3. Table 3 also shows additional error comparisons, including the root mean square error (RMSE), median absolute deviation (MAD), maximum and minimum error values, etc. It is important to note that the large differences between RMSE and MAD are indicative of the error density function [30].

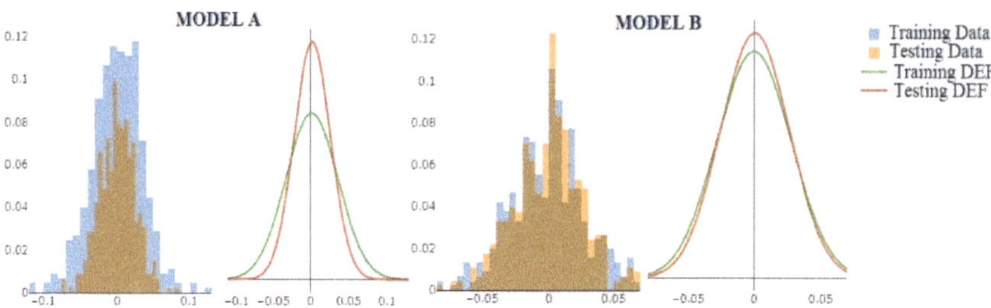

Figure 3. Example of error density function comparison.

Table 3. Error validation example.

Properties	Model A		Model B	
	Training Data	**Testing Data**	**Training Data**	**Testing Data**
Mean square error	11.9×10^{-4}	5.8×10^{-4}	7.8×10^{-4}	6.7×10^{-4}
Root mean square error	34.5×10^{-3}	24.2×10^{-3}	28.0×10^{-3}	25.9×10^{-3}
Median absolute deviation	22.4×10^{-3}	16.7×10^{-3}	18.2×10^{-3}	16.7×10^{-3}
Max. value	0.1275	0.0811	0.0698	0.0675
Min. value	-0.1167	-0.0791	-0.0828	-0.0828
Mean	19.4×10^{-5}	87.0×10^{-5}	-8.0×10^{-4}	-1.8×10^{-4}
Standard deviation	34.5×10^{-3}	24.2×10^{-3}	28.0×10^{-3}	26.0×10^{-3}

To achieve the best model for feed rate and cutting velocity, the following strategy was implemented to validate any possible architecture:

- Consider random architectures with one and two hidden layers with different neurons per layer.
- For each architecture, the training and testing error density functions are obtained. Hereinafter, these Gaussian functions are referred as m_1, m_2
- For each architecture, a Euclidean distance is evaluated between the functions of m_1 and m_2. This Euclidean distance is calculated with the properties described in Table 3. Equation (3) shows the implementation of this Euclidean distance:

$$d_{ED} = \sqrt{\sum(m_1 - m_2)} \tag{3}$$

- For each architecture, the maximum distance between the mean error values is considered to be zero. Equation (4) shows the implementation of this distance as the maximum absolute value of the mean property in testing and training error functions:

$$d_{mean} = \max(|mean_{m1}|, |mean_{m2}|) \tag{4}$$

- A maximum range of error functions must be calculated. This range is defined by the maximum and minimum properties for each error function (m_1, m_2).

$$d_{range} = \max(|\max_{m1} - \min_{m1}|, |\max_{m2} - \min_{m2}|) \tag{5}$$

- To consider the skew property for m_1 and m_2, the distance between RMSE to MAD is calculated. Equation (6) shows the implementation of this distance as the maximum absolute value for the difference between RMSE and MAD for both m_1 and m_2:

$$d_{skew} = \max(|RMSE_{m1} - MAD_{m1}|, |RMSE_{m2} - MAD_{m2}|) \tag{6}$$

- After this, the performance of a suggested model is given by Equation (7), which evaluates the Euclidean distance, mean distance to zero, the total error range and skew distance as a function of m_1 and m_2. The constant k allows the function to be adjusted to a certain range of values, which is shown as follows:

$$F_{perfo}(m_1, m_2) = \frac{k}{d_{ED} \cdot d_{mean} \cdot d_{range} \cdot d_{skew}} \tag{7}$$

To validate several random architectures, a heuristic optimization search was implemented by using a genetic algorithm for the feed rate and cutting velocity models. The architectures differ from each other in layer and neuron numbers. Furthermore, this genetic algorithm is defined by Equation (7) as the fitness function. Table 4 shows the reached architectures for the feed rate and cutting velocity models. For the feed rate, a neural network model was established with two hidden layers. Furthermore, this model has 15 and 4 neurons per layer. The cutting velocity model was also established with two hidden layers, which has 25 and 7 neurons per layer. The graphical performance for both models is shown in Figures 4 and 5 (feed rate and cutting velocity, respectively). In these figures, it is possible to appreciate the error density function in training and testing evaluations. These figures also show the reached genetic algorithm performance used to converge both models. It is important to note that the optimization by the genetic algorithm allows the achievement of good performance for both models in fewer iterations.

Table 4. Error validation for feed rate and cutting velocity models.

Models	Feed Rate		Cutting Velocity	
Datasets	**Training**	**Testing**	**Training**	**Testing**
Architecture	15–4		15–5	
Mean square error	4.95×10^{-4}	3.45×10^{-4}	57.71	66.66
Root mean square error	2.22×10^{-2}	1.8×10^{-2}	7.59	8.16
Median absolute deviation	1.41×10^{-2}	1.10×10^{-2}	3.84	3.91
Max. value	5.82×10^{-2}	5.82×10^{-2}	102.82	95.51
Min. value	-6.97×10^{-2}	-6.71×10^{-2}	-52.76	-75.10
Mean	-1.73×10^{-4}	1.24×10^{-4}	-3.60×10^{-2}	-3.61×10^{-2}
Standard deviation	2.22×10^{-2}	1.86×10^{-2}	7.59	8.16
d_{ED}	6.55×10^{-3}		25.16	
d_{mean}	1.73×10^{-3}		3.61×10^{-2}	
d_{range}	1.28×10^{-1}		170.61	
d_{skew}	8.14×10^{-3}		4.25	
k	1×10^{-5}		1×10^{5}	
F_{perfo}	845.26		151.68	

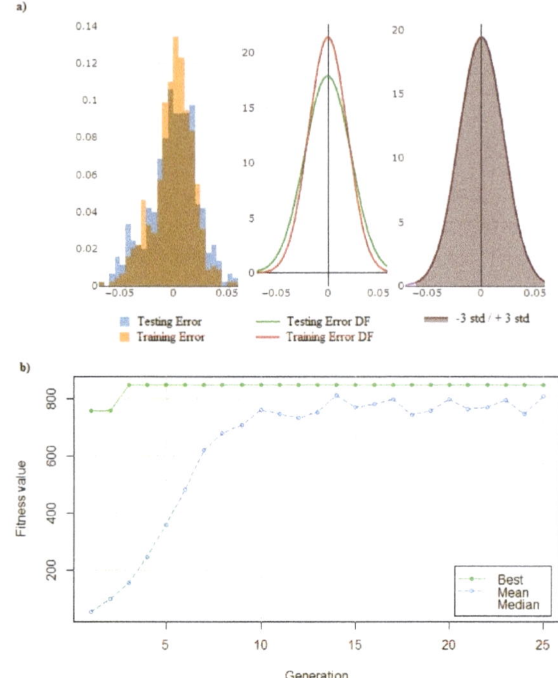

Figure 4. Feed rate model performance; (**a**) Error density function comparison; and (**b**) Genetic algorithm performance.

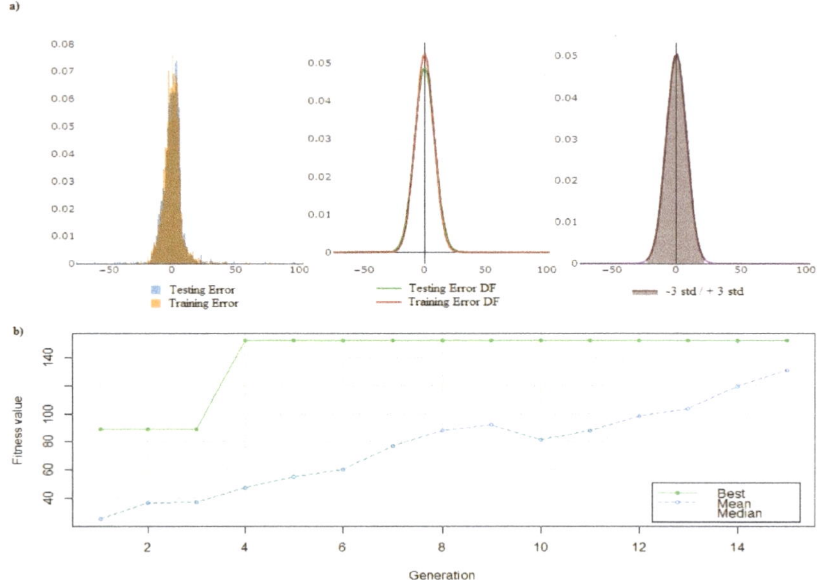

Figure 5. Cutting velocity model performance; (**a**) Error density function comparison; (**b**) Genetic algorithm performance.

4. Genetic Algorithm Optimization

A genetic algorithm (GA) is a heuristic search algorithm inspired by evolutionary mechanisms and biological natural selection. This algorithm defines a searching space of solutions and allows us to find individual solutions, which satisfy a certain condition. This condition is usually referred to as the fitness function [33]. This study aims to define a heuristic search based on GA, which allows us to find an optimal insert-tool for certain working specifications. This insert-tool search is governed by a fitness function, which is defined by a combination of the lowest power consumption, the shortest machining time and acceptable surface roughness among others. Figure 6 shows a general scheme of the proposed GA implementation for this research. Initially, it can be appreciated that the feed rate model evaluates the GA chromosome. After this, the feed rate and working conditions can be used as the inputs for the cutting velocity model. The resulting feed rate and cutting velocity are evaluated by the GA fitness function, which assess the current chromosome along with the working specifications. Finally, the GA operations transform the current population to the next generation. According to their characteristics, the individuals in the population improve with each iteration.

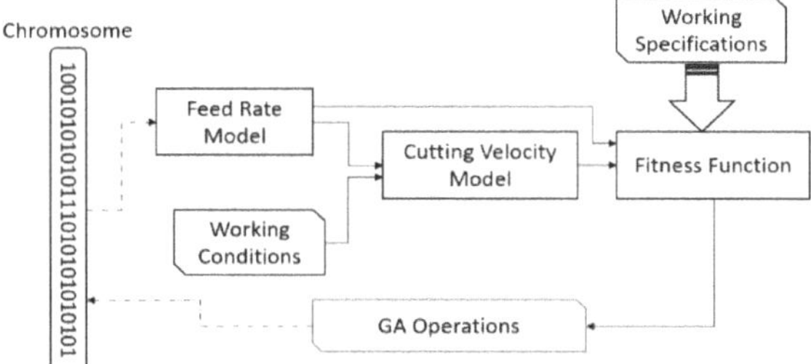

Figure 6. GA (genetic algorithm) working scheme.

4.1. Working Specifications

The working specifications for the GA represent the framework, which defines the optimization problem. For this research, the working specifications consider three groups: material specifications, machine specifications and general specifications. Table 5 introduces these categories for these working specifications. The material specifications represent the conditions related to the working piece, such as the initial and final diameters, the maximum allowed surface roughness and the machining length. The mechanical properties are also added to this group, including material harness and specific cutting forces. On the other hand, the machine specifications represent the relevant information of the CNC machine in a turning operation. For this approach, the total power available and the maximum spindle speed are considered. The general specifications define additional specifications, which also define a turning operation. For instance, the machine operation is considered in this group of specifications. This specification defines the CTPT feature, which sets the algorithm and considers if the insert-tool solution should be for finishing, medium or roughing operations. Furthermore, a stability parameter is also added to this group of specifications. This stability feature is related to the suitability for the expected insert-tool solution. Finally, the tool life and threshold distances are defined as turning constants, which allow for an adjustment of the accuracy of the algorithm.

Table 5. Working specifications.

Material specifications		
Initial diameter	D_i	Numeric values
Final diameter	D_f	Numeric values
Machining length	L_m	Numeric values
Hardness	HB	Numeric values
Specific cutting force	K_c	Numeric values
Max. surface roughness	R_{a_max}	Numeric values
Machine specifications		
Main motor power	P_{net}	Numeric values
Max. spindle speed	n_{max}	Numeric values
General specifications		
Machine operation	CTPT	Finishing–Medium–Roughing
Stability	/	Excellent–Good–Poor
Toot life	T_{life}	Numeric values
Threshold distance	Th_d	Numeric values

4.2. Encode-Decode Chromosomes

A binary chromosome refers to a string of zeros and ones, such as a binary number, and represents a certain bunch of features. Table 1 shows the description of the features of the chromosome structure. On the other hand, each feature has its own encoding length that allows it to be represented in the chromosome string. Table 6 shows the features used to build the GA chromosome and their binary length.

Table 6. Insert features descriptions for the decoding procedure.

Name	Type	Range	Length of Binary Numbers
ICmm	Numeric	15.875–3.970	4
LEmm	Numeric	21.20–5.65	4
REmm	Numeric	1.19–0.02	7
WEP	Logical	1 or 0	1
Smm	Numeric	5.56–1.98	7
AN	Logical	7 or 5	1
Angle	Numeric	90–35	6
Stability	Factor	not defined	not defined
ap_rec	Numeric	not defined	7

The length properties in Table 6 represent the lengths of the binary numbers used to define each feature. Furthermore, for the described numeric features, there is a specific decoding procedure called linear transformation, which is represented by Equation (8). For this equation, the binary number is evaluated by the function $int(X_{binary})$, which returns the integer value for a binary number. This integer value is then mapped by the maximum, minimum and length values to decode the corresponding features.

$$X_{real} = X_{min} + \frac{X_{max} - X_{min}}{2^{length} - 1} \cdot int(X_{binary}) \tag{8}$$

The logical features described in Table 6 are represented by one digit with only two possible values (1 or 0 and 5 or 7 for WEP and AN features, respectively). The stability feature has decoding and encoding procedures that differ from the rest of features. The stability feature refers to the suitability feature of an insert-tool. This feature can be set as Excellent, Good or Poor. This parameter is already given by the working specification described in Table 5. However, this information restricts the GA searching space because the grade of an insert-tool is defined under a certain range of application areas. This means that there is only a defined number of grades for a defined stability feature. Table 7 shows the stability condition associated with the insert-tool grades in the present dataset. Under this premise, a grades list (indexed list) and a binary number define the proposed encoding procedure for the features' stability. The binary number indexes the position for each grade on such list. For instance, we considered the stability "Good" described in Table 8. Since there are only four possible grades in this condition, the binary number to encode this feature is a string of three digits (000, 001, 010, 011, 111, etc.). Thus, the number of ones in such binary number represents the index in the grade list. For two ones in the binary number, the grade is 4225 and for 0 ones in the binary number, the grade is 1125, etc.

Table 7. Grade stability.

GRADE	Suitability	Stability	GRADE	Suitability	Stability
1125	75%	Good	4305	100%	Excellent
1515	75%	Good	4315	83.3%	Excellent
1525	100%	Excellent	4325	50%	Good
235	0%	Poor	4335	16.6%	Poor
4215	83.3%	Excellent	5015	100%	Excellent
4225	50%	Good	GC15	100%	Excellent
4235	20%	Poor	GC30	0%	Poor

Table 8. Indexed list for good stability.

Index	Grade	Stability
0	1125	Good
1	1515	Good
2	4225	Good
3	4325	Good

The ap_rec cutting parameter is a special feature for the chromosome definition. The range of this parameter was defined previously as an input to the model. This range is customized by the total and minimum depth of cut values for a certain turning process. Such values are defined with respect to operation issues and specific requirements. The chromosome structure for the GA optimization is shown in Figure 7. This structure defines at least 30 defined positions. It is important to note that each feature has its binary position and length in the chromosome structure.

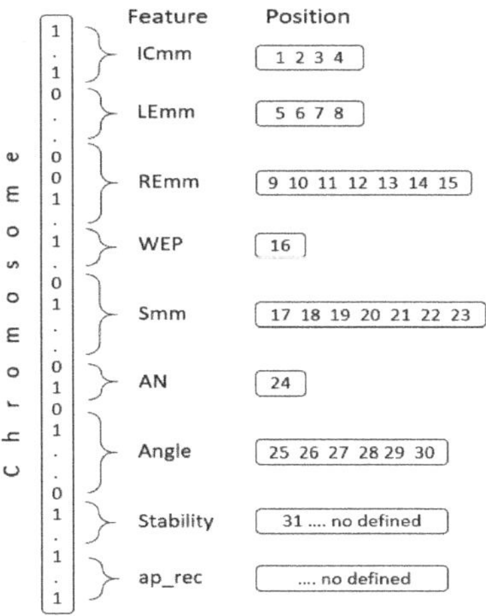

Figure 7. Chromosome structure.

4.3. Fitness Function Definition

The fitness function is the main concern for the development of the GA and its definition establishes the expected output of the algorithm. For this research, this function was defined as a function that evaluates the lowest power consumption, the shortest machining time and an acceptable surface roughness among others. Equation (9) defines the fitness function for this genetic algorithm optimization. This equation is a summation of goal functions g_i, which evaluate power consumption, machining time, surface roughness, etc. Furthermore, each g_i function is weighted by a constant ω_i, which allows the adjustment of the importance of each g_i function.

$$fitness = \sum_i \omega_i \cdot g_i \tag{9}$$

The goal function, g_1, described in Equation (10), evaluates the power consumption ratio in the turning process. This function uses Equation (11) to evaluate the theoretical power consumption, P [20]. Furthermore, the P_{net} constant is the total available power in the machine (introduced in Table 5). The constant γ allows for the consideration of the friction losses in the transmission motor [20] although this constant is equal to 1 for this research. Equation (12) shows the cutting force, F_c, as a function of the specific cutting force, K_c, related to the working material, depth of cut (*ap_rec*) and feed rate (*fn_rec*) [25]. It is important to note that the specific cutting force, K_c, is a parameter that was also introduced in Table 5.

$$g_1 = \frac{\gamma \cdot P_{net}}{P} \tag{10}$$

$$P = \frac{v_c \cdot F_c}{6 \times 10^4} \tag{11}$$

$$F_c = K_c \cdot ap\_rec \cdot fn\_rec \tag{12}$$

The goal function, g_2, described in Equation (13), defines the machining time ratio, which relates the tool life to the machining time. The tool life, T_{life}, is a tool supplier parameter, which defines the approximate tool life for a tool-insert. The machining time is an objective variable evaluated in Equation (14). This defines the machining time in the turning process with a certain group of cutting parameters. Equation (15) presents the spindle speed, n, as a function of the cutting velocity, v_c and the current machining diameter, D. It is important to note that n varies according to D. Thus, for each machining pass (diameter decreasing), the spindle speed increases.

$$g_2 = \frac{T_{life}}{T_m} \tag{13}$$

$$T_m = \frac{L_m}{fn\_rec \cdot n} \tag{14}$$

$$n = \frac{v_c \times 1000}{\pi \cdot D} \tag{15}$$

The goal function, g_3, defined in Equation (16), relates the maximum surface roughness, R_{a_max}, to the theoretically reached surface roughness, R_a. The parameter R_{a_max} is a working specification of material that defines the maximum allowed roughness. R_a is an objective variable defined in Equation (17), which estimates the surface roughness for a certain finishing operation [20].

$$g_3 = \frac{R_{a\_max}}{R_a} \tag{16}$$

$$R_a = \frac{125 \cdot fn\_rec^2}{REmm} \tag{17}$$

The goal function, g_4, defined in Equation (18), relates the objective variable, $d_{euclidean}$, to the parameter Th_d (threshold distance). The variable $d_{euclidean}$ is obtained by Equation (19). This equation calculates the Euclidean distance between the features x_i and y_i, which are defined by the suggested GA tool-insert features and the nth tool-insert in the dataset, respectively.

$$g_4 = \frac{Th_d}{d_{euclidean}} \tag{18}$$

$$d_{euclidean} = \sqrt{\sum_i (x_i - y_i)^2} \tag{19}$$

The goal function, g_5, shown in Equation (20), evaluates the spindle speed ratio, which is used in the turning process. This objective function relates the maximum spindle speed of the machine to the theoretical spindle speed reached in the turning process.

$$g_5 = \frac{n_{max}}{n} \tag{20}$$

The goal function, g_6, shown in Equation (21), evaluates the suitability range for a certain insert grade. This function assesses if a suggested grade is more appropriate for a specific application area compared to other solutions. Grades that are more specific are preferred instead of the general grades.

$$g_6 = \frac{1}{MC\_H - MC\_L} \tag{21}$$

Overall, there are two fitness functions regarding the turning processes: the finishing or roughing operations. In fact, for the roughing operation, the surface roughness is not considered in the fitness function, but the power consumption, machining time, Euclidean distance, spindle speed ratio and suitability are considered. For the finishing process, the fitness function considers the surface roughness, Euclidean distance and suitability grade. Equations (22) and (23) show the proposed fitness function for the roughing and finishing operations.

$$f_{roughing} = \omega_1 \frac{P_{net}}{P} + \omega_2 \frac{T_{life}}{T_m} + \omega_3 \frac{Th_d}{d_{eucliden}} + \omega_4 \frac{n_{max}}{n_5 \frac{1}{MC\_H - MC\_L}} \tag{22}$$

$$f_{finishing} = \omega_1 \frac{R_{a\_max}}{R_{a2} \frac{Th_d}{d_{eucliden}} 3 \frac{1}{MC\_H - MC\_L}} \tag{23}$$

4.4. Boundary Constraints

The boundary constraints for the GA optimization define the searching space for the algorithm. These boundary constraints allow us to control the chromosome evolution and ensure that the algorithm is converging towards a suitable solution. The first mechanism for controlling the population evolution is called the feasible solution control. This boundary control prevents infeasible individuals from appearing in the population [20]. This mechanism evaluates the goal function, g_4, by the constraint mentioned in Equation (24). Thus, only similar insert-tools are chosen as part of the population. The infeasible solutions are not considered.

$$g_4 = \frac{Th_d}{d_{euclidean}} > 1 \tag{24}$$

Some boundary constraints are established to prevent solutions that evaluate a parameter outside of the machine capabilities. For example, the expression mentioned in Equation (25) avoids GA individuals when evaluating power consumption, which exceeds the maximum power of the machine.

The constraint mentioned in Equation (26) avoids solutions that need spindle speeds greater than the maximum defined speed for the machine.

$$g_1 = \frac{\gamma \cdot P_{net}}{P} > 1 \tag{25}$$

$$g_5 = \frac{n_{max}}{n} \tag{26}$$

5. Application Examples

This section aims to explain some applications for the algorithm described in this research. Three application examples are used to show the performance and results of the proposed approach. These examples are defined under different working conditions and fitness functions.

5.1. Light Roughing Operation

A light roughing machining application can be defined as a turning operation with a small turning depth. In fact, these turning operations could be machined with only one pass. The described algorithm allows us to find a tool-insert solution that balances both power consumption and the number of passes. On the other hand, the machining time does not have the same importance as the power consumption as the number of passes is low. Table 9 introduces the working specifications for this light roughing operation. It sets an initial and a final diameter of 50 mm and 40 mm, respectively. This means a total depth of cut of 5 mm. Additionally, the total machining length is set to 100 mm. Figure 8 illustrates this information. The workpiece material is an unalloyed steel of 180 HB. This material has a specific cutting force of 600 N/mm². The machine specifications define a small lathe with a power capacity of 10 kW and a maximum spindle speed of 3000 rpm. Equation (27) shows the fitness function for this operation. The power consumption has been weighted to 100 times while the machining time was set to 0.5. This fitness function considers that the impact of power consumption is greater than that of the machining time during the turning process.

$$fitness = 100\frac{P_{net}}{P} + 0.5\frac{T_{life}}{T_m} + \frac{Th_d}{d_{eucliden}} + \frac{n_{max}}{n\frac{100}{MC\_H - MC\_L}} \tag{27}$$

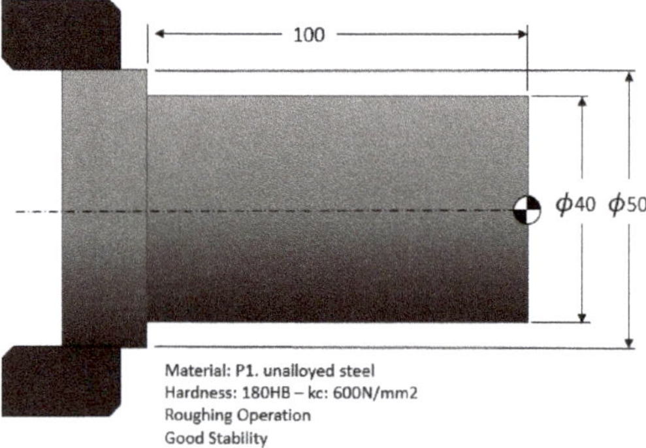

Figure 8. Example of a light roughing operation.

Table 9. Working specifications of the light roughing operation example.

Material Specifications		
Initial diameter	D_i	50 mm
Final diameter	D_f	40 mm
Machining length	L_m	100 mm
Hardness	HB	180 HB
Specific cutting force	H_c	600 N/mm^2
Machine Specifications		
Main motor power	P_{net}	10 kW
Maximum spindle speed	n_{max}	3000 rpm
General Specifications		
Machine operation	CTPT	Roughing
Stability	/	Good
Toot life	T_{life}	15 min
Threshold distance	Th_d	10
Range ap_rec	ap_rec	0.01 mm–5 mm

Figure 9 provides a complete description of the GA optimization. It begins with the definition of the working specifications, which are presented in Table 9. These specifications set the framework for this example. It is important to note that the working specifications are related to the GA model with respect to some stages ahead, such as the random population, feasible solution control, feed rate and cutting velocity models, boundary constraints and fitness evaluation stages. The stability condition establishes some insert grades as possible solutions since this specification was set at "Good" in Table 9. Such grade solutions were already shown in Table 8, which indexes the list for good stability. With the stability features already defined, the total chromosome length is defined as 40 for this application. On the other hand, the insert features and working specifications form the chromosome structure. Figure 10 describes the chromosome structure for this application. Given the random nature of the GA model, some infeasible individuals can be part of the population. The feasible solution control prevents infeasible solutions from being part of the GA population. The implementation of this control is introduced in Equation (24). This control evaluates a Euclidean distance by adjusting the Th_d parameter. All individuals out of the range of distance are discarded from the GA population.

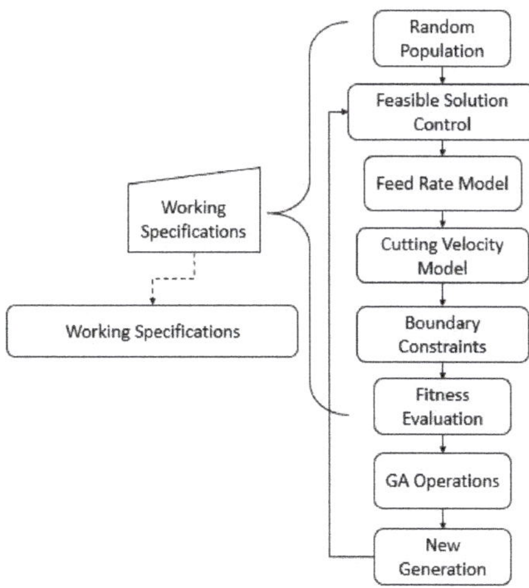

Figure 9. Flowchart of the GA implementation.

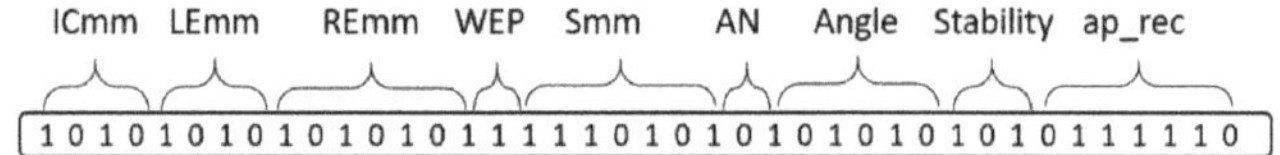

Figure 10. The GA chromosome structure.

Figure 11 shows the flowchart for obtaining the cutting parameters when given an individual chromosome in the population. It is important to note that some working specifications are used for both the feed rate and cutting velocity neural network models. In addition, the ap_rec cutting parameter (depth of cut) is defined by the GA chromosome and it is used in the feed rate model as an input. Once the cutting parameters, which are namely ap_rec, fn_rec and vc_rec, are obtained, it is possible to evaluate the performance of the current chromosome. Equation (27) shows the proposed fitness function to evaluate the performance of each GA individual. The variables, P, T_m, $d_{euclideanm}$ and n define the power consumption, machining time, Euclidean distance and spindle speed, respectively.

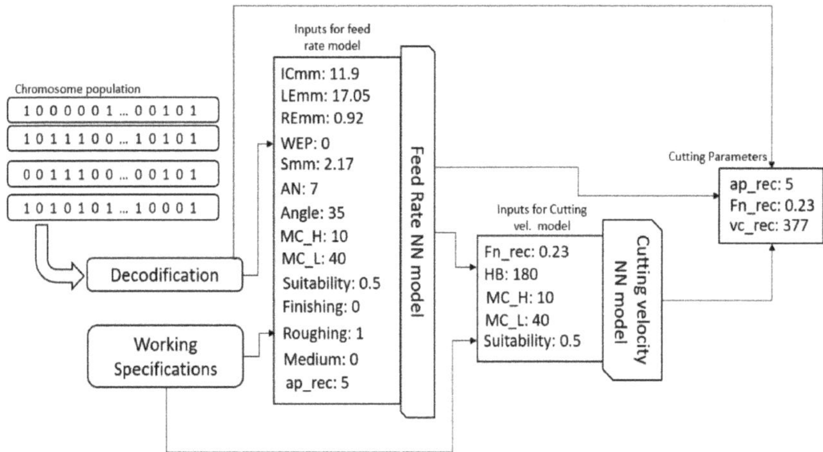

Figure 11. Neural network models in GA optimization.

Consistent with the flowchart for GA implementations, the boundary constraints establish population control again, which avoids individual GAs when evaluating parameters beyond the capabilities of the CNC machine. These constraints are related to the total power available and the maximum speed spindle defined in the CNC specifications. Once the entire population has been qualified by the fitness function, the GA operation crosses, selects and mutates the chromosomes to evolve them toward the next generation. In this way, after a certain number of iterations, the population is evolved enough towards a suitable insert-tool solution with its cutting parameters, as shown in Figure 12. Table 10 shows the reached non-existent insert by the GA evolution next to the defined closer tool-insert in the dataset. The reached tool-insert and its cutting parameters are presented as well. In Table 11, the evaluation for the goal variables, which define the performance for the suggested tool-insert, is introduced.

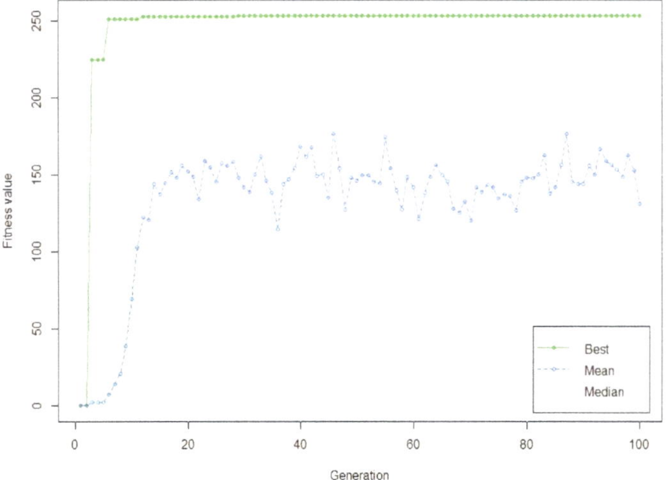

Figure 12. Example of GA performance for light roughing operation.

Table 10. Results of the light roughing operation example.

Features	GA Results	Closer Insert
ICmm	8.73	6.35
LEmm	10.5	10.3
REmm	0.68	0.39
WEP	false	false
Smm	2.37	2.38
AN	7°	7°
Angle	65°	60°
GRADE	4325	4325
Cutting Parameters		
Closer insert	TCMT 11 02 04-UR 4325	
Depth of cut	5 mm	
Feed rate	0.288 mm/r	
Cutting velocity	346 m/min	

Table 11. Goal variables evaluation for light roughing operation example.

Goal Variable	Variable	Evaluation
Power consumption	P	5 kW
Machining time	T_m	0.34 min
Euclidean distance	$d_{euclidean}$	6.307
Spindle speed	n	2761 rpm
High machining condition	MC_H	40
Low machining condition	MC_L	10
Fitness function	$fitness$	225.69

5.2. Heavy Roughing Operation

A heavy roughing application can be defined as a turning operation with a large depth turning. These turning operations cannot be machined in a single pass unlike light roughing machining. For this reason, the fitness function for these turning operations considers the machining time evaluation more than the power consumption. Table 12 introduces the working specifications for this roughing operation. It sets an initial diameter and a final diameter to 100 mm and 40 mm, respectively, which defines a total depth turning of 30 mm. Additionally, the total machining length is set to 100 mm. The geometrical information for this example is presented in Figure 13. The workpiece material is an unalloyed steel of 250 HB with a specific cutting force of 725 N/mm^2. Moreover, for this operation, the stability condition is set to "Good." The machine specifications define a small lathe with a total power capability of 10 kW and a maximum spindle speed of 3000 rpm. Equation (28) shows the fitness function for this operation. The machining time has been weighted 10 times, while the power consumption has been weighted 0.5 times. In this way, the fitness function establishes that the solution must prioritize the machining time instead of the power consumption.

$$fitness = 0.5\frac{P_{net}}{P} + 10\frac{T_{life}}{T_m} + \frac{Th_d}{d_{eucliden}} + \frac{n_{max}}{n\frac{100}{MC\_H - MC\_L}} \tag{28}$$

Table 12. Working specifications of the heavy roughing operation example.

Material Specifications		
Initial diameter	D_i	100 mm
Final diameter	D_f	40 mm
Machining length	L_m	100 mm
Hardness	HB	250 HB
Specific cutting force	K_c	725 N/mm^2
Machine Specifications		
Main motor power	P_{net}	10 kW
Maximum spindle speed	n_{max}	3000 rpm
General Specifications		
Machine operation	CTPT	Roughing
Stability	/	Good
Toot life	T_{life}	15 min
Threshold distance	Th_d	10
Range ap_rec	ap_rec	0.01–5 mm

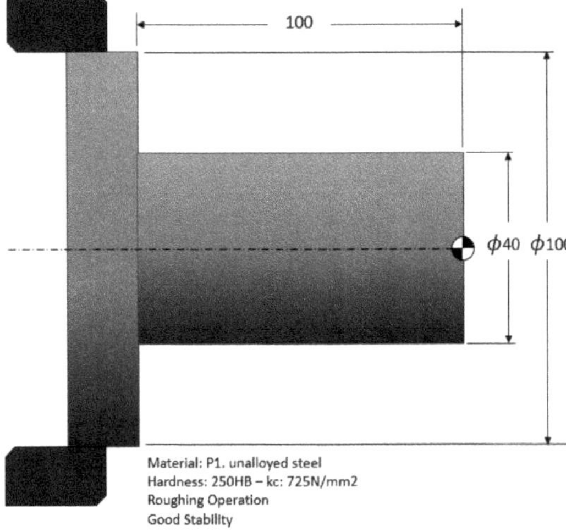

Figure 13. Example of heavy roughing operation.

Table 13 shows the results for the GA evolution, which are the obtained features for the GA algorithm next to the closer tool-insert in the dataset. This table also presents the suggested tool-insert and its cutting parameters for this operation. Furthermore, in Table 14, the evaluation for the goal variables, which define the performance for the reached tool-insert, is introduced.

Table 13. Results of the heavy roughing operation example.

Features	GA Results	Closer Insert
ICmm	15.87	12.07
LEmm	21.2	11.7
REmm	0.609	1.19
WEP	false	false
Smm	5.56	4.76
AN	7°	7°
Angle	90°	90°
GRADE	4325	4325
Cutting Parameters		
Closer insert	CCMT 12 04 12-PR 4325	
Number of passes	6	
Depth of cut	5 mm	
Feed rate	0.325 mm/r	
Cutting velocity	261 m/min	

Table 14. Goal variables evaluation of the heavy roughing operation example.

Goal variable	Variable	Evaluation
Power consumption	P	5.13 kW
Total power consumption	passes $\times$ P	30.83 kW
Machining time	T_m	1.84 min
Euclidean distance	$d_{euclidean}$	4.7830
Spindle speed	n	2080 rpm
High machining condition	MC_H	40
Low machining condition	MC_L	10
Fitness function	$Fitness$	86.261

5.3. Finishing Operation

Finishing operations differ from roughing operations mainly with regards to the fitness function. In this operation, the stability and surface roughness performance define the tool-insert solution. The stability features must be defined in such a way that the selected grade belongs to the lower ISO area application, which is related to the wear resistance performance. This area application is specified for finishing operations. Furthermore, the surface roughness performance must be lower than the maximum roughness defined by the working specifications of the process. Table 15 introduces the working specifications of this finishing operation. It sets a final diameter of 40 mm and a maximum surface roughness of ISO N8. This surface specification defines a roughness of 3 μm. The geometrical information for this example is presented in Figure 14. The workpiece material is an unalloyed steel of 180 HB with a specific cutting force of 600 N/mm^2. For this operation, the stability condition is set to "Good." The machine specifications define a medium lathe with a total power of 15 kW and a maximum spindle speed of 6000 rpm. Equation (29) sets the fitness function for this operation. This function evaluates the goal functions related to surface roughness and stability grade.

$$fitness = \frac{R_{a\_max}}{R_a \frac{Th_d}{d_{eucliden}} \frac{100}{MC\_H - MC\_L}} \tag{29}$$

Table 15. Working specifications of the finishing operation example.

Material Specifications		
Final diameter	D_f	40 mm
Machining length	L_m	50 mm
Hardness	HB	180 HB
Specific cutting force	K_c	600 N/mm^2
Max. surface roughness	R_{a_max}	3 µm
Machine Specifications		
Main motor power	P_{net}	15 kW
Maximum spindle speed	n_{max}	6000 rpm
General Specifications		
Machine operation	CTPT	Finishing
Stability	/	Good
Toot life	T_{life}	15 min
Threshold distance	Th_d	20
Range ap_rec	ap_rec	0.01–5 mm

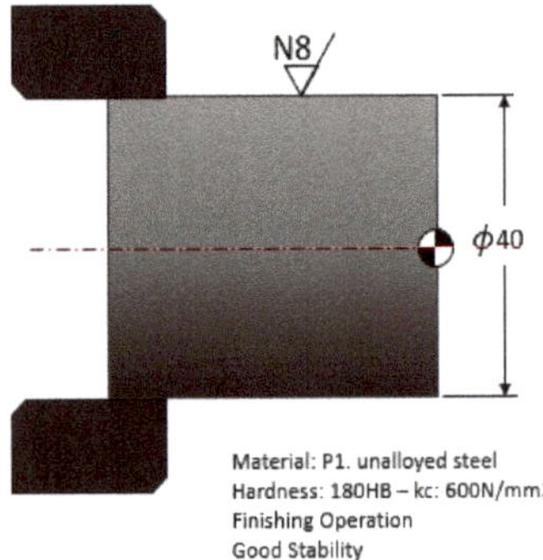

Figure 14. Example of a finishing operation.

Table 16 shows the results for the GA evolution, which are the obtained features of the GA algorithm next to the closer tool-insert in the dataset. The information related to the suggested tool-insert is also presented. The reached tool was VBMT 11 03 12-PF 4325. Table 17 introduces the evaluation for the goal variables, which define the performance of the suggested tool-insert.

Table 16. Results of the finishing operation example.

Features	GA Results	Closer Insert
ICmm	5.55	6.35
LEmm	5.65	9.87
REmm	0.99	1.19
WEP	False	false
Smm	1.98	3.18
AN	5°	5°
Angle	35°	35°
GRADE	4325	4325
Cutting Parameters		
Closer insert	VBMT 11 03 12-PF 4325	
Depth of cut	0.3 mm	
Feed rate	0.14 mm/r	
Cutting velocity	434 m/min	

Table 17. Goal variables evaluation of the finishing operation example.

Goal Variable	Variable	Evaluation
Power consumption	P	0.189 kW
Machining time	T_m	0.343 min
Euclidean distance	$d_{euclidean}$	6.48
Spindle speed	n	3456 rpm
Surface roughness	R_a	2.22 μm
High machining condition	MC_H	40
Low machining condition	MC_L	10
Fitness function	$fitness$	48.33

6. Conclusions

Neural network models for the selection of cutting inserts and cutting parameters were designed. These models were applied to roughing and finishing operations. This research uses the information from a tool supplier to embed knowledge for the tool-insert selection and optimization system developed in this research. The proposed system is based on artificial neural networks (ANNs) and a genetic algorithm (GA). These represent the modeling and optimization part of this research, respectively. For the modeling, two ANNs were implemented. These ANNs are able to infer the feed rate and cutting velocity parameters. The feed rate model is defined as a function of insert features and a set depth of the cut. The insert features represent the macro-geometries of a tool-insert, which include the cutting length, thickness, nose angle, nose radius, size and grade among others. For the cutting velocity model, the inputs are the material specifications and a set feed rate. The material specifications are defined as inherent features of a working material for turning processes. They include hardness, specific cutting force and ISO material group.

For the neural network validation, an error comparison based on density functions was implemented. This approach proposed an alternative solution for the error validation of regression models based on the recommended data by a supplier. To evaluate some architectures of the ANN models, a heuristic search based on GA was used. This approach evaluated the possible architectures and evolved them toward the most feasible one. For the proposed research, a heuristic searching method for a feasible tool-insert based on its characteristics in a certain environment was also introduced. The algorithm used for this search was a GA. The introduced GA optimization searches for an optimal tool-insert, which adapts to a working specification and evaluates an acceptable performance given by a customized objective function. This objective function evaluates the performance under certain working conditions, such as the lowest power consumption, the shortest machining time and an acceptable surface roughness. In this present study, different goal functions referring to different force models or tool life models can be used when designing the fitness function in order to obtain the optimal tool-insert with specific considerations. This research presents a model to simulate knowledge and expertise, which embeds the information from a tool supplier. It returns the most suitable insert-tool as a result, given certain working conditions. This research did not use a lookup table approach in the database as it instead modeled the mechanical relations between the geometrical features of an insert-tool and its recommended cutting parameters. This tool-insert selection and optimization system can embed the data from other models, different tool suppliers, previous machining works and expertise from machine-shop workers. However, because of the inherent discrepancy among different datasets, additional factors, such as input variation and data types, must be considered when using the developed system and approaches. Furthermore, because the developed system successfully represents the complex relationship between the working condition and the cutting parameters, the developed system can be a plug-in for CAD/CAM software or can be integrated with the controller of a CNC machine tool to be an auxiliary function for the automatic selection of the cutting parameters based on the preset working conditions presented in the part program. For this research, a fully connected ANN model was used due to the representation of a multi-dimensional space using the training data obtained from a database. Other networks with diverse characteristics, such as the convolutional networks, could also be used in future researches. Moreover, methods, such as the support vector machine and probabilistic regression, could be used for regression modeling. This research has used a GA for optimization. However, methods, such as the particle swarm optimization and gradient descent technique, can also be used to carry out optimization.

Author Contributions: Investigation, B.S.-P., D.H. and S.-S.Y.; Supervision, S.-S.Y.

Funding: This research was funded in part by the Ministry of Science and Technology, Taiwan, R.O.C., under Contract MOST104-2221-E-027-132 and MOST103-2218-E-009-027-MY2.

Acknowledgments: The authors would like to thank Hao-Wei Nien (Advantech-LNC Technology Co. Ltd., Taiwan, R.O.C.) and Chao-Choung Mai (ITRI Intelligent Machinery Technology Center, Taiwan, R.O.C.) for their suggestions.

Conflicts of Interest: The authors declare no conflict of interest.

References

1. Weyer, S.; Schmitt, M.; Ohmer, M.; Gorecky, D. Towards industry 4.0—Standardization as the crucial challenge for highly modular, multi-vendor production systems. *IFAC-Pap.* **2015**, *48*, 579–584. [CrossRef]
2. Leo Kumar, S.P. State of the art—Intense review on artificial intelligence systems application in process planning and manufacturing. *Eng. Appl. Artif. Intell.* **2017**, *65*, 294–329. [CrossRef]
3. Zarkti, H.; El Mesbahi, A.; Rechia, A.; Jaider, O. Towards an automatic-Optimized tool selection for milling process, based on data from Sandvik Coromant. In Proceedings of the Xème Conférence Internationale: Conception et Production Intégrées, Tanger, Morocco, 3–4 December 2015.
4. Arrazola, P.J.; Özel, T.; Umbrello, D.; Davies, M.; Jawahir, I.S. Recent advances in modelling of metal machining processes. *CIRP Ann.* **2013**, *62*, 695–718. [CrossRef]
5. Ganesh, N.; Kumar, M.U.; Kumar, C.V.; Kumar, B.S. Optimization of cutting parameters in turning of EN 8 steel using response surface method and genetic algorithm. *Int. J. Mech. Eng. Robot. Res.* **2014**, *3*, 75–86.
6. Yang, Y.; Li, X.; Gao, L.; Shao, X. A new approach for predicting and collaborative evaluating the cutting force in face milling based on gene expression programming. *J. Netw. Comput. Appl.* **2013**, *36*, 1540–1550. [CrossRef]
7. Cabrera, F.M.; Beamud, E.; Hanafi, I.; Khamlichi, A.; Jabbouri, A. Fuzzy logic-based modeling of surface roughness parameters for CNC turning of PEEK CF30 by TiN-coated cutting tools. *J. Thermoplast. Compos. Mater.* **2011**, *24*, 399–413. [CrossRef]
8. Joshi, S.R.; Ganjigatti, J.P. Application of general regression neural networks for forward and reverse modeling of aluminum alloy AA5083; H111 TIG welding process and comparison with feed forward back propagation, Elman back propagation neural networks. *Int. J. Eng. Technol. Sci. Res.* **2017**, *4*, 16–26.
9. Li, H.Z.; Guo, S.; Li, C.J.; Sun, J.Q. A hybrid annual power load forecasting model based on generalized regression neural network with fruit fly optimization algorithm. *Knowl. -Based Syst.* **2013**, *37*, 378–387. [CrossRef]
10. Cairns, J.; McPherson, N.; Galloway, A. Using artificial neural networks to identify and optimise the key parameters affecting geometry of a GMAW fillet weld. In Proceedings of the 18th International Conference on Joining Materials, JOM-18, Helsingor, Denmark, 26–29 April 2015.
11. Arezoo, B.; Ridgway, K.; Al-Ahmari, A.M.A. Selection of cutting tools and conditions of machining operations using an expert system. *Comput. Ind.* **2000**, *42*, 43–58. [CrossRef]
12. Dereli, T.; Filiz, I.H.; Baykasoglu, A. Optimizing cutting parameters in process planning of prismatic parts by using genetic algorithms. *Int. J. Prod. Res.* **2001**, *39*, 3303–3328. [CrossRef]
13. Benkedjouh, T.; Medjaher, K.; Zerhouni, N.; Rechak, S. Health assessment and life prediction of cutting tools based on support vector regression. *J. Intell. Manuf.* **2015**, *26*, 213–223. [CrossRef]
14. Özel, T.; Hsu, T.K.; Zeren, E. Effects of cutting edge geometry, workpiece hardness, feed rate and cutting speed on surface roughness and forces in finish turning of hardened AISI H13 steel. *Int. J. Adv. Manuf. Technol.* **2004**, *25*, 262–269. [CrossRef]

15. Xiong, J.; Zhang, G.; Hu, J.; Wu, L. Bead geometry prediction for robotic GMAW-based rapid manufacturing through a neural network and a second-order regression analysis. *J. Intell. Manuf.* **2012**, *25*, 157–163. [CrossRef]

16. Babu, K.V.; Narayanan, R.G.; Kumar, G.S. An expert system based on artificial neural network for predicting the tensile behavior of tailor welded blanks. *Expert Syst. Appl.* **2009**, *36*, 10683–10695. [CrossRef]

17. Özel, T.; Karpat, Y. Predictive modeling of surface roughness and tool wear in hard turning using regression and neural networks. *Int. J. Mach. Tools Manuf.* **2005**, *45*, 467–479. [CrossRef]

18. Malinov, S.; Sha, W.; McKeown, J.J. Modelling the correlation between processing parameters and properties in titanium alloys using artificial neural network. *Comput. Mater. Sci.* **2001**, *21*, 375–394. [CrossRef]

19. Kuo, R.J.; Chen, C.H.; Hwang, Y.C. An intelligent stock trading decision support system through integration of genetic algorithm based fuzzy neural network and artificial neural network. *Fuzzy Sets Syst.* **2001**, *118*, 21–45. [CrossRef]

20. Quiza Sardiñas, R.; Rivas Santana, M.; Alfonso Brindis, E. Genetic algorithm-based multi-objective optimization of cutting parameters in turning processes. *Eng. Appl. Artif. Intell.* **2006**, *19*, 127–133. [CrossRef]

21. Cus, F.; Balic, J. Optimization of cutting process by GA approach. *Robot. Comput. -Integr. Manuf.* **2003**, *19*, 113–121. [CrossRef]

22. Suresh, P.V.S.; Rao, P.V.; Deshmukh, S.G. A genetic algorithmic approach for optimization of surface roughness prediction model. *Int. J. Mach. Tools Manuf.* **2002**, *42*, 675–680. [CrossRef]

23. Yang, W.H.; Tarng, Y.S. Design optimization of cutting parameters for turning operations based on the Taguchi method. *J. Mater. Process. Technol.* **1998**, *84*, 122–129. [CrossRef]

24. Thamizhmanii, S.; Saparudin, S.; Hasan, S. Analyses of surface roughness by turning process using Taguchi method. *J. Achiev. Mater. Manuf. Eng.* **2007**, *20*, 503–506.

25. Sandvik Coromant. Available online: https://www.sandvik.coromant.com/en-gb/pages/default.aspx (accessed on 9 October 2017).

26. Murata, N.; Yoshizawa, S.; Amari, S. Network information criterion-determining the number of hidden units for an artificial neural network model. *IEEE Trans. Neural Netw.* **1994**, *5*, 865–872. [CrossRef] [PubMed]

27. Arnaiz-González, Á.; Fernández-Valdivielso, A.; Bustillo, A.; López de Lacalle, L.N. Using artificial neural networks for the prediction of dimensional error on inclined surfaces manufactured by ball-end milling. *Int. J. Adv. Manuf. Technol.* **2016**, *83*, 847–859. [CrossRef]

28. Jain, S.P.; Ravindra, H.V.; Ugrasen, G.; Prakash, G.V.N.; Rammohan, Y.S. Study of surface roughness and AE signals while machining titanium grade-2 material using ANN in WEDM. In *Materials Today: Proceedings*; Elsevier: Amsterdam, The Netherlands, 2017; pp. 9557–9560.

29. Anastasiadis, A.D.; Magoulas, G.D.; Vrahatis, M.N. New globally convergent training scheme based on the resilient propagation algorithm. *Neurocomputing* **2005**, *64*, 253–270. [CrossRef]

30. Twomey, J.M.; Smith, A.E. Validation and verification. In *Artificial Neural Networks for Civil Engineers: Fundamentals and Applications*; Expert Systems and Artificial Intelligence Committee, Ed.; American Society of Civil Engineers: Reston, VA, USA, 1997; pp. 44–64.

31. Olden, J.D.; Jackson, D.A. Illuminating the "black box": A randomization approach for understanding variable contributions in artificial neural networks. *Ecol. Model.* **2002**, *154*, 135–150. [CrossRef]

32. Bishop, C.M. *Pattern Recognition and Machine Learning*; Springer: New York, NY, USA, 2011.

33. Scrucca, L. GA: A package for genetic algorithms in R. *J. Stat. Softw.* **2013**, *53*, 1–37. [CrossRef]

Article

YOLOv3-Lite: A Lightweight Crack Detection Network for Aircraft Structure Based on Depthwise Separable Convolutions

Yadan Li [1,2], Zhenqi Han [2,*], Haoyu Xu [3], Lizhuang Liu [2,*], Xiaoqiang Li [1] and Keke Zhang [4]

[1] School of Computer Engineering and Science, Shanghai University, Shanghai 200444, China; liyadan@sari.ac.cn (Y.L.); xqli@shu.edu.cn (X.L.)
[2] Shanghai Advanced Research, Chinese Academy of Sciences, Shanghai 201210, China
[3] Lenovo Research, Shanghai Branch, Shanghai 201210, China; xuhaoyu@sina.com
[4] Shanghai Engineering Center for Microsatellites, Shanghai 201210, China; liyadan5201@gmail.com
*   Correspondence: hanzq@sari.ac.cn (Z.H.); liulz@sari.ac.cn (L.L.)

Received: date; Accepted: date; Published: date

Abstract: Due to the high proportion of aircraft faults caused by cracks in aircraft structures, crack inspection in aircraft structures has long played an important role in the aviation industry. The existing approaches, however, are time-consuming or have poor accuracy, given the complex background of aircraft structure images. In order to solve these problems, we propose the YOLOv3-Lite method, which combines depthwise separable convolution, feature pyramids, and YOLOv3. Depthwise separable convolution is employed to design the backbone network for reducing parameters and for extracting crack features effectively. Then, the feature pyramid joins together low-resolution, semantically strong features at a high-resolution for obtaining rich semantics. Finally, YOLOv3 is used for the bounding box regression. YOLOv3-Lite is a fast and accurate crack detection method, which can be used on aircraft structure such as fuselage or engine blades. The result shows that, with almost no loss of detection accuracy, the speed of YOLOv3-Lite is 50% more than that of YOLOv3. It can be concluded that YOLOv3-Lite can reach state-of-the-art performance.

Keywords: depthwise separable convolution; YOLOv3; feature pyramid; aircraft structure crack detection

---

1. Introduction

Identifying crack defects during the inspection of aging aircraft is of vital importance to the safety of the aircraft. The main reason for the China Airlines Flight 611 [1] air crash was the presence of cracks, which had not been completely flattened and became more and more serious.. They eventually caused the body of the planeto disintegrate in mid-air. This paper focuses on the detection of cracks in the aircraft structures, such as their engines or fuselage surface.

The method used at present for aircraft crack detection is visual inspection. This method involves a great amount of human labor, during which technicians must be fully focused in order to accurately find all the damaged areas. With the eruption in development of airlines in recent years, traditional visual inspection methods cannot fulfill the vast demand, as well as the high accuracy requirements, for crack inspection, due to the possibility of missing or false detection caused by human fatigue. Therefore, an automatic and intelligent method which can extract crack information from images or videos is necessary in order to reduce human labor and to help servicing engineers to speed up the inspection process and improve accuracy simultaneously.

Therefore, the study of aircraft crack detection has attracted the interest of many researchers. These works will be illustrated in two aspects. One is the traditional detection method, which depends on special hardware facilities, and the other is the visual detection method, based on deep learning.

The first kind of crack detection methods depends on specialized hardware devices. A resonant ultrasound spectroscopy apparatus was provided for detecting crack-like flaws in components in [2]. Searle et al. applied an aircraft structural health-monitoring system to detect damaged areas in a full-scale aircraft fatigue tests [3]. Kadam et al. [4] used a self-diagnosis technique to detect the cracking and de-bonding of the permanently embedded lead zirconate titanate (PZT). Cracks are reported as key in [5], which use the first-order reliability method (FORM + Fracture) to alleviate the computational cost of probabilistic defect-propagation analysis [6–10], which is commonly used to detect cracks using a special hardware device.

Some algorithms for crack detection are based on deep learning. Recently, deep learning has achieved great improvements in various visual tasks such as object classification [11–13], and object detection [14–16]. Deep learning fundamentally changes the ways to tackle some traditionally hard or intractable visual tasks and has produced many successful applications. Therefore, the application of deep learning in crack detection has emerged. Recent studies have designed deep learning-based crack recognition methods: [17] proposed a 5-layers Convolutional Neural Network (CNN) to detect cracks, while [18] evaluated five CNN architectures for corrosion detection from input images, in all of which large input images were cropped into small images of a fixed size. The CNNs were applied to classify whether cracks or corrosion were contained in each small input image. Ref. [19] proposed an aircraft engine borescope crack detection and segmentation system based on deep learning. Ref. [20] proposed an algorithm which learned hierarchical convolutional features for crack detection, which was actually an edge detection method. Although the crack features were extracted well, their detection scenarios are very simple, so the method could not be applied to the complex scene detection of aircraft structural cracks.

All of the crack detection methods mentioned above did not meet the needs of assistant technicians in order to complete rapid and accurate crack detection in aircraft structures. First, because the crack detection of aircraft structure is typically conducted in an outdoor environment, the detection equipment must be portable. However, the methods mentioned above require specialized professional and complex equipment for crack detection, which is inconvenient in aircraft inspection. Second, aircraft structures, especially the internal background of the engine, are very complex, and the performance of the algorithm in complex environments is very demanding. However, the background of the crack images used in the methods mentioned above is very clean, as is shown in Figure 1. Most of the cracks in other works are similar to those shown in the last two figures. Third, crack detection in aircraft structures needs to be completed accurately in a limited time. So, the offline detection speed of the algorithm needs to be close to real-time and guarantee high accuracy as well. However, when the efficiency of the above methods was improved, the accuracy decreased significantly. These two indicators cannot be taken into account simultaneously by the above methods.

To solve the problems mentioned above, we propose an aircraft structure crack detection algorithm: YOLOv3-Lite. This method is a deep learning-based framework, which can accurately and efficiently detect crack damage on an aircraft structure. Compared with the previous crack detection method, YOLOv3-Lite shows that it can reach state-of-the-art performance. It can run on a mobile device due to its light-weight characteristics. On the premise of guaranteeing a certain accuracy rate, the speed has been greatly increased. In addition, the speed improvement of YOLOv3-Lite refers to the shortening of offline detection time.

The algorithm pipeline can be described by three parts. Firstly, set is divided into the training set, the validation set, and the test set. The data is processed in a form that can be received by YOLOv3-Lite. Secondly, the YOLOv3-Lite architecture is built, which adopt depthwise separable convolution, feature pyramid, and YOLOv3 [21] methods. Additionally, transfer learning is introduced to reduce the required amount of data and yield high accuracy. We evaluate the model by comparing with baseline performance and achieve comparable AP scores and competitive efficiency as well. Finally, the model is applied to test data. The pipeline is shown in Figure 2.

Figure 1. A comparison of aircraft structure cracks and crack images used in other works. (**a**) crack in fuselage, (**b**) crack in engine, (**c**) cracks in the pavement, (**d**) cracks in concrete.

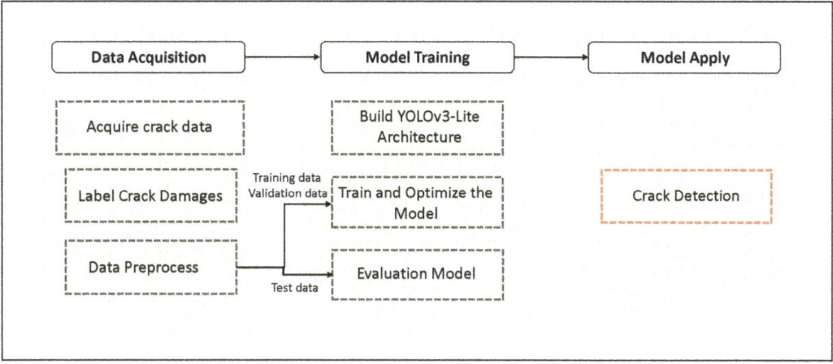

Figure 2. The overall architecture of the proposed YOLOv3-Lite algorithm.

The special contribution of this paper are as follows:

1. The proposed algorithm is a lightweight network with fewer parameters, such that it can be migrated to mobile devices.
2. It is a novel aircraft structure crack detection network which can detect the different parts of an aircraft and can detect different types of an aircraft, which means it has good generalization performance.
3. The crack detection network is combined with depthwise separable convolution and feature pyramids so it is fast and accurate.

The rest of the paper is organized as follows: Section 2 explains the YOLOv3-Lite algorithm. Section 3 shows some experiments using YOLOv3-Lite and discusses the experimental results. Finally, in Section 4, some future works are presented and the paper is concluded.

2. Aircraft Structural Crack Detection Method

In this section, we start with depthwise separable convolution, which is used by YOLOv3-Lite for a backbone network. Then, YOLOv3 is introduced, which is adopted for bounding box regression. Finally, the architecture of the aircraft structural crack detection network is proposed.

2.1. Depthwise Separable Convolution

Deep separable convolution is a convolution method proposed by MobileNet [22], which can greatly reduce the parameters and achieve the same effects as standard convolution. Depthwise separable convolutions are a form of factorized convolutions. They factorize a standard convolution into a depthwise convolution and a 1×1 convolution called a pointwise convolution. In depthwise convolution applies a single filter to each input channel and, then, a 1×1 convolution is applied to combine the outputs of the depthwise convolution by the pointwise convolution: compare this to a standard convolution whose input is convoluted and combined into a new set of outputs in one step the depthwise separable convolution splits this process into two layers, a separate layer for filtering and a separate layer for combining. This factorization has the effect of drastically reducing computation time and model size. The convolution principle of depthwise separable convolution and the standard convolution is shown in Figure 3.

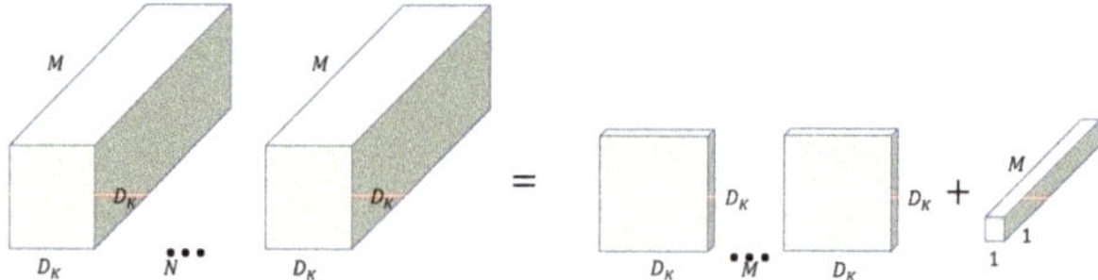

Figure 3. Standard convolution kernels: $N \times D_K \times D_K \times M$ (**left side**) and $M \times D_K \times D_K$ depthwise convelution kernels and $1 \times 1 \times M$ pointwise convolution kernels (**right side**).

A standard convolution layers takes a feature map of $D_F \times D_F \times M$ as input, where D_F is the spatial width and height of a square feature map and M is the number of input channels, which is parameterized by a convolution kernel K of size $D_K \times D_K \times M \times N$ where D_K is the spatial dimension of the kernel, M is the number of input channels, and N is the number of output channels. The output feature map for standard convolution, assuming a stride of one and padding computed as: $G_N = \sum K_{(N,M)} \times F_M$. Thus, the computation cost of the standard convolution is $D_K \cdot D_K \cdot M \cdot N \cdot D_F \cdot D_F$.

According to this formula, we can draw the conclusion that the number of input and output channels, the kernel size, and the feature map size contribute a great deal to the computational cost. However, depthwise separable convolutions break the interaction between the number of output channels and the size of the kernel. It is made up of two layers which are depthwise convolutions d_w and pointwise convolutions p_w. A single filter is applied to each input channel depthwise and then by pointwise convolution, a simple 1×1 convolution, which is conducted to create a linear combination of the output of the depthwise layer. The output feature map for depthwise convolution is computed as:

$$\hat{G}_M = \sum K_{(1,M)} \times F_M. \tag{1}$$

The computation cost of the depthwise separable convolution is $D_K \cdot D_K \cdot M \cdot D_F \cdot D_F$.

By decomposing the convolution process into two steps, the computational complexity is reduced by:

$$\frac{D_K \cdot D_K \cdot M \cdot D_F \cdot D_F + M \cdot N \cdot D_F \cdot D_F}{D_K \cdot D_K \cdot M \cdot N \cdot D_F \cdot D_F} = \frac{1}{N} + \frac{1}{D_K^2}. \tag{2}$$

Therefore, the computational effort of 3×3 depthwise separable convolutions will be reduced by 8 to 9 times, compared to standard convolution.

2.2. YOLOv3

YOLOv3 is a deep convolutional architecture designed for object detection, which uses the darknet as the feature extract network and then using dimension clusters as anchor boxes for predicting bounding boxes of the system. Four co-ordinate (t_x, t_y, t_w, t_h) values for each bounding box and their confidence scores are output from the input image directly through the regression operation, as well as class probabilities. Confidence scores represent the precision of the predicted bounding box when the grid contains an object.

At the training stage, the three feature maps ($13 \times 13, 26 \times 26, 52 \times 52$) output from the feature extract network. Taking the feature map of size 13×13 as an example, the proposed method divides the feature map into 13×13 grids. Each grid takes charge of the object detection in case the ground truth is contained in it. the predictions will be obtained as:

$$
\begin{aligned}
b_x &= \sigma(t_x) + C_x \\
b_y &= \sigma(t_y) + C_y \\
b_w &= p_w e^{t_w} \\
b_h &= p_h e^{t_h},
\end{aligned}
\tag{3}
$$

where the (C_x, C_y) denote that the center of an object is detected in a grid which is offset from the top left corner of the feature map; (p_w, p_h) denotes the width and height of the anchor box prior, respectively; and (t_x, t_y, t_w, t_h) are the four offset co-ordinate predicted by the network. Using a sigmoid to compress t_x and t_y to $[0, 1]$, the target center can be effectively ensured to be in the grid cell executing prediction. A bounding box, with dimension priors and location prediction, is shown in Figure 4.

The loss function of YOLOv3 is composed of three parts: a co-ordinate prediction error (terms 1 and 2), a confidence score (terms 3 and 4) which is the intersection over union (IoU) error, and a classification error (the last term). This loss is defined as follows:

$$
\begin{aligned}
Loss =\;&\lambda_{coord} \sum_{i=0}^{S^2} \sum_{j=0}^{B} 1_{ij}^{obj} [l(x_i, \hat{x}_i) + l(y_i, \hat{y}_i)] \\
&+\lambda_{coord} \sum_{i=0}^{S^2} \sum_{j=0}^{B} 1_{ij}^{obj} [(\sqrt{w_i} - \sqrt{\hat{w}_i})^2 + (\sqrt{h_i} - \sqrt{\hat{h}_i})^2] \\
&+ \sum_{i=0}^{S^2} \sum_{j=0}^{B} 1_{ij}^{obj} l(C_i, \hat{C}_i) \\
&+ \lambda_{noobj} \sum_{i=0}^{S^2} \sum_{j=0}^{B} 1_{ij}^{noobj} l(C_i, \hat{C}_i) \\
&+ \sum_{i=0}^{S^2} 1_{i}^{obj} \sum_{c \in classes} l(p_i(c), \hat{p_i}(c)),
\end{aligned}
\tag{4}
$$

where 1_{ij}^{obj} indicates that the target is detected by the j_{th} bounding box of grid i. In order to increase the loss from bounding box coordinate predictions and decrease the loss for confidence predictions for boxes that do not contain objects, the parameters λ_{coord} and λ_{noobj} are introduced and both set to 5. Then, $\hat{x}_i, \hat{y}_i, \hat{w}_i, \hat{h}_i$ are the predicted bounding box parameters of center co-ordinates and box size. x_i, y_i, w_i, h_i are the actual parameters, $\hat{C}_i$ is the prediction of the confidence score, C_i is the true data; $p_i(c)$ indicates the true value of the probability of the object in grid i belonging to class C; and $\hat{p}_i(c)$ is the predicted value. Except for the box size error, which uses the mean square error, the others use the binary cross-entropy loss $l(a, \hat{a})$ which is defined as follows:

$$l(a, \hat{a}) = -a_i log\hat{a}_i + (1 - a_i)log(1 - \hat{a}_i).$$ (5)

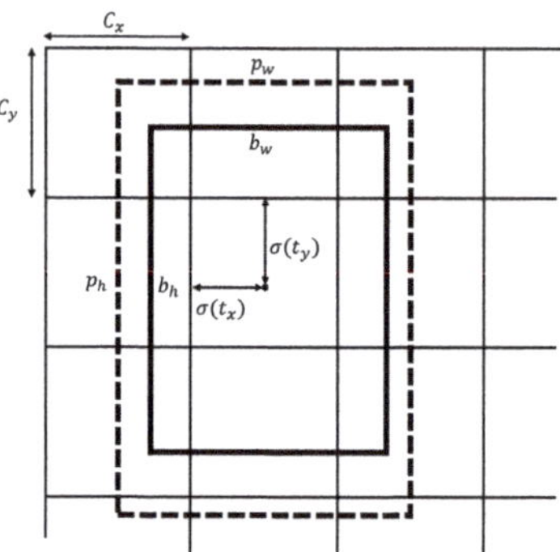

Figure 4. YOLOv3 predicts the width and height of the box as offsets from cluster centroids and predicts the center coordinates of the box relative to the location of filter application using a sigmoid function.

2.3. YOLOv3-Lite

In YOLOv3-Lite, the backbone network is designed, inspired by depthwise separable convolution. As the crack sizes vary greatly (i.e., there are centimeter-scale and decimeter-scale cracks), we extract features from three scales, using a similar concept to the feature pyramid network [23]. In order to realize the fusion of low-resolution and semantically strong features with more effective high-resolution features, we concatenate the feature pyramids. From our base feature extractor, we add one convolutional layer, which can improve the effectiveness of the network, as determined experimentally. The YOLOv3-Lite can greatly reduce the number of parameters and also achieve higher accuracy in portable equipment. The overall architecture of the network is shown in Figure 5.

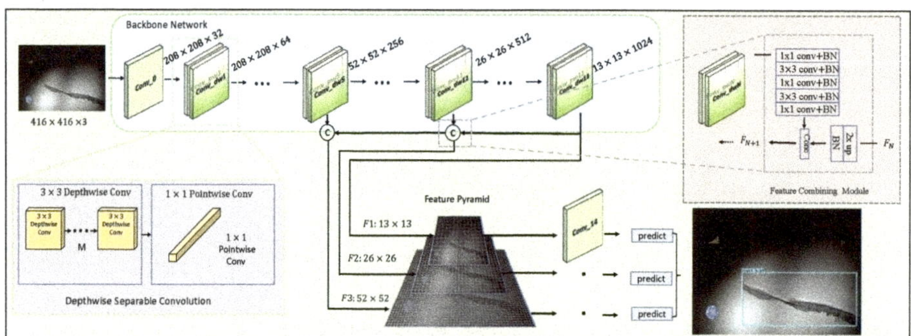

Figure 5. Network structure diagram of the YOLOv3-Lite model. The input is $416 \times 416 \times 3$. Crack featurs are extracted by the backbone network, designed using depthwise separable convolution. The feature pyramid is calculated for detecting cracks at different scales. Finally, the network outputs the detection results.

2.3.1. Backbone Network

Considering the need to migrate the algorithm to mobile devices, the backbone of the detection network is built on depthwise separable convolution to compose the basic convolution unit. The whole feature extraction network structure is defined in Table 1; there are 13 depthwise separable convolutional layers, where each layer contains one depthwise convolution and one pointwise convolution layer. All layers are followed by a batchnorm [24] and ReLU non-linearity layer. Downsampling is handled with stride convolution in the depthwise convolutions, as well as in the first layer.

Table 1. Backbone Architecture.

Type/Stride	Filter Shape	Output Size
Conv/s2	$3 \times 3 \times 3 \times 32$	$208 \times 208 \times 32$
Conv_dw/s1	$3 \times 3 \times 32$	$208 \times 208 \times 32$
Conv_dw/s1	$1 \times 1 \times 32 \times 64$	$208 \times 208 \times 64$
Conv_dw/s2	$3 \times 3 \times 64$	$104 \times 104 \times 64$
Conv_dw/s1	$1 \times 1 \times 64 \times 128$	$104 \times 104 \times 128$
Conv_dw/s1	$3 \times 3 \times 128$	$104 \times 104 \times 128$
Conv_dw/s1	$1 \times 1 \times 128 \times 128$	$104 \times 104 \times 128$
Conv_dw/s2	$3 \times 3 \times 128$	$52 \times 52 \times 128$
Conv_dw/s1	$1 \times 1 \times 128 \times 256$	$52 \times 52 \times 256$
Conv_dw/s1	$3 \times 3 \times 256$	$52 \times 52 \times 256$
Conv_dw/s1	$1 \times 1 \times 256 \times 256$	$52 \times 52 \times 256$
Conv_dw/s2	$3 \times 3 \times 256$	$26 \times 26 \times 256$
Conv_dw/s1	$1 \times 1 \times 256 \times 512$	$26 \times 26 \times 512$
$5\times$ Conv_dw/s1	$3 \times 3 \times 512$	$26 \times 26 \times 512$
Conv_dw/s1	$1 \times 1 \times 512 \times 512$	$26 \times 26 \times 512$
Conv_dw/s2	$3 \times 3 \times 512$	$13 \times 13 \times 512$
Conv_dw/s1	$1 \times 1 \times 512 \times 1024$	$13 \times 13 \times 1024$
Conv_dw/s1	$3 \times 3 \times 1024$	$13 \times 13 \times 1024$
Conv_dw/s1	$1 \times 1 \times 1024 \times 1024$	$13 \times 13 \times 1024$

We designed the backbone network that generates three-layer feature pyramid which contains three scales of feature maps in order to detect cracks of different sizes. The size of the network input image is 416×416, to which the multilayer depthwise separable convolution is applied. The last layer of the network outputs the 13×13 feature map, which is marked as $f1$. Layer 11 concatenates with $f1$ after up-sampling, then outputs the 26×26 feature map, which is marked as $f2$. Finally, the 52×52 feature map is computed by layer 5, concatenated with $f2$ after up-sampling. The concatenated parts are built by the residual network, which can combine low-level information and high-level Semantic Information. In this connection mode, the network can learn the crack features effectively. Finally, the pyramid feature map contains these three layers. Due to the large receptive field of the 13×13 feature map, it can detect large-sized cracks. On the contrary, the 52×52 feature map has smaller receptive field, such that small-sized crack can be detected by this layer.

2.3.2. Bounding Box Prediction

After extracting the features of the cracks, we regress the bounding box for crack detection, based on YOLOv3. There are nine appropriate prior anchor boxes which are clustered with training data labels by a K-means clustering algorithms. Each of the three feature maps has three anchor boxes. As multilayer convolution layers can reduce the resolution of an image, the last layer of the network has the lowest resolution. Thus, small crack features may be lost, while the large crack features still exist. Therefore, the last layer of the network is more sensitive to large cracks. At the same time, its receptive field is also the largest, so three larger anchor boxes are allocated to the 13×13 feature map obtained by the last layers for detecting large cracks. Three medium-size anchor boxes are for the 26×26 feature map and the smaller anchor boxes are for the 52×52 feature map, for detecting small cracks. Four co-ordinate values for each bounding box and their confidence scores output by YOLOv3. As we only have a crack class, the class probability is 1. Confidence scores represent the precision of the predicting bounding box when the grid contains the crack. Only the anchor box which has the highest confidence scores will be selected for regression; the other two anchors will be adopted by the non-maximum suppression algorithm. Then, the selected anchor box will gradually regress to the position and size of ground truth at the training stage.

3. Experimental Results and Analyses

Our experiments are aimed at crack detection in aircraft structures, including in the surface of the fuselage, engine blades, and in other parts. In this section, firstly, we will describe the data set composition and some training settings. Secondly, the effect of YOLOv3-Lite is demonstrated by several experimental results. Thirdly, the detection performance among YOLOv3-Lite, MobileNet-SSD [25], YOLO-Tiny and YOLOv3 will be compared.

3.1. Dataset Composition and Characteristic

Our dataset consists of two parts, one of which is derived from aircraft structures, such as fuselage, wing, aircraft tail, and engine interior. The length of cracks ranges from 1 to more than 10 cm. About 580 images with cracks were collected by ourselves and servicing engineers which contains different cases of structural cracks in real aircraft from our partner aviation companies. The other is from the data of industrial equipment with similar cracks, as shown in Figure 6. The dataset contains 960 pictures in total. We use 800 samples as a training set and use 80 samples as a validation set for selecting a well-generalized performance model. Another 80 samples independent of the training and validation set are used as a test set.

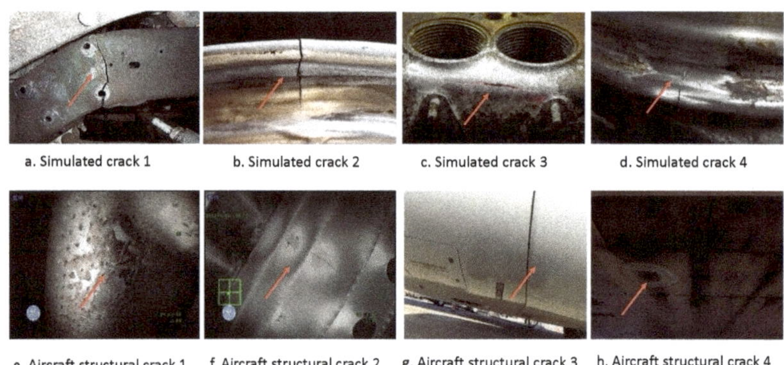

a. Simulated crack 1   b. Simulated crack 2   c. Simulated crack 3   d. Simulated crack 4

e. Aircraft structural crack 1   f. Aircraft structural crack 2   g. Aircraft structural crack 3   h. Aircraft structural crack 4

Figure 6. Different industrial equipments with cracks which have background interferences (**a–d**), aircraft structures with cracks (**e–h**).

As shown in Figure 6, the cracks on the fuselage and wings of an aircraft are small and light in color, so the difference between the cracks and the background is small. Meanwhile, strong illumination also increases the difficulty of crack detection in these areas. It is more difficult to detect cracks because of the complex internal structure and background of the aircraft engine.

3.2. Training Methodology

We implement YOLOv3-Lite using the Tensorflow 1.7.0 [26] and Python 3.7 running on an NVIDIA Tesla K20 GPU in the programming environment Linux 16.04. In YOLOv3-Lite, batch normalization is used after each convolutional layer, in both the deep separable convolution and in several standard convolution layers, which is used to speed up convergence in the training process. Firstly, we use the mechanism of transfer learning to train the network. For the backbone network, we adopt the parameters pre-trained in ImageNet. In fine-tuning, the initial learning rate is set to $1e-3$ and is divided by 10 after every epoch. The Adam [27] is employed to update the network parameters, with a batch size of 10 in each iteration. We train the network with 300 epochs in total. Secondly, all layers are unfrozen to train in detail for 50 epochs with a batch size of 4. The initial learning rate is set to $1e-4$ and the decay rate is the same as the fine-tuning stage. Adam is also used in this phase. The initialization parameters of the proposed model are shown in Tables 2 and 3. Using a pre-training weight training network can greatly reduce training time and experimental resources and can converge faster. Then we can adjust the weights of the whole network, through training,to make the network model more suitable in the context of aircraft structure crack detection.

Table 2. The initial parameters of YOLOv3-Lite in stage 1 of training.

Size of Input Images	Batch Size	Initial Learning Rate	Decay	Training Steps
416×416	10	0.001	0.1	300

Table 3. The initial parameters of YOLOv3-Lite in stage 2 of training.

Size of Input Images	Batch Size	Initial Learning Rate	Decay	Training Steps
416×416	4	0.0001	0.1	50

3.3. Evaluation Metrics

For each image, the intersection over union (IoU) between the bounding box of the detected crack and ground truth can be calculated as: $I_{oU} = \frac{A_o}{A_u}$, where I_{oU} is the intersection over union, A_o is the area of overlap, and A_u is the area of union.

When the IoU of the predicted bounding box and ground truth is greater than a certain threshold value (e.g., 0.5), it is considered to be a true positive; otherwise, it is a false positive. A false negative is obtained by missing a crack. Then we can calculate the precision and recall. Finally, average precision (AP) which is equivalent to the area under the precision-recall curve [28] can be computed. In addition, in order to measure the efficiency of the algorithm, the offline detection time for one picture is also an important evaluation index. We compare the detection time (i.e., the offline time) and analyze the accuracy performance of several models.

3.4. Detection Performance of YOLOv3-Lite

Depthwise separable convolution is used to achieve the goal of the lightweight network. At the same time, the feature pyramid method is used to detect crack defects of different sizes. In order for each layer of the feature pyramid to fuse the information of the high-level and low-level layers, the feature pyramids are concatenated. The experiments show that YOLOv3-Lite can achieve high accuracy and greatly improve the detection speed. The input size of the image for our aircraft structural crack detection network is adjusted to 416×416 pixels, in order to increase the model performance. The image is detected by YOLOv3-Lite in 0.1s with an AP of 38.7%. Finally, the network outputs the detection result. Our method can quickly and accurately detect cracks in various parts of the aircraft, such as the surface of the aircraft fuselage, engine blades, aircraft tires, wing tail, and other parts. The detection results are shown in Figure 7. Whether it is obvious large-sized cracks or small-sized cracks which are difficult to find with a nosisy background, they can be precisely detected by YOLOv3-Lite.

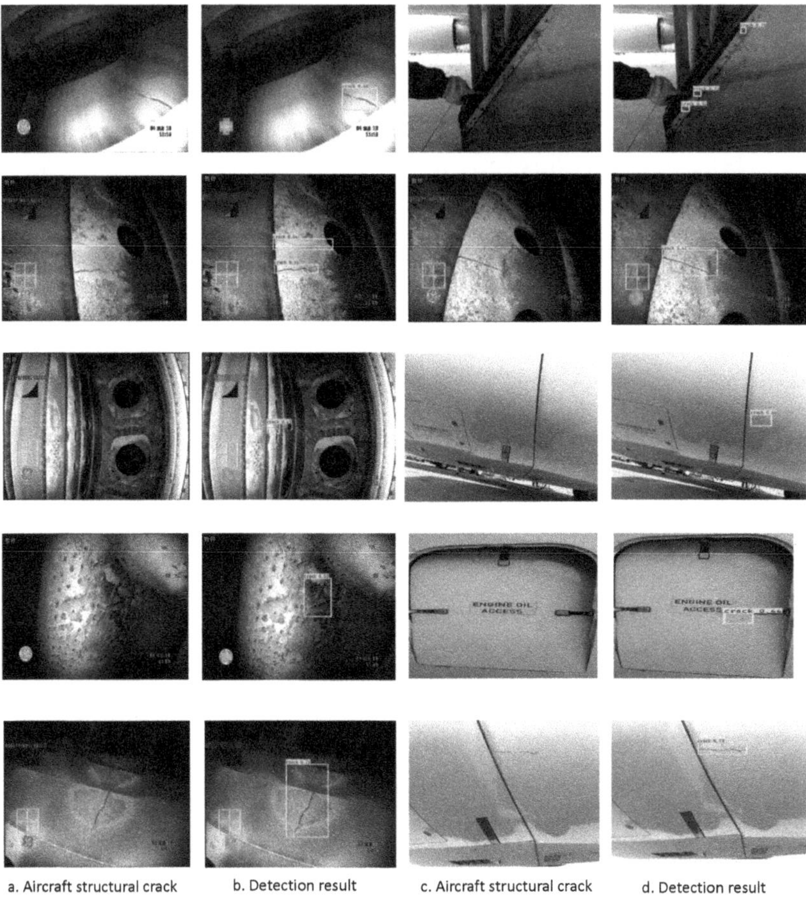

a. Aircraft structural crack   b. Detection result   c. Aircraft structural crack   d. Detection result

Figure 7. The experimental results of YOLOv3-Lite.

3.5. Comparison of YOLOv3-Lite with Three Modern Methods

To better verify the superiority of the improved method, we compare YOLOv3-Lite with YOLOv3, YOLO-Tiny and MobileNet-SSD. These models are evaluated in terms of detection of average precision (AP) and offline detection time. The results of the contraction experiment are shown in Table 4. The parameter quantity of YOLOv3-Lite is about 31 million, YOLOv3 is twice that of ours, at about 61 million, and YOLO-Tiny is minimum. The number of parameters of SSD-mobilenet is roughly the same as ours. However, the parameters of YOLOv3-Lite are small, the detection speed is faster than most of the algorithms, and its average precision is the highest. We adopt deep separable convolution, which reduces the number of parameters by 8–9 times, on average, with an effect similar to that of standard convolution. Therefore, the detection speed of YOLOv3-Lite is 50% faster than that of YOLOv3. As YOLO-Tiny only uses 13 standard convolution layers, its feature extraction ability is poor and AP is very low. Although we have fewer network layers, the proposed method is still much better than MobileNet-SSD as we use the way of connecting feature maps at different levels, each layer of feature pyramid containing not only high-level semantic information, but also low-level information. Therefore, YOLOv3-Lite not only has high detection accuracy but also achieves the goal of being lightweight, which greatly improves the network detection speed.

Table 4. Average precision, offline detection time and the number of parameters of each network.

	AP	Time (s)	The Number of Parameters (Million)
YOLOv3-Lite	38.7%	0.125	31
SSD-Mobilenet	17.1%	0.128	31
YOLOv3	43.1%	0.225	61
YOLO-Tiny	2.5%	0.09	9

Although the detection time of YOLO-Tiny is the shortest, its accuracy is too low to detect cracks at all. The accuracy of YOLOv3 is the highest, but its detection speed is 50% lower than YOLOv3-Lite, so it is not a good choice. We compare the detection results of YOLOv3-Lite and SSD-Mobilenet, which are shown in Figure 8. The bounding boxes of YOLOv3-Lite can accurately frame the location of cracks, while SSD-Mobilenet not only shows a large error in the location of the boxes, but also has a very low confidence level. In addition, there are many cracks that can not be detected by the SSD-Mobilenet method. Therefore, YOLOv3-Lite performs best in these modern methods.

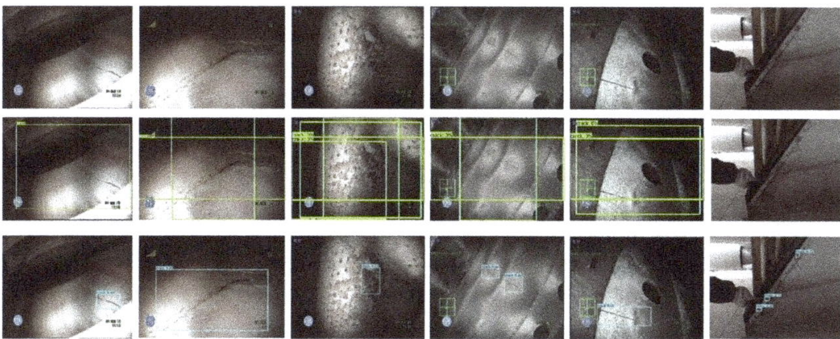

Figure 8. The first line is pictures of the original aircraft structures with cracks. The second line is the detection results of SSD-Mobilenet. The last line is the detection results of YOLOv3-Lite.

4. Conclusions

Every year, there are aircraft faults caused by crack defects. Thus, the research on crack inspection in aircraft structures is of far-reaching significance. However, The current related methods either depend on specialized hardware devices or adopt edge detection methods, which can not effectively resist the influence of background noise. We proposed the YOLOv3-Lite method to address these problems:

- We use deep separable convolution to design a feature extraction network. Using depthwise convolution and 1×1 pointwise convolution, instead of standard convolution, reduced lots of parameters.
- We adopt the idea of a feature pyramid network which combines low- and high-resolution information. This feature pyramid has rich semantic information at all levels and can be built quickly from a single input image scale.
- We use YOLOv3 for bounding box regression. The results show that the offline detection speed of YOLOv3-Lite is 50% faster than YOLOv3, and the detection accuracy and speed are better than SSD-MobileNet and YOLO-Tiny.

Therefore, YOLOv3-Lite is a light-weight, fast (50% faster), and accurate (38.7%) crack detection network, which shows that it can reach state-of-the-art performance. In order to extend YOLOv3-Lite to a general algorithm, our future work will aim optimize YOLOv3-Lite and apply it to more scenarios.

Author Contributions: Project administration, Y.L.; Validation, Y.L.; investigation, Y.L., Z.H. and X.L.; resources, Y.L., K.Z., L.L. and H.X.; visualization, Y.L.

Funding: This research was funded by Science and Technology Commission of Shanghai Municipality grant number 17511106400,17511106402 and Aerospace System Department grant number 30508020301.

Acknowledgments: Thanks for part of data collecting and labeling works by Chong Wang of our research group.

Conflicts of Interest: The authors declare no conflict of interest. The founding sponsors had no role in the design of the study; in the collection, analysis, or interpretation of data; in the writing of the manuscript, and in the decision to publish the results.

References

1. China Airlines Flight Cl-611 Crash Report Released. Available online: http://www.iasa.com.au/folders/Safety_Issues/FAA_Inaction/CI-611finalreport.html (accessed on 30 September 2010).
2. Migliori, A.; Bell, T.M.; Rhodes, G.W. Crack Detection Using Resonant Ultrasound Spectroscopy. U.S. Patent 5,351,543, 4 October 1994.
3. Searle, I.R.; Ziola, S.M.; Seidel, B. Crack detection on a full-scale aircraft fatigue test. In *Smart Structures and Materials 1997: Smart Sensing, Processing, and Instrumentation*; International Society for Optics and Photonics: San Diego, CA, USA, 1997; Volume 3042, pp. 267–277.
4. An, Y.K.; Kim, M.K.; Sohn, H. Airplane hot spot monitoring using integrated impedance and guided wave measurements. *Struct. Control Health Monit.* **2012**, *19*, 592–604. [CrossRef]
5. Coro, A.; Macareno, L.M.; Aguirrebeitia, J.; López de Lacalle, L.N. A Methodology to Evaluate the Reliability Impact of the Replacement of Welded Components by Additive Manufacturing Spare Parts. *Metals* **2019**, *9*, 932. [CrossRef]
6. Zhang, D.; Han, X.; Newaz, G. Sonic IR Crack Detection of Aircraft Turbine Engine Blades with Multi-Frequency Ultrasound Excitations. In *AIP Conference Proceedings*; AIP: College Park, MD, USA, 2014; Volume 1581, pp. 1644–1651.
7. Shin, H.J.; Lee, J.R. Development of a long-range multi-area scanning ultrasonic propagation imaging system built into a hangar and its application on an actual aircraft. *Struct. Health Monit.* **2017**, *16*, 97–111. [CrossRef]
8. Wang, X.; Wong, B.S.; Tan, C.; Tui, C.G. Automated crack detection for digital radiography aircraft wing inspection. *Res. Nondestruct. Eval.* **2011**, *22*, 105–127. [CrossRef]
9. Coro, A.; Abasolo, M.; Aguirrebeitia, J.; Lopez de Lacalle, L. Inspection scheduling based on reliability updating of gas turbine welded structures. *Adv. Mech. Eng.* **2019**, *11*, 1687814018819285. [CrossRef]
10. Llanos, I.; Aurrekoetxea, M.; Agirre, A.; de Lacalle, L.L.; Zelaieta, O. On-machine Characterization of Bulk Residual Stresses on Machining Blanks. *Procedia CIRP* **2019**, *82*, 406–410. [CrossRef]
11. Szegedy, C.; Ioffe, S.; Vanhoucke, V.; Alemi, A.A. Inception-v4, inception-resnet and the impact of residual connections on learning. In Proceedings of the Thirty-First AAAI Conference on Artificial Intelligence, San Francisco, CA, USA, 4–10 February 2017.
12. Iandola, F.; Moskewicz, M.; Karayev, S.; Girshick, R.; Darrell, T.; Keutzer, K. Densenet: Implementing efficient convnet descriptor pyramids. *arXiv* **2014**, arXiv:1404.1869.
13. Sun, Y.; Liang, D.; Wang, X.; Tang, X. Deepid3: Face recognition with very deep neural networks. *arXiv* **2015**, arXiv:1502.00873.
14. Girshick, R. Fast r-cnn. In Proceedings of the IEEE International Conference on Computer Vision, Araucano Park, Las Condes, Chile, 11–18 December 2015; pp. 1440–1448.
15. Ren, S.; He, K.; Girshick, R.; Sun, J. Faster r-cnn: Towards real-time object detection with region proposal networks. In Proceedings of the Advances in Neural Information Processing Systems, Montreal, QC, Canada, 7–12 December 2015; pp. 91–99.
16. Redmon, J.; Divvala, S.; Girshick, R.; Farhadi, A. You only look once: Unified, real-time object detection. In Proceedings of the IEEE Conference on Computer Vision and Pattern Recognition, Las Vegas, NV, USA, 26 June–1 July 2016; pp. 779–788.

17. Cha, Y.J.; Choi, W.; Büyüköztürk, O. Deep learning-based crack damage detection using convolutional neural networks. *Comput. Aided Civ. Infrastruct. Eng.* **2017**, *32*, 361–378. [CrossRef]

18. Atha, D.J.; Jahanshahi, M.R. Evaluation of deep learning approaches based on convolutional neural networks for corrosion detection. *Struct. Health Monit.* **2018**, *17*, 1110–1128. [CrossRef]

19. Shen, Z.; Wan, X.; Ye, F.; Guan, X.; Liu, S. Deep Learning based Framework for Automatic Damage Detection in Aircraft Engine Borescope Inspection. In Proceedings of the 2019 International Conference on Computing, Networking and Communications (ICNC), Honolulu, HI, USA, 18–21 February 2019; pp. 1005–1010.

20. Zou, Q.; Zhang, Z.; Li, Q.; Qi, X.; Wang, Q.; Wang, S. Deepcrack: Learning hierarchical convolutional features for crack detection. *IEEE Trans. Image Process.* **2018**, *28*, 1498–1512. [CrossRef] [PubMed]

21. Redmon, J.; Farhadi, A. Yolov3: An incremental improvement. *arXiv* **2018**, arXiv:1804.02767.

22. Howard, A.G.; Zhu, M.; Chen, B.; Kalenichenko, D.; Wang, W.; Weyand, T.; Andreetto, M.; Adam, H. Mobilenets: Efficient convolutional neural networks for mobile vision applications. *arXiv* **2017**, arXiv:1704.04861.

23. Lin, T.Y.; Dollár, P.; Girshick, R.; He, K.; Hariharan, B.; Belongie, S. Feature pyramid networks for object detection. In Proceedings of the IEEE Conference on Computer Vision and Pattern Recognition, Honolulu, HI, USA, 21–26 July 2017; pp. 2117–2125.

24. Ioffe, S.; Szegedy, C. Batch normalization: Accelerating deep network training by reducing internal covariate shift. *arXiv* **2015**, arXiv:1502.03167.

25. Li, Y.; Huang, H.; Xie, Q.; Yao, L.; Chen, Q. Research on a surface defect detection algorithm based on MobileNet-SSD. *Appl. Sci.* **2018**, *8*, 1678. [CrossRef]

26. Abadi, M.; Barham, P.; Chen, J.; Chen, Z.; Davis, A.; Dean, J.; Devin, M.; Ghemawat, S.; Irving, G.; Isard, M.; et al. Tensorflow: A system for large-scale machine learning. In Proceedings of the 12th {USENIX} Symposium on Operating Systems Design and Implementation ({OSDI} 16), Savannah, GA, USA, 2–4 November 2016; pp. 265–283.

27. Kingma, D.P.; Ba, J. Adam: A method for stochastic optimization. *arXiv* **2014**, arXiv:1412.6980.

28. Dollár, P.; Zitnick, C.L. Fast edge detection using structured forests. *IEEE Trans. Pattern Anal. Mach. Intell.* **2014**, *37*, 1558–1570. [CrossRef] [PubMed]

Article

Development of Operator Theory in the Capacity Adjustment of Job Shop Manufacturing Systems

Ping Liu [1,*], Qiang Zhang [1] and Jürgen Pannek [2,*]

[1] International Graduate School for Dynamics in Logistics, Faculty of Production Engineering, University of Bremen, 28359 Bremen, Germany; zha@biba.uni-bremen.de
[2] BIBA Bremer Institut für Produktion und Logistik GmbH, Faculty of Production Engineering, University of Bremen, 28359 Bremen, Germany
* Correspondence: liu@biba.uni-bremen.de (P.L.); pan@biba.uni-bremen.de (J.P.)

Received: 30 April 2019; Accepted: 27 May 2019; Published: 31 May 2019

Abstract: With the development of industrial manufacture in the context of Industry 4.0, various advanced technologies have been designed, such as reconfigurable machine tools (RMT). However, the potential of the latter still needs to be developed. In this paper, the integration of RMTs was investigated in the capacity adjustment of job shop manufacturing systems, which offer high flexibility to produce a variety of products with small lot sizes. In order to assist manufacturers in dealing with demand fluctuations and ensure the work-in-process (WIP) of each workstation is on a predefined level, an operator-based robust right coprime factorization (RRCF) approach is proposed to improve the capacity adjustment process. Moreover, numerical simulation results of a four-workstation three-product job shop system are presented, where the classical proportional–integral–derivative (PID) control method is considered as a benchmark to evaluate the effectiveness of RRCF in the simulation. The simulation results present the practical stability and robustness of these two control systems for various reconfiguration and transportation delays and disturbances. This indicates that the proposed capacity control approach by integrating RMTs with RRCF is effective in dealing with bottlenecks and volatile customer demands.

Keywords: capacity control; job shop systems; RMTs; operator theory

---

1. Introduction

As customer demand is changing quickly, production and logistic systems become more complex and dynamic. Moreover, various delays, stochastic disturbances, and bottlenecks in the production system induce additional challenges for manufacturers. To deal with these problems, numerous advanced technologies, such as reconfigurable machine tools (RMT), Internet of Things (IoT), radio-frequency identification (RFID), and cyber-physical systems (CPS), have been proposed for a more automatic, accurate, and reliable monitoring and controllable manufacturing system [1–3]. This displays that the manufacturing industry is on its way toward a new generation, named Industry 4.0 [4]. However, to use these technologies efficiently, effective integration plays a vital role, which has attracted researchers' increasing attention. Capacity adjustment is one such approach for manufacturing systems, especially for job shop systems, to deal with volatile customer demands. This paper focuses on developing an effective capacity control strategy for the job shop system by means of RMTs.

A job shop system is one production mode with flexible producing paths for a variety of products. Compared with traditional flow shop manufacturing systems with fixed producing paths, job shop systems have the advantage of being able to satisfy changing demand. However, they also have several drawbacks, such as high work-in-process (WIP) levels, high cost, and low productivity [5]. In the literature, some researchers applied scheduling methods to deal with these problems. For example, Wang and Rosenshine [6] discussed the scheduling for combination of made-to-stock and made-to-order jobs in a job shop environment considering mean flowtime. In [7], simulation-based multiobjective optimization as a hyper-heuristic was utilized in the automatic design of scheduling rules for complex job shop systems. Further, other algorithms, such as tabu search [8], genetic algorithm (GA) [9], genetic programming based hyper-heuristic approach (GA-HH) [10], and particle swarm optimization (PSO) [11], were applied to the flexible job shop scheduling problem. The aim of these works was to improve productivity as well as to minimize cost and completion time. Another possibility to solve these problems is capacity adjustment in short- or mid-term horizons [12]. Related work is introduced in the next section.

1.1. Related Research

The methods of capacity adjustment can be classified into two types: Labor-oriented approaches and machinery-oriented approaches. Labor-oriented approaches mainly modify capacity through adjusting the working time. In contrast to these, machinery-oriented approaches adjust capacity through the flexibility of machines. Reconfigurable machine tools (RMT), as one advanced piece of technology of Industry 4.0, provide a new opportunity for machinery-oriented capacity adjustment, which cannot be achieved using classical dedicated machine tools (DMT) only. In [12], RMTs were used in harmonizing the throughput-time capacity control approach to plan the delivery dates and analyze the inventory range of each workstation considering reconfiguration delays. The proportional–integral–differential (PID) control method [13,14] was applied to adjust the capacity, and a mathematical model of job shop systems was developed, including the new application degree of freedom of RMTs. Furthermore, the respective model was continuously extended, including WIP and planned WIP level of each workstation, and a model predictive control (MPC) approach was applied considering time-varying input orders [15]. However, job shop systems are not single-input–single-output (SISO) systems but instead show nonlinear dynamics as well as a multi-input–multi-output (MIMO) structure with strong coupling between the workstations (subsystems). Additionally, these systems also suffer from various disturbances and delays, which are unaccounted for in the literature. Operator-based robust right coprime factorization (RRCF) [16] represents one option to deal with these issues.

Operator theory is an advanced control theoretical approach to effectively control and analyze the dynamic and stability performances of a class of nonlinear systems [17]. Moreover, it is accessible to ensure the stability of nonlinear systems using the Bezout identity [18]. The robust stability and traceability of the nonlinear affine systems with unknown bounded disturbances were analyzed in [19]. The design of robust and tracking controllers of nonlinear feedback systems was considered in [20–23]. Additionally, this method has been applied to deal with complex problems, such as disturbances, delays, couplings, and so on. In [24], this method was used to deal with internal and external disturbances to ensure robust stability. Deng et al. [25,26] analyzed the delays and faults detection in a thermal process control system using operator theory. In [27], a tracking operator was designed to cope with unknown time-varying delays in nonlinear uncertain systems. Bi and Deng [28,29] solved the coupling problem in MIMO systems and designed robust controllers for MIMO systems. The literature illustrates the effectiveness of the RRCF approach in dealing with nonlinearities, disturbances, delays, and couplings, which occur in job shop systems.

1.2. Research Work

In this paper, the RRCF approach is utilized to design controllers for the capacity adjustment of job shop systems with RMTs and analyze the dynamic and stability performance of the system. In the design of the capacity control system, the WIP level is one key performance indicator, which highly influences cost, throughput, and delivery date reliability [30]. Therefore, our aim was to ensure that the WIP of each workstation is on a planned level. However, the complex material flow, couplings, various delays, and disturbances, including transportation delays between workstations, reconfiguration delays of RMTs, and stochastic demands, make this problem more difficult, especially for large-scale companies. To deal with these challenges, a decoupling controller based on operator theory was applied to transform the complex MIMO system into multiple SISO systems to decrease the complexity for a quick response and less involvement with other workstations [30]. Then, the capacity of each SISO system (workstation) was controlled to ensure the WIP on a planned level. Additionally, to verify the effectiveness of the RRCF control system, PID was considered as the benchmark.

The structure of the paper is organized as follows: In Section 2, the mathematical preliminaries aew introduced. Then, the model of the job shop system with RMTs ia described in Section 3. Thereafter, Section 4 concentrates on the capacity control design, and simulation results are shown in Section 5. Finally, the conclusions and an outlook are provided in Section 6.

2. Mathematical Preliminaries

In this paper, general nonlinear input–output systems of the form:

$$P : U \to Y, \tag{1}$$

are considered, where the input and output spaces U and Y are two normed linear spaces over the field of complex numbers, endowed, respectively, with norms $\| \cdot \|_U$ and $\| \cdot \|_Y$. The set of all (nonlinear) operators is denoted by $\mathcal{N}(U, Y)$ and call $\mathcal{D}(P)$ and $\mathcal{R}(P)$ the domain and range of P. A (semi)-norm on (a subset of) $\mathcal{N}(D_s, Y)$ is defined via:

$$\|P\| := \sup_{x, \tilde{x} \in D_s \& x \neq \tilde{x}} \frac{\|P(x) - P(\tilde{x})\|_Y}{\|x - \tilde{x}\|_U}.$$

Given such a system, our aim is to show the stability of the system, which is formally defined as follows:

Definition 1 (Finite-Gain Input–Output Stability). *An operator $P \in \mathcal{N}(U_s, Y_s)$ with $U_s \subseteq U$ and $Y_s \subseteq Y$ is called finite-gain input–output stable if:*

1. *it is input–output stable, i.e., $P(U_s) \subseteq Y_s$, and if;*
2. *the norm $\|P\|$ is well defined and finite, i.e., $\|P\| < \infty$.*

Here, U_s and Y_s are called the stable input subspace and the stable output subspace of the operator P, respectively. Moreover, an operator P is called causal, stabilizable or unimodular if:

1. for the projection: (causal)

$$Q_T(x(t)) = \begin{cases} x(t), & 0 \leq t \leq T \\ 0, & T \leq t \leq \infty \end{cases},$$

it has $Q_T \circ P \circ Q_T = Q_T \circ P$ for all $x(t) \in U$ and all $T \in [0, \infty)$;
2. there exists an operator $Q : \mathcal{D}(Q) \to \mathcal{D}(Q)$ such that $P \circ Q$ is input–output stable; (stabilizable)
3. P is stabilizable and $P^{-1} \in \mathcal{N}(Y_s, U_s)$. (unimodular)

These three properties allow us to introduce our main tool to show finite-gain input–output stability.

Definition 2 (Right Coprime Factorization (RCF)). *Let $P : \mathcal{D}(P) \to \mathcal{R}(P)$ be a causal and stabilizable operator. We say that P has a right coprime factorization illustrated in Figure 1, if there exist finite-gain input–output stable and causal operators $D : \mathcal{D}(P) \to \mathcal{D}(P)$, $N : \mathcal{D}(P) \to \mathcal{R}(P)$ as well as $A : \mathcal{R}(N) \to \mathcal{D}(P)$ and $B : \mathcal{R}(D) \to \mathcal{D}(P)$ such that:*

1. *D is causal, invertible and $P = N \circ D^{-1}$ holds on $\mathcal{D}(P)$, and:*

2.   *for the unimodular operator $M : \mathcal{D}(P) \to \mathcal{D}(P)$, the Bezout identity is:*

$$A \circ N + B \circ D = M. \tag{2}$$

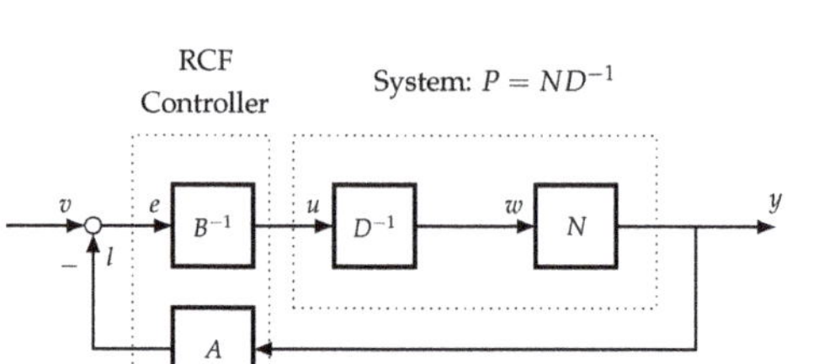

Figure 1. Nonlinear feedback control system.

Here, y, w, and u represent the output, quasi-state, and input signal respectively, cf. [16] for details. v is the reference input, l is the feedback state, and e is the error between v and l. Then, the latter definition allows us to convert the control system (1) to a dynamical system.

Theorem 1. *If a causal and stabilizable operator $P : \mathcal{D}(P) \to \mathcal{R}(P)$ has right coprime factorization, then the respective closed loop is finite-gain input–output stable. Moreover, for any reference v, the closed loop simplifies to $y = N \circ M^{-1}(v)$.*

Proof. As P has a right coprime factorization, Figure 1 can be utilized to obtain $l = A \circ N(w)$ and $e = B \circ D(w)$. Therefore, we have $v = l + e = (A \circ N + B \circ D)(w)$. Again, by the right coprime factorization property, the Bezout identity (2) can be applied to obtain $v = M(w)$ and $w = M^{-1}(v)$.

Hence, by $y = N(w) = N \circ M^{-1}(v)$, cf. Figure 2, the second assertion follows.

Regarding finite-gain input–output stability, unimodularity of M and finite-gain input–output stability of N are first utilized to obtain:

$$P(U_s) = N \circ M^{-1}(U_s) \subseteq N(U_s) \subseteq Y_s,$$

which shows input–output stability. Similarly, we obtain by unimodularity of M:

$$\|P\| = \|N \circ M^{-1}\| := \sup_{x,\tilde{x} \in U_s \& x \neq \tilde{x}} \frac{\|N \circ M^{-1}(x) - N \circ M^{-1}(\tilde{x})\|_Y}{\|x - \tilde{x}\|_U}$$
$$= \sup_{x,\tilde{x} \in U_s \& x \neq \tilde{x}} \frac{\|N(x) - N(\tilde{x})\|_Y}{\|M(x) - M(\tilde{x})\|_U}.$$

Now, finite-gain input–output stability of N is utilized to conclude $\|P\| < \infty$, which completes the proof. $\square$

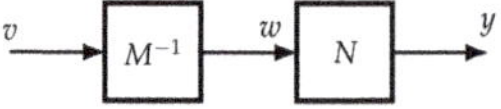

Figure 2. Input–output system equivalent to Figure 1.

In order to include model uncertainties, the mapping P is modified respectively, i.e., an unknown but bounded operator $\triangle N$ in parallel is integrated to N.

Definition 3 (Robust Right Coprime Factorization (RRCF)). *Consider $P : \mathcal{D}(P) \to \mathcal{R}(P)$ to be a causal and stabilizable operator with right coprime factorization and suppose a bounded model disturbance to act as shown in Figure 3. Then, P has robust right coprime factorization if the two operators A and B satisfy the Bezout identity $A \circ (N + \triangle N) + B \circ D = \tilde{M}$, where $\tilde{M}$ is a unimodular operator.*

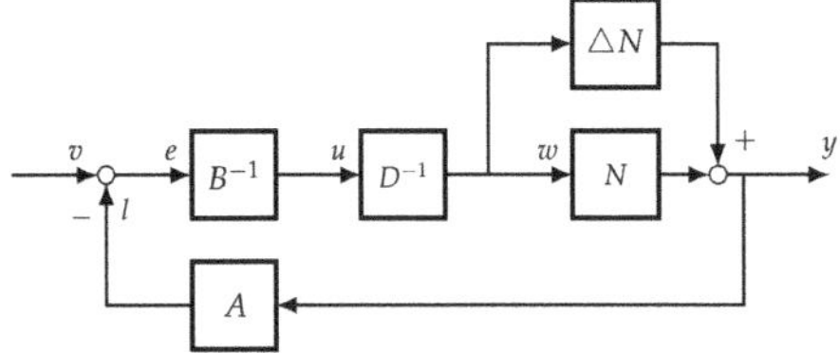

Figure 3. Nonlinear feedback system with disturbances.

Similar to Theorem 1, the closed loop in Figure 3 can be simplified to a dynamical system, cf. Figure 4.

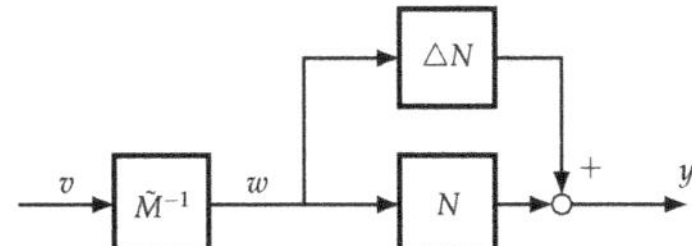

Figure 4. Equivalent of Figure 3.

Corollary 1. *If a causal and stabilizable operator $P : \mathcal{D}(P) \to \mathcal{R}(P)$ has robust right coprime factorization, then the respective closed loop is finite-gain input–output stable. Moreover, for any reference v, the closed loop simplifies to $y = (N + \triangle N) \circ \tilde{M}^{-1}(v)$, with $\tilde{M}$ from Definition 3.*

Proof. Completely analog to the proof of Theorem 1, where N is replaced by $(N + \triangle N)$. $\square$

Using these definitions reveals an effective approach to control and analyze the stability and performance of a class of nonlinear control systems, which include job shop systems. Before designing a respective controller, the model of such a system is shortly recapped.

3. Mathematical Model

Job shop manufacturing systems have a high flexibility and can produce numerous types of products, as these may flow flexibly between all workstations. The variables in the model are given in Table 1. In general, for a job shop system with n workstations, the simplified model of the jth ($j = 1, \ldots, n$) workstation is illustrated in Figure 5. The input rate of the workstation is the sum of output rate from all workstations, including the workstation itself and a possible initial stage. The output rate of the workstation is the current capacity.

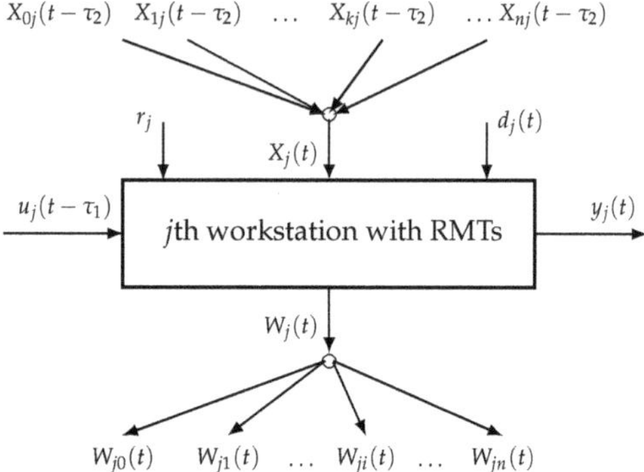

Figure 5. Model of jth workstation with reconfigurable machine tools (RMTs).

Table 1. Variables within the workstation with RMTs.

Variable	Description
$X_{kj}(t)$	Orders input rate from workstation k to workstation j for $k, j \in \{0, \ldots, n\}$
$X_j(t)$	Orders input rate of workstation $j \in \{1, \ldots, n\}$
$W_{jk}(t)$	Orders output rate of workstation j to workstation k for $j, k \in \{1, \ldots, n\}$
$W_j(t)$	Orders output rate of workstation $j \in \{1, \ldots, n\}$
$u_j(t)$	Input signal of workstation $j \in \{1, \ldots, n\}$, which is equal to the number of RMTs
$y_j(t)$	Output signal of workstation $j \in \{1, \ldots, n\}$, which is the WIP level of the workstation
$c_j(t)$	Current capacity of workstation $j \in \{1, \ldots, n\}$
$\bar{c}_j(t)$	Maximum capacity of workstation $j \in \{1, \ldots, n\}$
p_{jk}	Flow probability that the output orders from workstation j to workstation k for $j, k \in \{0, \ldots, n\}$
n^{RMT}	Number of RMTs in the system
n_j^{RMT}	Number of RMTs in workstation $j \in \{1, \ldots, n\}$
n_j^{DMT}	Number of DMTs in workstation $j \in \{1, \ldots, n\}$
v_j^{DMT}	Production rate of DMTs in workstation $j \in \{1, \ldots, n\}$
v_j^{RMT}	Production rate of RMTs in workstation $j \in \{1, \ldots, n\}$
$d_j(t)$	Disturbances in workstation $j \in \{1, \ldots, n\}$
τ_1	Reconfiguration delay
τ_2	Transportation delay
n	Number of workstations in the system

The workstation receives orders from the initial stage ($k = 0$) and workstations $k \in \{1, \ldots, n\}$ and delivers its products to a final stage ($i = 0$) and workstation $i \in \{1, \ldots, n\}$. Thus, for the jth workstation, considering the transportation delay τ_2 between each workstations, the orders input rate is:

$$X_j(t) = \sum_{k=0}^{n} X_{kj}(t - \tau_2) = X_{0j}(t - \tau_2) + \sum_{k=1}^{n} W_{kj}(t - \tau_2), \tag{3}$$

where $X_{0j}(t)$ is the order input from the initial stage. The order output rate of the jth workstation is given by:

$$W_j(t) = \sum_{i=0}^{n} p_{ji} \cdot W_j(t) = \sum_{i=0}^{n} W_{ji}(t), \tag{4}$$

where the flow probabilities satisfy $\sum_{i=0}^{n} p_{ji} = 1$ for all $j \in \{1, \ldots, n\}$. The current WIP as the output signal of the jth workstation is the integral difference between the order input and output rate plus disturbances over time:

$$y_j(t) = y_j(0) + \int_0^t (X_j(\tau) + d_j(\tau) - W_j(\tau))d\tau. \tag{5}$$

The latter variable is of particular importance as for a high level of WIP. The order output rate is equal to the capacity of the workstation, that is:

$$W_j(t) = c_j(t) = \sum_{i=0}^{n} p_{ji} c_j(t).$$

In order to reflect the functionality of RMTs for capacity adjustment, in [31], an extended model of a job shop system with DMTs and RMTs was proposed, cf. Figure 5 for a sketch of workstation j with an assigned number of RMTs. Due to the high productivity of DMTs, this kind of machines will also be adopted in the system. The overall system includes n^{RMT} reconfigurable and n^{DMT} dedicated machine tools. Here, all RMTs can be used within all workstations but only perform one operation in the specific period. Moreover, each DMT can only perform one operation and is assigned to a specific workstation. Hence, each workstation consists of a fixed number of DMTs and a variable number of RMTs. Therefore, the maximal capacity of a workstation is given by:

$$\bar{c}_j(t) = n_j^{DMT} \cdot v_j^{DMT} + n_j^{RMT} \cdot v_j^{RMT}. \tag{6}$$

The number of RMTs in each workstation is considered to be our new degree of freedom. If the association of an RMT to a workstation is changed over time via $u_j(t)$, this renders the maximal capacity time-variant. Assuming a high WIP, each workstation is operating at its maximal capacity, and its output rate is given by:

$$W_j(t) = n_j^{DMT} \cdot v_j^{DMT} + u_j(t - \tau_1) \cdot v_j^{RMT}, \tag{7}$$

where τ_1 is the reconfiguration delay when RMTs change the operation between workstations. Note that if an RMT is reconfigured from workstation j to workstation k, the capacity of workstation k increases only after a lag δ (e.g., 2 h), while the capacity of workstation j decreases immediately. This reveals the description:

$$\tau_1 = \begin{cases} \delta, & u_j(t^+) \geq u_j(t^-), \\ 0, & \text{else.} \end{cases} \tag{8}$$

Then, the WIP level can be controlled via the function $u_j(\cdot)$ for all workstations in (5). The input–output model of the system, which is the plant operator P_j for $j = 1, \ldots, n$, is given by:

$$\begin{aligned} P_j : y_j(t) = &y_j(0) + \int_0^t X_{0j}(\tau - \tau_2) + d_j(\tau) \\ &+ \sum_{k=1}^{n} p_{kj} \cdot (n_k^{DMT} \cdot v_k^{DMT} + u_k(\tau - \tau_1 - \tau_2) \cdot v_k^{RMT}) \\ &- (n_j^{DMT} \cdot v_j^{DMT} + u_j(\tau - \tau_1) \cdot v_j^{RMT})d\tau, \end{aligned} \tag{9}$$

where $u_j \subseteq U$ and $y_j \subseteq Y$, for all $j = 1, \ldots, n$.

Additionally, the number of RMTs in the system is assumed to be limited by n^{RMT}, and each workstation contains at least 0 RMTs. This reveals the control constraints:

$$0 \leqslant u_j(t) \quad \text{and} \quad \sum_{j=1}^{n} u_j(t) \leqslant n^{RMT}. \tag{10}$$

Note that similar to n_j^{DMT} but in contrast to the input and output rate values $X_j(\cdot)$ and $W_j(\cdot)$, our control $u_j(\cdot)$ is integer instead of continuous. To compensate for the issue, the truncation $\lfloor \cdot \rfloor$ is utilized. Moreover, for the constraint $\sum_{j=1}^{n} u_j(t) \leq n^{RMT}$, the fractional approach [32] is utilized, which is given as:

$$\hat{u}_j(t) = \begin{cases} \lfloor u_j(t) \rfloor, & \text{if } \sum_{j=1}^{n} u_j(t) \leq n^{RMT} \\ \left\lfloor \dfrac{n^{RMT}}{\sum_{k=1}^{n} u_k(t)} u_j(t) \right\rfloor, & \text{else.} \end{cases}$$

Note that if the number of utilized RMTs does not exceed the total number of RMTs, then $\hat{u}_j(t)$ corresponds to the truncated value of u_j. Otherwise, the fractional discrete value of RMTs in the sum over all workstations is considered.

Note that model (9) only applies if the system is working at a high WIP level. In this case, the order output rate equals the maximum capacity, i.e., the WIP level can be controlled via the assignment of RMTs $u_j(\cdot)$ for all workstations.

4. Capacity Control

Based on the mathematical preliminaries and model from the previous sections, RRCF was used to control the capacity of the job shop systems. According to [26], the right factorization of the job shop system for $P = (N + \triangle N) \circ D^{-1}$ is obtained as:

$$\begin{aligned} w_j(t) &= D_j^{-1}(u_k(t - \tau_1 - \tau_2))(u_j(t - \tau_1)) \\ &= X_{0j}(t - \tau_2) + \sum_{k=1}^{n} p_{kj} \cdot (n_k^{DMT} \cdot v_k^{DMT} + u_k(t - \tau_1 - \tau_2) \cdot v_k^{RMT}) \\ &\quad - (n_j^{DMT} \cdot v_j^{DMT} + u_j(t - \tau_1) \cdot v_j^{RMT}) \end{aligned} \tag{11}$$

$$y_j(t) = N_j(w_j(\cdot)) + \triangle N_j(d_j(t)) = y_j(0) + \int_0^t w_j(\tau) + d_j(\tau)d\tau. \tag{12}$$

In this description, couplings, delays, and disturbances render the system more complex. In practice, operators have to consider all workstations to assign RMTs, which impacts on efficiency. In order to decrease complexity, a decoupling design allows responding quickly with less involvement from other workstations. Firstly, a decoupling controller was designed to split the MIMO system into multiple SISO systems. Then, the RRCF and tracking controllers for each SISO system were derived to ensure the WIP of workstations remains on the predefined levels.

4.1. Decoupling Control

In (11), there exist couplings between all workstations, which can be described as n linear equations. Solving the n linear systems:

$$u_j(\cdot) = \sum_{k=1}^{n} D_{jk}(w_k)(\cdot), \qquad j = 1, \ldots, n,$$

is obtained. To avoid the difficult computation of an RRCF control for the MIMO system, decoupling as proposed in [29] was utilized to transfer it into multiple SISO systems. To obtain n independent SISO systems, the decoupling operators H and G as shown in Figure 6 need to satisfy the following theorem, cf. Theorem 1 in [29].

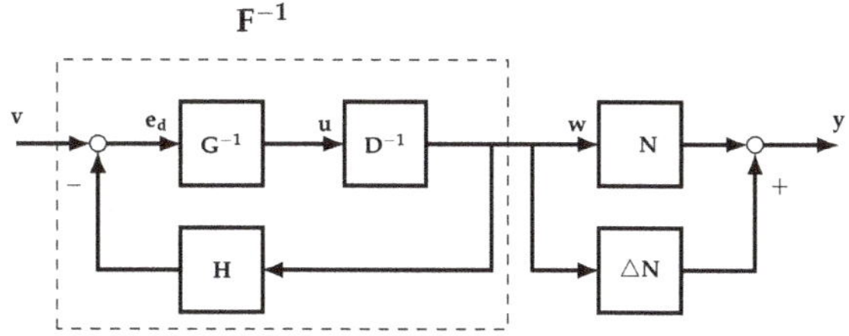

Figure 6. Decoupling control of multi-input–multi-output (MIMO) system.

Theorem 2 (Decoupled RRCF Control). *If G_j is linear and:*

$$\sum_{k=1,k\neq j}^{n} [H_{jk}(w_j)](w_k) + G_j \circ D_{jk}(w_k) = 0 \tag{13}$$

$$H_{jj}(w_j) + G_j \circ D_{jj}(w_j) = F_j(w_j), \tag{14}$$

then the MIMO system is decoupled and F_j is stable and invertible. Here, $\mathbf{F} = (F_1,\ldots,F_n)$ represents the decoupling operator with $v_j = F_j(w_j)$.

Applying the latter to the job shop system, the following proposition can be concluded.

Proposition 1 (Decoupled RRCF Control for Job Shop System). *Consider a plant (9) as well as decoupling parameters h_j with $\frac{1}{v_j^{RMT}} \neq |h_j| < \infty$ for $j = 1,\ldots,n$ and let $\mathbf{G} = (G_1,\ldots,G_n)$ be the identity operator, H_{jj} unimodular for $j = 1,\ldots,n$, such that:*

$$\sum_{k=1,k\neq j}^{n} [H_{jk}(w_j)](w_k) = - \sum_{k=1,k\neq j}^{n} G_j \circ D_{jk}(w_k),$$

holds. Then:

$$F_j(w_j) = \left(h_j - \frac{1}{v_j^{RMT}}\right) \cdot w_j - \frac{v_j^{DMT} n_j^{DMT}}{v_j^{RMT}}, j = 1,\ldots,n, \tag{15}$$

holds. Additionally, if n^{RMT} is sufficient large, then $F_j(w_j)$ is stable and invertible.

Proof. From (13) and (14), we obtain:

$$F_j(w_j) = H_{jj}(w_j) + G_j \circ D_{jj}(w_j).$$

As $\mathbf{G} = (G_1,\ldots,G_n)$ are the identity operators and H_{jj} unimodular operators $H_{jj} = h_j \cdot w_j$ for $j = 1,\ldots,n$, then plant (9) has:

$$F_j(w_j) = H_{jj}(w_j) + D_{jj}(w_j)$$
$$= \left(h_j - \frac{1}{v_j^{RMT}}\right) \cdot w_j - \frac{v_j^{DMT} n_j^{DMT}}{v_j^{RMT}}, j = 1,\ldots,n.$$

As $h_j \neq \frac{1}{v_j^{RMT}}$, it is a linear operator. Additionally, considering n^{RMT} is sufficient large, F_j is invertible. Its norm is:

$$\|F_j\| = |h_j - \frac{1}{v_j^{RMT}}|.$$

As $h_j < \infty$ and the production rate of RMTs v_j^{RMT} is a positive constant, hence, $\|F_j\| < \infty$. Thus, from Definition 1, this operator is stable, showing the assertion. $\square$

4.2. RRCF Control

After the decoupling, the RRCF control for each SISO system was considered separately. Based on Definition 3, the RRCF operators A_j and B_j (cf. Figure 7) can be designed following the Bezout identity:

$$A_j \circ (N_j + \triangle N_j) + B_j \circ F_j = \tilde{M}_j.$$

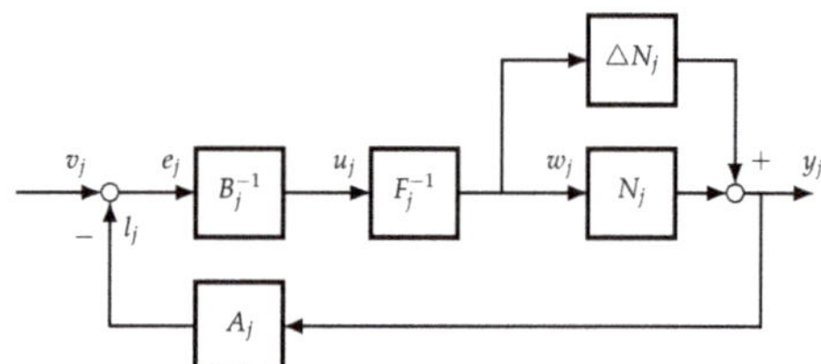

Figure 7. Robust right coprime factorization (RRCF) control of decoupled MIMO systems.

Here, for simplicity of exposition, the form:

$$A_j(s(\cdot)) = (1 - K_j) \cdot s(\cdot)', \tag{16}$$

is chosen to satisfy the sufficient condition with $K_j \in (0,1)$, where K_j is the RRCF control parameter for $j = 1, \ldots, n$. Then, let $\tilde{M}_j = I_j$, and from (12) and (15), we obtain:

$$B_j^{-1}(s(\cdot)) = \frac{(h_j v_j^{RMT} - 1)s(\cdot)}{v_j^{RMT} K_j} - \frac{v_j^{DMT} n_j^{DMT}}{v_j^{RMT}}. \tag{17}$$

Then, Corollary 1 can be applied to show the finite-gain input–output stability of the system.

4.3. Tracking Control

The RRCF control system has been designed to ensure finite-gain input–output stability but does not consider the tracking performance of the job shop system. In order to ensure the WIP can track a given level, a tracking controller C_j as proposed in [24] was integrated, cf. Figure 8 for a sketch. Here, for simplicity of exposition, the tracking controller was designed as:

$$C_j(s(\cdot)) = C_j^0 \cdot s(\cdot) + C_j^1 \cdot \int s(\cdot) \tag{18}$$

where C_j^0 and C_j^1 are tracking parameters. By setting the parameters appropriately, the tracking controller can minimize the error between the desired and actual WIP level.

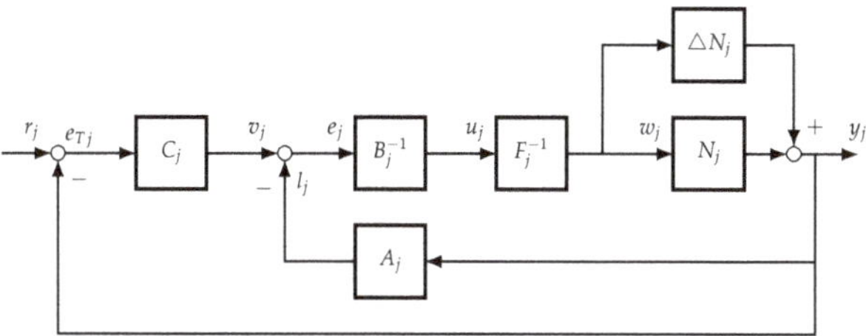

Figure 8. Nonlinear feedback tracking control of MIMO system.

In all, the capacity control of a job shop system $\mathbf{P} = (\mathbf{N} + \triangle\mathbf{N}) \circ \mathbf{D}^{-1}$ includes three parts: Firstly, the decoupling controllers $\mathbf{H}$ and $\mathbf{G}$ work on dealing with couplings between each workstation and transferring the complex MIMO system into multiple SISO systems. The second part is RRCF control of each SISO system (workstation), where RRCF controllers $\mathbf{A}$ and $\mathbf{B}$ are designed to ensure finite-gain input–output stability. Based on the stability, tracking controllers $\mathbf{C}$ are considered to guarantee that the WIP levels of all workstations track predefined levels. Including these adaptations of the decoupling, RRCF, and tracking controllers, this method can be utilized as a capacity adjustment method using RMTs in our job shop setting.

5. Case Study

To evaluate our proposed RRCF controller, a four-workstation three-product job shop system with $n^{RMT} = 10$ RMTs was considered. This case was defined in [33,34] in mould making. The flow probabilities for the three different products A_1, A_2, A_3 given by p^i_{jk} present the order output rate of product i from workstation j to workstation k (for $k = 1, 2, 3, 4$) and the final stage for $k = 0$, cf. Figure 9. It changes dynamically with the varying input rates. The parameter settings are shown in Table 2. Additionally, the PID control approach, cf. [32], was chosen as a benchmark for the RRCF control system. Here, two scenarios were considered in the simulation implemented in MATLAB. The performance of RRCF and PID control systems for constant and stochastic demands was analyzed sequentially.

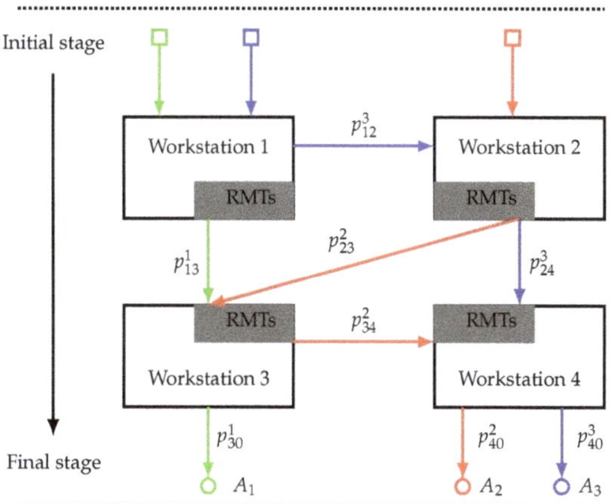

Figure 9. Four-workstation job shop manufacturing system with RMTs.

Table 2. Parameter settings of the four-workstation system.

Number of Workstation	1	2	3	4
Initial WIP level	400	400	300	200
Planned WIP level	240	400	400	240
Number of DMTs	4	2	2	4
Production rate of DMTs	20	40	40	20
Production rate of RMTs	10	20	20	10

5.1. Simulations for Constant Demands

Considering constant customer demands, the resulting performances of all workstations with delays are shown in Figures 10–12. The allocation of respective RMTs at each workstation is illustrated in Figure 10. The error between the planned and current WIP level of each workstation is given in Figure 11. The red and blue lines represent the variables with PID and RRCF control, respectively. The performance of WIP levels and distribution of RMTs of each workstation in this dynamical capacity control process is analyzed as follows.

At workstation 1, due to rush orders from the initial stage, the WIP level in workstation 1 increased and induced bottlenecks in the first 5 h, so the error in Figure 11 was highly decreased. Then both PID and RRCF controllers started assigning RMTs to this workstation, cf. Figure 10. Because of the reconfiguration delay, RMTs did not work for the first two hours, so the WIP was still increasing. However, after 2 h, PID and RRCF assigned 5 and 7 RMTs to this workstation, respectively, so the WIP in the RRCF case decreased more than in the PID one. After 10 h, the WIP of workstation 1 was lower than the planned level, so the number of RMTs was decreased. After around 50 h, the WIP in the RRCF case was practically stabilized (cf. [35], Chapter 2 for details), and the respective number of RMTs was also controlled between 2 to 3. The number of RMTs as the input signal is an integer, so the value of the WIP showed scattering but bounded behavior. For the PID case, the WIP level was practically stabilized after 130 h with higher overshoot.

At workstation 2, the initial WIP was lower than the planned level, so the error between planned and current WIP at the beginning was positive. In the first two hours, the error increased a bit due to fewer input orders from Workstation 1. After 2 h, owing to more orders flowing in from Workstation 1 but limited capacity, the WIP level was quickly increased and the error decreased quickly in the first 10 h in Figure 11. Therefore, bottleneck also moved to this workstation. After 10 h, both controllers started assigning RMTs to workstation 2 to solve the bottleneck in Figure 10. However, the RRCF had a quicker response with fewer than 2 RMTs to practically stabilize the WIP levels after 30 h, while the PID showed a slower response and took 4 RMTs to resolve the bottleneck. Additionally, the respective WIP level displayed higher overshoot and longer settling time than the RRCF control.

At workstations 3 and 4, the initial WIPs were less than the planned levels, so the initial error was positive in Figure 11. In the first two hours, as there was less order input and more output rate, the WIP level decreased, so the errors increased a bit. Then, due to increasing order inputs but limited capacity, the WIPs increased quickly over the planned levels and induced bottlenecks. After around 30 and 40 h at Workstations 3 and 4, respectively, both PID and RRCF controllers assigned RMTs to these two workstations. Then, the capacities increased, while the WIP levels decreased. However, the WIPs in RRCF were quickly stabilized and close to the planned levels. Same as for workstation 2, the RRCF control system showed a quicker response and used fewer RMTs than the PID control system to solve the bottlenecks. Moreover, the WIP levels of these two workstations within the RRCF control tended to be practically stable after 50 h, while the PID control system took around 150 h for the practical stabilization of the WIP levels. Additionally, the PID control showed higher overshoots and more oscillations. In Figures 10 and 11, both the PID and RRCF control approaches could practically stabilize the WIP level of all workstations. Nonetheless, RRCF showed better transient and steady state performance than the PID control system.

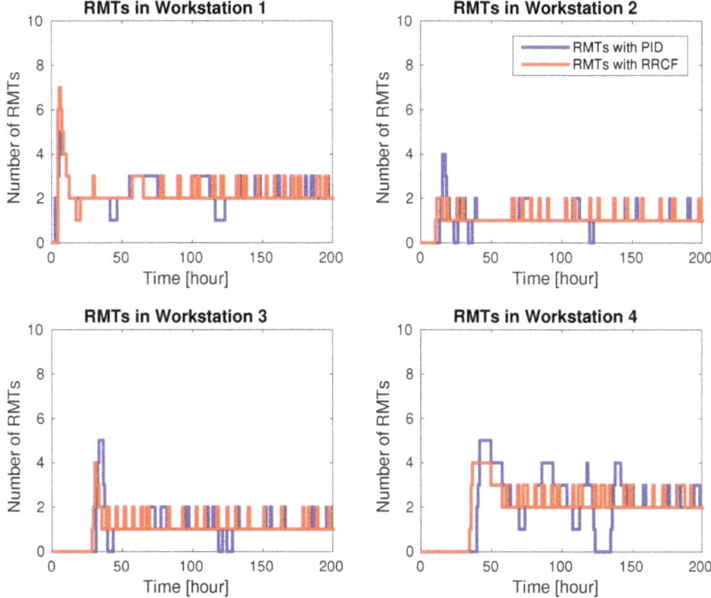

Figure 10. Number of RMTs in each workstation for constant demands.

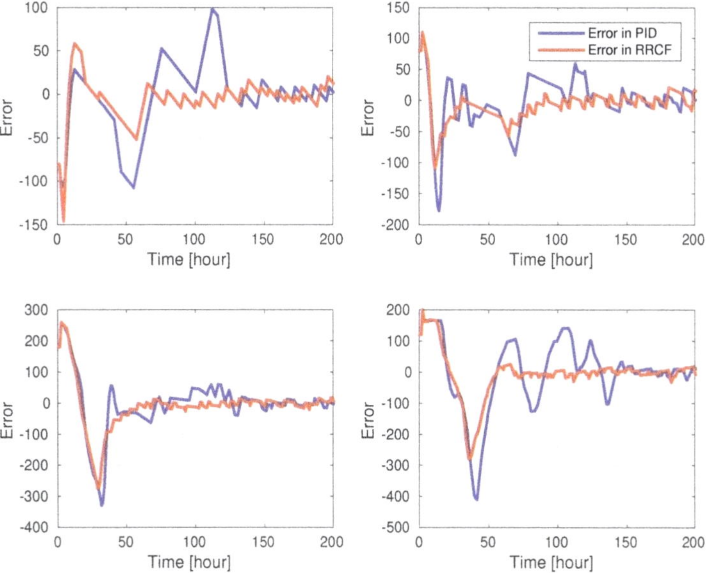

Figure 11. Error between planed and current WIP of each workstation for constant demands.

After the performance analysis of each workstation, the performance of each product was further analyzed. Figure 12 illustrates the order input and output rate of each product. The order input rate of each product was 51 per hour in red line. In Figure 9, all products were finished at workstation 3 or 4. Due to a capacity shortage, cf. Figure 10, the output rates of all products were mostly lower than the input rates for the first 30 h. Later, the order output rate of each product also showed scattering but bounded behavior for both PID and RRCF control. However, the RRCF control in green lines showed fewer overshoots and shorter settling time than the PID control in blue lines. This indicates that both the PID and the proposed RRCF method can practically stabilize the constant demand. However, compared to the PID control, the RRCF control has a quicker response to practically stabilize the system with fewer overshoots.

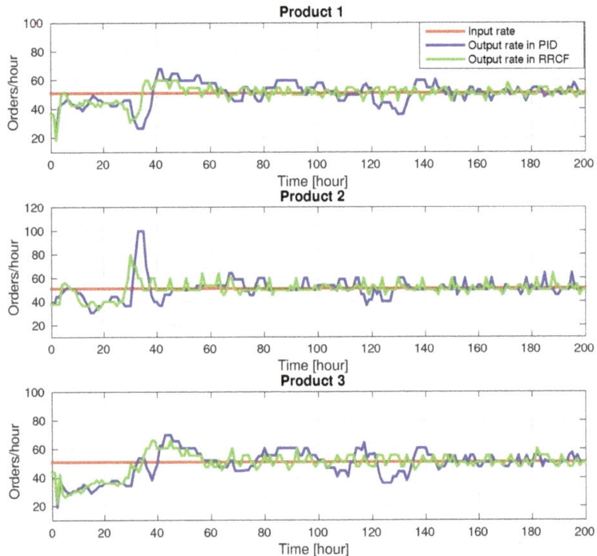

Figure 12. Order input and output rate of each product for constant demands.

5.2. Simulations for Stochastic Demands

Here, the demands of each product were assumed to be normally distributed with a mean value of 51 and variance of 5. Then, the practical stabilized closed loop was simulated 1000 times, and then the distribution of errors between the planned and current WIP at each workstation was obtained. The respective results are shown in Figure 13. This demonstrated that the mean errors of each workstation within the PID and RRCF control were almost close to zero. However, the distribution of the error in the RRCF control system is smaller than that for PID. Additionally, as to the number of RMTs X_j and the absolute errors Y_j of each workstation, the mean values $E(X)$, $E(Y)$ and the standard deviation $\sigma(X), \sigma(Y)$ are given in Table 3. At this time, the mean and standard deviation of the absolute errors in the RRCF control are less than the PID control most time. This illustrated that the RRCF control system is more robust than the PID one. Concerning the assignment of RMTs, there is no difference between both control systems on the mean value of RMTs, while RRCF showed higher values on standard deviation. This indicated that the proposed RRCF capacity control method can deal with the stochastic demands and showed higher robustness than the PID control method.

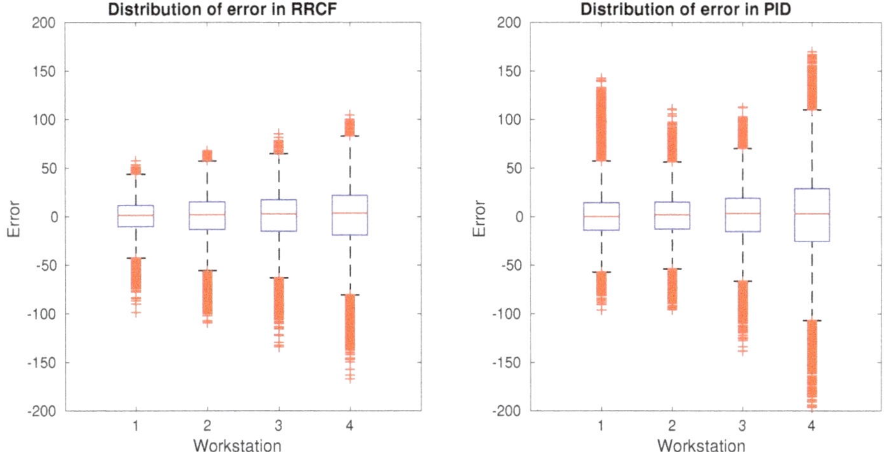

Figure 13. Distribution of errors at each workstation for stochastic demands.

Table 3. Performances of system for stochastic demands.

Controller	PID				RRCF			
Workstation	1	2	3	4	1	2	3	4
$E(X)$	2.20	1.10	1.10	2.19	2.20	1.10	1.11	2.19
$\sigma(X)$	0.56	0.54	0.63	0.97	1.12	0.76	0.86	1.96
$E(Y)$	17.58	17.00	21.26	32.76	13.11	17.33	19.48	24.58
$\sigma(Y)$	14.41	13.46	17.00	25.56	10.28	13.71	15.25	19.03

6. Conclusions and Outlook

In this paper, an operator-based approach incorporating RMTs was developed to control the capacity of job shop systems. A mathematical model integrating the flexibility of RMTs was presented for the capacity control. The design of the RRCF controller was discussed in the capacity control of a general job shop system. To evaluate the effectiveness of the RRCF control approach, a PID control was applied as a benchmark in the simulation of a four-workstation job shop manufacturing system. Then, conclusions are summarized as follows:

- The proposed RRCF capacity control method can deal with the constant demands and solve bottlenecks to ensure practical stability of the system in the WIP levels of all workstations and the output rates of all products. Additionally, compared to the PID method, the RRCF showed better transient and steady state performances, with shorter settling times and lower overshoots.
- For the stochastic demands, Monte Carlo simulation was utilized to evaluate the robustness of these two control systems. The simulation results illustrated that RRCF was more robust than PID with less mean and standard deviation of the absolute errors between planned and current WIP levels.
- The simulation results also supported the applicability and effectiveness of the proposed capacity control approach by integrating of RMTs with the RRCF method.
- As this capacity control approach is in a decentralized architecture, it also can be used in large-scale job shop systems.

In future work, the following research directions can be considered:

- The mathematical model can be further extended and integrated with more performance indicators, e.g., backlog and inventory for more complex problems with various perspectives.

- The proposed capacity control approach is designed from the customer perspective. Another work can be to develop an effective reconfiguration rule to optimize the performance, at the same time satisfying the demands.

Author Contributions: Validation, P.L., Q.Z. and J.P.; Writing—Original Draft Preparation, P.L. and Q.Z.; Writing—Review & Editing, P.L., Q.Z. and J.P.; Supervision, J.P.

Funding: This research was funded by the FUSION and gLINK programs of ERASMUS MUNDUS grant number 2013-2541/001-011-EM and 2014-0861/001-001-EM, respectively.

Acknowledgments: The publication of this paper was supported by Staats- und Universitaetsbibliothek Bremen (SuUB), Germany.

Conflicts of Interest: The authors declare no conflict of interest.

References

1. Landers, R.G.; Min, B.K.; Koren, Y. Reconfigurable machine tools. *CIRP Ann. Manuf. Technol.* **2001**, *50*, 269–274. [CrossRef]
2. Huang, G.Q.; Zhang, Y.; Jiang, P. RFID-based wireless manufacturing for real-time management of job shop WIP inventories. *Int. J. Adv. Manuf. Technol.* **2008**, *36*, 752–764. [CrossRef]
3. Atzori, L.; Iera, A.; Morabito, G. The Internet of Things: A survey. *Comput. Netw.* **2010**, *54*, 2787–2805. [CrossRef]
4. Almada-Lobo, F. The industry 4.0 revolution and the future of manufacturing execution systems (MES). *J. Innov. Manag.* **2015**, *3*, 16–21. [CrossRef]
5. Georgiadis, P.; Michaloudis, C. Real-time production planning and control system for job-shop manufacturing: A system dynamics analysis. *Eur. J. Oper. Res.* **2012**, *216*, 94–104. [CrossRef]
6. Wang, M.F.; Rosenshine, M. Scheduling for a combination of made-to-stock and made-to-order jobs in a job shop. *Int. J. Prod. Res.* **1983**, *21*, 607–616. [CrossRef]
7. Freitag, M.; Hildebrandt, T. Automatic design of scheduling rules for complex manufacturing systems by multi-objective simulation-based optimization. *CIRP Ann. Manuf. Technol.* **2016**, *65*, 433–436. [CrossRef]
8. Shen, L.; Dauzère-Pérès, S.; Neufeld, J.S. Solving the flexible job shop scheduling problem with sequence-dependent setup times. *Eur. J. Oper. Res.* **2018**, *265*, 503–516. [CrossRef]
9. Lu, P.H.; Wu, M.C.; Tan, H.; Peng, Y.H.; Chen, C.F. A genetic algorithm embedded with a concise chromosome representation for distributed and flexible job-shop scheduling problems. *J. Intell. Manuf.* **2015**, *29*, 1–16. [CrossRef]
10. Park, J.; Mei, Y.; Nguyen, S.; Chen, G.; Zhang, M. An investigation of ensemble combination schemes for genetic programming based hyper-heuristic approaches to dynamic job shop scheduling. *Appl. Soft Comput. J.* **2018**, *63*, 72–86. [CrossRef]
11. Nouiri, M.; Bekrar, A.; Jemai, A.; Niar, S.; Ammari, A.C. An effective and distributed particle swarm optimization algorithm for flexible job-shop scheduling problem. *J. Intell. Manuf.* **2018**, *29*, 603–615. [CrossRef]
12. Scholz-Reiter, B.; Lappe, D.; Grundstein, S. Capacity adjustment based on reconfigurable machine tools—Harmonising throughput time in job-shop manufacturing. *CIRP Ann. Manuf. Technol.* **2015**, *64*, 403–406. [CrossRef]
13. Kim, J.H.; Duffie, N.A. Design and analysis of closed-loop capacity control for a multi-workstation production system. *CIRP Ann. Manuf. Technol.* **2005**, *54*, 455–458. [CrossRef]
14. Liu, P.; Zhang, Q.; Pannek, J. Capacity adjustment of job shop manufacturing systems with RMTs. In Proceedings of the 10th International Conference on Software, Knowledge, Information Management & Applications (SKIMA), Chengdu, China, 15–17 December 2016. [CrossRef]
15. Zhang, Q.; Liu, P.; Pannek, J. Modeling and predictive capacity adjustment for job shop systems with RMTs. In Proceedings of the 25th Mediterranean Conference on Control and Automation, Valletta, Malta, 3–6 July 2017. [CrossRef]
16. Chen, G.; Han, Z. Robust right coprime factorization and robust stabilization of nonlinear feedback control systems. *IEEE Trans. Automat. Control* **1998**, *43*, 1505–1509. [CrossRef]

17. de Figueiredo, R.J.P.; Chen, G. *Nonlinear Feedback Control Systems: An Operator Theory Approach*; Academic Press Professional, Inc.: San Diego, CA, USA, 1993.

18. Deng, M. *Operator-based Nonlinear Control Systems: Design and Applications*; John Wiley & Sons: Hoboken, NJ, USA, 2014.

19. Wen, S.; Deng, M.; Ohno, Y.; Wang, D. Operator-based robust right coprime factorization design for planar gantry crane. In Proceedings of the IEEE International Conference on Mechatronics and Automation, Xi'an, China, 4–7 August 2010; pp. 1–5. [CrossRef]

20. Bi, S.; Deng, M. Operator-based robust control design for nonlinear plants with perturbation. *Int. J. Control* **2011**, *84*, 815–821. [CrossRef]

21. Bu, N.; Deng, M. System design for nonlinear plants using operator-based robust right coprime factorization and isomorphism. *IEEE Trans. Automat. Control* **2011**, *56*, 952–957. [CrossRef]

22. Wen, S.; Deng, M.; Inoue, A. Operator-based robust non-linear control for gantry crane system with soft measurement of swing angle. *Int. J. Model. Identif. Control* **2012**, *16*, 86–96. [CrossRef]

23. Wen, S.; Liu, P.; Wang, D. Optimal tracking control for a peltier refrigeration system based on PSO. In Proceedings of the International Conference on Advanced Mechatronic Systems, Kumamoto, Japan, 10–12 August 2014; pp. 567–571. [CrossRef]

24. Deng, M.; Inoue, A.; Ishikawa, K. Operator-based nonlinear feedback control design using robust right coprime factorization. *IEEE Trans. Automat. Control* **2006**, *51*, 645–648. [CrossRef]

25. Deng, M.; Inoue, A.; Edahiro, K. Fault detection in a thermal process control system with input constraints using a robust right coprime factorization approach. *Proc. Inst. Mech. Eng. Part J. Syst. Control Eng.* **2007**, *221*, 819–831. [CrossRef]

26. Deng, M.; Inoue, A. Networked non-linear control for an aluminum plate thermal process with time-delays. *Int. J. Syst. Sci.* **2008**, *39*, 1075–1080. [CrossRef]

27. Bi, S.; Deng, M.; Wen, S. Operator-based output tracking control for non-linear uncertain systems with unknown time-varying delays. *IET Control Theory Appl.* **2011**, *5*, 693–699. [CrossRef]

28. Deng, M.; Bi, S. Operator-based robust nonlinear control system design for MIMO nonlinear plants with unknown coupling effects. *Int. J. Control* **2010**, *83*, 1939–1946. [CrossRef]

29. Bi, S.; Xiao, Y.; Fan, X. Operator-based robust decoupling control for MIMO nonlinear systems. In Proceedings of the 11th World Congress on Intelligent Control and Automation, Shenyang, China, 29 June–4 July 2014; pp. 2602–2606. [CrossRef]

30. Lödding, H.; Yu, K.W.; Wiendahl, H.P. Decentralized WIP-oriented manufacturing control (DEWIP). *Prod. Plan. Control* **2003**, *14*, 42–54. [CrossRef]

31. Liu, P.; Pannek, J. Operator-based capacity control of job shop manufacturing systems with RMTs. In *Dynamics in Logistics*; Springer: Cham, Switzerland, 2018; pp. 264–272. [CrossRef]

32. Liu, P.; Chinges, U.; Zhang, Q.; Pannek, J. Capacity control in disturbed and time-delayed job shop manufacturing systems with RMTs. *IFAC-PapersOnLine* **2018**, *51*, 807–812. [CrossRef]

33. López de Lacalle, L.N.; Lamikiz, A.; Salgado, M.A.; Herranz, S.; Rivero, A. Process planning for reliable high-speed machining of moulds. *Int. J. Prod. Res.* **2002**, *40*, 2789–2809. [CrossRef]

34. López De Lacalle, L.N.; Lamikiz, A.; Muñoa, J.; Sánchez, J.A. The CAM as the centre of gravity of the five-axis high speed milling of complex parts. *Int. J. Prod. Res.* **2005**, *43*, 1983–1999. [CrossRef]

35. Grüne, L.; Pannek, J. *Nonlinear Model Predictive Control: Theory and Algorithms*; Springer International Publishing: Cham, Switzerland, 2017. [CrossRef]

Article

An Approach to Supporting the Selection of Maintenance Experts in the Context of Industry 4.0

Justyna Patalas-Maliszewska * and Sławomir Kłos

University of Zielona Góra, Faculty of Mechanical Engineering, Zielona Góra 65-417, Poland;
S.Klos@iizp.uz.zgora.pl
* Correspondence: J.Patalas@iizp.uz.zgora.pl; Tel.: +48-6828-2685

Received: 24 March 2019; Accepted: 1 May 2019; Published: 5 May 2019

Abstract: (1) Background: In recent years, many studies regarding the issues of improving the management and effectiveness of the maintenance department of manufacturing companies, in the context Industry 4.0, have been published. This makes it necessary to establish a research gap in the approach to obtaining support in realising management tasks in the maintenance area in the selection of appropriate employees to perform the given activities. (2) Methods: This article uses literature studies and empirical research results from manufacturing companies, in order to determine the approach in supporting the selection of maintenance experts. In the approach, the method used—which is based on rules should there be future any formalisation of the data—is also the Fuzzy Analytic Hierarchy Process (FAHP), which analyses the importance of a given competence, within a manufacturing resource, to undertake repairs. (3) Results: The innovative approach towards the selection of expert workers in a maintenance department is created, in part, in the form of an implemented web-application. The novelty of the "maintenance expert selection map", so-called, is the provision of formal procedures for describing the competence of each maintenance worker and defining the best "state of nature". (4) Conclusions: In the research that is presented here, the practicality for maintenance managers in the "maintenance expert selection map" was established. This map describes the competence of workers for selecting them for repair work within a given manufacturing resource; the scope of employee training was also determined in this research.

Keywords: maintenance expert; competence; decision support

---

1. Introduction

The implementation of the Industry 4.0 concept within manufacturing enterprises was, and still is, the objective of many research papers [1–4]. Maintenance in manufacturing companies plays a crucial role in improving their competitiveness [5]. Moreover, companies should develop and implement those models that have already been employed and that can predict reliable production in operation, according to the Industry 4.0 paradigm [6].

Human operators are key resources within a smart manufacturing company [7], since such workers are aware of specific technological processes; however, one can still observe the need to obtain support in the selection of appropriate employees, in order for them to undertake given activities. Maintenance department managers expect that, in the event of a breakdown, they will receive a list of those employees whose competence will guarantee that the machine will be repaired in the shortest time in real time.

The approach to the selection of expert workers needs to be computerised and codified, while using data of a specified format, in order that it may prove to be useful. According to [8], the formal representation of the competence of workers is the key factor of the model's effectiveness. Moreover, they stated that there exists a lack of the representation of competence for "Diagnosis" as well as

"Management" tasks that were carried out in maintenance departments based on the literature review of 74 research papers; it is clear that these competences are primarily associated with employees performing these processes in the company.

Our proposed approach focusses on supporting management tasks that were carried out in the maintenance department and it includes the following elements: (1) Defining the type of failure for each manufacturing resource, (2) Defining the parameters of each type of failure for each maintenance worker, (3) Defining the competences of each maintenance worker, (4) Defining the importance of competence while using the Fuzzy Analytic Hierarchy Process (FAHP) method, (5) Defining the "state of nature" for each manufacturing resource regarding the employee, and (6) Selecting the maintenance expert for the repair of the manufacturing resource. Our approach is also partly investigated in the form of a web-application and is presented, based on a case study. The problem in this paper is how to assess the competences of maintenance workers in manufacturing companies and how to select the employees with the appropriate competences to undertake the repair of a given machine in companies.

2. Supporting the Selection of Expert Maintenance Workers in the Context of Industry 4.0

In the literature, many examples of the system supporting work in the maintenance department within the context of Industry 4.0 are to be found.

Ni et al. [9] studied the extra hidden opportunities for preventive maintenance (PM) during production time without violating the requirements of system throughput. The authors created a mathematical prediction model to identify PM opportunity windows for large production systems based on real-time factory information system data. Ni and Jin [10] presented new decision support tools that are based on mathematical algorithms and simulation tools for effective maintenance operations. The system enables the short-term identification of throughput bottlenecks, estimates the windows of opportunity for maintenance, prioritises maintenance tasks, jointly produces and maintains scheduling systems, and maintains staff management. The system was implemented in an automotive manufacturing area. Xiao et al. [11] developed an optimisation model in order to minimise the total costs, namely, production costs, preventive maintenance costs, minimal repair costs for unexpected failures, and delay costs. They used genetic algorithms to illustrate the proposed model. Jin et al. [12] proposed an analytical, option-based cost model for scheduling joint production and preventive maintenance when demand is uncertain. They obtained the optimum number of preventive maintenance work-orders within a production system using the model.

Many articles deal with the application of augmented reality (AR) or virtual reality, to support maintenance activities. Massoni et al. [13] present an application for remote maintenance, which is based on off-the-shelf mobile and augmented reality (AR) technologies [14]. The application enables a skilled operator, in a control room, to be remotely connected to an unskilled operator, located where maintenance has to be performed. Technological limitations problems and the incorrect use of AR technology in the maintenance area were analysed. Securati et al. [15] created and adopted a controlled and exhaustive vocabulary of graphical symbols, to be used in augmented reality, to represent maintenance instructions. They identified the most frequent maintenance actions that were used in manuals and converted them into graphical symbols. Roy et al. [16] analysed the foundations and technologies that are required to offer the maintenance service for years to come.

In the literature, the adoption of the Condition-based Maintenance (CBM) approach, within the context of Industry 4.0, is to be found. CBM can be treated as the decision making strategy that is based on observation of the system within a manufacturing company and/or its components [17], as part of the main "Detect-Predict-Decide-Act" paradigms. The subject of current research is the "Decide" phase [18]. In the CBM approach, decisions are taken based on information that is collected by monitoring the condition [19] using various kinds of techniques, such as AI technologies, comprising ANN, the rule-based, expert system, and the Bayesian Network [20].

Therefore, the approach to Supporting the Selection of Maintenance Experts, which contributes the method used—which is based on rules should there be any future formalisation of the data—is also the Fuzzy Analytic Hierarchy Process (FAHP). This approach analyses the importance of a given its competence to undertake repair work within a manufacturing resource and it is defined and developed for the phase: "Decide" in the CBM approach.

Moreover, Belkadi et al. [21] performed a comparative analysis of decision support systems that are dedicated to maintenance departments, such as the 'Knowledge' based system for industrial maintenance [22] and the 'Intelligent' system for predicting breakdowns and monitoring industrial machines [23], which have the advantage of providing their solutions in the form of functionality, or, to put it another way, the transformation and adaptation of expert knowledge.

The management of competence in Industry 4.0 aims to identify not only the competences required within a company, but also the critical gaps in competences within a company. According to [24] and our previous research [25] and, as based on the survey and data obtained from 85 German and Polish Manufacturing Enterprises, the core competences, which are needed in manufacturing companies, in terms of Industry 4.0, were defined as technical, methodological, social, and personal.

Our proposed approach allows for managers to select maintenance department expert workers; the main functionalities of these innovations are:

- Integrating, with the data already collected, details from the information system implemented, of the time spent by each worker in repairing each type of failure in each manufacturing resource.
- Providing formal procedures for describing the competence of each maintenance worker.
- Defining the best natural state—meaning indicating those workers, the selection of whom will guarantee the maximum availability of the manufacturing resource.
- Assisting in the selection of maintenance expert.

3. An Approach to Selecting a Maintenance Expert

The proposed approach to presenting the selection of maintenance experts—based on their competences—for repairs within a manufacturing resource provides an opportunity to denote a particular worker as the expert worker within the maintenance department.

The approach to selecting a maintenance expert (Figure 1) is in line with the concept of reliability-centered maintenance (RCM). RCM can be treated as the reactive, preventive, and proactive maintenance practices that are introduced within a company [26]. It is also the approach to capturing the reason of downtime using two stages: (1) determine the critical components of the system and (2) application of decision rules to define categories of predictive maintenance (PM) [27]. The construction of the proposed approach corresponds to five stages defined in RCM process [28]:

- Selection of subsystem: maintenance competence management.
- Identification of component: defining the types of failure for each manufacturing resource and each competence, which has a considerable influence on reliability (stages 1-3, Figure 1).
- Analysis: defining the importance of each competence for the repair of each type of failure (stage 4, Figure 1).
- Optimal maintenance strategy selection: defining the "state of nature" and the implementation of maintenance expert selection map (stages 5-6, Figure 1).
- Analysis: the selection of this employee to repair a given resource, who guarantees an increase in the reliability level of a given manufacturing resource.

Each stage of the proposed approach must be formalised so that it can be computerised according to the Industry 4.0 concept.

The construction of the proposed approach is possible due to the acquisition and gathering of knowledge from the database of the information systems implemented within a company (stages 1–2, Figure 1) and the unique knowledge of employees performing activities in the maintenance department (stages 3–4, Figure 1). However, for acquired expert knowledge, so-called, to be useful, it needs to be represented and codified by data with a specified format. Accordingly, it is stated, at each stage of our approach, knowledge is defined, then acquired, and, finally, is then stored. Acquired knowledge must be converted into extracted and explicit knowledge, so that it can be computerised. This can be done [29] by using the frames based systems [30], frame logic [31], semantic networks [32], and conceptual graphs, with these being methods based on concept dictionaries, viz., ontologies [29], and methods that are based on established rules. The rule-based method was selected in order to create a formalised base for the approach (Figure 1).

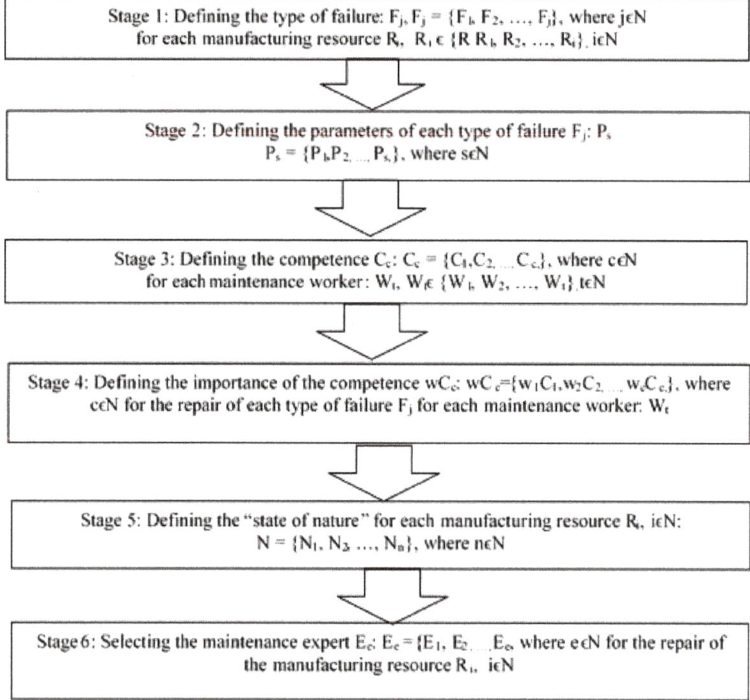

Figure 1. The proposed approach.

Stages one to four were based on the literature research results [4,24,33–35] and empirical research results [36] from the maintenance departments of manufacturing companies. In stage 1, (Figure 1) the types of failure for each manufacturing resource R_i, $i\epsilon N$ are defined: $F = \{F_1, F_2, \ldots, F_5\}$ (Table 1), where:

- F_1—failure of the control system.
- F_2—failure of the power system.
- F_3—failure of the cooling system.
- F_4—failure of the hydraulic system.
- F_5—failure of the material transfer system.

Table 1. Types of failure for each manufacturing resource.

Manufacturing Resources/Type of Failure	F_1	F_2	F_3	F_4	F_5
R_1	$1 \vee 0$	$1 \vee 0$	$1 \vee 0$	$1 \vee 0$	$1 \vee 0$
R_2	$1 \vee 0$	$1 \vee 0$	$1 \vee 0$	$1 \vee 0$	$1 \vee 0$
R_3	$1 \vee 0$	$1 \vee 0$	$1 \vee 0$	$1 \vee 0$	$1 \vee 0$
...	$1 \vee 0$	$1 \vee 0$	$1 \vee 0$	$1 \vee 0$	$1 \vee 0$
R_i, $i\epsilon N$	$1 \vee 0$	$1 \vee 0$	$1 \vee 0$	$1 \vee 0$	$1 \vee 0$

The rule for formalising the acquired data from the information system implemented, is defined in Table 1: If there is a failure in the manufacturing resource then the value of Fj = 1, if not Fj = 0, j = 1, ... , 5.

In stage 2, (Figure 2) the following parameters of each type of failure F_j, j = 1, ... , 5 for each manufacturing resource: R_i, i,jϵN are defined: P = {P_1, P_2, P_3} (Table 2), where kϵN

- P_1—time for diagnosing and finding the solution.
- P_2—maintenance operation time.
- P_3—time for testing.

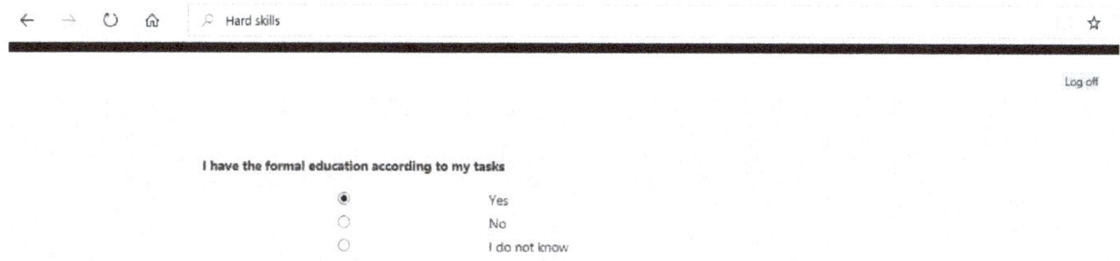

Figure 2. An extract from the web questionnaire, for employees facilitating the determining of competence: hard skills.

Table 2. The formalised data of parameters of each type of failure.

Parameters of Each Type of Failure	Description	Rules for Determining the Value of Parameters
P_1 – time for diagnose and finding solution	$P_1 \epsilon$ <10;30> [min]	if $P_1 \epsilon$ <10;15) [min] then P_1 = 1point if $P_1 \epsilon$ <15;20) [min] then P_1 = 2points if $P_1 \epsilon$ <20;25) [min] then P_1 = 3points if $P_1 \epsilon$ <25;28) [min] then P_1 = 4points if $P_1 \epsilon$ <28;30> [min] then P_1 = 5points
P_2 – maintenance operation time	$P_2 \epsilon$ <30;210> [min]	if $P_2 \epsilon$ <30;50) [min] then P_2 = 1point if $P_2 \epsilon$ <50;90) [min] then P_2 = 2points if $P_2 \epsilon$ <90;120) [min] then P_2 = 3points if $P_2 \epsilon$ <120;180) [min] then P_2 = 4points if $P_2 \epsilon$ <180;210> [min] then P_2 = 5points
P_3 – time for testing	$P_3 \epsilon$ <20;30> [min]	if $P_3 \epsilon$ <20;23) [min] then P_3 = 1point if $P_3 \epsilon$ <23;25) [min] then P_3 = 2points if $P_3 \epsilon$ <25;26) [min] then P_3 = 3points. if $P_3 \epsilon$ <26;28) [min] then P_3 = 4points if $P_3 \epsilon$ <28;30> [min] then P_3 = 5points

The rules for formalising the acquired data from the information system implemented is defined (Table 2):

In stage 3, (Figure 1) the competences of each maintenance worker: W_t, $W_t \epsilon$ {W_1, W_2, ... , W_t}, tϵN are defined: C = {C_1, C_2, C_3, C_4, C_5}, where:

- C_1—Hard skills.
- C_2—Knowledge-based.
- C_3—Methodical.
- C_4—Soft Skills.
- C_5—Experience.

Assessing workers' knowledge is not an easy task; moreover, the quality and scope of this knowledge is crucial to it being able to be repeatedly used. The following method for assessing knowledge has been distinguished [37]: questions with a defined set of choices [38,39], rating grids or rules [40], questions with open answers [41], and questions regarding domain variables [42]. In the proposed approach, the sub-model for selecting workers, based on their competences, is developed (Table 3).

Table 3. The competence of each maintenance worker.

Competence	Description	Rules for Determining the Value of Competence
Hard skills (C_1) [34]	Completed engineering studies, references, certificate, certificate for the completion of specialised training in the handling of resources: R_i, where $i \epsilon N$,	If a worker has no references, or has not completed engineering studies and possesses neither a certificate nor a certificate for the completion of specialised training, then $C_1 = 0$points. If a worker has references, but has not completed engineering studies, has no certificate and has no certificate for the completion of specialised training, then $C_1 = 1$point. If a worker has completed studies but has neither references, nor a certificate nor a certificate for the completion of specialised training, then $C_1 = 2$points If the worker has completed engineering studies and has references but has neither a certificate nor a certificate for the completion of specialised training, then $C_1 = 3$points. If the worker has completed engineering studies, has references and also has a certificate but has no certificate, for the completion of specialised training, then $C_1 = 4$points. If the worker has a certificate for the completion of specialised training, then $C_1 = 5$points.
Knowledge-based (C_2) [24,35]	A 15-question test about resources: R, where $i \epsilon N$	If up to 7 answers are correct, then: $C_2 = 0$points. If 7–8 answers are correct, then $C_2 = 1$point. If 9 answers are correct, then $C_2 = 2$points. If 10–11 answers are correct, then $C_2 = 3$points. If 12–13 answers are correct, then $C_2 = 4$points. If 14–15 answers are correct, then $C_2 = 5$points.
Methodical (C_3) [35,36]	A 15-question test about comparing and classifying information and the use of available resource: R_i, where$i \epsilon N$	If up to 7 answers are correct, then: $C_3 = 0$points. If 7–8 answers are correct, then $C_3 = 1$point. If 9 answers are correct, then $C_3 = 2$points. If 10–11 answers are correct, then $C_3 = 3$points. If 12–13 answers are correct, then $C_3 = 4$points. If 14–15 answers are correct, then $C_3 = 5$points.
Soft Skills (C_4) [4,34]	A 15-question test about the ability to organise work, the ability to work in a team, communication skills and the ability to undertake task-oriented work and working under pressure	If up to 7 answers are correct, then: $C_4 = 0$points. If 7–8 answers are correct, then $C_4 = 1$point. If 9 answers are correct, then $C_4 = 2$points. If 10–11 answers are correct, then $C_4 = 3$points. If 12–13 answers are correct, then $C_4 = 4$points. If 14–15 answers are correct, then $C_4 = 5$points.
Experience (C_5) [24]	Number of years in the current company (L) Number of years, generally, in the profession (Z)	If $L \leq 3$ years and $Z \leq 3$ years, then: $C_5 = 0$points. If $L \leq 3$ years and $3 < Z \leq 5$ years, then $C5 = 1$point. If $3 < Z \leq 5$ years and $5 < Z \leq 8$ years, then $C_5 = 2$points. If $5 < Z \leq 8$ years and $8 < Z \leq 10$ years, then $C_5 = 3$points. If $8 < Z \leq 10$ years and $Z > 10$ years, then $C_5 = 4$points. If $Z > 10$ years and $Z > 10$ years, then $C_5 = 5$points.

For each resource, R_i, where $i \epsilon N$, the value of each competence for each worker is determined according to the rules (Table 3).

In the fourth stage, the Fuzzy Analytic Hierarchy Process (FAHP) was implemented. It was possible to determine the relative predominance of a particular factor of the core-competence model, from those elements of the framework that could not be calculated using FAHP; furthermore, it was possible to evaluate these factors; therefore, this means that the importance of competence for the repair of each type of failure Fj for each maintenance worker: W_t, $W_t \epsilon$ {W_1, W_2, ... , W_t}, $t \epsilon N$ is defined. According to Nydick and Hill, a fuzzy number $\tilde{a} = (l, m, u)$ with a triangular fuzzy-membership function can describe a linguistic variable. The triangular fuzzy number is defined in the set $[l, u]$ and its membership function takes a value that is equal to 1 at point m. The fuzzy scale of preferences is strictly defined by [43]. Maintenance managers assess the validity of each competence for the purpose of repairing a given machine:

- C_1—equally important, or moderately more important, or of greater importance, or of the most importance, compared with C_2 or with C_3 or with C_4 or with C_5.
- C_2—equally important, or moderately more important, or of greater importance, or of the most importance, compared with C_1 or with C_3 or with C_4 or with C_5.
- C_3—equally important, or moderately more important, or of greater importance, or of the most importance, compared with C_1 or with C_2 or with C_4 or with C_5.
- C_4—equally important, or moderately more important, or of greater importance, or of the most importance, compared with C_1 or with C_2 or with C_3 or with C_5.
- C_5—equally important, or moderately more important, or of greater importance, or of the most importance, when compared with C_1 or with C_2 or with C_3 or with C_4.

The importance of each competence C_c, $c\epsilon N$ for the repair of each type of failure F_j, $j\epsilon N$ for each maintenance worker: W_t, $t\epsilon N$, is determined while using the FAHP method: wC: $wC = \{w_1 C_1, w_2 C_2, w_3 C_3, w_4 C_4, w_5 C_5\}$.

In the fifth stage, the "state of nature" N_{Ri}: $N = \{N_1, N_2, \ldots, N_{Ri}\}$, $i\epsilon N$ for each manufacturing resource R_i, $i\epsilon N$, is defined (Table 4), according to the formula:

for each $W_t, t \in N$

$$NRi \equiv \frac{w_1 C_1 + w_2 C_2 + w_3 C_3 + w_4 C_4 + w_5 C_5}{\bar{P}1 + \bar{P}2 + \bar{P}3},$$ where $i\epsilon N$ and, $s = \{1,2,3\}$ means the average time of the all-time measurements.

Table 4. The value of each "state of nature" for each manufacturing resource R_i, $i\epsilon N$.

Workers/"State of Nature"	N_{R1}	N_{R2}	...	N_{Ri} $i\epsilon N$
W_1	$N_{R1W1}\epsilon<0;1.66>$	$N_{R2W1}\epsilon<0;1.66>$	...	$N_{RiW1}\epsilon<0;1.66>$
...	...	...	...	...
W_t, $t\epsilon N$	$N_{R1Wt}\epsilon<0;1.66>$	$N_{R2Wt}\epsilon<0;1.66>$	...	$N_{RiWt}\epsilon<0;1.66>$

The higher the value of the "state of nature" ($maxN_{Ri} =1.66$), the greater is the certainty that the selection of this employee, to repair a given resource, guarantees an increase in the reliability level of a given manufacturing resource.

Our approach (stages 1-4) is partly investigated in the form of a web-application; this is presented below and it is based on a case study.

4. A Model for Supporting the Selection of Maintenance Experts

In order to illustrate the possibility of answering our research questions, let us consider the situation. The problem that is being considered entails selecting employees with the appropriate competence to undertake the repair of the given machine in companies, involving Industry 4.0. The research was carried out in the automotive industry company. Production, being partly automated, is carried out using a two-shift system. The maintenance manager supervises the work of four employees who service 18 machines. Below is an extract from the web-application for identifying the Industry 4.0, maintenance expert, based on the approach (Figure 1).

According to stage 1, (Figure 1) the data on the types of failure for each manufacturing resource from the information system is received and formalised, according to the rules that are included in Table 1 (Table 5)

Then, according to stage 2, the formalised data of the parameters of each type of failure (Table 2) is identified (Table 6).

According to the third stage, for each competence: C_1, C_2, C_3, C_4, C_5, a knowledge web-questionnaire is defined. The extracts from the web-questionnaires for workers facilitating the obtaining of values for each competence are presented (Figures 2–5).

Table 5. Data about the types of failure for each manufacturing resource.

Manufacturing Resources/Type of Failure	F_1	F_2	F_3	F_4	F_5
R1	1	1	1	1	1
R2	1	1	1	1	1
R3	1	1	1	1	1
R4	1	1	0	1	0
R5	1	1	0	1	0
R6	1	1	0	0	0
R7	1	1	1	0	0
R8	1	1	1	0	1
R9	1	1	1	0	1
R10	1	1	0	1	1
R11	1	1	1	1	1
R12	1	1	0	1	1
R13	1	1	0	1	0
R14	1	1	0	1	0
R15	1	1	0	0	0
R16	1	1	1	0	0
R17	1	1	1	0	1
R18	1	1	0	1	0

Table 6. Parameters of failure: F1—failure of the control system.

F_1 – Failure of Control System	F_{11}			F_{12}			F_{13}			F_{14}		
	P_1	P_2	P_3	P_1	P_2	P_3	P_1	P_2	P_3	P_1	P_2	P_3
R1	13	56	21	26	183	25	18	203	23	30	122	26
R2	15	104	22	11	71	20	23	102	24	18	145	25
R3	20	51	23	14	107	28	10	60	21	26	66	29
R4	15	182	24	27	100	30	24	117	28	20	45	24
R5	29	106	30	18	174	24	30	208	27	22	196	28
R6	18	102	23	20	150	22	26	76	20	20	149	20
R7	26	65	22	23	203	20	27	209	30	14	41	30
R8	13	203	25	17	51	22	29	116	28	17	107	29
R9	21	169	20	27	55	20	13	187	25	24	114	27
R10	19	202	24	25	139	26	27	166	24	20	81	23
R11	15	195	20	28	163	23	22	152	20	11	157	21
R12	29	53	22	19	159	25	29	163	21	15	188	21
R13	28	174	27	22	198	20	12	188	21	21	103	25
R14	14	158	27	25	61	22	14	132	23	29	188	26
R15	29	30	30	14	80	28	26	105	21	24	99	27
R16	14	30	23	17	52	30	16	168	21	28	41	30
R17	25	206	28	19	84	25	13	51	25	15	193	22
R18	11	106	23	17	50	26	17	91	28	27	115	27

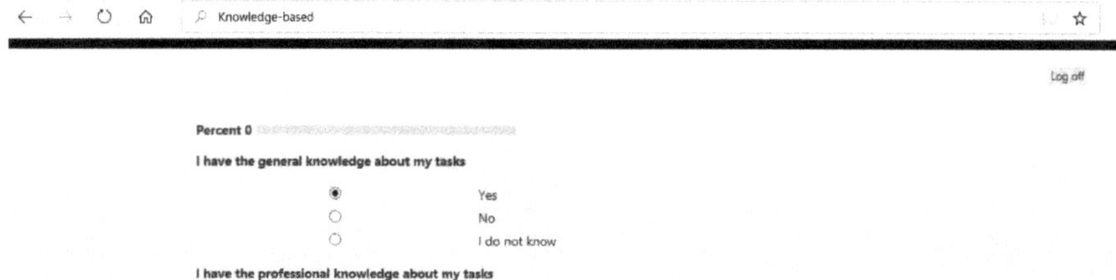

Figure 3. An extract from the web questionnaire, for employees facilitating the determining of competence: knowledge-based.

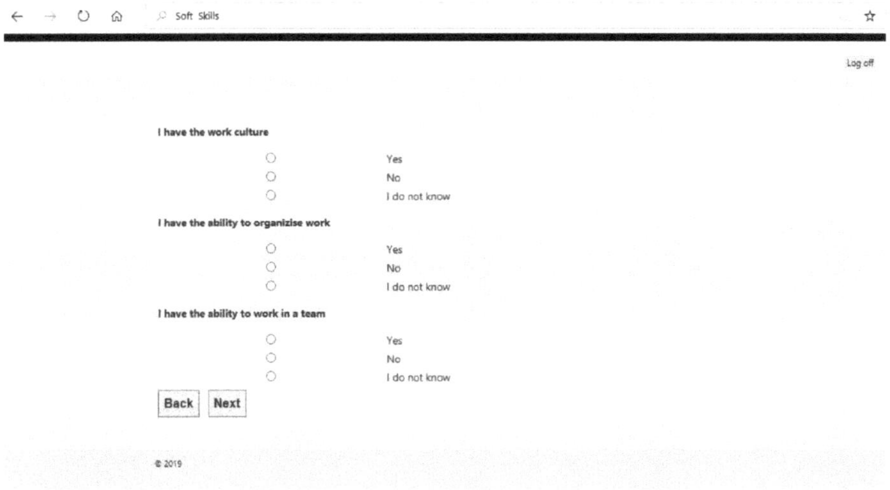

Figure 4. An extract from the web questionnaire, for employees facilitating the determining of competence: methodical.

Figure 5. An extract from the web questionnaire, for employees facilitating the determining of competence: soft skills.

Each of the four employees completed the web-forms of questionnaires and, based on their responses as well on the rules included in the Table 3, the following values of each competence for each worker are received (Table 7).

Table 7. The values of competence for each maintenance worker.

Workers/the Values of Competence	C_1	C_2	C_3	C_4	C_5
W_1	2	2	4	1	1
W_2	1	1	3	3	1
W_3	2	1	5	0	0
W_4	1	2	4	1	1

According to the fourth stage, the FAHP method was implemented and used. The fuzzy weightings matrix of competences for the repair of each defined failure—see Table 8 as the example for the F1—failure of control system.

Table 8. The values of elements of the comparison matrix, using the Fuzzy Analytic Hierarchy Process (FAHP) method as the example for the F_1—failure of control system.

Competence	C_1	C_2	C_3	C_4	C_5
C_1	(1,1,1)	(1/3,1,1)	(3,5,7)	(3,5,7)	(3,5,7)
C_2	(1,1,3)	(1,1,1)	(3,5,7)	(3,5,7)	(3,5,7)
C_3	(1/7,1/5,1/3)	(1/7,1/5,1/3)	(1,1,1)	(1/5,1/3,1)	(1/5,1/3,1)
C_4	(1/7,1/5,1/3)	(1/7,1/5,1/3)	(1,3,5)	(1,1,1)	(1/3,1,1)
C_5	(1/7,1/5,1/3)	(1/7,1/5,1/3)	(1,3,5)	(1,1,3)	(1,1,1)

Using the FAHP method, the importance of the each competence for the repair of the defined failure F_1 was obtained, respectively:

- The importance of C1: w1 = 0.4014.
- The importance of C2: w2 = 0.3429.
- The importance of C3: w3 = 0.1060.
- The importance of C4: w4 = 0.0904.
- The importance of C5: w5 = 0.0593.

Accordingly, for each of the four employees, the following values for each competence, dedicated to the F1 (failure of the control system), were obtained (Table 9).

Table 9. The values of the competence of each maintenance worker.

The Importance of Competence	w_1C_1	w_2C_2	w_3C_3	w_4C_4	w_5C_5
W_1	0.8028	0.6858	0.424	0.0904	0.0593
W_2	0.4014	0.3429	0.318	0.2712	0.0593
W_3	0.8028	0.3429	0.53	0	0
W_4	0.4014	0.6858	0.424	0.0904	0.0593

According to the fifth stage, the values of each natural state for each manufacturing resource R18 for the F_1 (failure of control system) were defined. Table 10 presents the formalised data from Table 6.

Accordingly, based on the data from Tables 9 and 10, the "states of nature" values are received (Table 11).

Table 10. The values of the parameters for each manufacturing resource R18 for the F1 - failure of control system.

e Employees/ Manufacturing Resource	R_1	R_2	R_3	R_4	R_5	R_6	R_7	R_8	R_9	R_{10}	R_{11}	R_{12}	R_{13}	R_{14}	R_{15}	R_{16}	R_{17}	R_{18}
W_1	4	6	7	9	13	7	7	9	8	9	8	8	14	10	11	4	14	6
W_2	12	4	9	12	8	8	9	5	7	12	11	9	9	5	8	9	7	8
W_3	9	8	4	11	14	7	14	13	9	11	8	10	7	7	8	7	6	10
W_4	13	9	11	6	13	8	7	10	10	7	6	8	9	14	10	11	8	11

Table 11. The values of the "states of nature".

	R_1	R_2	R_3	R_4	R_5	R_6	R_7	R_8	R_9	R_{10}	R_{11}	R_{12}	R_{13}	R_{14}	R_{15}	R_{16}	R_{17}	R_{18}
W_1	0.5156	0.343717	0.294614	0.229144	0.158638	0.294614	0.294614	0.229144	0.257788	0.229144	0.257788	0.257788	0.147307	0.20623	0.187482	0.515575	0.147307	0.343717
W_2	0.116067	0.3482	0.154756	0.116067	0.1741	0.1741	0.154756	0.27856	0.198971	0.116067	0.126618	0.154756	0.154756	0.27856	0.1741	0.154756	0.198971	0.1741
W_3	0.186189	0.209463	0.418925	0.152336	0.119693	0.239386	0.119693	0.1289	0.186189	0.152336	0.209463	0.16757	0.239386	0.239386	0.209463	0.239386	0.279283	0.16757
W_4	0.127762	0.184544	0.150991	0.276817	0.127762	0.207613	0.237271	0.16609	0.16609	0.237271	0.276817	0.207613	0.184544	0.118636	0.16609	0.150991	0.207613	0.150991

The "state of nature" means the relation of the competence level of a given employee, whose validity has been correctly assessed for the purpose of repairing a given resource in the enterprise, as compared to the time that is taken to repair a given resource. The following rules for the definition of the "maintenance expert selection map" are defined:

for $N_{RiWt} \in (1.2;1.66>$ very strongly recommended for the repair of a given resource

for $N_{RiWt} \in (0.8.;1.2>$ strongly recommended for the repair of a given resource

for $N_{RiWt} \in (0.5;0.8>$ recommended for the repair of a given resource

for $N_{RiWt} \in (0.27;0.5>$ weakly recommended for the repair of a given resource

for $N_{RiWt} \in <0;0.27>$ not recommended for the repair of a given resource

Table 12 presents the "maintenance expert selection map" that is based on the value of the "States of nature" (Table 11).

Table 12. The "maintenance expert selection map".

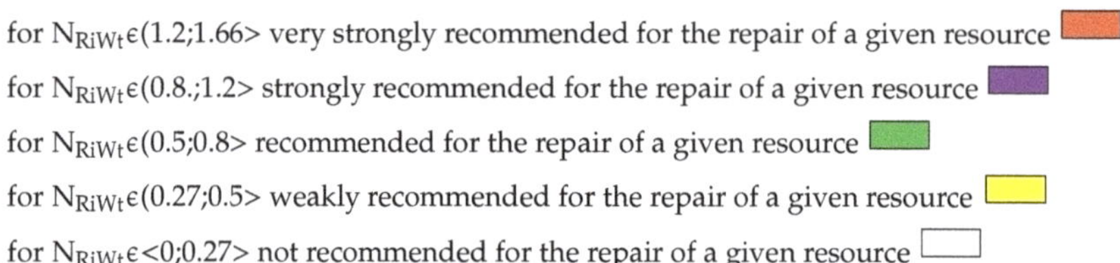

	R$_1$	R$_2$	R$_3$	R$_4$	R$_5$	R$_6$	R$_7$	R$_8$	R$_9$	R$_{10}$	R$_{11}$	R$_{12}$	R$_{13}$	R$_{14}$	R$_{15}$	R$_{16}$	R$_{17}$	R$_{18}$
W$_1$	green	yellow	yellow			yellow	yellow									green		yellow
W$_2$		yellow						yellow						yellow				
W$_3$			yellow														yellow	
W$_4$				yellow							yellow							

Hence, the maintenance expert remains undefined in our case study, since no "state of nature" has been marked in red. The maintenance manager received support to select W$_1$ to repair resources R$_1$ and R$_{16}$, W$_2$ was selected to repair resources R$_2$,R$_8$ and R$_{16}$, W$_3$ was selected to repair resources R$_3$, and R$_{17}$, W$_4$ was selected to repair resources R$_4$ and R$_{11}$. It should provide a concise and precise description of the experimental results, their interpretation, as well as the experimental conclusions that can be drawn.

5. Discussion

The proposed approach makes it possible to identify the competences of maintenance department employees and it also makes it possible to give them validity to assign repair work correctly, depending on the type of failure.

The model also assumes that data is obtained regarding the repair times of a given machine, from IT systems that the enterprise has implemented. Thanks to such defined elements, it is possible to define the so-called "state of nature" for each enterprise resource vis-à-vis each employee. The higher the value of a given natural state (max = 1.66), the better the matching of a given employee, to a given failure, on a given resource. In addition, formalisation rules were used for each element in the proposed approach in order to implement the IT system. The proposed solution allows for maintenance managers to increase the availability of the resources of the enterprise.

Quantitatively defining the competences of employees in our research was a particularly difficult task.

Therefore, formalisation rules were strictly defined for each type of competence, with the IT implementation of the given approach then being presented.

The use of the proposed approach allows for decision making to be supported when selecting an expert with the best ratio of competences, in relation to the time that is required to repair a failure in a given resource. The use of the maintenance expert selection map is helpful in:

- Selecting a maintenance expert, from among available employees, to repair a given resource.
- Selecting the scope of employee training, in order to improve the competences of employees in relation to the effective repair of resources, by shortening the elimination time of failures and by reducing the downtime of failures.

- Defining a motivating system for all maintenance workers based on the value of natural states.

The proposed approach is currently implemented in the form of a decision support system when selecting the most effective maintenance employees for repairing failures in manufacturing resources. The current implementation work has been partly completed in the form of a web-application, related to the definition of competences. The five competence questionnaires were defined and implemented. This application also implemented the FAHP method. Our web-application will be extended by the algorithm for automatic data, for the repair times of each failure and also extended by the acquisition and automatic creation of a maintenance expert selection map as part of further research.

6. Conclusions

In the research presented, an innovative approach to selecting expert workers in a maintenance department is created, in part, also in the form of a web-application. The "maintenance expert selection map" innovation, so-called, provides the formal procedures for describing the competences of each maintenance worker and defining the best "state of nature". It was established that the "maintenance expert selection map" is useful in:

- Describing the competence of workers.
- Delecting workers according to competence for repairing a given manufacturing resource.
- Determining the scope of employee training.

In sustainable manufacturing, the right employee, with the proper competence and being employed in the right workplace is crucial [33]. The proposed approach can help to increase the sustainability of the company in all three of its dimensions:

- Economic and environmental—the proposed approach allows for managers to assign a particular worker to repair a given resource; selecting this worker will guarantee the maximum availability of the manufacturing resource. The right assignment of highly-qualified maintenance staff to repair a resource results in lower downtime costs, lower additional costs due to defective products, and a reduction in the risk of the possibility of total damage and of the risk of loss of warranty.
- Social—the proposed approach allows not only for the core competences to be determined, but also the need for new competences and the demand for training programmes for low-qualified maintenance department workers. By using the proposed approach, the manager may decide to assign a given employee to a place of work that is more appropriate to his or her qualifications, which will ultimately translate into the achievement of better working conditions.

We are planning to expand the approach with the dynamic measurements idea in our further works [44,45]. In the next stage, a method for inspection for the obtained improvement of availability of the manufactured resource thanks to the use of the proposed "maintenance expert selection map" will be developed. Subsequently, the model for updating the assessment of availability based on the formulation of additional events, which may affect the extension of the repair time of the resource, regardless of the competences of the employees, will be established.

Although this study is an attempt at dealing with the several aspects of decision making, to be supported when selecting an expert with the best ratio of competences, there are still some limitations, which could be considered in future research.

Firstly, there is the limitation in creating a standard for our approach to supporting the selection of maintenance experts, namely, the development of an application. Secondly, there may be some limitations when it comes to providing integration with the CBM standard for developing the application of open-software, since no standard currently exists. Thirdly, the approach should be a prominent inscribed feature in the company's development strategy and it should also be part of the evaluation of employees and the system by which they are motivated.

The proposed "maintenance expert selection map" approach will be useful for maintenance department managers and will allow them to define not only core competences, but also to enjoy maximum availability regarding manufactured resources and preparing training programmes that are adequate to the needs both of current and of new employees, despite the above limitations.

Author Contributions: Conceptualization, J.P.-M. and S.K.; Formal analysis, J.P.-M.; Investigation, J.P.-M.; Methodology, J.P.-M.; Resources, J.P.-M. and S.K.; Software, J.P.-M.; Validation, J.P.-M.; Writing – original draft, J.P.-M. and S.K.; Writing – review & editing, J.P.-M.

Acknowledgments: This work is supported by program of the Polish Minister of Science and Higher Education under the name "Regional Initiative of Excellence" in 2019 - 2022, project no. 003/RID/2018/19, funding amount 11 936 596.10 PLN).

Conflicts of Interest: The authors declare no conflict of interest.

References

1. Naskar, S.; Basu, P.; Sen, A.K. A literature review of the emerging field of IoT using RFID and its applications in supply chain management. In *The Internet of Things in the Modern Business Environment*; IGI Global: Hershey, USA, 2017.
2. Wang, S.; Wan, J.; Zhang, D.; Li, D.; Zhang, C. Towards smart factory for Industry 4.0: A self-organized multi-agent system with big data based feedback and coordination. *Comput. Netw.* **2016**, *101*, 158–168. [CrossRef]
3. Wollschlaeger, M.; Sauter, T.; Jasperneite, J. The future of industrial communication: Automation networks in the era of the internet of things and industry 4.0. *IEEE Ind. Electron. Mag.* **2017**, *11*, 17–27. [CrossRef]
4. Kłos, S.; Patalas-Maliszewska, J. Using a Simulation Method for Intelligent Maintenance Management. In *International Conference on Intelligent Systems in Production Engineering and Maintenance: ISPEM 2017*; Advances in Intelligent Systems and Computing; Springer International Publishing: Cham, Switzerland, 2018; Volume 637, pp. 85–95.
5. Holmberg, K.; Adgar, A.; Arnaiz, A.; Jantunen, E.; Mascolo, J.; Mekid, S. *E-Maintenance*, 1st ed.; Springer: London, UK, 2010.
6. Bommer, S.C.; Fendley, M. A theoretical framework for evaluating mental workload resources in human systems design for manufacturing operations. *Int. J. Ind. Ergon.* **2018**, *63*, 7–17. [CrossRef]
7. Kaasinena, E.; Schmalfuß, F.; Özturkc, C.; Aromaa, S.; Boubekeur, M.; Heilala, J.; Heikkilä, P.; Kuula, T.; Liinasuo, M.; Mach, S.; et al. Empowering and engaging industrial workers with Operator 4.0 solutions. *Comput. Ind. Eng.* **2019**. [CrossRef]
8. del Amoa, I.F.; Erkoyuncua, J.A.; Roy, R.; Palmarini, R.; Onoufriou, D. A systematic review of Augmented Reality content-related techniques for knowledge transfer in maintenance applications. *Comput. Ind.* **2018**, *103*, 47–71. [CrossRef]
9. Ni, J.; Gu, X.; Jin, X. Preventive Maintenance Opportunities for Large Production Systems. *CIRP Ann. Manuf. Technol.* **2015**, *64*, 447–450. [CrossRef]
10. Ni, J.; Jin, X. Decision Support Systems for Effective Maintenance, Operations. *CIRP Ann. Manuf. Technol.* **2012**, *61*, 411–414. [CrossRef]
11. Xiao, L.; Song, S.; Chen, X.; Coit, D.W. Joint optimization of production scheduling and machine group preventive maintenance. *Reliab. Eng. Syst. Saf.* **2016**, *146*, 68–78. [CrossRef]
12. Jin, X.; Li, L.; Ni, J. Option model for joint production and preventive maintenance system. *Int. J. Prod. Econ.* **2009**, *119*, 347–353. [CrossRef]
13. Masoni, R.; Ferrise, F.; Bordegoni, M.; Gattullo, M.E.; Uva, A.E.; Fiorentino, M.; Carrabba, E.; Donatoe, M. Supporting remote maintenance in industry 4.0 through augmented reality. *Procedia Manuf.* **2017**, *11*, 1296–1302. [CrossRef]
14. Palmarini, R.; Erkoyuncu, J.A.; Roy, R.; Torabmostaedi, H. A systematic review of augmented reality applications in maintenance. *Robot. Comput. Integr. Manuf.* **2018**, *49*, 215–228. [CrossRef]
15. Scurati, G.W.; Gattullo, M.; Fiorentino, M.; Ferrisea, F.; Bordegonia, M.; Uvab, A.E. Converting maintenance actions into standard symbols for Augmented Reality applications in Industry 4.0. *Comput. Ind.* **2018**, *98*, 68–79. [CrossRef]
16. Roy, R.; Stark, R.; Tracht, K.; Takata, S.; Mori, M. Continuous maintenance and the future—Foundations and technological challenges. *CIRP Ann. Manuf. Technol.* **2016**, *65*, 667–688. [CrossRef]

17. Kothamasu, R.; Huang, S.; Verduin, W.H. System health monitoring and prognostics-are view of current paradigms and practices. *Int. J. Adv. Manuf. Technol.* **2006**, *28*, 1012–1024. [CrossRef]

18. Bousdekis, A.; Papageorgiou, N.; Magoutasa, B.; Apostolouab, D.; Mentzasa, G. Enabling condition-based maintenance decisions with proactive event-driven computing. *Comput. Ind.* **2018**, *100*, 173–183. [CrossRef]

19. Wu, F.; Wang, T.; Lee, J. An online adaptive condition-based maintenance method for mechanical systems. *Mech. Syst. Signal Process.* **2010**, *24*, 2985–2995. [CrossRef]

20. Shin, J.H.; Jun, B.J. On condition based maintenance policy. *J. Comput. Des. Eng.* **2015**, *2*, 119–127. [CrossRef]

21. Belkadia, F.; Dhuieb, M.A.; Aguadoc, J.V.; Larochea, F.; Bernard, A.; Chinesta, F. Intelligent Assistant System as a context-aware decision-making support for the workers of the future. *Comput. Ind. Eng.* **2019**. [CrossRef]

22. Toro, C.; Sanín, C.; Vaquero, J.; Posada, J.; Szczerbicki, E. Knowledge based industrial maintenance using portable devices and augmented reality. In *Knowledge-Based Intelligent Information and Engineering Systems*; Springer: Berlin/Heidelberg, Germany, 2007; pp. 295–302.

23. Espíndola, D.B.; Fumagalli, L.; Garetti, M.; Pereira, C.E.; Botelho, S.S.; Henriques, R.V. A model-based approach for data integration to improve maintenance management by mixed reality. *Comput. Ind.* **2013**, *64*, 376–391. [CrossRef]

24. Hecklaua, F.; Galeitzkea, M.; Flachsa, S.; Kohlb, H. A holistic approach to human-resource management in Industry 4.0. *Procedia CIRP* **2016**, *54*, 1–6. [CrossRef]

25. Patalas-Maliszewska, J.; Kłos, S. An Intelligent System for Core-Competence Identification for Industry 4.0 Based on Research Results from German and Polish Manufacturing Companies. In *International Conference on Intelligent Systems in Production Engineering and Maintenance: ISPEM 2017*; Advances in Intelligent Systems and Computing; Springer International Publishing: Cham, Switzerland, 2018; Volume 637.

26. Igba, J.; Alemzadeh, K.; Anyanwu-Ebo, I.; Gibbons, P.; Friis, J. A Systems Approach Towards Reliability-Centred Maintenance (RCM) of Wind Turbines. *Procedia Comput. Sci.* **2013**, *16*, 814–823. [CrossRef]

27. Selvik, J.T.; Aven, T. A framework for reliability and risk centered maintenance. *Reliab. Eng. Syst. Saf.* **2011**, *96*, 324–333. [CrossRef]

28. Gupta, G.; Mishra, R.P. Identification of Critical Components Using ANP for Implementation of Reliability Centered Maintenance. *Procedia CIRP* **2018**, *69*, 905–909. [CrossRef]

29. Bekkaoui, M.; Karray, M.-H.; Sari, Z. Knowledge formalization for experts' selection into a collaborative maintenance platform. *IFAC-PapersOnLine* **2015**, *48*, 1445–1450. [CrossRef]

30. Potes Ruiz, P.; Kamsu Foguem, B.; Grabot, B. Generating knowledge in maintenance from Experience Feedback. *Knowl.-Based Syst.* **2014**, *68*, 4–20. [CrossRef]

31. Angele, J.; Lausen, G. Ontologies in f-logic. In *Handbook on Ontologies*; Staab, S., Studer, R., Eds.; Springer: Berlin, Germany, 2004; pp. 29–50.

32. Yao, H.; Etzkorn, L. Automated conversion between different knowledge representation formats. *Knowl.-Based Syst.* **2006**, *19*, 404–412. [CrossRef]

33. Melosi, F.; Campana, G.; Cimatti, B. Competences Mapping as a Tool to increase Sustainability of Manufacturing Enterprises. *Procedia Manuf.* **2018**, *21*, 806–813. [CrossRef]

34. Decius, J.; Schaper, N. The Competence Management Tool (CMT)—A new instrument to manage competences in small and medium-sized manufacturing enterprises. *Procedia Manuf.* **2017**, *9*, 376–383. [CrossRef]

35. Patalas-Maliszewska, J. *Reference Models of Knowledge Management for Manufacturing Companies*; PWN: Warsaw, Poland, 2019.

36. Patalas-Maliszewska, J.; Skrzeszewska, M. An Evaluation of the Effectiveness of Applying the MES in a Maintenance Department—A Case Study. *Found. Manag.* **2018**, *10*, 257–270. [CrossRef]

37. Madhusudanan, N.; Chakrabarti, A. A questioning based method to automatically acquire expert assembly diagnostic knowledge. *Comput. Aided Des.* **2014**, *57*, 1–14. [CrossRef]

38. Gruber, T.R. The acquisition of strategic knowledge. In *Perspectives in Artificial Intelligence*; Academic Press: Cambridge, MA, USA, 1989; Volume 4.

39. Preston, P.; Edwards, G.; Compton, P. A 1600 Rule Expert System Without Knowledge Engineers. In Proceedings of the Second World Congress on Expert Systems, Moving Towards Expert Systems Globally in the 21st Century, New York, NY, USA; 1993; pp. 220–228.

40. Boose, J.H.; Bradshaw, J. Expertise transfer and complex problems: Using AQUINAS as a knowledge-acquisition workbench for knowledge-based systems. *Int. J. Man Mach. Stud.* **1987**, *26*, 3–28. [CrossRef]

41. Winter, G.B. An automated knowledge acquisition system for model-based diagnostics. In Proceedings of the AUTOTESTCON'92, IEEE Systems Readiness Technology Conference, Conference Record, Metropolitan, NY, USA, 12–14 May 1992.

42. Cheah, W.P.; Kim, Y.S.; Kim, K.-Y.; Yang, H.J. Systematic causal knowledge acquisition using FCM constructor for product design decision support. *Expert Syst. Appl.* **2011**, *38*, 15316–15331. [CrossRef]

43. Nydick, R.L.; Hill, R.P. Using the analytic-hierarchy process to structure the supplier selection procedure. *Int. J. Purch. Mater. Manag.* **1992**, *28*, 31–36. [CrossRef]

44. Lopez de Lacalle, L.N.; Viadero, F.; Hernandez, J.M. Applications of dynamic measurements to structural reliability updating. *Probabilistic Eng. Mech.* **1996**, *11*, 97–105. [CrossRef]

45. Coro, A.; Abasolo, M.; Aguirrebeitia, J.; Lopez de Lacalle, L.N. Inspection scheduling based on reliability updating of gas turbine welded structures. *Adv. Mech. Eng.* **2019**, *11*, 1–2. [CrossRef]

Article

Pushing Digital Automation of Configure-to-Order Services in Small and Medium Enterprises of the Construction Equipment Industry: A Design Science Research Approach

Christoph Paul Schimanski [1,2,*], **Gabriele Pasetti Monizza** [2], **Carmen Marcher** [1,2] **and Dominik T. Matt** [1,2,*]

[1] Free University of Bozen-Bolzano, Faculty of Science and Technology, Piazza Università 5, 39100 Bolzano, Italy
[2] Fraunhofer Italia Research, Via A.-Volta 13A, 39100 Bolzano, Italy
*   Correspondence: christophpaul.schimanski@natec.unibz.it (C.P.S.); dominik.matt@unibz.it (D.T.M.)

Received: 9 August 2019; Accepted: 5 September 2019; Published: 9 September 2019

Featured Application: Our developed artefacts can help practitioners from the construction equipment industry understand the added value of the Building Information Modeling method and its role as an Industry 4.0 enabler in construction. We further demonstrate new ways to exploit Building Information Models to achieve a higher degree of digital automation in small and medium enterprise organizations.

Abstract: In order to efficiently transform business processes (such as product design, product engineering, production, logistics, sales, deliveries, etc.) into digitally automated processes, new concepts have been introduced in both the manufacturing and construction industries. Under the term Industry 4.0, promising possibilities for high-performance production processes are emerging based on e.g., digital twins and cyber-physical systems. However, the construction industry lags behind in adapting these ideas, and is still facing severe productivity deficits. This paper addresses these deficits by assessing the hypothesis of Building Information Modeling—as the digital twinning methodology in construction—representing a key driver for digital automation and thus enabling more productive processes. To this end, we apply a design science research approach to develop artefacts using computational methods for the automation of business processes in a configure-to-order industry partner. The evaluation is done in the context of a pilot project together with this industry partner. The findings obtained in the pilot project revealed time savings in the phases of bid estimation and work preparation. Based on the findings, the applicability and utility of the suggested approach are discussed and allow for the conclusion that Building Information Model data can usefully streamline and automate many processes at the interface between design and production, if structured and preprocessed purposefully.

Keywords: Industry 4.0; configure-to-order; BIM; construction equipment; digital information flow; lean assembly; digital twins; cyber-physical production systems

---

1. Introduction

According to previous research activities [1], the construction industry is characterized as one of the less efficient industries on the market, and the productivity gap with other industries has been growing over the last decade. The construction industry is characterized by highly customized products and components that increase the complexity of the information and material flow [2,3]. With a development toward an escalating number of individual product variants and product configurations, the need for adaptable production and the complexity of manufacturing processes and systems increase simultaneously.

Manufacturing systems in a mass-customization environment should be able to produce small quantities in a highly flexible way and should be rapidly reconfigurable [4,5]. The latest trends in mass customization are the concepts of "Industry 4.0" or "cyber-physical systems" (CPS). Industry 4.0 can be understood as the development toward more digitization and automation in production [6]. The term CPS describes particularly the continuously growing integration of information and communication technology (ICT) with physical real-world entities. The large potential of Industry 4.0 will be a key enabler for further developments in mass-customization manufacturing [7].

This paper extends the work published in [8] to investigate strategies for the digital transformation of the information flow and business processes of a configure-to-order (CtO) service provider as a local industry partner. This industry partner designs and configures temporary container facilities according to individual customer needs. We extend our previous work by presenting more findings obtained in the pilot project and an in-depth discussion on applicability and utility with respect to practical implementations.

As outlined by Girmscheid [9], the information flow in construction projects has to be considered the core element for planning and controlling the production in an industrialized environment. The industry partner involved in this study belongs to the construction equipment industry, which, compared to the main construction industry, has had even less contact with Industry 4.0 concepts.

Even though research [10] shows that the discussion about the adaptation of Industry 4.0 in construction has not yet been finished, in this sector the method of Building Information Modeling (BIM) plays the role of enabling Industry 4.0 ideas [1,3,10]. BIM describes a method of cross-linked planning, construction, and operation of buildings, infrastructures, and industrial facilities using software [11]. Its significance as an Industry 4.0 enabler is mainly due to the reason that BIM, as the representative of digital twinning in the construction industry [12], can provide for the digital characterization and hence the information basis of all relevant physical and functional properties of construction projects [13]. This in turn, comprises the starting point for any digital construction site support.

In particular, BIM is crucial for the interplay of digital and nondigital entities, and thus essential for the establishment of cyber-physical production systems in construction [10], through which manufacturing processes are already positively affected [14]. Taking advantage of the BIM approach, the authors suggest a digital process flow using computational design and digital fabrication techniques to enhance the automation and the efficiency of the information flow in selected business processes. Digital fabrication refers to production processes controlled by computers. The business processes considered here refer to a customized product in a container construction pilot project and comprise all interrelated activities from capturing the customer requirements to the bid estimation and work preparation phase. Figure 1 shows the technique applied in this study (b) in comparison to the schema of conventional computational design (a) by Bohnacker et al. [15]:

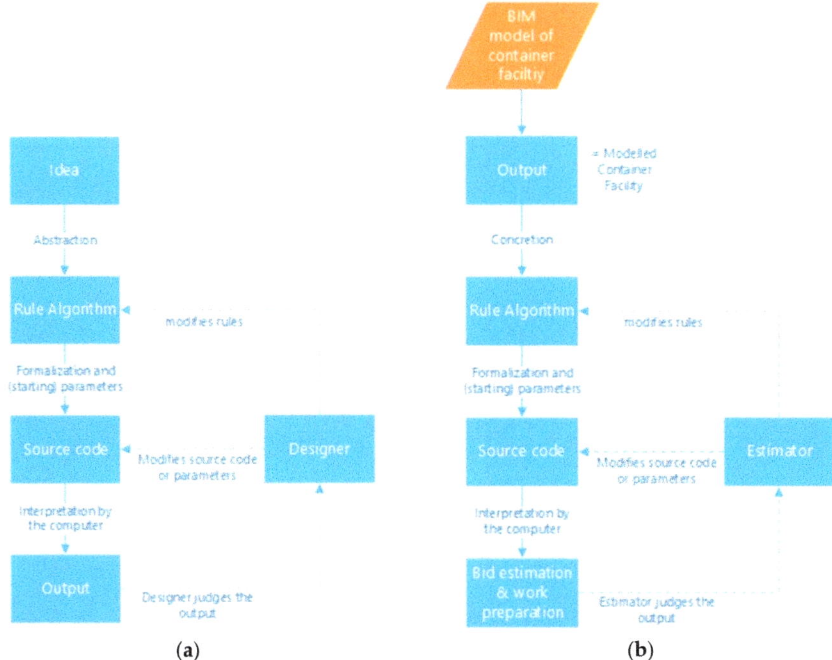

Figure 1. Computational design: (**a**) Conventional schema by Bohnacker et al. [15] vs. (**b**) Application in this study.

Despite the conventional technique, we will begin our considerations with an accurately designed BIM model of the container facility containing all relevant customer requirements in terms of metadata. This BIM model serves as a starting point to formulate algorithms that can process the concrete elements of the model to the bid estimation and work preparation phase. The final assembly, which is based on the work preparation phase, is then assumed as digital fabrication, since it is entirely based on digital information, even if this is done with the help of physical manpower. Analogous to conventional computational design, these algorithms are translated into code and interpreted by the computer. The user must validate the result, which is the outcome of an automated and BIM-based procedure.

Even though BIM is becoming increasingly important in the construction industry [16], and affecting most of the construction stakeholders, little attention is paid in the scientific community to the benefits for the construction equipment industry. This raises the research questions of:

- How to use BIM to benefit the construction equipment industry?
- What is the impact of BIM implementation in the construction equipment industry on process automation?

In the context of the pilot project presented here, a total of four artefacts applying a design science research (DSR) approach have been developed in order to assess the hypothesis that BIM can streamline automation and information flow in the construction equipment industry.

The body of this article is organized as follows. Section 2 draws a line to related work from previous scientific studies regarding BIM-based automation in the construction equipment industry. Section 3 provides a deeper explanation of the applied research strategy. Sections 4 and 5 describe the development (4) and evaluation (5) of the artefacts with respect to DSR. In terms of evaluation, we will show how the joint application of the artefacts can enable a modular and automized configuration of a container system regarding the customer's demand. Also, in Section 5, we will quantify the benefits of the proposed approach in terms of time-savings, and discuss its applicability and utility in real-world scenarios. Section 6 provides a final discussion, and Section 7 concludes this study.

2. Related Work

The discussion surrounding BIM and construction equipment in the literature is multifaceted, but, from our point of view, it is not yet sufficiently exhausted. On the one hand, the richness of facets is shown by an ambiguous understanding of the notion of *equipment* in construction. As an example, Lau et al. [17] presented a tool for a BIM-based cost analysis of equipment in buildings. However, they refer to equipment as components relevant for facility management in the operation phase rather than machinery supporting the construction processes themselves, which is our focus. Other research has shown the importance of studying construction tools [18]. However, we want to narrow down the scope of our considerations to BIM being applied in processes on the construction site that support the actual value-adding main production processes leveraging machinery and/or temporary facilities. Thus, our definition of construction equipment is in line with the list provided by Jahr and Borrmann [19], who classify a total of seven construction equipment groups. These groups include—besides the earlier mentioned machineries and temporary facilities—storage and traffic areas, supply of water/power and waste disposal, site securities such as fences and scaffoldings, as well as excavation support.

The biggest share of research in the light of BIM and construction equipment deals with applications for the automated or semi-automated selection of construction machineries such as cranes or hoists, and site layout planning with respect to certain project characteristics defined by the BIM model and/or external variables such as e.g., the maximum crane lifting capacity. With this regard, Ji et al. [20] used expert interviews and technical standards to define optimization models and criteria for tower crane planning. After this optimization model proposes an optimal crane constellation, BIM models are taken into account to visually confirm and check the plausibility of the result, leveraging the 3D representation, but not the metadata. Abbott et al. [21] addressed a selection support for mobile cranes, which are characterized by their varying locations of application on-site. Here, the focus is on avoiding accidents with other agents that are present in the current operating area. In this context, they propose deploying BIM information, particularly the 3D visualizations.

Ji and Leite [22] presented a framework for reviewing tower crane planning. This framework incorporates BIM 4D as input information to a software prototype that applies a rule-based checking engine for evaluating the crane plans. Another example for the rule-based checking of BIM models for site layout optimization problems can be found in [23]. The authors make use of Business Rule Management Systems (BRMS) to check BIM models represented by Industry Foundation Classes (IFC) files against predefined rules that could support humans when dimensioning resources such as construction equipment.

The earlier mentioned work of Jahr and Borrmann [19] also addresses the utilization of BIM for equipment selection and configuration as part of a rule-based knowledge inference system. However, as opposed to the studies mentioned above, they exclusively consider the BIM model's metadata, not accounting for 3D representations of the equipment in question when deriving the decision.

Kan et al. [24] made use of CPS for planning and controlling the movements of mobile cranes on-site that exploit sensor data for localization and use a 3D model for the control interface. However, this virtual model in the interface is based on a video game engine, and does not provide semantic data as BIM models do.

Han et al. [25] proposed a decision support system for selecting the best matching cranes in the heavy civil industry making use of 3D simulations. Li et al. [26] made use of mathematical algorithms to optimize crane positions and material supply drop-off points for trucks in high-rise constructions, also taking into account BIM models. Here, the BIM models are mainly used for a 4D simulation to verify the plausibility of the defined points on-site. Al-Hussein et al. [27] are addressing optimal crane selections by means of simulations as well. However, their simulations are based on 3D representations, whilst the metadata of BIM models are not considered.

Besides crane selection and layout planning, other scientific work in the light of BIM and construction equipment must be mentioned: another related field of research deals with decision-making processes in the work preparation phase, where BIM can be considered a source of knowledge. A question that arises in this context is for example which technological solution to decide for when several solutions are possible. Practical examples for this problem are given in terms of comparison of conventional and modular construction [28], and assessment of building stocks [29]. In addition to that, Kim et al. [30] presented a decision support system for the automatic generation of scaffolding plans in different scenarios, which are evaluated according to safety, cost, and installation durations based on BIM information as input. Likewise, in the field of scaffolding planning, Feng and Hsu [31] proposed a framework for automatized planning using BIM authoring software which is based on an ontology model from the perspective of material management, taking into account BIM-based quantity take-offs.

In conclusion of this literature review, to our knowledge, there have been very few approaches leveraging the actual metadata available in BIM models to automate and streamline the internal processes of a construction equipment supplying company. Most of the studies account solely for 3D visualization as an inherent part of BIM models with respect to optimization and layout problems. In particular, we have found little scientific attention to the assembly and installation of temporary construction containers themselves, which are the focus of this study.

3. Research Strategy

The approach in this study follows a design science research (DSR) approach. DSR foresees obtaining new knowledge by designing artefacts (which can be e.g., processes or products) and evaluating their utility and applicability in practice [32]. The latter is directed to an a priori identified problem for which a hypothesis with respect to a potential solution is formulated. Eventually, the artefacts and their application aim at assessing this hypothesis.

As a starting point for the DSR approach in this study, we have identified the lack of digital automation in the construction equipment industry (PROBLEM), whilst we see BIM as a potential key for driving digital automation in this sector (HYPOTHESIS). We have developed a total of four artefacts (DEVELOPMENT), which are namely (1) a new BIM-based process definition, (2) a BIM object library, (3) a container management tool, and (4) a database infrastructure representing a virtual container warehouse. The development phase is described in Section 4.

The evaluation of the artefacts according to DSR has been conducted in a real-world pilot project together with an industry partner from Northern Italy. The findings are given in Section 5.2. The impacts on the findings for the evaluation of applicability and utility according to DSR are discussed in Section 5.3. The applied research approach is illustrated in Figure 2.

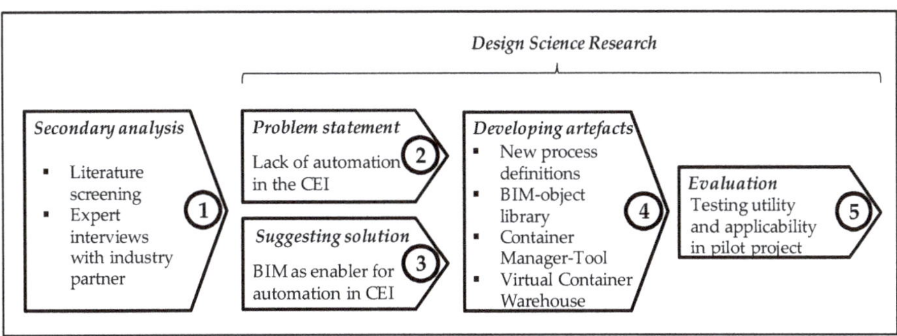

Figure 2. Research strategy.

4. Development Phase According to Design Science Research

As mentioned above, the DSR approach is characterized by an initial problem statement. The problem statement here has evolved through a literature screening and process analysis of the involved industry partner. To understand the potential improvement measures in this CtO container service environment, it is important to understand how the containers dealt with can be principally composed and configured with respect to customer demand: The containers are categorized into different typologies (BM10′, BM16′, BM20′, BM24′, and BM30′) varying in size. The typology denominations refer to the length of the longitudinal side in the unit of feet. Furthermore, the containers are modular components whose external walls are composed of panels with equal geometric dimensions. The longer the longitudinal side, the more panels can be placed there (panel slots). The short side (NORTH + SOUTH) of all container types always has two panel slots (Figure 3). The panels themselves are subdivided into five types differentiated by their functions: (1) full panel, (2) window panel, (3) sanitary window panel, (4) glazed door panel, and (5) door panel.

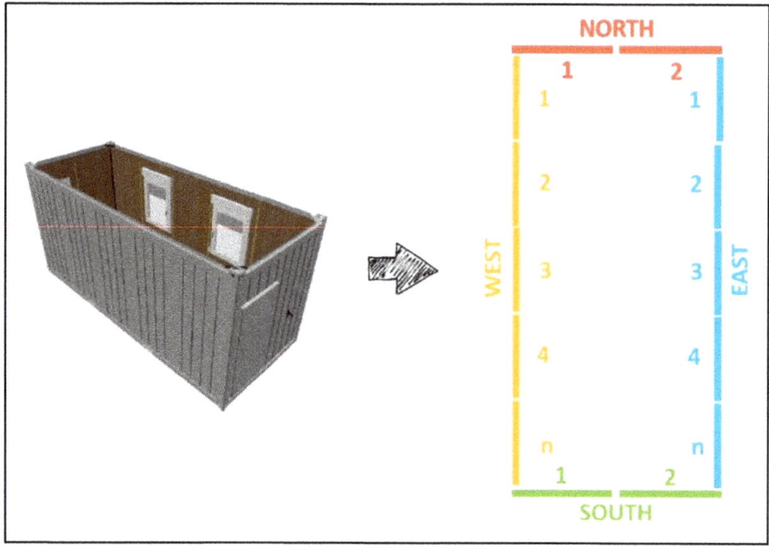

Figure 3. General container and panel layout.

This means that, depending on its longitudinal dimension, a container can have **n** panels on the longitudinal side and exactly two panels on the short side. For unambiguous identification, we have classified the possible panel positions by their sky orientation and sequence (Figure 3). The single panels can be modified arbitrarily according to the customer's requirements to form various types of container facilities such as e.g., temporary schools or construction site offices.

At the time of our investigations, the main internal processes of the industry partner whose core business is to plan and install container facilities as rental properties were essentially characterized by (1) requirement analysis of customer demand regarding the container facilities to be erected; (2) translation of these requirements into 2D floor plans using computer-aided design (CAD) software; (3) physical lookup in the local container warehouse to identify which currently present container corresponds the most to the drawing, and hence requires the least modification effort; (4) manual entry of the container's rental ID number of the containers selected onto the floor plan printout (Figure 4); (5) manual transfer of rental ID from printout to CAD software; (6) estimation of modification effort of selected containers with respect to the drawing based on experience; (7) making bid to customer. Figure 5 shows the simplified process from customer inquiry to contract award.

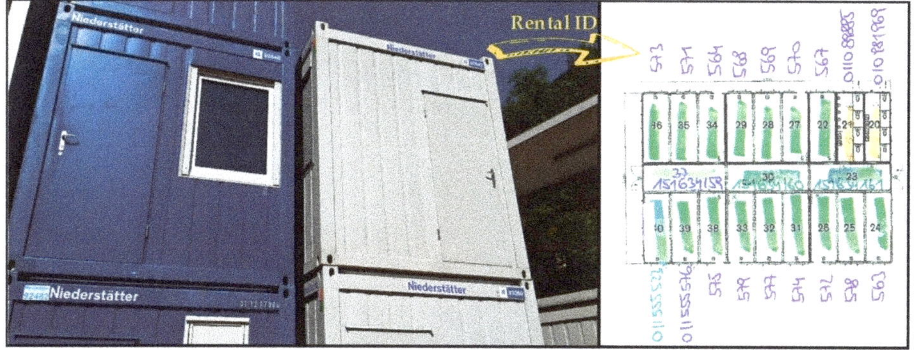

Figure 4. Manual assignment of rental IDs.

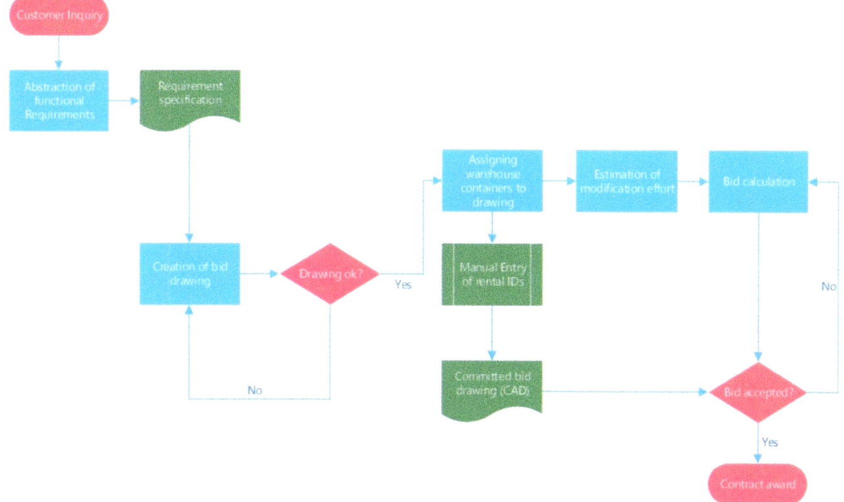

Figure 5. Current overall process.

In addition, the involved industry partner maintained a sort of 2D plan repository of PDF plans for every container in its current configuration. As a consequence, after panel modification for the current project, every 2D plan with respect to both floor plans and views had to be updated manually by means of CAD, which the industry partner considered very time consuming.

Considering the above, the identified potentials for improvement can be stated as:

- Manually maintained 2D drawing repository, which is not linked to the container warehouse
- No documentation about the current container warehouse inventory
- Container assignment (= assigning rental ID to drawing) requires an employee to physically inspect the warehouse
- Effort for panel modifications is estimated based on experience
- Rental IDs are entered manually on 2D plan printouts.

The following actions are derived from these improvement potentials:

- Development of a virtual container warehouse that represents the current availability of the physical containers

- Development of a parametrized BIM object library, which contains the different container typologies and panel types
- Accessibility of the customer requirements stored in the BIM model
- Creation of a link between BIM models and the virtual container warehouse
- Automatic transfer of the assigned rental IDs to the BIM model
- Automatic determination of panel modification and assembly effort for the selected containers
- Semi-automated bid estimation based on the BIM model and assigned containers from the warehouse
- Automatic maintenance of the 2D drawings repository as derivatives of the BIM models.

Based on that, we developed the following artefacts with respect to DSR:

4.1. BIM-Based Process Definitions

4.1.1. Creation of Bid Drawing

The bid drawing used to be created exclusively in 2D using a CAD software system. For this purpose, a library of so-called *drawing blocks* is available, which contains different container typologies. When creating the bid drawing, the customer requirements are reflected exactly as requested. This means that all containers are drawn with the desired panel configuration, without considering the actual warehouse inventory.

This principle approach was kept in the new BIM-based process of bid drawing creation. However, in the new process, the working methodology changes from a two-dimensional drawing to a three-dimensional, BIM-based approach, which replaces the CAD blocks with BIM objects. The BIM objects are part of the BIM object library containing all necessary container typologies. After the requested container typology has been set in the model, the panel configuration can be adjusted by selecting the dedicated panel slot and choosing a panel type from a dropdown list. Then, the information about the container typology and panel configuration can be exported from the BIM authoring tool to the so-called *Container Manager* (MS Excel workbook). There, the container information is available for further processing.

4.1.2. Container Allocation

The container allocation in the new BIM-based process differs fundamentally from the current process in that no physical inspection of the warehouse will be necessary. The required container typologies and panel configurations according to customer requirements are exported from the BIM modeling software to the *Container Manager*. The *Container Manager* provides direct access to the *Virtual Container Warehouse* in the form of an MS Access database. This merging of the two information sources and data streams in the *Container Manager* enables a target-oriented allocation of the most suitable containers according to the drawing requirements. For this purpose, a selection support has been developed in the *Container Manager*, which allows for filtering according to the currently available inventory containers with the lowest effort in terms of panel modification.

If the physical containers are assigned from the warehouse, their status in the virtual container warehouse automatically changes to *"lent"*. In addition, the current panel configuration (as metadata) is updated to that of the bid drawing, so that the correct status is already stored in the virtual container warehouse on return.

As a further functionality within the new BIM-based process, it is also possible to import the assigned rental IDs back into the BIM model, so that manual input is no longer necessary. In addition, at the push of a button, all 2D drawings of individual containers comprising the entire facility are updated according to the bid drawing, and plan files are generated. These are then to be saved as PDF files so that the 2D drawing repository can be updated within a very short time without any additional effort.

4.1.3. Bid Estimation

The bid estimation in the new BIM-based process will change in such a way that numerous calculation steps will be automated, and calculation becomes thoroughly BIM-based. For example, this means that the overall effort for panel modification and assembly will be no longer estimated based on experience, but deterministically calculated based on the exact number of panels to be modified according to the BIM model and actual inventory. The comparison of as-planned and as-stocked is done within the *Container Manager*. When the user assigns a certain physical container from the warehouse to a specific drawn container of the bid drawing, the *Container Manager* reveals information on how many panels must be installed, removed, or exchanged in this particular match. In addition, it becomes possible to attribute the effort for modifications with cost rates to calculate accurate bidding prices.

4.1.4. Assembly Planning

The new BIM-based assembly planning provides for an improvement in the determination of assembly and delivery times: Based on the detailed knowledge of real costs and the coupling to the BIM model and the virtual container warehouse, it is possible to calculate the required time for the modification of the panel configuration as well as the assembly on site. This enables the *Container Manager* to calculate the required start dates using a backward calculation based on the agreed delivery and erection dates, and thus also supports the control of the site logistics.

4.2. BIM Object Library

The BIM object library consists of a total of five different container typologies, which have been preconfigured for immediate project use. The containers follow a modular design and can be differentiated according to panel configuration and length (Table 1).

Table 1. Selected container typologies.

Typology	Length (ext.) [m]	Height (ext.) [m]	# Panels (long.) *	Width (ext.) [m]	Weight ** [kg]
BM 10′	2.989	2.591	2.5	2.435	from 1.490
BM 16′	4.885	2.591	4	2.435	from 2.000
BM 20′	6.055	2.591	5	2.435	from 2.490
BM 24′	7.355	2.591	6	2.435	from 3.000
BM 30′	9.120	2.591	7	2.435	from 3.490

* Short side always consists of two panels. ** Depends on detailed panel configuration and outfitting.

The modularity is characterized by multiple possible panel configurations. The different panel types are also part of the object library (Figure 6).

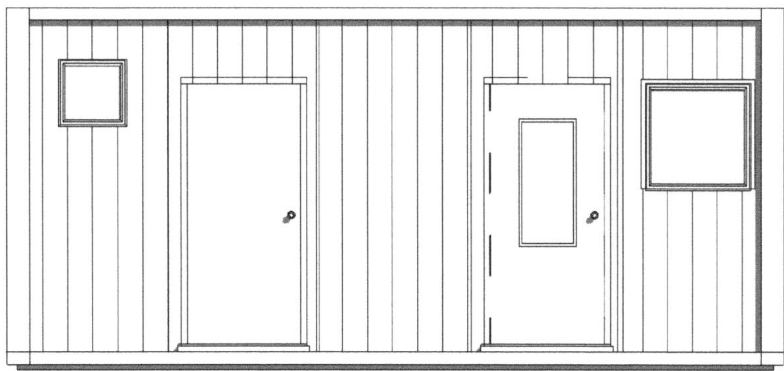

Figure 6. Panel types (from left to right: Sanitary window panel, door panel, full panel, glazed door panel, window panel).

This means that the components of the BIM object library can be—analogous to the blocks previously used in CAD—selected and put down predefined into the bid drawings. In addition, the exact panel configuration can be set within the BIM authoring tool for each of the five container typologies. Figure 7 shows the developed BIM object library in its basic configuration.

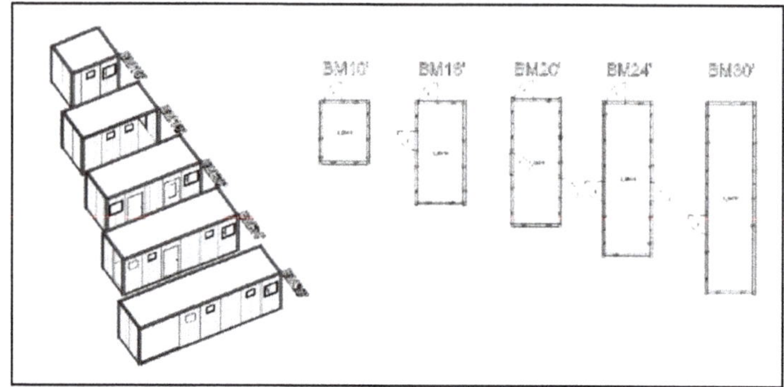

Figure 7. Building Information Modeling (BIM) object library in basic configuration.

The modular external walls of the single container typologies were modeled within in the BIM authoring tool as "curtain walls" with a grid division corresponding to the number of panels. For each panel, there is a slot in the container outer wall for which the desired panel type can be chosen (e.g., door panel, window panel or full panel).

In the BIM object library, a new parameter called "N_SidePanel" (Figure 8) was created, which constitutes a property parameter that divides the external walls of the containers according to the possible panel configuration.

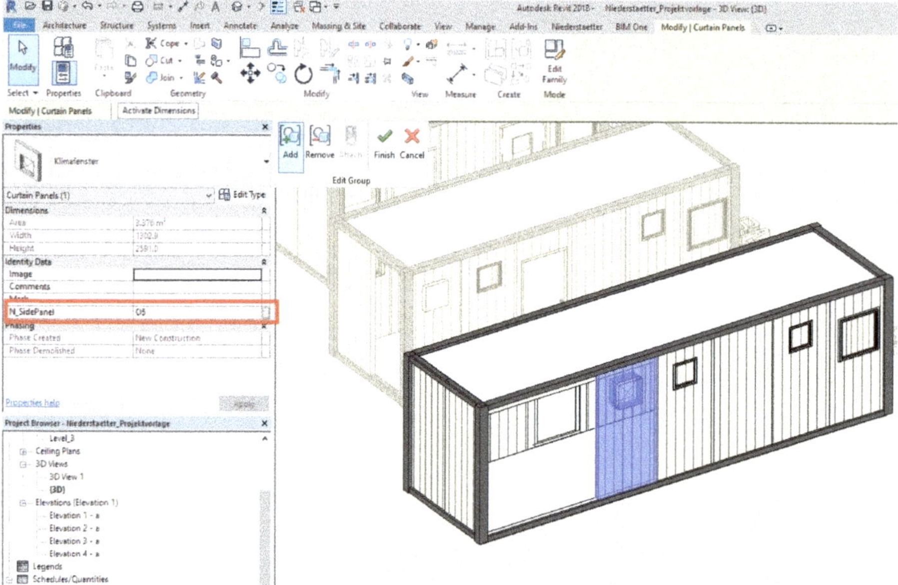

Figure 8. Parameter N_SidePanel in the BIM object library.

The value for N_SidePanel allows a clear comparison of drawing containers and warehouse containers with regard to the modification effort of the individual panels. Therefore, this parameter decides whether a panel has to be installed, removed, or exchanged. A prerequisite for this is the digitization of the current warehouse inventory into the virtual container warehouse with regard to the exact and current panel configuration.

The value for N_SidePanel is already prefilled within the BIM object library for each container group. The possible values for the parameter N_SidePanel depend on the sky orientation:

- in north direction: N1, N2
- in east direction: E1, E2, E3, E4, E5, E6, E7
- in south direction: S1, S2
- in west direction: W1, W2, W3, W4, W5, W6, W7.

4.3. Container Manager

We have developed an MS Excel-based tool that we have called the *Container Manager*. The *Container Manager* retrieves and manipulates BIM data to automate the container management process in the phases of bid estimation and work preparation according to the new BIM-based process definition (artefact #1). The information flow between the BIM model and the *Container Manager* is established through programmed algorithms making use of computational design techniques. The functionality of the individual algorithms is described below.

4.3.1. Algorithm: Export to Container Manager

In principle, the algorithm depicted in Figure 9 has the function to write the container typology and the panel types as well as their position into an Excel table (= *Container Manager*).

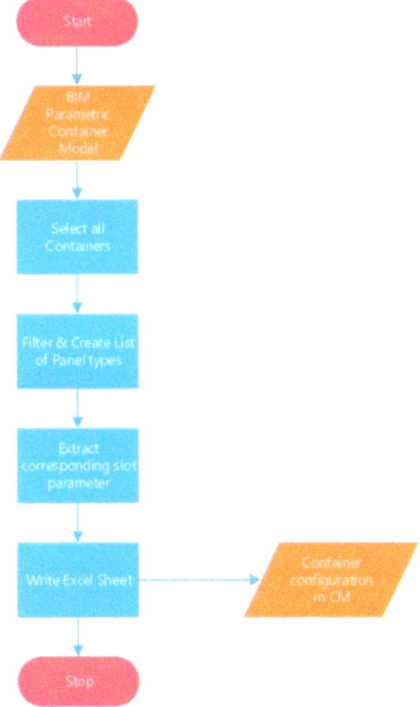

Figure 9. Export to *Container Manager* algorithm.

Description of algorithm steps:

(1) *Select all containers:* In the first step, all containers available in the project are selected. (2) *Filter and create list of panel types:* The elements of interest are extracted from the container element groups, in this case, the so-called "CurtainPanels", which represent the BIM elements of the modeled panels.

(3) *Extract corresponding slot parameter:* The position parameters (N_SidePanel) and the name of the corresponding panel type are extracted from a list of all CurtainPanels. (4) *Write Excel Sheet:* Finally, the extracted data is exported in processed form to the *Container Manager* as an Excel workbook, and the *Container Manager* starts automatically.

The algorithm leads to this representation in the *Container Manager* (Figure 10), which shows the drawn panel type for every slot in tabular form:

CONTAINERTYP	N1	N2	O1	O2	O3	O4	O5	O6	O7	S1	S2	W1	W2	W3	W4	W5	W6	W7
BM 24:										sanitärfer	Fenster	Voll	Tür	Voll	Voll	Voll		
BM 24'			Voll	Voll	Voll	Voll	Voll			Fenster	sanitärfer							
BM 30'	sanitärfer	Voll	Voll	Voll	Voll	Voll	Voll			Tür	Voll	Voll	Voll	Voll	Voll	Voll		
BM 30'	sanitärfer	Voll	Voll	Voll	Voll	Voll	Voll			Tür	Voll	Voll	Voll	Voll	Voll	Voll		
BM 30'	sanitärfer	Voll	Voll	Voll	Voll	Voll	Voll			Tür	Voll	Voll	Voll	Voll	Voll	Voll		
BM 30'	Tür	Voll	Voll	Tür	Voll	Voll	Voll			Fenster	Fenster	Voll	Voll	Voll	Voll	Voll		
BM 30'										sanitärfer	Fenster							

Figure 10. Screenshot of *Container Manager*: Export panel configuration from BIM model.

4.3.2. Algorithm: Assign Container Rental IDs

This algorithm depicted in Figure 11 imports the rental IDs of the physical containers assigned in the *Container Manager* back into the BIM model, and renames the corresponding drawn containers.

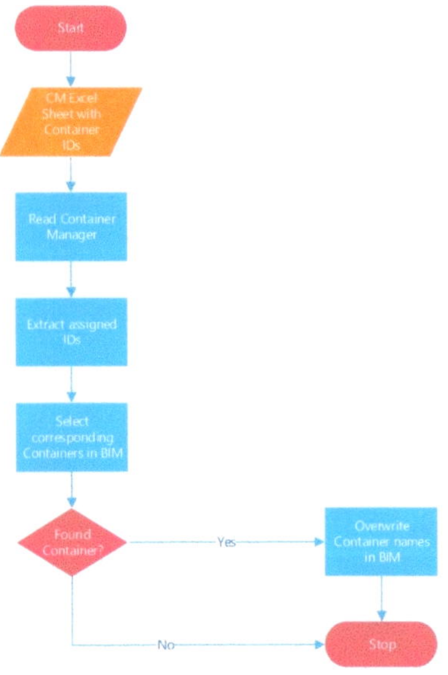

Figure 11. Assign container rental IDs algorithm.

Description of algorithm steps:

(1) *Read Container Manager and extract assigned IDs:* This part reads the *Container Manager* information and extracts the assigned rental IDs. (2) *Select corresponding containers in BIM:* This part searches for the BIM objects in the drawing to which the rental IDs have been assigned in the *Container Manager*. (3) *Overwrite container names in BIM:* In the last step, both sets of information are compared, and the container name is overwritten in the BIM model according to the rental ID.

4.3.3. Algorithm: Create 2D plans

The *Create 2D plans* algorithm depicted in Figure 12 generates plans for all drawn containers taking into account the panel configuration and assigned rental ID. The plans are output in A3 format and contain the floor plan representation as well as two view perspectives of each container. In addition, the rental ID is displayed in the plan header, so that the plan documents generated in this way are saved as PDF files and can be used directly to update the 2D drawing inventory with respect to the current container state.

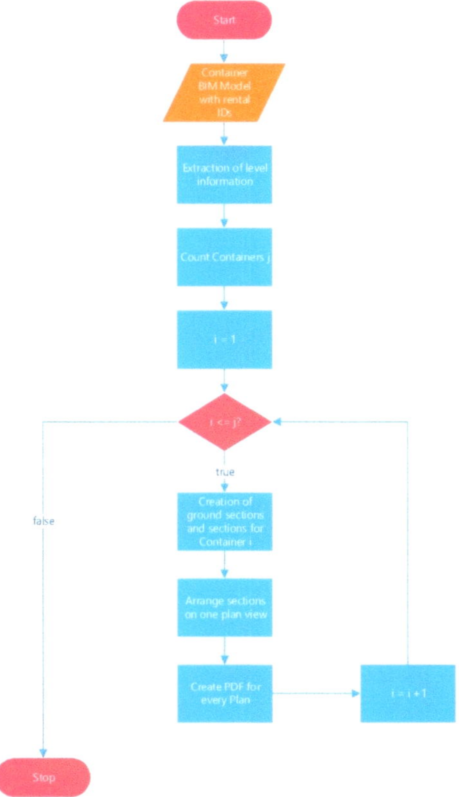

Figure 12. *Create 2D plans* algorithm.

Description of algorithm steps:

(1) *Extraction of level information:* This part analyzes how many floors are present in the drawn container layout, and transfers this information to the next algorithm step for creating floor plan and view perspectives. (2) *Create ground sections and sections for every container:* The floor plans and views of the isolated containers are generated. The north and west sides were selected for the two view perspectives. (3) *Create sections on one plan view:* Here, it is selected on which plan template the floor plan and views are placed. In the next step, they are positioned centrally on the plan. (4) *Create PDF for every plan:* A PDF for every plan is generated and saved.

All the algorithms can be fired directly from the BIM authoring tool, and have been realized through add-ins in the user interface, as shown in Figure 13.

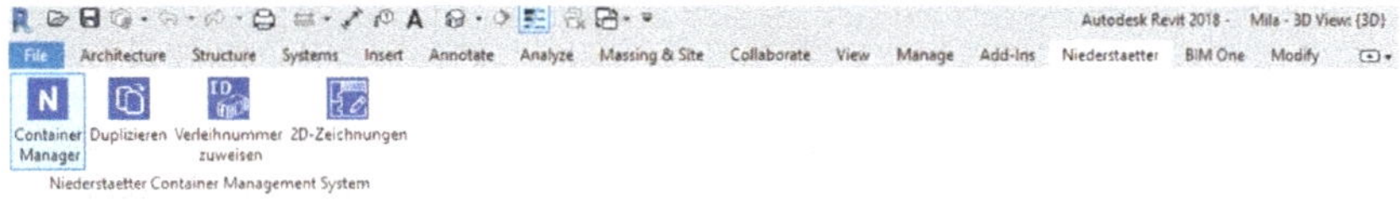

Figure 13. Algorithms in the BIM authoring user interface.

4.4. Database for a Virutal Container Warehouse

The industry partner's physical container warehouse has a capacity of more than 300 containers, which can be considered for assignment to a new container facility project. To digitize this inventory, a database system has been developed that accounts for the container typology, current panel configuration, rental ID, and cost attributes, as well as other secondary parameters. This database system is considered a reliable virtual container warehouse and is linked to the *Container Manager*. There, it serves as the basis for smart container assignment when comparing drawing information with actual availabilities, since best matches are proposed to the user. As soon as the container assignment to the bid drawing has been completed, the assigned containers and their rental ID as well as the new panel configuration can be imported back into the virtual container warehouse. Accordingly, these containers are no longer available until the status is set to *"not lent"* again.

4.5. Interplay of the Artefacts

The BIM object library (artefact #2) forms the basis for creating the bid drawing considering the customer requirements to the container facility. Thanks to the application of the BIM approach, these requirements can be become part of the model through metadata. This metadata in terms of container typology and panel configuration can be exported to the *Container Manager* **(artefact #3)**, where the available containers in the physical inventory are imported from the virtual container warehouse (artefact #4).

Within the *Container Manager*, available containers are contrasted with the drawn containers, and based on the metadata comparison, the best match is suggested, aiming for the lowest modification effort. Once the container assignment has been completed, the *Container Manager* automatically calculates the bid for the customer, taking into account the effort for panel modification and container installation as well as rental costs, which comprise attributes of the selected containers. Additional items that are not based on the drawing (e.g., transportation costs or the use of cranes) can be added manually to the bid estimation sheet. After that, the rental IDs of the assigned containers are fed back into the BIM model. This is the starting point for the automated creation of up-to-date plans of all the containers for the 2D plan repository. The process steps for these activities constitute artefact #1 and are depicted in Figure 14.

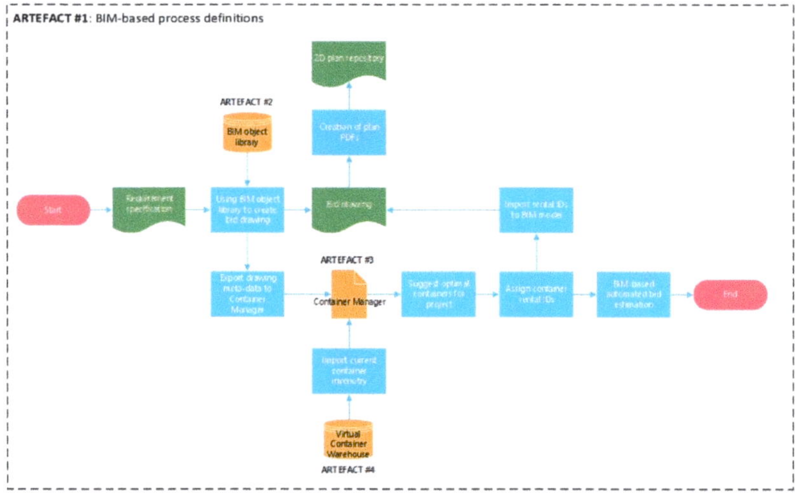

Figure 14. Interplay of the artefacts (= artefact #1).

5. Evaluation Phase According to Design Science Research

5.1. Description of the Pilot Project

In order to test the artefacts and to evaluate their applicability and utility according to DSR, a pilot project together with the industry partner has been conducted. The pilot project consisted of the planning phase for a two-story container facility composed of 28 containers of two different typologies (BM 20′ and BM 24′). It was supposed to be erected as a temporary office building for a medium-sized company in Northern Italy. Figure 15 shows the floor plans of the two floors.

Figure 15. Pilot project: Floor plans.

For this research study, the pilot project was worked on in the phases of planning, bid estimation, and work preparation in two ways: (1) a traditional way applying the current process, and (2) a BIM-based way applying the newly developed artefacts. Approach (2) resulted in a BIM model representing the customer requirements in terms of container typologies and container arrangement, as well as panel configuration. This resulting BIM model (a) and the eventually realized project (b) are shown in Figure 16.

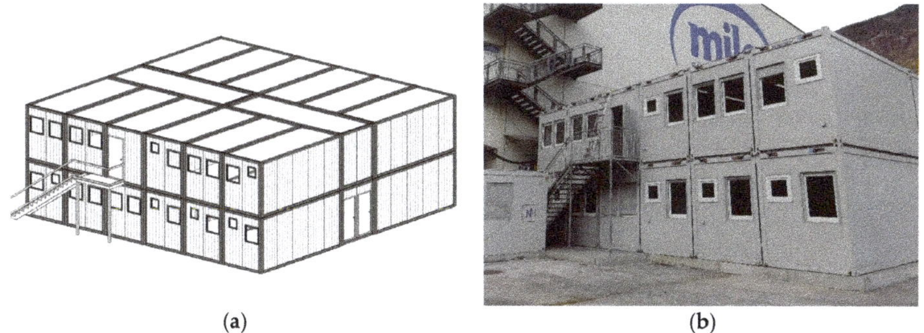

(a)　　　　　　　　　　　　　　(b)

Figure 16. Pilot project: (a) BIM model and (b) Realization.

Both approaches have been applied for comparison and gaining information regarding the process durations for (1) creation of the bid drawing, (2) assignment of physical containers to the drawing, and (3) the final creation of 2D plans and updating the repository.

Additionally, the results of the bid estimation were also considered. Here, the traditionally estimated times for panel modification and container installation were compared with the automatically calculated durations based on the BIM model.

5.2. Findings

The findings of the pilot project revealed time savings in the process steps of container assignment and transmission of the assigned rental IDs to the bid drawing as well as in the final creation of up-to-date 2D plans of the containers. In the latter-mentioned process step, a reduction from approximately 2 h to 5 min could be achieved, which corresponds to a time saving of ~95%. In comparison, the time savings in the former-mentioned process amount to ~83% when comparing the current process to the BIM-based process. For the creation of the bid drawing itself, no time difference between the current process, which is characterized by using CAD and predefined blocks to the BIM approach making use of a BIM authoring tool and the developed BIM object library, could be observed (Table 2).

Table 2. Comparison of process durations: Current process vs. BIM-based.

Sub-Process	Current Process	BIM-Based	Time Savings [%]
Creation of bid drawing	2 h	2 h	0
Container selection and ID assignment	3 h	0.5 h	83.3
Creation of 2D plans	2 h	0.1 h	95.8

Regarding the assembly times as part of bid estimation, the following findings have been obtained considering a crew of three fitters when applying the current estimation process based on experience and an automated calculation based on BIM information: The estimated time for the panel modifications of the assigned containers was 2 weeks, whilst the BIM-based calculations resulted in an estimation of 2.5 weeks. The container installation duration has been estimated in the traditional way at 5 days, whilst the BIM-based calculation resulted in 7 days. The results are given in Table 3. The BIM-based calculations were built on performance factors provided by the industry partner. These performance factors describe the time required for the modification of one single panel respectively for the installation of one single container.

Table 3. Comparison of assembly times: Current process vs. BIM-based.

	Time Needed According to Applied Approach	
Bid Estimation Item	**Current Process**	**BIM-based**
Panel modifications	2 weeks	2.5 weeks
Container installation	5 days	7 days

5.3. Assessing Utility and Applicability

The applicability with respect to DSR has been proven through successful practice in the pilot project. Experts from the industry partner who applied this approach had no or little transition problems following the new process and making use of the artefacts. On the one hand, this was confirmed by the interviews of experts involved. On the other hand, the findings show that the time required to prepare the bid drawing using the BIM-based approach was identical to the previous CAD approach, although this was a first experiment. With increasing experience in handling the BIM authoring tool and the artefacts, an even easier handling in the course of learning effect and thus a higher applicability can be expected.

In terms of utility with respect to DSR, Table 2 shows that the BIM-based approach provides for benefits in the time required to allocate the physical containers. This makes a direct application of the new BIM process for the industry partner recommendable, since its utility is characterized by substantial time savings. On the other hand, Table 3 indicates that the automated calculation of the time required for panel modifications and container installation is greater than the traditionally estimated time. Therefore, the BIM-based bid estimation could lead to less economical bids, and thus brings with it an increased risk of being outbid by competitors in future tenders. This mismatch is due to the absence of an accurate effort function for panel modifications, which takes into account non-linearities when panels are modified in a continuous fashion during the container preparation phase rather than constituting discrete events.

In order to fully reap the benefits of an accurate BIM-based calculation, it is therefore necessary to formulate a proper effort function for panel modification, since the total effort is not a simple linear proportionality of the number of panels to be modified. Rather, the situation is that, depending on the total number of containers, a minimum effort is required if at least one panel must be modified on the container in question. However, there is a difference in terms of effort per panel whether, e.g., all five or six panels on one wall must be modified, or just one. In this case, the effort per panel decreases the more panels that need to be modified.

Hence, to increase the utility of the suggested approach, an effort function must be formulated which accounts for non-linear phenomena and interactions between effort predictors. For that purpose, we conducted interviews with experts from the industry partner, asking them to estimate the effort for panel modification in different hypothetical project scenarios according to their experience. This data set given in Table 4.

Table 4. Data set for statistical analyses.

Project	Number Containers	Dominant Container Typology	Overall Number Panels to be Modified	Number of Full Walls to be Modified *	Panel Modification Effort (Estimated by Experts) [d]
Project1	33	BM 16′	32	7	7
Project2	25	BM 16′	32	4	7
Project3	20	BM 20′	40	4	7.5
Project4	102	BM 20′	40	10	13
Project5	44	BM 20′	40	12	8
Project6	55	BM 16′	32	7	9
Project7	25	BM 16′	32	7	6
Project8	30	BM 20′	40	10	7
Project9	16	BM 16′	32	8	5.5
Project10	2	BM 30′	7	1	2
Project11	4	BM 30′	4	0	2
Project12	42	BM 20′	40	8	8.5
Project13	3	BM 16′	12	3	1.5
Project14	8	BM 10′	20	3	2
Project15	10	BM 16′	32	3	5

* Walls with more than three panels to be modified are considered as full walls.

Given this data set, we performed statistical analyses based on a response surface methodology (RSM) approach. RSM is a collection of statistical and mathematical methods to investigate the impact of several predicting variables (predictors) on one or more responding variables (response) [33]. Depending on the dominant container size, the statistical analyses led to different regression equations providing for a response, which is the effort of panel modifications in days. The stepwise backward elimination of statistically non-relevant predictors in the regression procedure revealed that besides the dominant container typology (D), a total of three predictors have significant impact on the response: (A) Number of containers, (B) Overall number panels to be modified, and (C) Number of entire walls to be modified. These results are visualized in the Pareto chart in Figure 17.

The different regression equations in function of the dominant container typology are presented in Table 5.

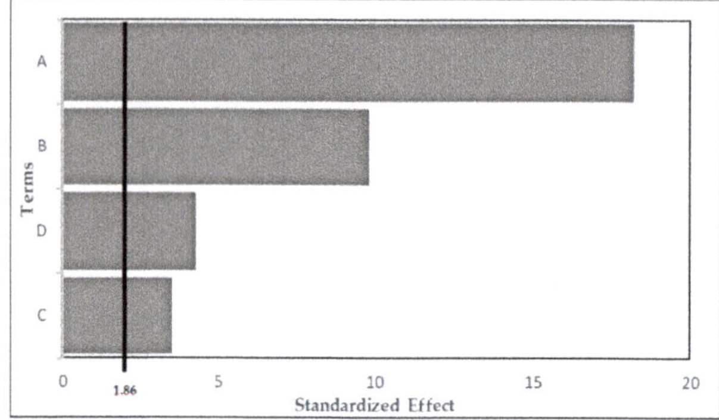

Figure 17. Pareto chart of the standardized effects for the determination of the total panel modification effort, $\alpha = 0.1$. Factor abbreviations: (**A**) Number containers, (**B**) Overall number panels to be modified, (**C**) Number of entire walls to be modified, and (**D**) Container typology.

Table 5. Regression equations.

Dominant Container Typology *	y = Effort Panel Modification [d]	Predictors **
BM 10′	y =	−1.689 + 0.08429 A + 0.1746 B − 0.1589 C
BM 16′	y =	−0.357 + 0.08429 A + 0.1746 B − 0.1589 C
BM 20′	y =	−0.797 + 0.08429 A + 0.1746 B − 0.1589 C
BM 30′	y =	0.866 + 0.08429 A+ 0.1746 B − 0.1589 C

** BM 24′ as part of the BIM object library never constitutes the dominant container typology, since it is usually used as a corridor in container facilities. **A = Number containers, B = Overall number panels to be modified, C = Number of entire walls to be modified.*

By applying the models determined by the RSM approach, a more accurate effort can be calculated than before, as the comparison to the pilot project findings shows (see Table 3), (with: dominant container typology: BM 20′; number containers: 28; overall number panels to be modified: 60; number of entire walls to be modified: 11):

$$y = -0.797 + 0.08429 \times 28 + 0.1746 \times 60 - 0.1589 \times 11 \tag{1}$$

$$y = 10.29 \; days \; \sim \; 2 \; working \; weeks \tag{2}$$

Thus, the utility according to DSR is demonstrated not only with regard to time savings for the processes of *assignment of physical containers to the drawing* and *final creation of 2D plans and updating the repository* (see Table 2), but also within the automated calculation of the bid estimation. However, these statistical analyses are based on expert knowledge when this study was carried out, and should be verified in the future by real historical project data.

6. Discussion

According to Kasanen et al. [34], design science research follows the steps of (1) finding a practically relevant problem with research potential; (2) obtaining knowledge about the topic in which the problem lies; (3) suggesting an innovative solution for that problem; (4) demonstrating the workability of the solution; (5) showing the solution's connection to theory and contribution to research; and finally, (6) evaluating the applicability within a defined scope.

In this study, all these six steps have been addressed: The practically relevant problem has been identified as a lack of digital automation in the construction equipment industry (1), which was revealed by the demonstrated absence of wide-ranging scientific studies dealing with this topic (2). This literature analysis disclosed particularly a gap of knowledge regarding the potential impact of digitization in the field of temporary construction facilities. We have shown that BIM can be considered as a starting point for digitization here. As a consequence, BIM became part of our solution suggestion for driving the digital automation in this field (3). The solution itself consists of a total of four BIM-based artefacts that can be practically applied to streamline and automate the existing processes of an industry partner from the container construction sector. We have demonstrated the workability in a pilot project scenario (4) and evaluated the utility and applicability within the scope of this pilot project, which can be considered a contribution to engineering practice.

The research strategy followed consisted of prescriptive research resulting in new artefacts that are supposed to improve parts of the real world within the considered scope. This kind of research occupies a space between descriptive theory and practical application [35]. A contribution to research (5) has been achieved through providing an in-depth pilot project that investigates the role of BIM—and particularly its underlying metadata—for the construction equipment industry from the perspective of temporary container facilities. As shown in the related work section, to our knowledge, little research has been conducted to develop new prescriptive artefacts for practical application in this sector that leverage BIM data systematically from both perspectives: 3D representation and metadata. In this context, the answer to the introductory stated research questions is twofold:

Firstly (i), metadata has to be input systematically and according to customer requirements into the BIM model. Moreover, metadata has to be extracted and further processed for a horizontal digital information flow. This information flow must be tailored to the characteristics of the respective current business processes. This study has shown how to design a BIM-based information flow making use of computational design techniques.

Secondly (ii), and assuming (i) as given, this study has demonstrated a positive impact of BIM on process automation in terms of time and labor savings. However, this impact cannot be generalized based on this study. Future evaluations require a context-specific consideration taking into account how the BIM metadata was used and what for.

7. Conclusions

This paper proposes the systematic capitalization of BIM technology in the construction equipment industry, leveraging the pilot project findings as groundings. The pilot project constitutes the evaluation phase of the applied design science research approach, collaborating with a CtO service provider as an industry partner to assess the hypothesis of BIM representing an important driver for digital automation in this sector. To this end, the utility and applicability according to design science research were evaluated.

- The results indicated that parts of the bid estimation and work preparation can be digitized and automated.
- Several currently personnel-intensive and time-demanding processes can be streamlined. This could free up internal resources for other activities, such as, e.g., increased and improved quality management.
- This study has initiated the vertical digitization and automation of the corporate structure of the industry partner. This can be extended into a horizontal direction across different corporate divisions, and facilitate striving for a continuous digital information flow. Practically conceivable, for example, would be interfaces for the *Container Manager* and the *Virtual Container Warehouse* to the company's Enterprise Resource Planning (ERP) system. Such research could eliminate redundancies in data storage and processing routines. Internal processes could be further streamlined and consistently converted to a BIM-based information cycle.
- To consolidate the status quo of development, it is necessary to collect further representative data by additional pilot projects regarding the real panel modification and assembly effort. Nonetheless, the statistical analyses carried out can already provide indications for this.

- Finally, this study shows the great potential of BIM data being preprocessed for supporting the construction site in an application-oriented way from the perspective of the construction equipment industry apart from crane setup applications, as presented in the state-of-the-art section.

Future research could aim at demonstrating how these data can be used in a structured way, even for continuous and longer-lasting production of the main construction industry beyond the phase of work preparation, which for example has already been pointed out by Bortolini et al. [36]. In addition, sustainability considerations in accordance with [37] can be a walking direction from here. In terms of BIM-based site management itself, a combination with lean management approaches at the data-processing level to create new integrated information systems could be a field of further research. More in detail, an integrated information system of BIM and the Last Planner® System as one of the most important Lean Construction methods for production planning and control can be a way to achieve this [38]. Through a continuous and systematic BIM reference also in the phase of production on-site, new applications in the light of Industry 4.0 will be possible in the future, and thus the basis for the promising cyber-physical production systems can also be formed in the construction sector.

Author Contributions: Conceptualization, C.P.S. and C.M.; methodology, C.P.S.; investigation, C.P.S.; writing—original draft preparation, C.P.S. and G.P.M.; supervision, D.T.M.

Funding: This work was supported by the Open Access Publishing Fund of the Free University of Bozen-Bolzano.

Acknowledgments: We sincerely would like to thank the company *Niederstätter*, located in South Tyrol, Italy, for giving us the opportunity to carry out the evaluation phase on a pilot project and providing us with valuable data for analysis. Special thanks are also directed to Giovanni Toller and Gian Luca Regis from *Fraunhofer Italia* for support during the programming of the algorithms as well as to Joseph Alexander Roberts for proofreading.

Conflicts of Interest: The authors declare no conflict of interest.

References

1. Pasetti Monizza, G.; Bendetti, C.; Matt, D.T. Parametric and Generative Design techniques in mass-production environments as effective enablers of Industry 4.0 approaches in the Building Industry. *Autom. Constr.* **2018**, *92*, 270–285. [CrossRef]
2. Matt, D.T.; Rauch, E. Implementing Lean in Engineer-to-Order Manufacturing. In *Handbook of Research on Design and Management of Lean Production Systems*; Modrák, V., Semančo, P., Eds.; IGI Global: Hershey, PA, USA, 2014; pp. 148–172. ISBN 97814666503981.
3. Dallasega, P.; Rauch, E.; Matt, D.T.; Fronk, A. Increasing productivity in ETO construction projects through a lean methodology for demand predictability. In Proceedings of the IEOM 2015-5th International Conference on Industrial Engineering and Operations Management, Hyatt Regency Dubai, UAE, 3–5 March 2015.
4. Qiao, G.; Lu, R.F.; McLean, C. Flexible manufacturing systems for mass customisation manufacturing. *Int. J. Mass Cust.* **2006**, *1*, 374. [CrossRef]
5. Thirumalai, S.; Sinha, K.K. Customization of the online purchase process in electronic retailing and customer satisfaction: An online field study. *J. Oper. Manag.* **2011**, *29*, 477–487. [CrossRef]
6. Oesterreich, T.D.; Teuteberg, F. Understanding the implications of digitisation and automation in the context of Industry 4.0: A triangulation approach and elements of a research agenda for the construction industry. *Comput. Ind.* **2016**, *83*, 121–139. [CrossRef]
7. Kull, H. Intelligent Manufacturing Technologies. In *Mass Customization-Opportunities, Methods, and Challenges for Manufacturers*; Apress: New York, NY, USA, 2015; pp. 9–20.
8. Schimanski, C.P.; Pasetti Monizza, G.; Marcher, C.; Toller, G.; Matt, D.T. Enhancing Automation in the Construction Equipment Industry through Implementation of BIM. In Proceedings of the CDVE2019 Proceedings, Mallorca, Spain, 6–9 October 2019.
9. Girmscheid, G. Industrialization in Building Construction Production Technology or Management Concept? In Proceedings of the 11th Joint CIB International Symposium Combining Forces-Advancing Facilities Management and Construction through Innovation, Helsinki, Finland, 13–16 June 2005; Volume 1, pp. 427–441.
10. Maskuriy, R.; Selamat, A.; Ali, K.N.; Maresova, P.; Krejcar, O. Industry 4.0 for the Construction Industry—How Ready Is the Industry? *Appl. Sci.* **2019**, *9*, 2819. [CrossRef]

11. Eastman, C.M.; Teicholz, P.; Sacks, R.; Liston, K. *BIM Handbook: A Guide to Building Information Modeling for Owners, Managers, Designers, Engineers and Contractors*, 1st ed.; Wiley: Hoboken, NJ, USA, 2008; ISBN 9780470541371.
12. Besenyői, Z.; Krämer, M.; Husain, F. Building Information Modelling in Agile Environments-an Example of Event Management at the Airport of Tempelhof. *MATEC Web Conf.* **2018**, *251*, 03064. [CrossRef]
13. Chougule, N.S. A Review of Building Information Modeling (BIM) for Construction Industry. *Int. J. Innov. Res. Adv. Eng.* **2015**, *2*, 98–102.
14. Rojas, R.A.; Rauch, E. From a literature review to a conceptual framework of enablers for smart manufacturing control. *Int. J. Adv. Manuf. Technol.* **2019**, *104*, 517–533. [CrossRef]
15. Bohnacker, H.; Gross, B.; Laub, J.; Lazzeroni, C.; Frohling, M. *Generative Design: Visualize, Program, and Create with Processing*; Abrams & Chronicle Books: London, UK, 2012; ISBN 9781616890773.
16. Logothetis, S.; Karachaliou, E.; Valari, E.; Stylianidis, E. Open source cloud-based technologies for BIM. In Proceedings of the International Archives of the Photogrammetry, Remote Sensing and Spatial Information Sciences ISPRS Archives, Riva del Garda, Italy, 4–7 June 2018; Volume 42, pp. 607–614.
17. Lau, C.; Yang, M.-X.; Allan, L.; Ku, C.-J. Cost Analysis of Equipment in a Building Using BIM-based Methods. In Proceedings of the 35th International Symposium on Automation and Robotics in Construction (ISARC), Berlin, Germany, 20–25 July 2018.
18. De Lacalle, L.N.L.; Fernández-Larrinoa, J.; Rodríguez-Ezquerro, A.; Fernández-Valdivielso, A.; López-Blanco, R.; Azkona-Villaverde, I. On the cutting of wood for joinery applications. *Proc. Inst. Mech. Eng. Part B J. Eng. Manuf.* **2015**, *229*, 940–952.
19. Jahr, K.; Borrmann, A. Semi-automated site equipment selection and configuration through formal knowledge representation and inference. *Adv. Eng. Informatics* **2018**, *38*, 488–500.
20. Ji, Y.; Sankaran, B.; Choi, J.; Leite, F. Integrating BIM and Optimization Techniques for Enhanced Tower Crane Planning. *Comput. Civ. Eng. 2017* **2017**, *3*, 67–74.
21. Abbott, E.L.S.; Peng, L.; Chua, D.K.H. Using building information modelling to facilitate decision making for a mobile crane lifting plan. In *8th International Conference on Engineering, Project, and Product Management (EPPM 2017)*; Springer: Berlin, Germany, 2018; pp. 77–89.
22. Ji, Y.; Leite, F. Automated tower crane planning: Leveraging 4-dimensional BIM and rule-based checking. *Autom. Constr.* **2018**, *93*, 78–90. [CrossRef]
23. Schwabe, K.; Teizer, J.; König, M. Applying rule-based model-checking to construction site layout planning tasks. *Autom. Constr.* **2019**, *97*, 205–219.
24. Kan, C.; Anumba, C.J.; Messner, J.I. A Framework for CPS-Based Real-Time Mobile Crane Operations. In *Advances in Informatics and Computing in Civil and Construction Engineering*; Springer: Berlin, Germany, 2018; pp. 653–660.
25. Han, S.; Hasan, S.; Bouferguene, A.; Al-Hussein, M.; Kosa, J. An integrated decision support model for selecting the most feasible crane at heavy construction sites. *Autom. Constr.* **2018**, *87*, 188–200. [CrossRef]
26. Li, R.; Fu, Y.; Liu, G.; Mao, C.; Xu, P. An Algorithm for Optimizing the Location of Attached Tower Crane and Material Supply Point with BIM. In Proceedings of the 35th International Symposium on Automation and Robotics in Construction (ISARC), Berlin, Germany, 20–25 July 2018.
27. Al-Hussein, M.; Athar Niaz, M.; Yu, H.; Kim, H. Integrating 3D visualization and simulation for tower crane operations on construction sites. *Autom. Constr.* **2006**, *15*, 554–562. [CrossRef]
28. Hammad, A.; Akbarnezhad, A. Modular vs Conventional Construction: A Multi-Criteria Framework Approach. In Proceedings of the ISARC 34th International Symposium on Automation and Robotics in Construction, Taipei, Taiwan, 28 June–1 July 2017; Department of Construction Economics & Property, Vilnius Gediminas Technical University: Vilnius, Lithuania, 2017; Volume 34, pp. 214–220.
29. Carbonari, A.; Corneli, A.; Di Giuda, G.; Ridolfi, L.; Villa, V. BIM-Based Decision Support System for the Management of Large Building Stocks. In Proceedings of the ISARC International Symposium on Automation and Robotics in Construction, Berlin, Germany, 20–25 July 2018; Ruhr-Universität Bochum: Bochum, Germany; Scotland, UK, 2018; Volume 35, pp. 348–355.
30. Kim, K.; Cho, Y.; Kim, K. BIM-Driven Automated Decision Support System for Safety Planning of Temporary Structures. *J. Constr. Eng. Manag.* **2018**, *144*, 04018072. [CrossRef]

31. Feng, C.W.; Hsu, T.F. Using BIM to Automate Information Generation for Assembling Scaffolding—A Material Management Approach. In Proceedings of the 34th International Symposium on Automation and Robotics in Construction (ISARC), Taipei, Taiwan, 28 June–1 July 2017; Department of Construction Economics & Property, Vilnius Gediminas Technical University: Vilnius, Lithuania, 2017; Volume 34, pp. 610–617.
32. Van Aken, J.E. Management research as a design science: Articulating the research products of mode 2 knowledge production in management. *Br. J. Manag.* **2005**, *16*, 19–36. [CrossRef]
33. Sredović Ignjatović, I.D.; Onjia, A.E.; Ignjatović, L.M.; Todorović, Ž.N.; Rajaković, L.V. Experimental Design Optimization of the Determination of Total Halogens in Coal by Combustion–Ion Chromatography. *Anal. Lett.* **2015**, *48*, 2597–2612. [CrossRef]
34. Kasanen, E.; Lukha, K.; Siitonen, A. The constructive approach in management accounting research. *J. Manag. Account. Res.* **1993**, *5*, 243–264.
35. Van Aken, J.E. Management Research Based on the Paradigm of the Design Sciences: The Quest for Field-Tested and Grounded Technological Rules. *J. Manag. Stud.* **2004**, *41*, 219–246. [CrossRef]
36. Bortolini, R.; Formoso, C.; Viana, D.D. Site logistics planning and control for engineer-to-order prefabricated building systems using BIM 4D modeling. *Autom. Constr.* **2019**, *98*, 248–264. [CrossRef]
37. Tokbolat, S.; Kaur Calay, R. Residential Construction Sustainability in the UK and Prospects of Knowledge Transfer to Kazakhstan. *J. Sustain. Dev.* **2015**, *8*, 14.
38. Sacks, R.; Radosavljevic, M.; Barak, R. Requirements for building information modeling based lean production management systems for construction. *Autom. Constr.* **2010**, *19*, 641–655.

Article

Marketing Innovations in Industry 4.0 and Their Impacts on Current Enterprises

Otakar Ungerman * and Jaroslava Dědková

Technical University of Liberec Faculty of Economics, Department of Marketing and Trade Studentska 2, 46117 Liberec, Czech Republic
* Correspondence: otakar.ungerman@tul.cz; Tel.: +42-48535-2417

Received: 13 August 2019; Accepted: 2 September 2019; Published: 5 September 2019

Abstract: This paper discussed the marketing innovations associated with Industry 4.0 and the effects that these innovative approaches cause. The main aim of the research was to discover the relationship between marketing innovations and their effects. Knowledge of this relationship can be used for the strategic planning of industrial companies in practice. The research methodology consisted of pilot research followed by primary research in industrial enterprises. The data were evaluated by descriptive statistics, statistical hypothesis, and correlation analysis. Through the research, the authors identified the importance of 17 innovative marketing tools and the strength of the use of 11 effects resulting from the implementation of these tools. The authors identified the relationships between tools and their implications in Industry 4.0 where a correlation was demonstrated. A list of 11 strategic objectives was created and, subsequently, a specific marketing mix proposal for each objective consisting of innovative marketing tools was as well. The results of this work enable enterprises involved in Industry 4.0 to better plan.

Keywords: Industry 4.0; marketing innovations; innovative marketing tools; impacts marketing innovations

---

1. Introduction

From the perspective of enterprises, innovations represent a key activity for their further development and increasing competitiveness within the current globalized market. When it comes to fundamental technological innovations that are linked to a sudden increase in labor productivity, this already concerns an industrial revolution. The first jump in labor productivity came in the 18th century in conjunction with the use of steam—the first industrial revolution. Subsequent labor productivity jumps were associated with the expansion of factory machine production—the beginning of the use of electricity and oil, automation, and subsequent digitization. Digitization is the main engine of the next expected jump in industry called Industry 4.0. According to the authors of References [1–3], Industry 4.0 involves complete digitization, robotization, and automation of most of today's human activities to ensure greater speed and efficiency of production, for more efficient use of materials, and for greener industry and human life. These are technological changes, such as the electronic transfer of information between machines (machine-to-machine), which solves problem situations without human involvement.

According to Reference [4], human capital is to be perfectly replaced in industry by autonomous robots and workers' professions will be abolished. The emergence of Industry 4.0 was first published in 2011 in Hanover, where in the publication "Industry Manifest 4.0", they presented the basic features of this industrial revolution [5,6]. Industry 4.0, however, is a broad term and is interpreted by various authors in different contexts. What the authors agree on is that Industry 4.0 will lead to fundamental changes in the economy, the working environment, and the development of skills [7]. At the time of the establishment of Industry 4.0 to the present day, many authors (e.g., [8,9]) have begun to call Industry 4.0 the Industrial Revolution 4.0.

The research results are significant for the whole European context, although data were collected only in the Czech Republic. The Czech economy is export based, where 80% of goods and services are exported [10]. At the same time Czech industry is strongly linked to multinationals property-wise. The share of foreign capital in Czech companies makes up half of the total capital. This interconnection of Czech companies with foreign corporations, e.g., Skoda auto versus Volkswagen leads to the conviction that the results can be generalized for the whole of developed Europe.

This paper deals with innovations, so it is necessary to define innovations correctly. According to the 2010 Eurostat update of the methodology used in the EU Business Innovation Activity Survey [11] (innovative enterprises are those which have introduced some of the innovations during the listed period:

(1)   Product innovation—marketing of a new or substantially improved product or service;
(2)   Process innovation—the introduction of a new or substantially improved method of production, provision of services, mode of supply, storage, distribution, introduction or substantial improvement of enterprise support activities;
(3)   Marketing innovation—introduction of a new method of promotion, valuation or sale of products/services, significant changes in the aesthetic design or packaging of the offered products;
(4)   Organizational innovation—introduction of a new way of organizing the supplier-customer relationship management, human resources or a new approach to the organization of external relations.

These four groups of innovations can be further divided into technological and non-technological. Technological innovations include product and process innovations and non-technological innovations include marketing and organizational innovations. This paper discusses marketing innovations that are categorized as non-technological innovations, nevertheless, it is impossible to ignore technology with them. There are big differences in the current characteristics of Industry 4.0-related marketing innovations and their impact. There are no current studies when it comes to examining the relationship between marketing innovation and impact. This situation led to the determination of the main research question:

Main research question:

How to use the current innovative marketing tools associated with Industry 4.0 for enterprise practice? This research question was divided into three research sub-questions:

1.   How are innovative marketing tools used in Industry 4.0?
2.   What are the implications of innovative marketing tools if enterprises implement them in their strategy?
3.   What marketing mix does an enterprise involved in Industry 4.0 have to build to achieve a strategic goal?

Several hypotheses were defined in the research and were tested using statistical induction tests. The research questions were conditioned by performing the primary research, where the main objective was to find out what the implementation of Industry 4.0 brings to enterprises in the field of marketing innovations and what are the effects of these new trends. The methodology of the primary research is presented in Section 3.

2. Literature Review

After the research questions were set, a thorough investigation of scientific databases (Web of Science, Scopus, ProQuest) and other specialized literature was carried out. The research focused on three areas that were the subject of secondary research: innovation in Industry 4.0, innovation in marketing, and the effect that introducing these innovations brings to enterprises.

2.1. Innovation in Industry 4.0

As mentioned in the introduction, the EU classifies marketing innovation as a non-technological innovation group. However, if we want to define innovations related to Industry 4.0 that are based on technological progress, technology cannot be omitted from marketing innovations. Innovation in the era of industry 4.0 is perceived differently by authors. Lu [12] and Witkowski [13] emphasize that innovation is based on the development of the Internet and the existence of big data.

They state that the main innovations include the Internet of Things (IoT), cyber–physical systems (CPS), information and communication technologies (ICT), enterprise architecture (EA), and enterprise integration (EI), while the utilization of cybernetic systems is also reported by Marešová [14], who defines innovations in marketing as a flexible linking of products and services over the Internet or other network applications such as blockchain or a peer-to-peer network system.

The authors Zezulka and Veselý [15] divide innovation in Industry 4.0 into three groups that are interconnected and mutually influencing areas:

- Digitization and the integration of any production–business relationship: All links in the production chain will be able to access all necessary data. This can be very useful because, for example, machine builders, manufacturers of software and other production chain manufacturers and the entire production chain will be able to develop their products with the knowledge of the latest components that component manufacturers are yet to develop and test. The increase in digitization has an impact on companies' business activities, including their business models, in that they enable new forms of cooperation and lead to new products and services, as well as new forms of relationships with customers and employees. This digitization also places pressure on enterprises to consider their strategies and to systematically explore new business opportunities [16].
- Digitizing production and services: Based on data available through the cloud, manufacturers will be able to predict, for example, the failure or imminent failure of any manufacturer of electronic components that is needed for "their" production, machinery or equipment. The digitization of production data enables the optimization of demand, increases productivity, and allows efficient creation of values at the company's own production sites. The implementation of Industry 4.0 requires a high computing power to plan, process, simulate, and monitor production lines and to optimize and analyze data generated during the product lifecycle [17].
- New business models: They arise from the digitization and utilization of big data and lead to a precise definition and subsequent addressing of a homogeneous target group.

The authors of References [18–21] agree that Industry 4.0 works on six basic principles:

1. Interoperability: the ability of cyber–physical systems, people, and all smart factory components to communicate with each other through the Internet of Things and Services.
2. Virtualization: the ability to link physical systems to virtual models and simulation tools.
3. Decentralization: decision-making and control is carried out autonomously and in parallel in individual subsystems.
4. Ability to work in real-time: real-time compliance is a key requirement for any communication, decision-making, and control in real-world systems.
5. Service orientation: the preference of the computing philosophy of offering and using standard services, which leads to SOAs (service-oriented architectures).
6. Modularity and reconfigurability: Industry 4.0 systems should be maximally modular and capable of autonomous reconfiguration based on automatic situation detection.

An overview of the Industry 4.0 core tools has been compiled and published in the National Industry Initiative 4.0, where a team of authors with the support of the Ministry of Industry has compiled an overview of ten innovative tools of Industry 4.0.

1. System integration—This is based on the interconnection of all links in the value chain from suppliers to the organizational structure of the manufacturing company itself to distribution to the end customer. The condition of the functionality of the interconnection is real-time data processing, information sharing, and continuous communication. At present, these connections are still inadequate and underdeveloped [22].

2. Big Data Analysis—Big data is usually considered to be data in the range of peta bytes (10^{15} bytes) or more that are currently at the edge of database technology capability. Examples are image data, text data from the Internet, business and security data, and combined multimodal data. Big data processing serves to optimize a company's production, related services, and distribution. The effort is to involve big data analysis for easier innovation, surpassing cheap mass production [23,24].

3. Autonomous robots—robotic devices that work independently and do not require human control which are controlled by a program. Very often this type of robot works in collaboration with a person, where both actors complement each other. Nowadays, robots that are able to learn by themselves are beginning to gain ground and are developing the program themselves.

4. Communication infrastructure—means using secure high-speed communication, primarily through wired and wireless networks. A link between products and networks is created where the necessary information is transferred among devices and machines throughout the production process.

5. Data storage and cloud computing—These are server networks, each with a different function. Cloud solutions are used to store "big data", such as unstructured data. Clouds also help with planning new production. Using cloud solutions opens up opportunities for productivity growth and cost optimization. The big advantage is the possibility to share information among hundreds of branches of one company, e.g., about customers or sales structure [25].

6. Additive production—it is the process of joining material according to 3D digital data, most often layer by layer. The product is produced quickly and precisely, even the most complex shapes such as printing a house. Additive technology makes it possible to produce diverse parts without the need for lengthy programming preparation [26]. Currently, 4D is being tested, which is a 3D product that can later change and reshape over time.

7. Augmented Reality—connects the physical and virtual worlds. It extends the human perception of the world with new information that is not easily and quickly recognizable. Current applications are focused on smartphones and tablets that enable visualization of virtual tours, composing product groups, etc. Applications can be found in warehouse and logistics operations (barcode reading) in transport (traffic information) and in service (component visualization).

8. Sensors—Sensors include methods and tools for measuring and sensing various variables that are important in an industrial automation system.

9. Cybernetics and artificial intelligence provide key technologies for Industry 4.0 system solutions. These are the principles of organization, management, and decision-making as well as procedures to integrate autonomous systems.

10. New technologies—unused technologies will find their place in the Industry 4.0 process and new technologies will emerge, especially in the areas of biotechnology, information technology, and genetic technologies [27].

2.2. Marketing Innovations

Following the introduction of Industry 4.0 innovations, researchers focused their attention on industry-related marketing innovations. An exhaustive definition was used by Kotler [28] (p. 104):

> "By innovative (lateral) marketing we mean a sequence of work that, when applied to existing products, leads to the creation of new products or new services to meet new needs, bring new areas of use, new situations or discover new target groups of consumers. It is therefore

a process, offering a significant opportunity to create entirely new product categories or to form entirely new markets."

The same author equates innovative marketing to a process that requires a methodical approach. Its application to existing products or services brings possible innovations, for example, in the form of a new market or product category [28]. A concise definition of innovative marketing can also be found in the study entitled "Marketing Innovation: The Unheralded Innovation Vehicle to Sustained Competitive Advantage". The definition is as follows: "generation and implementation of new ideas for creating, communicating, and delivering value to customers and for managing customer relationships in ways that benefit the organization" [29] (p. 5). According to Reference [11], marketing innovation is the implementation of a new marketing method involving significant changes in the product or packaging design, product placement, product promotion or pricing. According to References [30,31], marketing innovation is defined as an innovation and a new method by which firms can sell to potential or existing customers. It includes significant changes in the implementation of various marketing strategies to increase marketing efficiency [32] allowing enterprises to gain a competitive edge and create value for shareholders.

The secondary research compiled an overview of some of the most important marketing trends of today, as perceived by world authors:

- Digital marketing—includes all marketing communications operating on the basis of digital technologies. Digital marketing trends include:

  - Artificial intelligence—autonomously evaluates the behavior of users on social networks. On a website, content automatically adapts to who's on the page. Artificial intelligence is used to write newsletters or posts on Facebook, increasing the clickthrough rate by tens of percent [33,34].
  - Conversational marketing—allows you to engage people in natural communication. Conversational will strengthen the brand and ultimately increase sales. This is a real-time conversation with a customer using chatbots. Chatbots find out everything a customer wants and prepare specific communication for them [35].
  - Personal brand/influencer—most commonly associated with video and youtubers. People do not want to follow enterprises but want to follow other people who are somehow interesting. This creates an "influencer." Today, an "influencer" is rewarded for their influence and product placement [36].
  - Search engine optimization (SEO)—social networks seemed to have killed SEO, but SEO is still very important and its importance has started to grow again. Link building that is used for SEO is focused on content quality and corporate blogs with videos and comments and client responses. In voice search, only the first position leads to a conversion. [37].
  - Omnipresence—Customers use multiple channels at once, and enterprises must spread their communications across all types of communication channels. This is a coordinated connection of all channels: email, YouTube, Instagram, Facebook, WhatsApp, and blogs with professional networks such as LinkedIn [38].

- Internet marketing, unlike digital marketing, always requires an internet connection. Ren, Xie, and Krabbendam [39] point out that internet marketing and marketing relationships have recently become the main focus of marketing innovations that companies use to achieve a sustainable competitive advantage. Similarly, Prahalad and Ramaswamy [40] and Son et al. [41] argue that the internet has changed the ways in which people live.
- Relationship marketing—focuses on long-term results to provide customers with long-term value and create high customer loyalty through building relationships at many levels, whether economic, social, technical or legal. It focuses on product benefits in a long-term horizon. It prefers intensive contact with customers with an emphasis on high responsibility towards them. [42].

- Mobile marketing—a form of advertising that is usually displayed on mobile devices, mostly phones or tablets. The huge increase in the popularity of smart phones with large displays has also caused the growing importance of mobile marketing. For mobile marketing, the abbreviation MAGIC can be applied: Mobile, Anytime, Globally, Integrated, Customized; that means using all the possibilities of digital marketing on a mobile device. It is therefore possible to influence the customer 24/7 [43].

These and other marketing innovations are used by enterprises to develop a strategy that meets a specific goal. The goal is to create new products and services that fill the gaps in the market and meet customers' unsuspecting needs. By implementing marketing innovation, a company can discover new product and service utilization, new market opportunities, and new groups of potential customers [25].

The authors Ohtonen [44] and Cummins et al. [45] add that the use of these marketing innovations can fulfill many diverse goals such as the introduction of new sales promotion methods, improvements in product packaging, innovation in promotion or the new use of media. It is evident that the scale of innovation activities in companies is determined by financial resources, the difficulty of introducing innovation to human resources, time and external factors such as political conditions, and the public perception of potential investments related to the introduction of innovation [46].

2.3. Impacts of Marketing Innovation

The impact of the onset of Industry 4.0 is already being reflected in enterprises. The main effect is the economic performance of enterprises, which can be measured in terms of accounting indicators such as cash flow and profitability. In addition, O'Sullivan and Abela [47] report that marketing performance is measured by return on assets (ROA) and return on investment (ROI). However, marketing performance can be measured by sales volume, revenue growth, and market share, while financial performance can be measured by profitability, revenue percentage, return on investment, profit, and profit growth. However, these are not only economic effects, but a whole range of effects associated with Industry 4.0. Industry 4.0 and marketing innovation provide a number of new opportunities and, if utilized, will have a strong impact on whole value chains. Among the key effects, according to References [7,48–51], include:

- Increase in labor productivity—the number of products/services and the production time are important to calculate labor productivity. Industry 4.0 leads to a dramatic increase in productivity, significantly reducing work time for the same production volume. Above all, there is a complete interconnection of the production process, including development and subsequent service. In factories, machines will be controlled by sensors, readers. and cameras. Robots automatically report maintenance to the maintenance staff. On the whole, the production process will be sped up and refined, while productivity will increase overall.
- The emergence of new business models—this is the basic principle of business, the way an enterprise creates and gains value from providing its services or selling products. New business models are linked to autonomous robotization in engineering. Industry 4.0 leads to new business models primarily related to direct selling [52]. New business models can resolve customer problems more effectively and find brand new customer segments [53]. Those business models, based on new technologies and big data, are focused on new services, value-linked ecosystems, and the approach to the customer to enable production to better respond to user-focused design and to better align with the processes and contexts involved in creating value for the customer [54].
- The cessation of "classic" jobs and the creation of "new" jobs—with the advent of Industry 4.0, people who devote themselves to automated operations will lose their jobs. It is not only the workers in assembly line production, but all who work in a routine way. Yet massive unemployment is not imminent. This is because a number of new positions will be created in services or industries where it is necessary to produce "customized goods".

- New workflows—process robotization, combining with IoT and other Industry 4.0 tools will lead to major workflow changes. This primarily involves the simplification of production. The change in workflows is related to changes in work organization. All this will translate into changes in the organizational structures of companies and the abandonment of the classic line models [55].
- New communication systems—these are models that are not linear but are circular, network models. With social networks, credit cards, and other Internet footprints, enterprises have perfect information about each individual. By processing big data, they can get to know their customer's behavior intimately and tailor their communication mix to fit their needs and, thus, achieve greater satisfaction. [56].
- Increased occupational safety—this impact is associated with the abolition of blue-collar jobs where there is the greatest risk of occupational injury. Leaving the risky work to robots will bring about a significant increase in occupational safety [57].
- Increasing competitiveness—the competitive strategy of companies involved in Industry 4.0 is primarily quality associated with precision processing. On the other hand, Industry 4.0 leads to higher production efficiency and reduced overall costs. The competitive advantage may then be the price [58].
- Increase PR—the involvement of companies in Industry 4.0 is used by enterprises in marketing. In their communication, they present the application of innovation, thereby building a better employer brand, which leads to better human capital. The presentation of the application of Industry 4.0 in the media leads to a better image in relation to the general public [12,47].

3. Materials and Methods

The primary methodology was designed to meet the research goal of "Finding out what Industry 4.0's implementation of marketing innovation brings to enterprises and the impact of these new trends". For a better overview of the procedure, both the conceptual research framework and the research evaluation methods are presented in Figure 1.

The basis of the research was a pilot project carried out in 2017. The aim of the pilot research was to identify innovative marketing tools and the effects they bring. For this identification, qualitative research inquiries with an in-depth view and a focus group were used. The results of the pilot research are presented in papers published in the Web of Science database [59,60].

In the Czech Republic where the research was conducted, there are a total of 31,966 enterprises registered in the Commercial Register [10], with 10 or more employees. Those enterprises comprised the core set (population). We contacted enterprises with more than 10 employees; smaller enterprises were excluded on the assumption that it is harder for them to implement innovations and those microenterprises would thus distort the sample set. The authors tried to obtain a representative sample of respondent and, thus, the minimal sample set size was calculated. The sample, with a ±5% sampling error at a 95% confidence interval, comprised 380 enterprises. During our research, we contacted a total of 605 enterprises; the return rate was 34%, i.e., 210 completed responses. Although the return rate of the responses was relatively high due to the fact of personal contact, the objective was not fulfilled. The sampling error was recalculated and, in conclusion, it may be said that 210 enterprises made up the representative sample with a 6.1% sampling error at a 95% confidence interval [60].

The structure of the enterprises by size can be broken down into enterprises with 10–49 employees, of which there are 21,101, making up 66% of enterprises. Enterprises with 50–249 employees, of which there were 8443, making up 26% of enterprises. Enterprises with 250 or more employees, of which there were 2422, making up 8% of enterprises [10]. The composition of the sample set was as follows: 10–49 employees 55%, 50–249 employees 30%, and 250 or more employees 15%. The distribution of respondents roughly corresponded to the composition of the core set, indicating that the sample was representative. The research included a sorting question, but the answers were not included in the assessment. The authors wanted to obtain a comprehensive view across all types of enterprises.

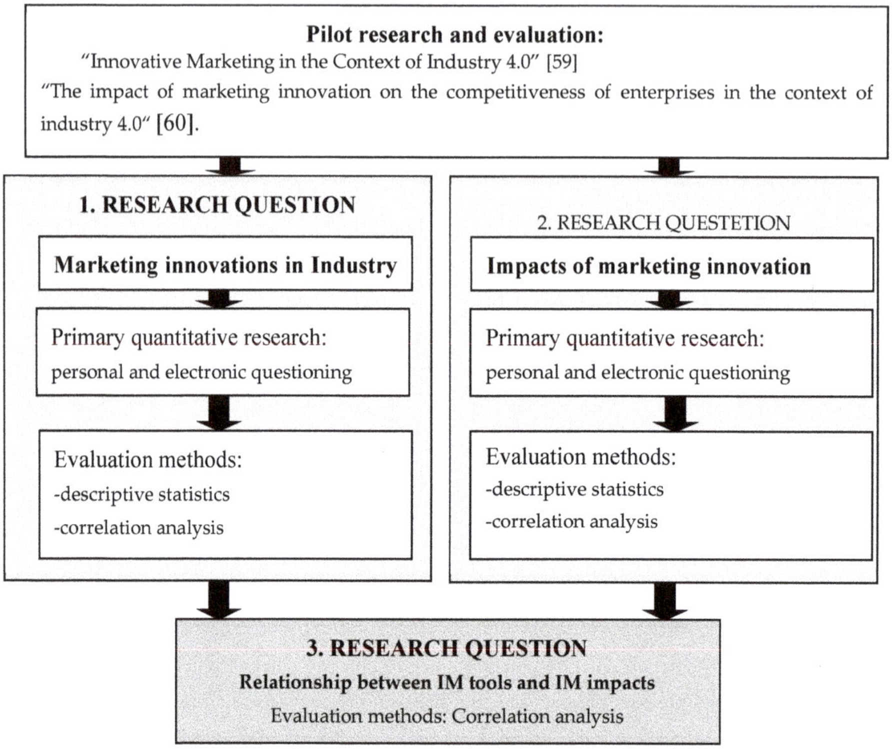

Pilot research and evaluation:
"Innovative Marketing in the Context of Industry 4.0" [59]
"The impact of marketing innovation on the competitiveness of enterprises in the context of industry 4.0" [60].

1. RESEARCH QUESTION	**2. RESEARCH QUESTETION**
Marketing innovations in Industry	**Impacts of marketing innovation**
Primary quantitative research: personal and electronic questioning	Primary quantitative research: personal and electronic questioning
Evaluation methods: -descriptive statistics -correlation analysis	Evaluation methods: -descriptive statistics -correlation analysis

3. RESEARCH QUESTION
Relationship between IM tools and IM impacts
Evaluation methods: Correlation analysis

Figure 1. Conceptual framework of this study. Source: own.

The respondents were selected using systematic random sampling, which had to be performed in order to enable the use of statistical induction tests [61]. The parent university had access to the Bisnode database of companies, where enterprises are sorted alphabetically only, and which contains the entire population. The system used to select the enterprises involved contacting every 50th enterprise, which complied with the conditions of random research and enabled further statistical processing.

The pilot project was followed by primary research, which took place between June 2018 and March 2019. Two hundred and ten companies participated in the research, out of a total of 605 respondents, which was a return of 34%. The return on responses was relatively high, as the enterprises were approached in person and were subsequently sent a questionnaire electronically. It was necessary to obtain answers from key people, either owners, top management or the head of marketing in the enterprise. Enterprises with more than 10 employees were contacted and smaller enterprises were excluded on the assumption that innovation is more difficult to implement and that these micro-enterprises would distort the sample.

The data were collected using the IBM SPSS Data Collection software; IBM SPSS Advanced Statistics was used for the statistical assessment.

In the conceptual framework, it can be seen that research innovations in marketing and the effect they cause take place in parallel. The two investigations are then interconnected to establish the relationship. Methods of evaluating the detected data included:

(a) *Descriptive statistics* to determine and summarize information, process it in the form of graphs and tables, and calculate their numerical characteristics. Data processing methods used in the research included: frequency, arithmetic mean, standard deviation, median confidence interval, and minimum and maximum scale values.

(b) *Correlation analysis* depicts the statistical dependence of two quantitative variables. The aim of the correlation analysis was to determine the strength of the linear dependence between the frequency of use and the arithmetic mean of innovative marketing and the effects they caused. Correlation was also used to identify the dependence between marketing innovations and impact [62].

4. Final Evaluation of Research

Three research sub-questions were determined for this paper. According to these research questions, this section is divided into three subchapters, each giving an answer to one research question.

4.1. Innovative Marketing Tools

The first research question was set to identify the current use of innovative marketing in practice: "How are innovative marketing tools used during industry 4.0?" The tools examined were identified in the pilot research carried out by the qualitative method of data collection. In the pilot project, enterprises identified 17 tools that they perceived as part of Industry 4.0.

1. Additive production—this is the formation of a physical product by gradual controlled addition of materials, such as metals, plastics, thermoplastics, glass. These include, in particular, casting and 3D printing [63]. According to Corsini, Aranda-Jan, and Moultrie [64] the standard metal and plastic machining industry cannot be replaced by additive manufacturing. Three-dimensional printing is not yet suitable for mass production and is only suitable for unique and complex products.
2. Augmented reality—this is a representation of the real environment and the subsequent addition of visual information using 3D graphics. Thanks to 3D animation it is possible to present not only the appearance of the product, but also a demonstration of the product cut and its functionality [65].
3. Virtual reality—allows the user to find himself in a simulated environment associated with user interaction. Virtual reality creates the illusion of a real world or a fictional world. In practice, it is used for the construction of buildings and cars, in the medical industry, or for computer games [66].
4. Virtual currency (cryptocurrency)—this is based on the principle of peer-to-peer networks (client–client). This currency system has no superior control to regulate the currency. Virtual currency cannot be falsified due to the complex encryption. All transactions and accounts are public, which acts as protection and prevention of financial crime. The use of payment systems in practice is currently limited and used more as an investment [67].
5. Autonomous distribution—in the consumer market, it is the delivery of the product directly to the customer's home, currently drones are the most used. In the industrial market, it is used in logistics, in transport among the intersections in the distribution channel. At the same time it is used in in-house logistics, where autonomous trucks provide production [68].
6. Organizing events—this is a method of marketing communication connected with a form of performance, an experience that is associated with affecting emotions. In practice, events are divided into external communication, building relationships with stakeholders, and internal communication focused on their own employees. According to the authors Biswas and Suar [69] and Dabirian, Kietzmann, and Diba [70] event marketing is now at its peak again and its effectiveness is primarily in building employer branding.
7. Relationship marketing—in contrast to transaction marketing, which is based on business needs, relationship marketing is based on customer needs. In practice, it is an effort to create, maintain, and expand strong and valuable relationships with stakeholders. Gillett [71] has proven that building customer loyalty and business success is strongly correlated.
8. Product placement on shared multimedia—product placement is the placement of a product or brand in a movie, series, video, or photo to make it visible. The highest efficiency and effectiveness

according to Reference [65] is to use product placement on shared multimedia such as YouTube, Instagram or Vimeo.

9. Mobile app marketing—Kaplan [43] defines mobile marketing as a marketing activity conducted over the Internet to which consumers are constantly connected via a personal mobile device. Currently, companies see the greatest mobile marketing opportunities in applications, but this is no longer possible without app store optimization that will ensure a good position [72].

10. Quality function deployment—the aim is to incorporate the requirements of end customers into the final product of the company. The best way to apply QFD is to engage customers directly in product development. Lam and Bai [73] have proven that QFD leads to increased business competitiveness because the company develops and manufactures only products that the customer expects.

11. Product and packaging ecodesign—a systematic process of product design and development that, in addition to classic features such as functionality, places great emphasis on achieving a minimum negative impact of the product on the environment in terms of its entire life cycle. In practice, this means that the company will reuse all parts of the product at the end of its life cycle [74].

12. Internet of Things—this is so-called machine-to-machine communication. The product must have a built-in communication device to receive information from another device, process it, and provide it to another device. In practice, enterprises use IoT both in manufacturing, for example, in supplying production lines, or producing products that communicate with each other [75].

13. Circular economy—The circular economy separates economic growth from the need to extract new and rare materials. In reality, enterprises focus on material savings, recycling, reuse, and refurbishment. Lewandowski [76] and Velenturf and Purnell [77] have proven that business involvement in the circular economy is economically beneficial in the long run.

14. Guerilla and viral marketing—guerilla marketing is an unconventional form of marketing intended to shock. The goal is to get the maximum effect from minimal sources. Its low-cost use is primarily used by smaller enterprises. These aggressive attacks are mostly associated with a viral spread through social networks [78].

15. Advergaming—creating computer games for presenting enterprises or products. This is a link between the gaming business and marketing. These may be virtual worlds such as The Sims worlds or augmented games that combine reality with fiction such as Pokémon, where real "sponsored" sites serve as part of the game [79].

16. Employer branding—creating an employer's brand is a long-term and continuous process and consists in systematically creating and sharing positive employee experience. The main tool is sophisticated personnel communication with current, future, and former employees of the company. According to Reference [69], enterprises use this strategic tool to prevent employee turnover and attract the best possible future candidates.

17. Individual marketing using social media—also known as one-to-one marketing, this is a marketing strategy by which companies leverage data analysis and digital technology to deliver individualized messages and product offerings to current or prospective customers. According to Reference [80], enterprises need to use big data in conjunction with social media to carry out individual marketing.

These tools have been subjected to research. First, Table 1 presents descriptive statistics. The tools are listed in order of importance, from the most important to the least important.

Table 1. Evaluation of innovative marketing tools.

Tool	$\bar{x}$	n	SD	95% Confidence Interval		Median
				Min.	Max.	
Tool 17	2.65	157	1.86	2.36	2.94	2
Tool 7	2.91	137	1.772	2.61	3.21	2
Tool 6	2.99	148	1.841	2.69	3.29	3
Tool 13	3.26	154	1.885	2.96	3.56	3
Tool 9	3.41	96	2.05	2.99	3.82	3
Tool 8	3.44	116	1.971	3.08	3.8	3
Tool 10	3.57	107	2.047	3.18	3.96	3
Tool 14	3.61	137	1.884	3.29	3.93	3
Tool 16	3.61	145	1.761	3.32	3.9	3
Tool 2	3.89	102	2.009	3.5	4.29	3
Tool 12	3.94	68	2.136	3.42	4.46	3
Tool 3	4.01	79	2.047	3.55	4.47	4
Tool 11	4.03	77	2	3.57	4.48	4
Tool 1	4.15	78	2.064	3.69	4.62	4
Tool 15	4.55	73	2	4.08	5.01	5
Tool 4	5	54	2.249	4.39	5.61	6
Tool 5	5.18	56	1.983	4.65	5.71	6

Source: own, (n = 210, 1 = maximum importance, 7 = minimum importance).

This section may be divided by subheadings. It should provide a concise and precise description of the experimental results, their interpretation as well as the experimental conclusions that can be drawn.

If we divide the importance rating into two intervals (<1; 4) and (4; 7>), we get eleven tools that are important in the first interval and six that are not important. The second column shows the number of enterprises using each tool. The resulting values show that those tools that are perceived as important are used the most. The calculation of the correlation between the measure of importance and the frequency of use was −0.87234, which confirms the negative dependence.

The enterprises agree on the evaluation, which was confirmed by the standard deviation, where there were no large deviations. The identified tools can be combined into three groups. Each group contains tools that are identical in their use. These are:

(A) Tools targeting narrow homogeneous segments or individuals—enterprises identify innovative marketing as the most important tools that enable accurately targeting the most homogeneous segment. Even the most important was the tool that targets directly to individuals. Structured but also unstructured data are used for this direction, accessible from social networks. Industry 4.0 is about processing large amounts of unstructured information (big data). Focusing directly on the individual is associated with engaging customers directly in creating product design, organizing events, or reaching out with viral content.

(B) Promotion tools based on technological innovations—the second group of tools related to Industry 4.0 are technological marketing innovations, which include augmented reality, virtual reality, virtual currency, autonomous distribution, Internet of Things or advergaming. These tools are assessed by contradictory companies, which is evident from the higher dispersion. This may be due to the different levels of business involvement in Industry 4.0. The importance of these tools is clearly demonstrated despite lower assessments.

(C) Corporate social responsibility—The third group of marketing innovations are tools associated with corporate social responsibility. Although enterprises have classified them as Industry 4.0 related tools, they are non-technological tools. Their use is very common, and their importance is also relatively high. These are ecodesign, circular economy, and employer branding.

4.2. Impacts of Innovative Marketing

The second research question was "What impact do innovative marketing tools cause when enterprises implement them in their strategy?" The solution to this research question was a follow-up to the previous section. Identifying marketing innovations and their importance was a precondition for researching the impacts they cause. Impacts were also identified in the pilot project through qualitative research. Overall, 11 impacts were included in primary research.

The list of impacts was also determined in the pilot project, using qualitative research. Fifty enterprises participated in the pilot research, which, according to the available information, use innovative marketing in their business and present themselves as implementing Industry 4.0. The research was conducted in the form of personal questions, using the focus group method. The outputs were subjected to a content analysis, involving marketing experts and a practical expert specializing in the implementation of Industry 4.0 in enterprises. The responses to the open questions were first coded and then grouped into clusters. This procedure identified 11 main impacts of innovative marketing that the firms we questioned considered important. The resulting list may be described as containing fundamental impacts in the application of innovative marketing 4.0.

1. Building PR and thus increasing the value of the enterprise. By implementing Industry 4.0 and innovative marketing in its program, the enterprise exhibits to the stakeholders and the surrounding area that it has a long-term vision. Implementation is usually linked to capital investment, which increases costs but also improves the image and value of the enterprise. This impact was most common in the replies half of all enterprises.
2. Higher demands on employees. Although marketing innovations are classified as non-technological, the implications of implementation are clearly linked to higher demands on employees. The Industry 4.0 philosophy will lead to a massive reduction in manual workers and a high demand for skilled people. Enterprises are already aware of the need to change the structure of their employees.
3. Improving communication with customers. The identified marketing innovations lead to better knowledge of customers. With social networks, credit cards, and other Internet footprints, enterprises will have perfect information about each individual. By processing big data, they can get to know their customer's behavior intimately and tailor their communication mix to fit their needs and, thus, achieve greater satisfaction. Improved communication with the target group leads to the acquisition of new customers.
4. Increasing the competitiveness of the enterprise. Respondents said that the implementation of innovative marketing and Industry 4.0 is in itself a competitive advantage which leads to assertion in a certain field compared to other enterprises. It is an increase in structural competitiveness resulting from ownership of assets or technology.
5. Change in the amount of costs. Enterprises have agreed that the implementation of innovative marketing and Industry 4.0 is associated with cost changes. The replies showed that it was not possible to unequivocally claim that there was an increase in costs because some enterprises stated that there was a reduction in costs. The time factor plays a large role in costs. In the short term, due to the introduction of Industry 4.0, this is a cost which, in the long run, leads to cost reductions.
6. Entering new markets. Thanks to an innovative approach, enterprises gain a competitive edge which aims to determine the growth strategy. These growth strategies often involve entering new markets. It is very often the internationalization of the enterprise, which concerns technologically developed countries. However, enterprises can expand their operations to other new segments where, for example, the expansion from the industrial to the consumer market has not yet functioned.
7. Increasing labor productivity. Labor productivity is increased as a result of the introduction of improved technologies. Marketing innovations increase overall output, divided by work input.

Higher labor productivity leads to higher profits. This profit can then be used in the form of free capital for investments leading to further innovation. This cycle is driven by innovations associated with Industry 4.0.

8. Change of distribution channels. It is primarily a systemic vertical integration that leads to property interconnection from production to sale. This situation leads to many acquisitions and mergers, which create large multinational corporations. Autonomous robotization will play a major role in distribution, especially in engineering. At the same time, Industry 4.0 will lead to the autonomous distribution of goods to the end customer, for example, using drones.

9. Improving product quality. Better technology clearly implies an increase in product quality, for example, 3D printing while maintaining maximum accuracy in product manufacturing. Thanks to Industry 4.0, new materials are used with new properties that lead to improved product quality. The impact of implementation is clearly with products at a higher price level.

10. Changes in strategic planning. The implementation of Industry 4.0 leads to changes in long-term business planning. The basics of long-term planning is a strategic plan, where the vision of the enterprise is changed. The changes also affect other parts of the strategic plan, which are strategy and tactics. Enterprises see the great importance of digitization in the control of a strategic plan where, thanks to big data processing, the enterprise has a perfect overview of all outputs and hard data in context.

11. Change of company culture. Corporate culture can be characterized as a way of doing work and dealing with people. These are symbols of the enterprise (abbreviations, slang, dress code, symbols), hero promotion (serves as a model of ideal behavior), rituals (informal activities, formal meetings), and values that represent the deepest level of corporate culture. Respondents stated it was necessary to adapt to these changing needs of the market and clients as a result of Industry 4.0.

The identified impacts were subjected to research and subsequent statistical evaluation. Descriptive statistics were used once again, and the results are listed in Table 2 arranged in order of the strongest effects.

Table 2. The effects of innovative marketing on enterprises.

Impact	$\bar{x}$	n	SD	95% IS		Median
				Min.	Max.	
Impact 6	2.58	192	1.59	2.35	2.8	2
Impact 4	2.74	191	1.675	2.5	2.98	2
Impact 1	2.78	175	1.58	2.54	3.01	2
Impact 9	3.07	171	1.696	2.81	3.33	3
Impact 5	3.13	183	1.664	2.88	3.37	3
Impact 7	3.26	160	1.982	2.95	3.57	3
Impact 2	3.39	163	1.783	3.12	3.67	3
Impact 8	3.43	172	1.648	3.18	3.68	3
Impact 10	3.52	174	1.733	3.26	3.78	3
Impact 11	3.83	156	1.911	3.52	4.13	3
Impact 3	3.86	170	1.718	3.6	4.12	4

Source: own, 1 = strongest impacts; 7 = weakest impacts.

If the effects of innovative marketing are again broken down by a threshold of 4, it can be stated that all of the effects were identified by enterprises as very strong when they were above the threshold. At the same time, the strength of the impacts correlated significantly with the number of enterprises. The correlation coefficient between the impact strength and the number of enterprises that were identified as having impact effects was −0.7990. At the same time, there was a broad consensus on the assessment, as evidenced by the low standard deviation for all impacts.

4.3. The Relationship of Innovative Marketing and Its Impact

After identifying current innovative marketing tools and their impacts, the authors focused on the relationship between the resulting innovative tools and impacts. The basic idea was that the identified effects from Section 4.2 can be viewed in reverse. Effects arising from the implementation of marketing innovations can be understood as the strategic goals of the enterprise. In practice, instead of impacts, an enterprise can include the identified impacts in its plan as possible targets. For example, the identified impact of "Building Public Relations and Branding" can be included by an enterprise in its strategic plan as a long-term goal.

If all the identified impacts change to goals, we can meet each goal with the innovative marketing tools identified. For each goal, it is possible to build an individual marketing mix, by calculating the correlation between tools and impacts. The correlation data were collected by a systematic random selection of respondents, as described in the methodology. The research question "What marketing mix does an enterprise involved in Industry 4.0 have to build to achieve a certain strategic goal" was identified.

To determine the correlation between tools and impacts, pairs of scaled vectors were first created. Pairs where respondents said they did not use the tool or impact were omitted. The created pairs were subjected to correlation analysis, where the influence of the evaluation of the importance of individual tools of innovative marketing on the evaluation of individual impacts was examined. In this way, 17×11 pairs of vectors were generated, and their mutual state was analyzed. The following model was investigated:

H0: $r_{Toolx, Impactx} = 0$

HA: $r_{Toolx, Impactx} \neq 0$

Table 3 summarizes the resulting p-values for each correlation test. The reliability coefficient is set to $\alpha = 0.05$, where the *p*-value is 0.05, a statistically significant relationship is indicated.

Table 3. Resulting *p*-values of the correlate analysis test of the pair tool × impact.

IM Tools	Impact											
	1	2	3	4	5	6	7	8	9	10	11	Σ
1	0.813	0.728	0.917	0.997	0.275	0.493	0.044	0.637	0.036	0.615	0.076	2
2	0.052	0.555	0.068	0.362	0.005	0.433	0.038	0.014	0.141	0.114	0.007	4
3	0.675	0.855	0.092	0.231	0.083	0.260	0.004	0.084	0.018	0.640	0.133	2
4	0.472	0.006	0.289	0.598	0.101	0.364	0.044	0.774	0.048	0.891	0.803	3
5	0.237	0.037	0.152	0.599	0.474	0.352	0.654	0.636	0.338	0.755	0.253	1
6	0.021	0.745	0.225	0.000	0.153	0.007	0.811	0.003	0.130	0.039	0.232	5
7	0.090	0.369	0.005	0.169	0.136	0.068	0.496	0.037	0.033	0.040	0.015	5
8	0.568	0.432	0.328	0.555	0.411	0.022	0.335	0.112	0.201	0.076	0.379	1
9	0.624	0.563	0.049	0.509	0.639	0.010	0.998	0.022	0.053	0.333	0.816	3
10	0.108	0.915	0.371	0.544	0.781	0.724	0.038	0.399	0.472	0.806	0.543	1
11	0.023	0.599	0.322	0.064	0.035	0.744	0.465	0.846	0.967	0.887	0.047	3
12	0.508	0.736	0.144	0.570	0.011	0.057	0.206	0.167	0.075	0.255	0.225	1
13	0.233	0.925	0.458	0.063	0.009	0.006	0.103	0.082	0.114	0.112	0.013	3
14	0.036	0.598	0.056	0.008	0.000	0.000	0.873	0.051	0.003	0.002	0.077	6
15	0.013	0.381	0.021	0.262	0.590	0.251	0.238	0.536	0.473	0.805	0.564	2
16	0.060	0.043	0.112	0.222	0.049	0.051	0.017	0.001	0.436	0.017	0.177	5
17	0.000	0.613	0.031	0.002	0.399	0.000	0.050	0.034	0.068	0.101	0.131	6
Σ	5	3	4	3	6	6	7	6	5	4	4	

Source: own. (IM = Innovative Marketing).

The resulting table shows that each impact was related to several tools. There were a minimum of three and a maximum of seven tools. The correlation analysis confirmed that all tools were associated with certain impacts. Overall, the two tools "guerilla and viral marketing" and "individual marketing using social media" correlated most with the six impacts. By contrast, four innovative tools correlated with only one impact.

If an enterprise had one of the identified impacts among its strategic objectives, it can successfully use the identified tools. These are the tools that companies had identified as an effective strategy leading to the set goals. If innovative marketing tools are divided into the three groups (A–C) when identifying innovative marketing tools, this breakdown does not apply when compiling individual marketing mixes, as shown in Table 4.

Table 4. Strategic objective and marketing mix. (SM: social media).

Strategic Objective	Marketing Mix of Innovative Tools		
	A. Tools Targeting Narrow Homogeneous Segments or Individuals	B. Promotion Tools Based on Technological Innovations	C. Corporate Social Responsibility
1. Building public relations and brand	Organizing events, guerilla and viral marketing, Individual marketing using SM	Advergaming	Ecodesign of product and packaging
2. Higher demands on employees		Virtual currency Autonomous distribution	Employer branding
3. Improving communication with customers	Relationship marketing Individual marketing using SM	Advergaming Mobile app marketing	
4. Increasing competitiveness	Organizing events, Guerilla and viral marketing, Individual marketing using SM		
5. Change in the amount of costs	Guerilla and viral marketing	Augmented reality Internet of Things	Circular economy Ecodesign of product and packaging
6. Entering new markets	Organizing events Product placement on shared multimedia Guerilla and viral marketing Individual marketing using SM	Mobile app marketing	Circular economy Ecodesign of product and packaging
7. Increasing labor productivity	Quality function deployment Individual marketing using SM	Virtual reality Additive production Virtual currency Augmented reality	Employer branding
8. Change of distribution channels	Organizing events Relationship marketing Individual marketing using SM	Mobile app marketing Augmented reality	Employer branding
9. Product quality improvement	Relationship marketing Guerilla and viral marketing	Virtual currency Virtual reality Additive production	
10. Changes in strategic planning	Organizing events Relationship marketing Guerilla and viral marketing		Employer branding
11. Change of company culture	Relationship marketing	Augmented reality	Circular economy Ecodesign of product and packaging

The results of the research have shown that it is not only tools from one group that can be used to implement the strategy, but a toolkit composed of all three groups. In conclusion, there is a clear link between the marketing tools associated with Industry 4.0 and the impacts associated with this time of major change. These relationships can be used by enterprises for accurate and successful strategy development.

5. Discussion

The findings refute the classic perception of innovation presented in the EU in the Oslo Manual of 2005. According to this material, marketing innovations are classified as non-technological innovations, but research has shown that current enterprises perceive marketing innovations as a combination of technological and non-technological innovations. Therefore, marketing innovations in Industry 4.0 cannot be described as non-technological, but they cannot be labeled as merely technological either.

An interesting difference was found in the perception of Industry 4.0 in the secondary research. Some authors, for example Wee et al. [9] and Cooper et al. [8], present Industry 4.0 as an industrial revolution associated with a step-up in productivity gains. The authors showed that new technologies, digitization, and robotization are leading to a revolution in industry. On the other hand, the authors of References [57,81,82] perceive Industry 4.0 as a natural evolution in industry, not revolution. From the presented research we cannot unequivocally favor either evolution or revolution.

If we evaluate the innovative marketing tools associated with Industry 4.0 that have been identified by this research, it is a combination of long-term tools and brand-new tools. Examples of marketing tools that were used before the idea of Industry 4.0 came into being include events, relationship marketing, employer branding, and product placement. Examples of current marketing tools include the Internet of Things, additive production, augmented reality, virtual reality, and virtual currency. When comparing the identified innovative marketing tools from secondary research with the findings from primary research, there was a strong consensus. The identified innovative marketing tools identified from secondary tools coincided with primary research. Research has shown that there is no boundary between Industry 4.0 and innovative marketing. The identified list of seventeen tools was a compilation of Industry 4.0 as perceived by the authors of References [34–37,40,41,43].

The fundamental local effect of any industrial revolution is the rapid increase in labor productivity, as confirmed by References [83–85]. In the presented research, enterprises also identified an increase in labor productivity, but this ranked sixth in the ranking of importance. The most important impacts were "entering new markets, increasing competitiveness", which are connected with growth strategies. The third strongest impact of Industry 4.0 implementation was "building PR and a brand". Many enterprises today present themselves as implementing Industry 4.0, but even the presentation of business involvement in Industry 4.0 is often just a marketing strategy. In this way, the enterprise presents itself is connected with innovation, but the reality is completely different.

The relationship of innovative marketing in Industry 4.0 and the impact that this implies is essential for practical use in strategic planning. Identified impacts that may be strategic goals in strategic planning as stated by Müller et al. [86] and Wolf and Floyd [87] can be fulfilled with a marketing strategy in the form of a precisely designed marketing mix. On the other hand, marketing innovations need to be seen only as part of the business strategy, along with product, process, and organizational innovations.

6. Conclusions

The reason for carrying out the research presented in this paper was to find out the current perception of Industry 4.0 in marketing practice. Theoretically, this topic has been written about quite often, as shown by the secondary research, but there is a lack of a practical perspective coming directly from enterprises. This paper provided this enterprise view in which the main research question was "How to use current Industry 4.0 innovative marketing tools for enterprises practice?" The authors conducted a pilot study, the results of which followed the primary research. We managed to identify innovative marketing tools and sort them by importance. The next step was to identify effects and to rank them by their impact on enterprises. Finally, the tools were confronted with impacts and dependent pairs were constructed using correlation analysis. This correlation was used in the preparation of specific marketing mixes for individual strategic objectives. Every enterprise plan and strategic goal are a part of long-term plans. For these goals, marketing mixes identified in the field of marketing, consisting of innovative tools, which according to companies belong to Industry 4.0, can be used. This conclusion can answer the main research question. Three other research questions were answered in the research process.

The main importance of the presented research is the possibility of applying the results in practice. The results obtained can be used as guidelines for drawing strategic plans. A large number of strategic objectives and the ways to achieve them were identified. Enterprise managers can use the marketing mix discovered by the research, which is likely to be effective, for achieving these goals. Of course, the use of only marketing to fulfill the enterprise strategy would be insufficient. The importance of the paper is also in the theoretical area, where a comprehensive overview of the current Industry 4.0, as perceived at present, was recorded.

This paper is highly extensive and comprehensive, in that it combines several studies and places them into logical contexts. Even so, there are certain limitations to the results it presents due to the scope. This paper does not present the differences resulting from the sorting parameters of the enterprises, such as the size of the enterprise, sector or ownership. This could form the subject of future research. The research was conducted in 2018/19; however, the situation in the sector in question is developing very rapidly, and so the same research method should be applied regularly, such as every two years. The research may serve as a basic record of situations that could be the subject of further research using the same methodology. The paper presents a combination of seventeen innovative marketing tools and eleven effects that they cause. Each of the tools and impacts is described briefly in the paper, but a more precise and detailed explanation could be another possible direction on which researchers could focus. The greatest benefit of the paper is that the results may be used in practice for strategic planning. Verifying the accuracy of the compiled proposals is the biggest challenge for marketing researchers in the era of Industry 4.0.

In conclusion, the research results showed that emerging opportunities from Industry 4.0 are positive drivers for growth strategies and increased business competitiveness.

Author Contributions: Conceptualization, O.U. and J.D.; methodology, O.U.; software, O.U.; validation, O.U. and J.D.; formal analysis, J.D.; investigation, J.D.; resources, O.U.; writing—original draft preparation, O.U.; writing—review and editing, O.U.; visualization, J.D.; supervision, O.U.; project administration, J.D.; funding acquisition, J.D.

Funding: This research received no external funding.

Acknowledgments: This article was written as part of a project to support research teams of excellence at the Faculty of Economics of the Technical University of Liberec, which is financed by institutional support for long-term conceptual development at the Faculty of Economics of the Technical University of Liberec.

Conflicts of Interest: The authors declare no conflict of interest.

References

1. Vojáček, A. Co se skrývá pod výrazy Industry 4.0/Průmysl 4.0?| Automatizace.HW.cz. Available online: https://automatizace.hw.cz/mimochodem/co-je-se-skryva-pod-vyrazy-industry-40-prumysl-40.html (accessed on 11 August 2019).
2. Oesterreich, T.D.; Teuteberg, F. Understanding the implications of digitisation and automation in the context of Industry 4.0: A triangulation approach and elements of a research agenda for the construction industry. *Comput. Ind.* **2016**, *83*, 121–139. [CrossRef]
3. Stock, T.; Seliger, G. Opportunities of sustainable manufacturing in industry 4.0. *Procedia CIRP* **2016**, *40*, 536–541. [CrossRef]
4. Hofmann, E.; Rüsch, M. Industry 4.0 and the current status as well as future prospects on logistics. *Comput. Ind.* **2017**, *89*, 23–34. [CrossRef]
5. Lee, J.; Kao, H.-A.; Yang, S. Service innovation and smart analytics for industry 4.0 and big data environment. *Procedia CIRP* **2014**, *16*, 3–8. [CrossRef]
6. Zhou, K.; Liu, T.; Zhou, L. Industry 4.0: Towards future industrial opportunities and challenges. In Proceedings of the IEEE 2015 12th International Conference on Fuzzy Systems and Knowledge Discovery (FSKD), Zhangjiajie, China, 15–17 August 2015; pp. 2147–2152.

7. Pereira, A.C.; Romero, F. A review of the meanings and the implications of the Industry 4.0 concept. *Procedia Manuf.* **2017**, *13*, 1206–1214. [CrossRef]
8. Cooper, S. Designing a UK industrial strategy for the age of industry 4.0. *Rethink Manuf.* **2017**, *5*, 1–27.
9. Wee, D.; Kelly, R.; Cattel, J.; Breunig, M. Industry 4.0-how to navigate digitization of the manufacturing sector. *McKinsey Co.* **2015**, *58*, 56–67.
10. ČSÚ Inovace. Available online: https://www.czso.cz/csu/czso/statistika_inovaci (accessed on 10 August 2019).
11. Manual, O. *Oslo Manual: Guidelines for Collecting and Interpreting Innovation Data*; Committee for Scientific and Technological Policy, OECD-OCDE: Paris, France, 2005.
12. Lu, Y. Industry 4.0: A survey on technologies, applications and open research issues. *J. Ind. Inf. Integr.* **2017**, *6*, 1–10. [CrossRef]
13. Witkowski, K. Internet of things, big data, industry 4.0–innovative solutions in logistics and supply chains management. *Procedia Eng.* **2017**, *182*, 763–769. [CrossRef]
14. Maresova, P.; Soukal, I.; Svobodova, L.; Hedvicakova, M.; Javanmardi, E.; Selamat, A.; Krejcar, O. Consequences of Industry 4.0 in Business and Economics. *Economies* **2018**, *6*, 46. [CrossRef]
15. Zezulka, F.; Veselý, I. Úvod do problematiky a základní modely Industry 4.0. Available online: https://www.systemonline.cz/rizeni-vyroby/uvod-do-problematiky-a-zakladni-modely-industry-4.0.htm (accessed on 11 August 2019).
16. Rachinger, M.; Rauter, R.; Müller, C.; Vorraber, W.; Schirgi, E. Digitalization and its influence on business model innovation. *J. Manu Tech. Manag.* **2018**, *30*, 18. [CrossRef]
17. Müller, J.M.; Buliga, O.; Voigt, K.-I. Fortune favors the prepared: How SMEs approach business model innovations in Industry 4.0. *Technol. Forecast. Soc. Chang.* **2018**, *132*, 2–17. [CrossRef]
18. Leyh, C.; Martin, S.; Schäffer, T. Industry 4.0 and Lean Production—A matching relationship? An analysis of selected Industry 4.0 models. In Proceedings of the IEEE 2017 Federated Conference on Computer Science and Information Systems (FedCSIS), Prague, Czech Republic, 3–6 September 2017; pp. 989–993.
19. Hermann, M.; Pentek, T.; Otto, B. Design principles for industrie 4.0 scenarios. In Proceedings of the IEEE 2016 49th Hawaii International Conference on System Sciences (HICSS), Koloa, HI, USA, 5–8 January 2016; pp. 3928–3937.
20. Basl, J. Pilot study of readiness of Czech companies to implement the principles of Industry 4.0. *Manag. Prod. Eng. Rev.* **2017**, *8*, 3–8. [CrossRef]
21. Mařík, V. *A kol. Národní iniciativa průmysl 4.0*; Ministerstvo průmyslu a obchodu ČR: Prague, Czech Republic, 2015.
22. Kopp, J.; Basl, J. Studie připravenosti českých podniků na trendy Průmysl 4.0. *Systémová Integrace* **2017**, *24*, 77–86.
23. Zillner, S.; Lasierra, N.; Faix, W.; Neururer, S.B. User needs and requirements analysis for big data healthcare applications. *Stud. Health Technol. Inform.* **2014**, *205*, 657–661.
24. Obitko, M.; Jirkovský, V. Big data semantics in industry 4.0. In Proceedings of the International conference on industrial applications of holonic and multi-agent systems, Valencia, Spain, 2–3 September 2015; Springer: Cham, Switzerland, 2015; pp. 217–229.
25. Wang, C.; Wang, Q.; Ren, K.; Lou, W. Privacy-preserving public auditing for data storage security in cloud computing. In Proceedings of the 2010 Proceedings IEEE Infocom, San Diego, CA, USA, 15–19 March 2010; pp. 1–9.
26. Burger, J.; Hasse, H. Multi-objective optimization using reduced models in conceptual design of a fuel additive production process. *Chem. Eng. Sci.* **2013**, *99*, 118–126. [CrossRef]
27. Basl, J.; Doucek, P. A Metamodel for Evaluating Enterprise Readiness in the Context of Industry 4.0. *Information* **2019**, *10*, 89. [CrossRef]
28. Kotler, P.; de Bes, F.T. *Inovativní Marketing: Jak Kreativním Myšlením Vítězit u Zákazníků*; Grada Publishing a.s.: Prague, Czech Republic, 2005.
29. Tinoco, J.K. Marketing innovation: The unheralded innovation vehicle to sustained competitive advantage. *Int. J. Sustain. Strateg. Manag.* **2010**, *2*, 168. [CrossRef]
30. Saxena, A. *Marketing Innovation in FMCG Industry*; In Proceedings of the PRIMA; Publishing India Group: Delhi, India, 2011; Volume 2, p. 9.
31. Gupta, S.; Malhotra, N.K.; Czinkota, M.; Foroudi, P. Marketing innovation: A consequence of competitiveness. *J. Bus. Res.* **2016**, *69*, 5671–5681. [CrossRef]

32. Moreira, J.; Silva, M.J.; Simoes, J.; Sousa, G. Marketing innovation: Study of determinants of innovation in the design and packaging of goods and services—Application to Portuguese firms. *Contemp. Manag. Res.* **2012**, *8*. [CrossRef]

33. Lu, H.; Li, Y.; Chen, M.; Kim, H.; Serikawa, S. Brain intelligence: Go beyond artificial intelligence. *Mob. Netw. Appl.* **2018**, *23*, 368–375. [CrossRef]

34. Tiago, M.T.P.M.B.; Veríssimo, J.M.C. Digital marketing and social media: Why bother? *Bus. Horiz.* **2014**, *57*, 703–708. [CrossRef]

35. Hancock, B.; Bordes, A.; Mazare, P.-E.; Weston, J. Learning from Dialogue after Deployment: Feed Yourself, Chatbot! *arXiv* **2019**, arXiv:1901.05415.

36. Gross, J.; Wangenheim, F.V. The Big Four of Influencer Marketing. A Typology of Influencers. *Mark. Rev. St. Gallen* **2018**, *2*, 30–38.

37. Baye, M.R.; De los Santos, B.; Wildenbeest, M.R. Search engine optimization: What drives organic traffic to retail sites? *J. Econ. Manag. Strategy* **2016**, *25*, 6–31. [CrossRef]

38. Schivinski, B.; Dabrowski, D. The effect of social media communication on consumer perceptions of brands. *J. Mark. Commun.* **2016**, *22*, 189–214. [CrossRef]

39. Ren, L.; Xie, G.; Krabbendam, K. Sustainable competitive advantage and marketing innovation within firms: A pragmatic approach for Chinese firms. *Manag. Res. Rev.* **2009**, *33*, 79–89. [CrossRef]

40. Prahalad, C.K.; Ramaswamy, V. Co-creation experiences: The next practice in value creation. *J. Interact. Mark.* **2004**, *18*, 5–14. [CrossRef]

41. Son, J.; Sadachar, A.; Manchiraju, S.; Fiore, A.M.; Niehm, L.S. Consumer adoption of online collaborative customer co-design. *J. Res. Interact. Mark.* **2012**, *6*, 180–197. [CrossRef]

42. Kotler, P.; Wong, V.; Saunders, J.; Armstrong, G. *Moderní Marketing*; Grada Publishing a.s.: Prague, Czech Republic, 2007.

43. Kaplan, A.M. If you love something, let it go mobile: Mobile marketing and mobile social media 4x4. *Bus. Horiz.* **2012**, *55*, 129–139. [CrossRef]

44. Ohtonen, J. The 8 Phases of an Innovation Management Process. BPM Leader, 2013. Available online: http://www.bpmleader.com (accessed on 2 September 2019).

45. Cummins, D.; Gilmore, A.; Carson, D.; O'Donnell, A. Innovative marketing in SMEs: A conceptual and descriptive framework. *Int. J. New Prod. Dev. Innov. Manag.* **2000**, *2*, 231–248.

46. Lovichová, M. Inovativní marketing ve střední firmě. *Bus. Trends* **2014**, *3*, 64–70.

47. O'sullivan, D.; Abela, A.V. Marketing performance measurement ability and firm performance. *J. Mark.* **2007**, *71*, 79–93. [CrossRef]

48. Erol, S.; Jäger, A.; Hold, P.; Ott, K.; Sihn, W. Tangible Industry 4.0: A scenario-based approach to learning for the future of production. *Procedia CIRP* **2016**, *54*, 13–18. [CrossRef]

49. Schlechtendahl, J.; Keinert, M.; Kretschmer, F.; Lechler, A.; Verl, A. Making existing production systems Industry 4.0-ready. *Prod. Eng.* **2015**, *9*, 143–148. [CrossRef]

50. Sala, H.; Trivín, P. The effects of globalization and technology on the elasticity of substitution. *Rev. World Econ.* **2018**, *154*, 617–647. [CrossRef]

51. Muangkhot, S.; Ussahawanitchakit, P. Strategic marketing innovation and marketing performance: An empirical investigation of furniture exporting businesses in Thailand. *Bus. Manag. Rev.* **2015**, *7*, 189.

52. Burmeister, C.; Lüttgens, D.; Piller, F.T. Business model innovation for Industrie 4.0: Why the "Industrial Internet" mandates a new perspective on innovation. *Die Unternehmung* **2016**, *70*, 124–152. [CrossRef]

53. Müller, J.M.; Däschle, S. Business Model Innovation of Industry 4.0 Solution Providers Towards Customer Process Innovation. *Processes* **2018**, *6*, 260. [CrossRef]

54. Ibarra, D.; Ganzarain, J.; Igartua, J.I. Business model innovation through Industry 4.0: A review. *Procedia Manuf.* **2018**, *22*, 4–10. [CrossRef]

55. Strange, R.; Zucchella, A. Industry 4.0, global value chains and international business. *Multinatl. Bus. Rev.* **2017**, *25*, 174–184. [CrossRef]

56. Wollschlaeger, M.; Sauter, T.; Jasperneite, J. The future of industrial communication: Automation networks in the era of the internet of things and industry 4.0. *IEEE Ind. Electron. Mag.* **2017**, *11*, 17–27. [CrossRef]

57. Barreto, L.; Amaral, A.; Pereira, T. Industry 4.0 implications in logistics: An overview. *Procedia Manuf.* **2017**, *13*, 1245–1252. [CrossRef]

58. Mrugalska, B.; Wyrwicka, M.K. Towards lean production in industry 4.0. *Procedia Eng.* **2017**, *182*, 466–473. [CrossRef]

59. Ungerman, O.; Dedkova, J. Innovative Marketing in the Context of Industrie 4.0. In *Proceedings of the 13th International Conference Liberec Economic Forum 2017*; Kocourek, A., Ed.; Technical Univ Liberec, Faculty Economics: Liberec, Czech Republic, 2017; ISBN 978-80-7494-349-2.

60. Ungerman, O.; Dedkova, J.; Gurinova, K. The impact of marketing innovation on the competitiveness of enterprises in the context of industry 4.0. *J. Compet.* **2018**, *10*, 132. [CrossRef]

61. Mazzocchi, M. *Statistics for Marketing and Consumer Research*; SAGE: Thousand Oaks, CA, USA, 2008; ISBN 978-1-4739-0352-4.

62. Rossi, P.E.; Allenby, G.M.; McCulloch, R. *Bayesian Statistics and Marketing*; John Wiley & Sons: Hoboken, NJ, USA, 2012; ISBN 978-0-470-86368-8.

63. Bin Hamzah, H.H.; Keattch, O.; Covill, D.; Patel, B.A. The effects of printing orientation on the electrochemical behaviour of 3D printed acrylonitrile butadiene styrene (ABS)/carbon black electrodes. *Sci. Rep.* **2018**, *8*, 9135. [CrossRef]

64. Corsini, L.; Aranda-Jan, C.B.; Moultrie, J. Using digital fabrication tools to provide humanitarian and development aid in low-resource settings. *Technol. Soc.* **2019**, *24*, 101–117. [CrossRef]

65. Thomas, D.J. Augmented reality in surgery: The Computer-Aided Medicine revolution. *Int. J. Surg.* **2016**, *36*, 25. [CrossRef]

66. Sherman, W.R.; Craig, A.B. *Understanding Virtual Reality: Interface, Application, and Design*; Morgan Kaufmann: Burlington, MA, USA, 2018.

67. Briere, M.; Oosterlinck, K.; Szafarz, A. Virtual currency, tangible return: Portfolio diversification with bitcoin. *J. Asset Manag.* **2015**, *16*, 365–373. [CrossRef]

68. Northcote-Green, J.; Wilson, R. *Control and Automation of Electrical Power Distribution Systems*, 1st ed.; Northcote-Green, J., Wilson, R.G., Eds.; CRC Press: Boca Raton, FL, USA, 2017; ISBN 978-1-315-22146-5.

69. Biswas, M.K.; Suar, D. Antecedents and Consequences of Employer Branding. *J. Bus. Ethics* **2016**, *136*, 57–72. [CrossRef]

70. Dabirian, A.; Kietzmann, J.; Diba, H. A great place to work!? Understanding crowdsourced employer branding. *Bus. Horiz.* **2017**, *60*, 197–205. [CrossRef]

71. Gillett, A.G. REMARKOR: Relationship marketing orientation on local government performance. *J. Serv. Res.* **2015**, *15*, 97.

72. Grundy, Q.H.; Wang, Z.; Bero, L.A. Challenges in assessing mobile health app quality: A systematic review of prevalent and innovative methods. *Am. J. Prev. Med.* **2016**, *51*, 1051–1059. [CrossRef]

73. Lam, J.S.L.; Bai, X. A quality function deployment approach to improve maritime supply chain resilience. *Transp. Res. Part E Logist. Transp. Rev.* **2016**, *92*, 16–27. [CrossRef]

74. Dalhammar, C. Industry attitudes towards ecodesign standards for improved resource efficiency. *J. Clean. Prod.* **2016**, *123*, 155–166. [CrossRef]

75. National University of Sciences and Technology, Islamabad, Pakistan; Joyia, G.J.; Liaqat, R.M.; Farooq, A.; Rehman, S. Internet of Medical Things (IOMT): Applications, Benefits and Future Challenges in Healthcare Domain. *JCM* **2017**, *12*, 240–247. [CrossRef]

76. Lewandowski, M. Designing the business models for circular economy—Towards the conceptual framework. *Sustainability* **2016**, *8*, 43. [CrossRef]

77. Velenturf, A.; Purnell, P. Resource Recovery from Waste: Restoring the Balance between Resource Scarcity and Waste Overload. *Sustainability* **2017**, *9*, 1603. [CrossRef]

78. Tam, D.D.; Khuong, M.N. The Effects of Guerilla Marketing on Gen Y's Purchase Intention—A Study in Ho Chi Minh City, Vietnam. *Int. J. Trade Econ. Financ.* **2015**, *6*, 191. [CrossRef]

79. Sharma, M. Advergaming—The Novel Instrument in the Advertsing. *Procedia Econ. Financ.* **2014**, *11*, 247–254. [CrossRef]

80. Jussila, I.; Tarkiainen, A.; Sarstedt, M.; Hair, J.F. Individual psychological ownership: Concepts, evidence, and implications for research in marketing. *J. Mark. Theory Pract.* **2015**, *23*, 121–139.

81. Yin, Y.; Stecke, K.E.; Li, D. The evolution of production systems from Industry 2.0 through Industry 4.0. *Int. J. Prod. Res.* **2018**, *56*, 848–861. [CrossRef]

82. Bagheri, B.; Yang, S.; Kao, H.-A.; Lee, J. Cyber-physical systems architecture for self-aware machines in industry 4.0 environment. *IFAC-PapersOnLine* **2015**, *48*, 1622–1627. [CrossRef]

83. Rüssmann, M.; Lorenz, M.; Gerbert, P.; Waldner, M.; Justus, J.; Engel, P.; Harnisch, M. Industry 4.0: The future of productivity and growth in manufacturing industries. *Boston Consult. Group* **2015**, *9*, 54–89.

84. Schuh, G.; Potente, T.; Wesch-Potente, C.; Weber, A.R.; Prote, J.-P. Collaboration Mechanisms to Increase Productivity in the Context of Industrie 4.0. *Procedia CIRP* **2014**, *19*, 51–56. [CrossRef]

85. Drath, R.; Horch, A. Industrie 4.0: Hit or hype? [industry forum]. *IEEE Ind. Electron. Mag.* **2014**, *8*, 56–58. [CrossRef]

86. Müller, J.M.; Kiel, D.; Voigt, K.-I. What drives the implementation of Industry 4.0? The role of opportunities and challenges in the context of sustainability. *Sustainability* **2018**, *10*, 247. [CrossRef]

87. Wolf, C.; Floyd, S.W. Strategic Planning Research: Toward a Theory-Driven Agenda. *J. Manag.* **2017**, *43*, 1754–1788. [CrossRef]

Article

Construction of an Industrial Knowledge Graph for Unstructured Chinese Text Learning

Mingxiong Zhao, Han Wang, Jin Guo, Di Liu, Cheng Xie *, Qing Liu and Zhibo Cheng

School of Software, Yunnan University, Yunnan 650500, China
* Correspondence: chengxie@sjtu.edu.cn; Tel.: +1-818-3897-655

Received: 26 May 2019; Accepted: 2 July 2019; Published: 5 July 2019

Abstract: The industrial 4.0 era is the fourth industrial revolution and is characterized by network penetration; therefore, traditional manufacturing and value creation will undergo revolutionary changes. Artificial intelligence will drive the next industrial technology revolution, and knowledge graphs comprise the main foundation of this revolution. The intellectualization of industrial information is an important part of industry 4.0, and we can efficiently integrate multisource heterogeneous industrial data and realize the intellectualization of information through the powerful semantic association of knowledge graphs. Knowledge graphs have been increasingly applied in the fields of deep learning, social network, intelligent control and other artificial intelligence areas. The objective of this present study is to combine traditional NLP (natural language processing) and deep learning methods to automatically extract triples from large unstructured Chinese text and construct an industrial knowledge graph in the automobile field.

Keywords: social network; industry 4.0; industrial knowledge graph; deep learning; industrial big data; intellectualization of industrial information

---

1. Introduction

Industry 4.0 is an intelligent era, which promotes industrial transformation through the use of information technology, such that traditional manufacturing and value creation will undergo revolutionary changes. Industry 4.0 is divided into two main parts: one is the intellectualization of industrial control, and the other is the intellectualization of industrial information. There has been much research on the intellectualization of industrial control that is now relatively mature [1–7]. However, the intellectualization of industrial information is still in the research stage and there are some difficulties, mainly because industrial data are heterogeneous and multisource, and most of them are unstructured data. Therefore, determining how to automatically extract useful information from these unstructured data and integrate them is an important part of the intelligence of industrial information. Taking the automobile industry as an example, services oriented toward the users' experience are an important part of value creation and are becoming increasingly more important; however, it is not just the automobile production information but also the valuable information that can be automatically extracted from user evaluations that can help enterprises improve products and serve users. A recent study of the car market found that China has been the world's largest seller of cars for nine consecutive years. In 2017, China's total vehicle sales reached 28.879 million, more than 11 million ahead of the United States and accounting for a third of global sales. The relevant automobile websites and BBS (bulletin board system) generate a large amount of user data, which are mainly unstructured data without a specific format. The main work of this paper is to extract the structured information automatically from these unstructured Chinese texts and build the knowledge graph of the automobile industry based on the extracted structured information. Data sources as well as NLP (natural language processing) or other methods with which to process the data are unique among languages, especially for those belonging to different language families. Currently, most projects are concerned with knowledge graph systems in the English language. Because Chinese belongs to a different language family, directly translating English knowledge graphs into Chinese is not always feasible; hence, Chinese knowledge graph construction is of great significance. Currently, much progress has been made for knowledge graphs in the English language. However, Chinese knowledge graph construction has more challenges because Chinese is significantly different from English from various linguistic perspectives [8].

In recent years, the knowledge graph, as a new technology to realize large-scale semantic integration and interactive operation, has attracted great attention and research interest from industry and academia. The knowledge graph is a structured knowledge base that is different from the traditional relational database in that a knowledge graph uses a statement composed of two nodes and one edge to represent a fact, which is specifically expressed as a triple (h, r, t) [9], where h represents the head entity, r represents the relationship between the two entities, and t represents the tail entity. A knowledge graph usually consists of a large number of triples. Knowledge graphs have been increasingly applied in the fields of deep learning, computer vision, intelligent control and other artificial intelligence areas. The construction of a knowledge graph is divided into two parts: entity extraction and relation extraction. Knowledge graph has gone through the process from manual construction, such as WordNet and CyC, to automatic acquisition using machine learning and information extraction technology. This paper proposes a novel method that combines entity extraction with relational extraction to realize the automatic extraction of triples that are shaped as "entity-relation-entity" from unstructured Chinese text, and a feasible approach that extracts user evaluation information in the form of "entity-attribute-evaluation" from unstructured Chinese text.

In summary, the contributions of our work are highlighted as follows:

(1) A feasible method is proposed to achieve automatic extraction of triples from unstructured Chinese text by combining entity extraction and relationship extraction.
(2) An approach is proposed to extract structured user evaluation information from unstructured Chinese text.
(3) A knowledge graph of the automobile industry is constructed.

The remainder of the paper is organized as follows: Section 2 reviews the related works. Section 3 describes the proposed method in detail. In Section 4, the complete experiment and the knowledge graph construction is presented. Section 5 concludes the paper.

2. Related Work

To construct the knowledge graph of the automobile industry, we need to extract triples, including entity extraction and relation extraction. The related works summarize the state-of-the-art studies about entity extraction, relation extraction, and the introduction of existing knowledge graphs.

2.1. Entity Extraction

Entity extraction is also called entity linking or entity annotation. It is a hot topic in knowledge accessing and web-based content processing. Much work has been conducted toward entity linking in recent years, which has resulted in several different solutions. By English entity extraction, Wikify! uses unsupervised keyword extraction techniques to extract entities from text [10]. Then, Wikipedia is applied to find the matching pairs with the extracted entities. Finally, two different disambiguation algorithms are employed to link the correct Wikipedia page with the entity. In a similar way, Tagme and Spotlight extract and link entities to a knowledge base [11–13]. The major difference is that Spotlight uses DBpedia as its knowledge base. For Chinese entity extraction, CMEL builds a synonym dictionary for Chinese entities from Microblog [14]. Then, Wikipedia is applied as the linking knowledge base. An SVM method is used to address disambiguation. Yuan et al. use SWJTU Chinese word segmentation in entity recognition [15]. Pinyin edit distance (PED) and LCS (longest common subsequence) are applied to entity linking. Additionally, Wikipedia is applied as the linking knowledge base. CN-EL uses a similar process for entity extraction, but the difference is that it uses CN-DBpedia as its knowledge base. It also provides a stable online interface for both research and commercial access. Table 1 summarizes the above methods in detail [16]. It is observed from Table 1 that Wikify! and TAGME are the recommendations for traditional wiki-page linking. Spotlight can be used for LOD linking for English entities and CN-DBpedia can be used for LOD linking for Chinese entities. Recently, entity extraction is transformed into sequence annotation problem, He et al. propose a method about Chinese entity extraction based on bidirectional LSTM networks [17]. Dash et al. use big data mechanics enhance entity extraction [18]. All have achieved good results.

Table 1. Summary of entity extraction approaches.

	Language	Online API (application programming interface)	Status	Commercial
Wikify!	English	Yes	Active	No
TAGME	English	Yes	Active	No
Spotlight	English	Yes	Active	No
CMEL	Chinese	no	update to 2014	No
Yuan et al.	Chinese	no	update to 2015	No
CN-EL	Chinese	Yes	Active	Yes

Because of the large number of unrelated entities that would be introduced using the above tools, in this paper, we extract named entities by dictionary matching. We first create a dictionary of the car, and then create a character iterator and identify the name of the car by string matching.

2.2. Relation Extraction

Relation extraction is one of the most important tasks in NLP (natural language processing). Many efforts have been invested in relation extraction. Relationship extraction is transformed into relationship classification [19]. One related work was proposed by Rink and Harabagiu [20] and utilizes many features derived from external corpora for a support vector machine (SVM) classifier. Recently, deep neural networks have been shown to learn underlying features automatically and have been used in the literature. The most representative progress was made by Zeng et al., who utilized convolutional neural networks (CNNs) for relation classification [1,21]. While CNNs are not suitable for learning long-distance semantic information, the RNN (recurrent neural network) is often used for text processing [22]. One related work was proposed by Zhang and Wang, which employed bidirectional RNN to learn patterns of relations from raw text data [23]. Although the bidirectional RNN has access to both past and future context information, the range of context is limited due to the vanishing gradient problem [24]. To overcome this problem, long short-term memory (LSTM) units were introduced by Hochreiter and Schmidhuber [25]. Moreover, the GRU (gated recurrent unit) proposed by Cho et al. is a good variant of the LSTM network [26]. It is simpler and more efficient than the LSTM network, so the method of this paper builds on the bidirectional GRU. Most of these methods are supervised relation extraction, which is time-consuming and labor intensive. To address this issue, Mintz et al. align plain text with free-base by distance supervision [27]. However, distance supervision inevitably encounters the wrong labeling problem. To alleviate the wrong labeling problem, Riedel et al. model distant supervision for relation extraction as a multi-instance single-label problem [28], and Hoffmann et al. adopt multi-instance multilabel learning in relation extraction [29,30]. However, all of the feature-based methods strongly depend on the quality of the features generated by NLP tools, which will suffer from the error propagation problem and the difficulty of applying the multi-instance learning strategy of conventional methods in neural network models. Therefore, Zeng et al. combine at-least-one multi-instance learning with a neural network model to extract relations on distant supervision data [31]. However, they assume that only one sentence is active for each entity pair, and it will therefore lose a large amount of rich information contained in those neglected sentences. Hence, Lin et al. propose sentence-level attention over multiple instances, which can utilize all informative sentences [32]. Since each word in a sentence has a different importance to the semantic expression of the sentence, this paper also uses the word-level attention. In recent years, the research of graph neural network has become a hot topic in the field of deep learning, Zhu et al. use graph neural newtwork extract relation, and achieve good results.

In this paper, we will extract the relation between cars from unstructured Chinese text. For example, given the Chinese text "Volkswagen's two classic b-class cars Magotan and Passat have been occupying a large share of domestic automobile sales", we can extract that the semantic relation between "Magotan" and "Passat" is "Same Level". In this experiment, we define four semantic relations: "Same Level", "Homology", "Subordinate" and "Unknown".

2.3. Knowledge Graph

Knowledge graphs can be divided into universal knowledge graphs and industry knowledge graphs. The universal knowledge graph is based on common knowledge and emphasizes the breadth of knowledge. The industry knowledge graph is based on industry-specific data and emphasizes the depth of knowledge. In the universal knowledge graph, Freebase, Wikidata, DBpedia, and YAGO are representative examples. DBpedia is a multilanguage comprehensive knowledge base that was created by researchers from the University of Leipzig and the University of Mannheim in Germany and is at the core of the LOD (linking open data) project [33]. DBpedia extracts structured information from a multilingual Wikipedia and publishes it as linked data on the Internet for online web applications, social networking sites, and other online knowledge bases [34]. YAGO is a comprehensive knowledge base that was built by researchers from the Max Planck institute (MPI) in Germany. YAGO integrates Wikipedia, WordNet, GeoNames and other data sources, and integrates the classification system in Wikipedia with that in WordNet to build a complex hierarchy of categories. Freebase knowledge base was originally created by Metaweb and later acquired by Google [35]. Freebase knowledge base has become an important part of the Google knowledge graph. The data in Freebase is mainly constructed by humans, while the other data are mainly from Wikipedia, IMDB, Flickr and other websites or corpora. Wikidata are a collaborative knowledge base that was designed to support Wikipedia, Wikimedia Commons, and other Wikimedia projects. It is the central repository for structured data in Wikipedia, Wikivoyage, and Wikisource and is free to use [36]. The data in Wikidata are primarily stored as documents and currently contain over 17 million documents. Most universal knowledge graphs are constructed to obtain knowledge from semistructured or structured web pages. In terms of processing semistructured data, the main task is to learn the extraction rules of semistructured data through wrappers. Because semistructured data have a large number of repetitive structures, a small amount of annotation data can allow the machine to learn certain rules and then use the rules to extract the same type of data in the whole site. The construction of an industry knowledge graph is different from the construction of a universal knowledge graph. At present, there is little research on the industry knowledge map and is limited to a few fields. Due to the complex data structure, most of it is unstructured data, which makes the construction of an industry knowledge graph more challenging.

An industry knowledge graph can also be called a vertical knowledge graph. The description target of this kind of knowledge graph is the specific industry domain, which usually relies on the data of a specific industry to build, so its description scope is very limited. In the automotive industry, there is no corresponding knowledge graph. In this article, we will crawl the unstructured data related to the automotive field from the vehicle websites and BBS, and extract the structured knowledge from the unstructured data by employing the method of the bidirectional GRU (gate recurrent unit) combined with an attention mechanism. We construct the knowledge graph of the automotive industry based on the structured knowledge. The construction of the knowledge graph is divided into two main parts: entity extraction and relationship extraction. Entity extraction is also known as named entity recognition (NER) [37] and refers to automatic recognition of named entities from the data set. In this experiment, we automatically extract specific automobile names from unstructured texts, such as "Chevrolet" and "Ford". After entity extraction of the text corpus, we obtain a series of discrete named entities. To obtain semantic information, we also need to extract the relationship between entities from the relevant corpus and form a network knowledge structure by connecting the entities through the relationship.

Figure 1 shows the pipeline of the method. The input of the method is unstructured Chinese text, where a large number of triples are obtained after processing, and we link the same entities together to form a knowledge graph.

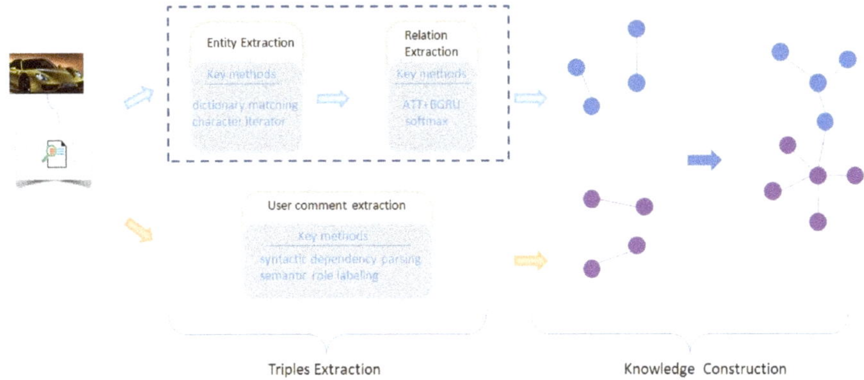

Figure 1. The pipeline of the method.

2.4. Automated Knowledge Base Management

A fundamental challenge in the intersection of Artificial Intelligence and Databases consists of developing methods to automatically manage Knowledge Bases which can serve as a knowledge source for computer systems trying to replicate the decision-making ability of human experts.

Although the challenge for dealing with knowledge is an old problem, it is perhaps more relevant today than ever before. The reason is that the joint history of Artificial Intelligence and Databases shows that knowledge is critical for the good performance of intelligent systems. In many cases, better knowledge can be more important for solving a task than better algorithms [38].

It is widely accepted that the complete life cycle for building systems of this kind can be represented as a three-stage process: creation, exploitation and maintenance [39]. These stages in turn are divided into other disciplines. In Table 2, we can see a summary of the major disciplines in which the complete cycle of knowledge (a.k.a. Knowledge Management) is divided [40].

Table 2. Summary of concepts in the Knowledge management field.

Knowledge Creation	Knowledge Exploitation	Knowledge Maintenance
Knowledge acquisition	Knowledge reasoning	Knowledge meta-modeling
Knowledge representation	Knowledge retrieval	Knowledge integration
Knowledge storage and manipulation	Knowledge sharing	Knowledge validation

3. Methods

3.1. Semantic Relation Extraction

The extraction of an entity relationship can be transformed into relation classification. An example is shown in Figure 2, the pipeline of semantic relation extraction mainly includes three steps.

Step one: We convert each word of the input sentence into a vector by an embedding matrix $\mathbf{V} \in \mathbb{R}^{d^w \times |V|}$, where V is a fixed-sized vocabulary and d^w is a hyperparameter to be chosen by the user. The purpose of providing two entities in input is to calculate the relative distance between each word and two entities, we connect the word vector and position vector to obtain the distributed representation of each word, which is the input of the model.

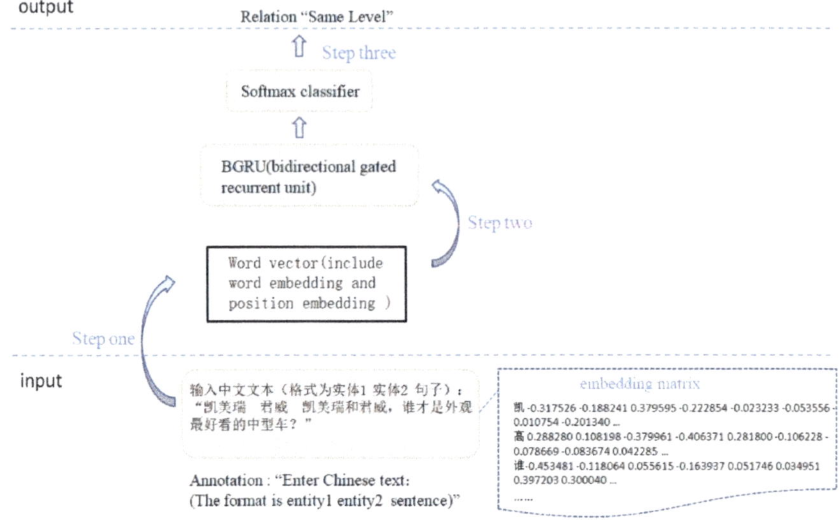

Figure 2. The pipeline of semantic relation extraction.

Step two: The model BGRU is able to exploit information both from the past and the future, and finally outputs the distributed representation of the whole sentence.

Step three: After going through the classifier, we can get the probability of each category and select the relationship of maximum probability as the final result.

When we use the model, we simply enter Chinese text (the format is "entity1 entity2 sentences"), and the model outputs relation. Take Figure 2 as an example, we enter the Chinese text "Camry Regal, which is the best-looking midsize car, Camry or Regal?", model output relation "Same Level".

3.1.1. Sentence Encoder

In this section, we transform the sentence x into its distributed representation X by the BGRU+Attention model. As shown in Figure 3, the model contains the following components:

1. Input layer,
2. Embedding layer,
3. BGRU layer,
4. Attention layer,
5. Output layer.

The inputs of the BGRU are raw words of the sentence x. We transform words into low-dimensional vectors by a word embedding matrix. In addition, we also use position embeddings for all words in the sentence to specify the position of each entity pair.

Word Embedding. Input a sentence x consisting of n words x $= (w_1, w_2, \cdots, w_n)$. This part aims to transform every word into distributed representations that capture syntactic and semantic meanings of the words by an embedding matrix $\mathbf{V} \in \mathbb{R}^{d^w \times |V|}$, where V is a fixed-sized vocabulary and d^w is a hyperparameter to be chosen by the user. As shown in Figure 4, we give a partial word embedding matrix, whose first column is a word, and the latter part is a 100-dimensional vector.

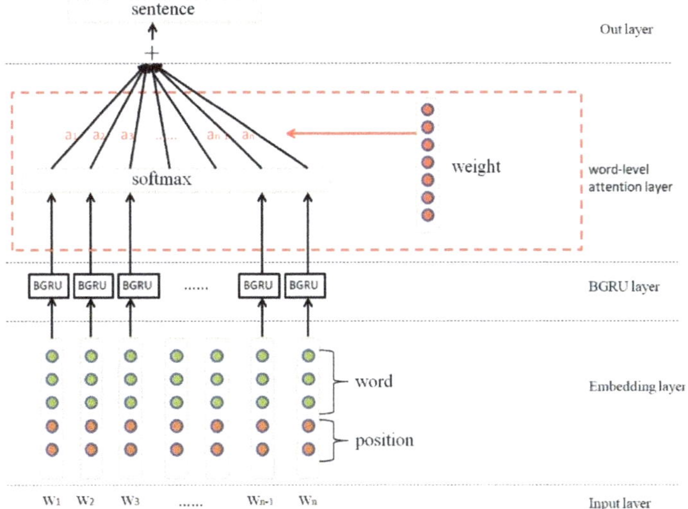

Figure 3. The architecture of BGRU (bidirectional gated recurrent unit)+Attention used for the sentence encoder.

Annotation:
kay	凯	-0.317526 -0.188241 0.379595 -0.222854 -0.023233 -0.053556 0.010754 -0.201340 ...
High	高	0.288280 0.108198 -0.379961 -0.406371 0.281800 -0.106228 -0.078669 -0.083674 0.042285 ...
Who	谁	-0.453481 -0.118064 0.055615 -0.163937 0.051746 0.034951 0.397203 0.300040 ...
Steam	汽	-0.118446 0.062470 0.194381 0.720970 0.308287 0.142574 -0.136900 -0.161334 -0.167365 ...
Ride	骑	-0.342828 0.018589 0.178869 0.183846 -0.162444 -0.102450 0.124333 -0.056581 0.192219 0.032627 ...
Man	郎	-0.230347 0.004020 0.005433 0.301461 0.043111 -0.317013 0.066877 0.193077 0.387688 0.084718 ...
Hugh	休	-0.154594 0.133991 0.385062 -0.008949 -0.140869 -0.150756 -0.511297 -0.331444 0.041822 0.432020 ...
Print	刊	-0.079298 -0.379804 -0.446937 -0.006616 -0.288559 -0.121265 0.303156 -0.150285 0.134322 -
Centre	中	0.041946 -0.008482 -0.159824 -0.049679 -0.076711 0.167492 0.130599 -0.032046 0.064531 -0.079213
Type	型	0.050682 -0.398424 0.092322 0.475399 0.222126 0.147334 -0.019527 -0.068095 -0.118040 ...
Most	最	-0.212371 -0.048879 -0.345541 -0.315009 -0.200895 0.250481 0.104016 -0.129055 -0.239268

......

Figure 4. Word embedding matrix.

Position Embedding. Contextual information at any location affects the extraction of entity relationships, and the words close to the target entities are usually informative to determine the relation between entities. Therefore, by defining the combination of the relative distances from the current word to the head or tail entities, the GRU can keep track of how close each word is to the head or tail entities.

Finally, we concatenate the word embedding and position embedding of all words to be a vector sequence S = $(w_1, w_2, \cdots, w_n)$, where $w_i \in \mathbb{R}^d$ $(d = d^w + d^p)$.

The GRU (gate recurrent unit) is a kind of recurrent neural network (RNN) that has also been proposed to solve problems such as the gradient vanishing in long-term memory [26]. Compared with LSTM, there are only two "gates" inside the GRU, and it has fewer parameters than LSTM but can also achieve the same function as the LSTM [41]. Considering the computing power and time cost of the hardware, we will often choose a more practical GRU. The architecture of the GRU block is shown in Figure 5.

Typically, the GRU-based recurrent neural networks contain an update gate z_t and reset gate r_t. The update gate is used to control the extent to which the status information of the previous moment is brought into the current state. The larger the value of the update gate is, the more the status information of the previous moment h_{t-1} is brought in. The reset gate is used to control the degree of ignoring the status information of the previous moment h_{t-1}. The smaller the value of the reset gate is, the more the status information of the previous moment is ignored, just as these following equations demonstrate:

$$z_t = \sigma\left(W_z \cdot [h_{t-1}, x_t]\right),$$
$$r_t = \sigma\left(W_r \cdot [h_{t-1}, x_t]\right),$$
$$\tilde{h}_t = \tanh\left(W \cdot [r_t * h_{t-1}, x_t]\right), \tag{1}$$
$$h_t = (1 - z_t) * h_{t-1} + z_t * \tilde{h}_t,$$

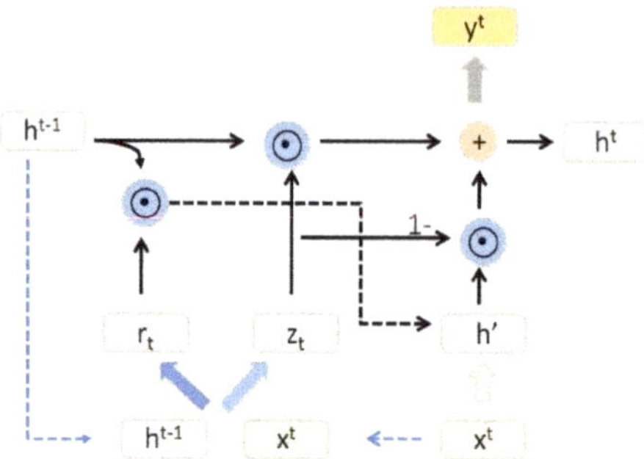

Figure 5. The architecture of the GRU(gate recurrent unit) block.

It is beneficial to have access to the future as well as the past context for many sequence modeling tasks. However, standard GRU networks process sequences in temporal order, and they ignore the future context. Bidirectional GRU networks are able to exploit information both from the past and the future by introducing a second layer that reverses the hidden connections flow. As shown in Figure 6, the output is represented as $h_i = \left[\overrightarrow{l_i} \oplus \overleftarrow{r_i}\right]$.

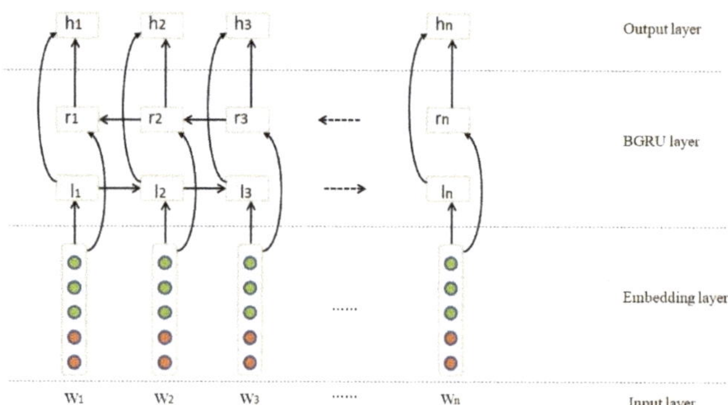

Figure 6. The architecture of the bidirectional GRU.

3.1.2. Relation Classification

After the embedding layer, the original sentence becomes the corresponding sentence vector. As shown in Figure 7, we use a softmax classifier to predict relation y from sentence set S, just as these following equations demonstrate:

$$p(y|S) = \mathrm{softmax}(WS + b),$$
$$y = \arg\max p(y|S), \tag{2}$$

where W is a trained parameter vector and b is a bias, and n indicates the number of sentence sets. The loss function is defined as J:

$$J(\theta) = -\frac{1}{m} \sum_{i=1}^{m} r_i \log (y_i, \theta),$$ (3)

where r is the one-hot representation of the truth relation and θ represents all parameters of the model.

The attention model was originally applied to image recognition, mimicking the focus of the eye moving on different objects when the person viewed the image [42–44]. Similarly, when recognizing an image or a language, a neural network is focused on a part of the feature each time, and the recognition is more accurate. This motivates determining how to measure the importance of features. The most intuitive method is to use a weight. Therefore, the result of the attention model is to calculate the weight of each feature first and then apply the weight to features.

Word-level attention. As shown in Figure 6, the output layer H can be represented as a matrix consisting of vectors $[h_1, h_2, \ldots, h_n]$, where n is the sentence length. The representation S of the sentence is formed by a weighted sum of these output vectors h_i:

$$M = \tanh(H),$$
$$\alpha = \text{softmax}\left(w^T M\right),$$ (4)
$$S = H\alpha^T,$$

where $H \in \mathbb{R}^{d^w \times n}$, d^w is the dimension of the word vector, and w is a trained parameter vector.

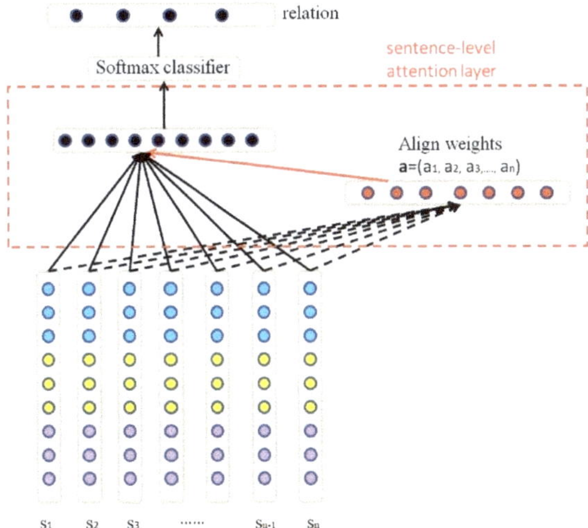

Figure 7. Relation classification.

Sentence-level attention. As shown in Figure 7, if we regard each sentence equally, the wrong labeling of sentences will introduce a massive amount of noise during training and testing. Therefore, sentence-level attention is important for relation extraction. The set vector X is computed as a weighted sum of these sentence vectors s_i:

$$e_i = x_i A r,$$
$$\alpha = \text{softmax}\left(e_i\right),$$ (5)
$$X = \sum_i \alpha_i S_i.$$

As shown in Figure 8, every line is a sentence (the annotations are in parentheses). Red denotes the sentence weight and blue denotes the word weight. We normalize the word weight by the sentence weight to make sure that only important words in important sentences are emphasized. Figure 8 shows that the model can select the words carrying strong sentiment like "middle-size", "MPV", "same price" and their corresponding sentences. Sentences containing many words like "common", "sales", "from" are disregarded. Note that the model can not only select words carrying strong sentiment; it can also deal with complex across-sentence context.

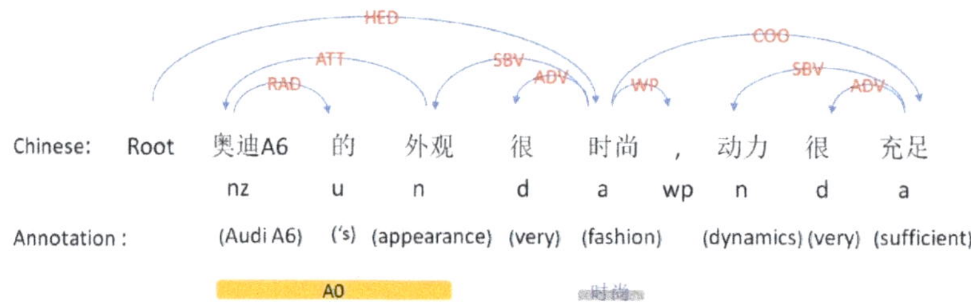

Figure 8. Example of visualization of attention.

3.2. User Comment Information Extraction

Syntax dependency parsing is one of the key techniques in natural language processing (NLP). Its basic task is to determine the syntactic structure of a sentence or the dependencies between words in a sentence. As shown in Figure 9, an example of syntactic dependency parsing and semantic role labeling is depicted in a Chinese sentence.

Figure 9. An example of syntax dependency parsing and semantic role labeling.

To facilitate subsequent structural understanding and extraction of content, we will organize the results of the above analysis into a dataframe, as shown in Table 3.

The "Word" column is the participle result of this sentence and the "Relation" column represents the relation between this word and the match word. Table 4 shows the corresponding syntactic relations. The "match word" column shows the match terms according to relationships, the "pos" column is the part of speech of each word, the "tuple word" column is a combination of two words, and the "match word n" column is the sequence number of the match word.

Semantic role labeling is a shallow semantic analysis of sentences, which centers on verbs to find the executor and acceptor of actions, as well as the components modified by adjectives. As shown in Figure 10, we find the component "A0" modified by the adjective "fashion" through semantic role labeling, and then find the main component "appearance" and the attribute "Audi A6" through the syntactic analysis of "A0". Finally, we can obtain a triple in the shape of "Audi A6-appearance-fashion".

Table 3. An example of dependency parsing.

	Match Word	Match Word n	pos	Relation	Word	Tuple Word
0	appearance	2	nz	ATT (attribute)	Audi A6	Audi A6 - appearance
1	Audi A6	0	u	RAD (right adjunct)	's	Audi A6 - 's
2	fashion	4	n	SBV (subject-verb)	appearance	appearance - fashion
3	fashion	4	d	ADV (adverbial)	very	very - fashion
4	root	root	a	HED (head)	fashion	root - fashion
5	fashion	4	wp	WP (punctuation)	,	fashion - ,
6	sufficient	8	n	SBV (subject-verb)	dynamics	dynamics - sufficient
7	sufficient	8	d	ADV (adverbial)	very	very - sufficient
8	fashion	4	a	COO (coordinate)	sufficient	sufficient - fashion

Table 4. The syntactic relations.

Tag of Relationship Types	Description
ATT	attribute
RAD	right adjunct
SBV	subject-verb
ADV	adverbial
HED	head
COO	coordinate

3.3. Automatic Triples Extraction

We extract named entities by dictionary matching. We first create a dictionary of the car, then create a character iterator, and we identify the name of the car by string matching. Finally, the identified two entities and the corresponding text constitute the input of the relation extraction model. The model outputs the possibility of four relations. We select the relation between two entities with the highest probability, and obtain the triples shaped as "entity-relation-entity". We also obtain the triples of user comments by syntactic dependency parsing and semantic role labeling. Figure 10 shows the flow of triples extraction.

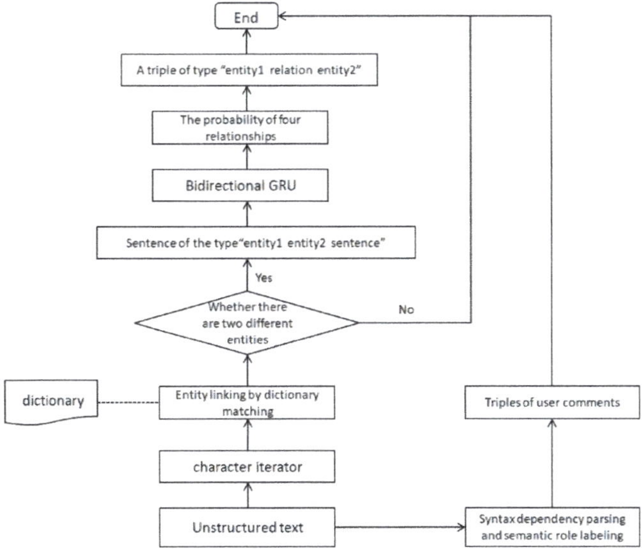

Figure 10. Flow chart of triples extraction.

4. Experiment

Semantic relation extraction is one of the tasks that is transformed into relation classification and implemented by the supervised learning method in the above section. First, we defined four relationship categories—"Same Level", "Homology", "Subordinate", and "Unknown"—and construct the corresponding data set for each relationship. Then, we train the relationship extraction model and realize the automatic extraction of triples by combining named entity recognition. Finally, we construct the knowledge graph of the automotive domain through the obtained triples.

4.1. Dataset

We need to find the corresponding semantic training text for each relationship. For example, the Chinese text "When we talk about French cars, we have to mention PSA group's two twin stars, Citroen and Peugeot" can be expected to be the training text of the semantic relationship of "Homology" between the two entities "Citroen" and "Peugeot". The relationship between two cars from the same country is "Homology". To find the corresponding training text, we first sort out several popular cars from nine countries and then combine the cars from the same country. Finally, we crawl the text in which two cars' names appear at the same time as the training text of the relation "Homologous". The method improves the efficiency of data processing but also introduces considerable noise data. The sentence-level attention mentioned in the previous section reduces the influence of noise data. Figure 11 shows the number of training texts for the relation "Homologous".

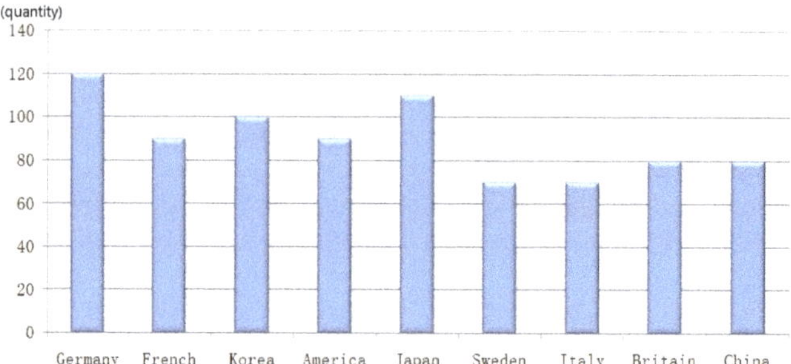

Figure 11. The number of training texts for the relation "Homology".

The "subordinate" relationship reflects the information of the superior and the subordinate characteristics. Figure 12 is the subordinate diagram of "Volkswagen", where the relationship between "FAW-Volkswagen" and "Jetta" is "Subordinate". Similarly, we sort out the combinations of other brands and find the corresponding training text for the relation "Subordinate".

Similarly, we sort out 11 levels of partial vehicles, then combine cars of the same level, and finally crawl the corresponding training text. Figure 13 shows the number of training texts for the relation.

Figure 14 shows the data statistics of training data of four kinds of relations, where "unknown" stands for no relation between entities.

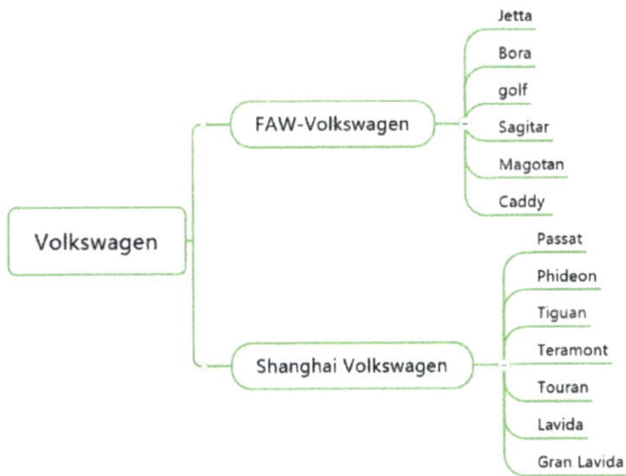

Figure 12. The number of training texts for the relation "Same Level".

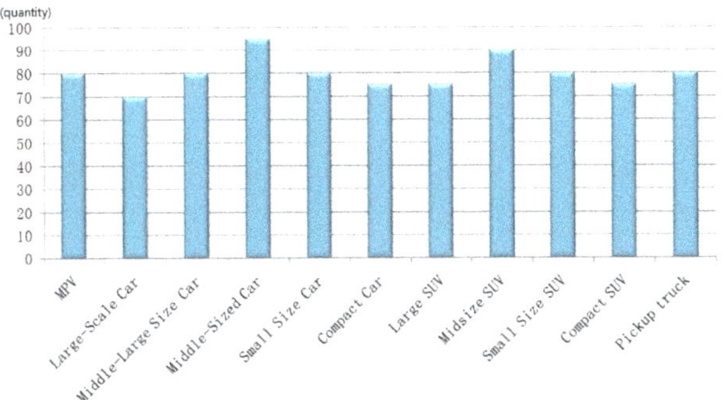

Figure 13. Volkswagen affiliate diagram.

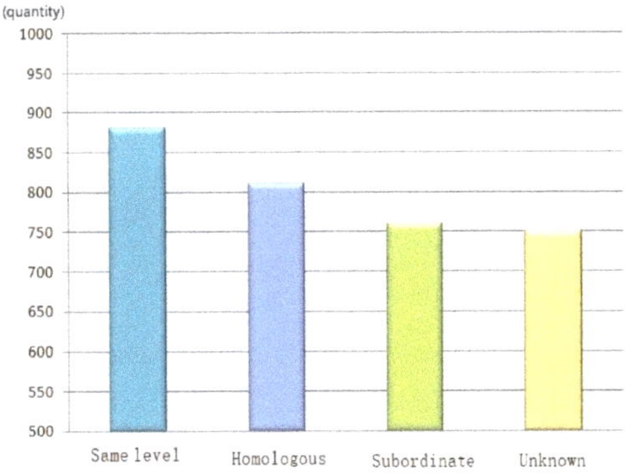

Figure 14. Data statistics of training data of four kinds of relations.

4.2. Model Training

We train the model with 3200 sentences and test it with 700 sentences. We use a grid search to determine the optimal parameters and select the batch size$\in\{10,20,...,50\}$, the neural network layer number$\in\{1,2,3\}$, and the number of neurons in each layer$\in\{200,250,300\}$. As shown in Figure 15, we form 36 different combinations based on different hyper-parameters, and obtain the average accuracy of each combination through experiments. Table 5 lists the specific experimental results. We select the hyper-parameter combination with the maximum average accuracy as the optimal parameter set. For other parameters, since they have little effect on the results, so we initialize common values. In Table 6, we show the hyper-parameters used in experiments.

According to whether the classification results are correct, TP, TN, FP, and FN can be determined. TP means that the classification result is a true positive, TN means true negative, FP means false positive, and FN means false negative. We use accuracy and recall rate to evaluate the effect of the model. The specific formula is as follows:

$$P = \frac{TP}{TP + FP},$$
$$R = \frac{TP}{TP + FN}. \tag{6}$$

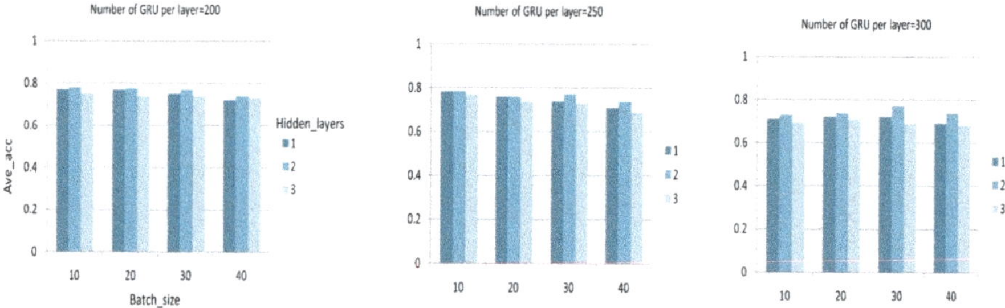

Figure 15. Comparison of experimental results of different parameter combinations.

Table 5. Average accuracy of different hyper-parameters.

Number of GRU per Layer		200	200	200	250	250	250	300	300	300
hidden layers		1	2	3	1	2	3	1	2	3
bach size	10	0.77	0.78	0.75	0.783	**0.79**	0.769	0.71	0.73	0.691
	20	0.768	0.776	0.74	0.76	0.77	0.74	0.72	0.74	0.71
	30	0.75	0.768	0.74	0.74	0.77	0.73	0.72	0.77	0.69
	40	0.72	0.74	0.73	0.71	0.74	0.69	0.69	0.74	0.68

Table 6. Hyper-parameter settings.

Word dimension	100
Position dimension	5
Dropout probability	0.5
Batch size	10
BGRU (bidirectional gated recurrent unit) layer number	2
GRU (gated recurrent unit) size of each layer	250

We randomly divide the data set into training set and test set. We train the model with the training set and evaluate the accuracy of the model with the test set. We divide the data set four times, and carry out experiments for each time. Finally, the average of the results of each experiment is used to represent the performance of the model. As shown in Figure 16, we make a comparative experiment between the two models, the blue curve represents the accuracy/recall rate curve of BGRU, the red curve represents the accuracy/recall rate curve of BLSTM, and the specific results of the four experiments are listed in Table 7.

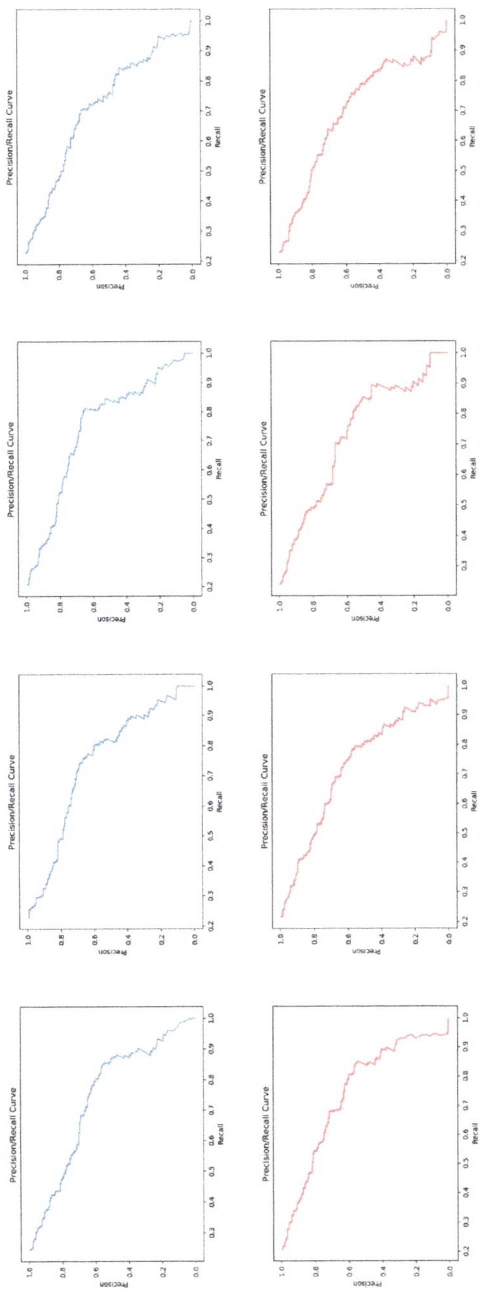

Figure 16. Precision/Recall Curve (blue is BGRU (bidirectional gated recurrent unit), and red for BLSTM (bidirectional long short-term memory)).

Table 7. The accuracy of four experiments of two models.

	1	2	3	4	Mean
BLSTM (bidirectional long short-term memory)	0.752	0.767	0.76	0.75	0.757
BGRU (bidirectional gated recurrent unit)	0.785	0.781	0.776	0.77	0.778

We compare the run-time performance of BGRU and BLSTM on a 3.6 GHz Intel Core i7-7700 Think Station P318 with a 32 G DDR4 memory. We calculate the mean values of the four experiments of the two models and comparing the mean values found that BGRU incurs 9.2% smaller run-time compared to BLSTM. Figure 17 shows the run-time of the two models in four experiments. BGRU train faster and perform better than BLSTM on less training data because BGRU has less parameters per "cell", allowing it in theory to generalise better from less examples, at the cost of less flexibility.

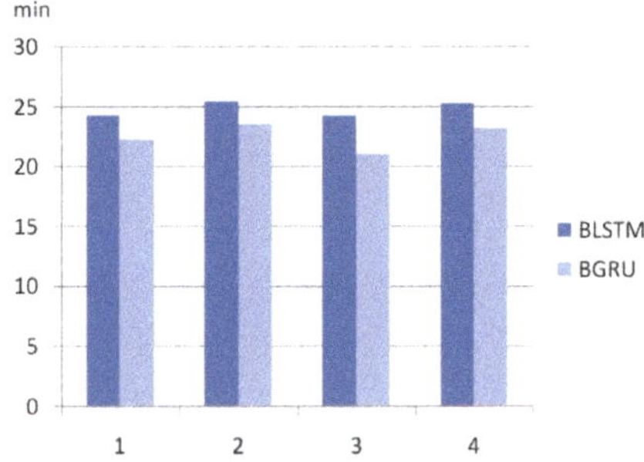

Figure 17. Run-time of four experiments of two models.

4.3. The Result of Triple Extraction

It is feasible to evaluate the correctness of triples extraction since the test set selected for the work is rather small. Table 8 is the evaluation of car entity extraction in the test data, and evaluation result is given in Table 9. In total, from Table 9, there are 700 texts that could be checked by human efforts. We check these 700 texts and annotate the correct triples as the ground truth. Based on the ground truth, the F1-measure criterion is applied.

Table 8. Evaluation of car entity extraction on the test set.

All Entities	Extracted	Correct	F1-Measure
1428	1400	1400	0.99

Table 9. Evaluation of triple extraction on the test set.

Relation	Ground Truth	Extracted	Correct	Precision	F1-Measure
Same Level	230	203	168	0.83	0.78
Homology	150	102	74	0.73	0.59
Subordinate	92	63	49	0.78	0.63
Unknown	228	192	145	0.76	0.69

Table 10. Statistics of triple extraction.

Texts	Triples
53,200	30,500

From the result shown in Table 8, almost all the entities in the test set are correctly identified, and the F1-measure achieves 99% calculated by the precision and recall (2*precision*recall/(precision+recall)). It means that most of the car entities in the unstructured Chinese text could efficiently be extracted. It is because the car in both the dictionary and the text share the same naming standard. From the result shown in Table 9, the extraction of triples can achieve more than 73% accuracy, indicating that the model can effectively identify the semantic relationship between entities and extract triples automatically. The triple extraction with the relation of "same level" can achieve a high accuracy rate because the Chinese text about cars often appears keywords that represent vehicle types, such as "SUV,MPV". We believe that cars of the same type satisfy the relation "same level", and these keywords are easy to be given a high weight by the model and easy to be recognized. Similarly, the text also contains some keywords of other relations, and the model can quickly and accurately identify the meaning relations of these texts. However, according to the experimental results, we find that the recall rate of triples extraction with different relations is generally low, which indicates that the extraction efficiency of the model is obviously insufficient for most Chinese texts about cars whose meaning is not clearly expressed, so the model needs more types of texts to train and improve its generalization ability. Finally, we crawl the 50,000+ texts and extract the 30,000+ triples through the model, and Table 10 shows the statistics of the quantity.

Figure 18 shows several examples of triples extraction. An unstructured Chinese text is used as input to the model, and the model automatically outputs two entities and their relation, as well as the triple in the form of "entity–relation–entity".

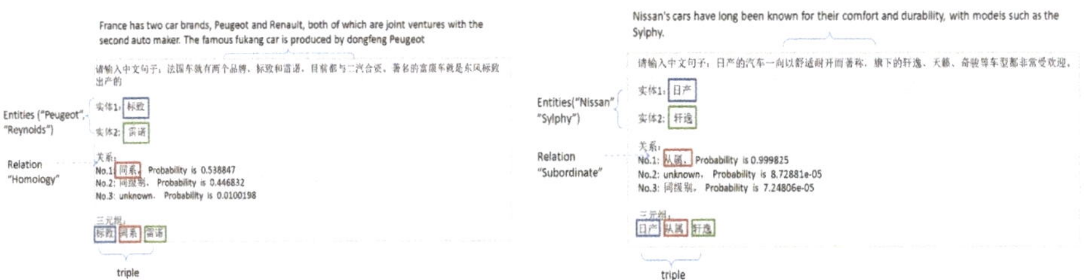

Figure 18. Some examples of triple extraction.

4.4. Knowledge Graph Construction

As shown in Figure 19, we extract a number of triples from an unstructured text and then link them by connecting entities with the same name. In Figure 20, a knowledge graph composed of partial triples is depicted. The nodes in the knowledge graph represent car entities, and the edges represent the relationship between the two entities. Some Chinese annotations are given in Figure 21.

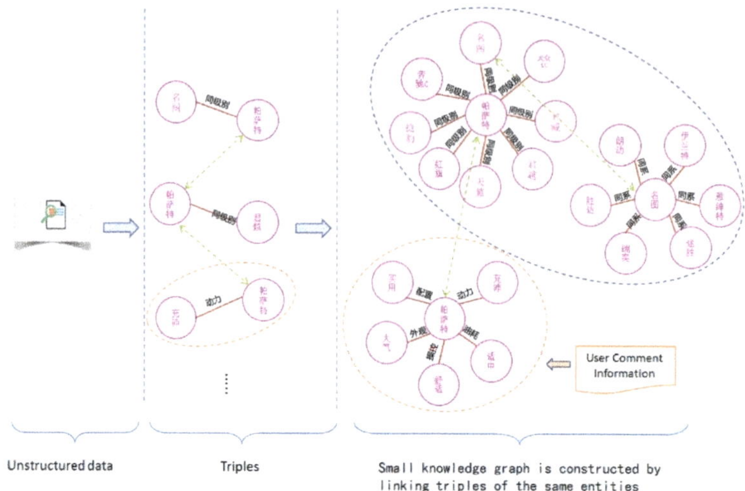

Figure 19. An example of the construction of knowledge graph.

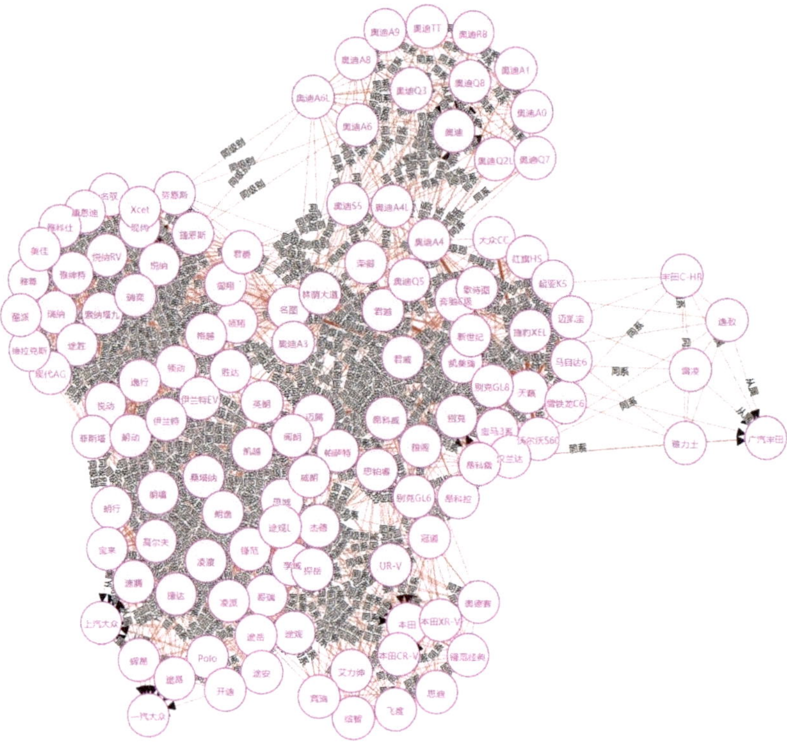

Figure 20. knowledge graph of the car.

node(entity)		edge(relation)	
Chinese	Annotation	Chinese	Annotation
名图	MISTRA	同级别	Same Level
帕萨特	Passat	同系	Homology
君越	LaCrosse	从属	Subordinate
君威	Regal		
大众CC	VWCC	edge(attribute)	
天籁	Teana		
红旗	Red Banner	动力	power
捷豹	Jaguar	配置	configuration
奔驰C	Mercedes C	外观	appearance
朗动	Elantra	油耗	fuel consumption
胜达	Santa Fe	...	...
瑞奕	Rui Yi		
途胜	Tucson		
雅绅特	Hyundai Accent		
伊兰特	Elantra		
...			

Figure 21. The partial annotations in the knowledge graph.

4.5. Discussion

From the real-world case study, we have learned that the unstructured data in the industrial field contain considerable useful information that can be effectively integrated by the powerful semantic association of the knowledge graph. The method proposed in this paper realizes efficient and accurate extraction of information. On the other hand, three major problems were also learned from the case study:

1. Relation Selection

As mentioned in Table 6, the accuracy and recall rate of triple extraction change with the change of the semantic relation, which indicates that an effective semantic relation setting can improve the efficiency of information extraction. Due to the diversity of industrial fields, entity relations in different fields need to have a special evaluation standard.

2. Entity Extraction

As shown in Table 5, almost all the entities in the test set are correctly identified, which indicates that the dictionary matching method can effectively identify entities in the text. However, this also brings about the same problems. One of them is that the contents of a dictionary need to be complete, and it will take considerable time and money to build a dictionary. In addition, due to the diversity of industrial fields, entity recognition in different fields needs to construct corresponding dictionaries, and this method has poor portability. Entity recognition based on deep learning is more generalized, which is worth studying.

5. Conclusions

The industrial 4.0 era is the fourth industrial revolution and is characterized by network penetration. Massive text data will be produced in different industrial fields, but the publication of data are not standardized, and the data quality is not high. The main work of this paper includes:

- A feasible method is proposed to achieve automatic extraction of triples from unstructured Chinese text by combining entity extraction and relationship extraction.
- An approach is proposed to extract structured user evaluation information from unstructured Chinese text.

- A knowledge graph of the automobile industry is constructed.

  In the future, we will explore the following directions:

(1) We mainly crawl data from BBS and automobile sales websites. We will expand our data in future work, such as unstructured objective data in the automobile manufacturing process or unstructured data in other industrial fields.

(2) In the process of constructing the industrial knowledge graph, we only aligned the entities with the same name and did not take into account the entities with ambiguity, that is, those with the same name but different meanings. Moreover, we did not merge the entities with different names but which had the same meanings. In the future, we will study the disambiguation and fusion of entities in the process of constructing knowledge graphs.

(3) We have constructed the knowledge graph of the automobile industry. In the future, we will design a corresponding application according to this knowledge graph. For example, the KBQA (knowledge base question answering) in the automobile field holds prospects.

Author Contributions: Conceptualization, C.X.; Data curation, J.G.; Investigation, D.L.; Methodology, M.Z. and H.W.; Resources, Q.L.; Supervision, Z.C.

Funding: Science Foundation of Yunnan University: No. 2017YDQN11.; Yunnan Provincial Science Research Project of the Department of Education: No.2018JS008.; Youth Talent Project of the China Association for Science and Technology: No. W8193209.

Conflicts of Interest: The authors declare no conflict of interest.

Abbreviations

The following abbreviations are used in this manuscript:

NLP	Natural Language Processing
LOD	Linking Open Data
GRU	Gate Recurrent Unit
CNN	Convolutional Neural Network
RNN	Recurrent Neural Network
LSTM	Long Short-Term Memory
BGRU	Bidirectional Gated Recurrent Unit
BLSTM	Bidirectional Long Short-Term Memory
ATT	Attribute
RAD	Right Adjunct
SBV	Subject-Verb
ADV	Adverbial
HED	Head
COO	Coordinate
KBQA	Knowledge Base Question Answering

References

1. Quintana, G.; Campa, F.J.; Ciurana, J. Productivity improvement through chatter-free milling in workshops. *Proc. Inst. Mech. Eng.* **2011**, *225*, 1163–1174. [CrossRef]

2. Li, B.; Hou, B.; Yu, W. Applications of artificial intelligence in intelligent manufacturing: A review. *Front. Inf. Technol. Electron. Eng.* **2017**, *18*, 86–96. [CrossRef]

3. Bahrin, M.A.K.; Othman, M.F.; Azli, N.H.N. Industry 4.0: A review on industrial automation and robotic. *J. Teknol.* **2016**, *78*, 137–143.

4. Zhong, R.Y.; Xu, X.; Klotz, E. Intelligent manufacturing in the context of industry 4.0: A review. *Engineering* **2017**, *3*, 616–630. [CrossRef]

5. Barrio, H.G.; Morán, I.C.; Ealo, J.A.; Barrena, F.S.; Beldarrain, T.O.; Zabaljauregui, M.C.; Zabala, A.M.; Arriola, P.J.A. Proceso de mecanizado fiable mediante uso intensivo de modelización y monitorización del proceso: Enfoque 2025. *DYNA* **2018**, *93*, 689–696. [CrossRef]

6.   Wang, J.; Ma, Y.; Zhang, L. Deep learning for smart manufacturing: Methods and applications. *J. Manuf. Syst.* **2018**, *48*, 144–156. [CrossRef]

7.   Bustillo, A.; Urbikain, G.; Perez, J.M.; Pereira, O.M.; de Lacalle, L.N.L. Smart optimization of a friction-drilling process based on boosting ensembles. *J. Manuf. Syst.* **2018**, *48*, 108–121. [CrossRef]

8.   Wang, C.; Gao, M.; He, X. Challenges in chinese knowledge graph construction. In Proceedings of the 2015 31st IEEE International Conference on Data Engineering Workshops, Seoul, Korea, 13–17 April 2015; pp. 59–61.

9.   Sintek, M.; Decker, S. TRIPLE—A query, inference, and transformation language for the semantic web. In Proceedings of the International Semantic Web Conference, Sardinia, Italy, 9–12 June 2002; pp. 364–378.

10.   Mihalcea, R.; Csomai, A. Wikify!: Linking documents to encyclopedic knowledge. In Proceedings of the Sixteenth ACM Conference on Conference on Information and Knowledge Management, Lisbon, Portugal, 6–10 November 2007; pp. 233–242.

11.   Ferragina, P.; Scaiella, U. Tagme: On-the-fly annotation of short text fragments (by wikipedia entities). In Proceedings of the 19th ACM international conference on Information and knowledge management, Toronto, ON, Canada, 26–30 October 2010; pp. 1625–1628.

12.   Hasibi, F.; Balog, K.; Bratsberg, S.E. On the reproducibility of the TAGME entity linking system. In Proceedings of the European Conference on Information Retrieval, London, UK, 10–12 April 2016; pp. 436–449.

13.   Mendes, P.N.; Jakob, M.; García-Silva, A. DBpedia spotlight: Shedding light on the web of documents. In Proceedings of the 7th International Conference on Semantic Systems, Graz, Austria, 7–9 September 2011; pp. 1–8.

14.   Meng, Z.; Yu, D.; Xun, E. Chinese microblog entity linking system combining wikipedia and search engine retrieval results. In *Natural Language Processing and Chinese Computing*; Springer: Berlin/Heidelberg, Germany, 2014; pp. 449–456.

15.   Yuan, J.; Yang, Y.; Jia, Z. Entity recognition and linking in Chinese search queries. In *Natural Language Processing and Chinese Computing*; Springer: Cham, Switzerland, 2015; pp. 507–519.

16.   Xie, C.; Yang, P.; Yang, Y. Open knowledge accessing method in IoT-based hospital information system for medical record enrichment. *IEEE Access* **2018**, *6*, 15202–15211. [CrossRef]

17.   He, Z. Chinese entity attributes extraction based on bidirectional LSTM networks. *Int. J. Comput. Sci. Eng.* **2019**, *18*, 65–71. [CrossRef]

18.   Enhanced Entity Extraction Using Big Data Mechanics. Available online: https://link.springer.com/chapter/10.1007/978-981-13-2673-8_8 (accessed on 5 July 2019)

19.   Daojian, Z.; Kang, L.; Siwei, L.; Guangyou, Z.; Jun, Z. Relation Classification via Convolutional Deep Neural Network. 2014. pp. 2335–2344. Available online: https://scholar.google.com.hk/scholar?hl=zh-CN&as_sdt=0%2C5&q=Relation+Classification+via+Convolutional+Deep+Neural+Network.&btnG= (accessed on 5 July 2019)

20.   Rinkm, B.; Harabagium, S. Utd: Classifying semantic relations by combining lexical and semantic resources. In Proceedings of the 5th International Workshop on Semantic Evaluation, Los Angeles, CA, USA, 15–16 July 2010; pp. 256–259.

21.   Relation Classification via Convolutional Deep Neural Network. Available online: http://ir.ia.ac.cn/bitstream/173211/4797/1/Relation%20Classification%20via%20Convolutional%20Deep%20Neural%20Network.pdf (accessed on 5 July 2019)

22.   Tomas, M.; Martin, K.; Lukás, B.; Jan, C.; Sanjeev, K. Recurrent neural network based language model. In Proceedings of the 11th Annual Conference of the International Speech Communication Association, Chiba, Japan 26–30 September 2010; pp. 1045–1048.

23.   Zhang, D.; Wang, D. Relation classification via recurrent neural network. *arXiv* **2015** arXiv:1508.01006.

24.   Zhang, S.; Zheng, D.; Hu, X. Bidirectional long short-term memory networks for relation classification. In Proceedings of the 29th Pacific Asia Conference on Language, Information and Computation, Shanghai, China, 30 October–1 November 2015; pp. 73–78.

25.   Hochreiter, S.; Schmidhuber, J. Long short-term memory. *Neural Comput.* **1997**, *9*, 1735–1780. [CrossRef] [PubMed]

26.   Chung, J.; Gulcehre, C.; Cho, K.H. Empirical evaluation of gated recurrent neural networks on sequence modeling. *arXiv* **2014**, arXiv:1412.3555.

27. Mintz, M.; Bills, S.; Snow, R. Distant supervision for relation extraction without labeled data. In Proceedings of the Joint Conference of the 47th Annual Meeting of the ACL and the 4th International Joint Conference on Natural Language Processing of the AFNLP, Singapore, 2 August 2009; pp. 1003–1011.

28. Riedel, S.; Yao, L.; McCallum, A. Modeling relations and their mentions without labeled text. In Proceedings of the Joint European Conference on Machine Learning and Knowledge Discovery in Databases, Barcelona, Spain, 20–24 September 2010; pp. 148–163.

29. Hoffmann, R.; Zhang, C.; Ling, X. Knowledge-based weak supervision for information extraction of overlapping relations. In Proceedings of the 49th Annual Meeting of the Association for Computational Linguistics: Human Language Technologies, Portland, OR, USA, 19–24 June 2011; pp. 541–550.

30. Surdeanu, M.; Tibshirani, J.; Nallapati, R. Multi-instance multi-label learning for relation extraction. In Proceedings of the 2012 Joint Conference on Empirical Methods in Natural Language Processing and Computational Natural Language Learning, Jeju Island, Korea, 13 July 2012; pp. 455–465.

31. Zeng, D.; Liu, K.; Chen, Y. Distant supervision for relation extraction via piecewise convolutional neural networks. In Proceedings of the 2015 Conference on Empirical Methods in Natural Language Processing, Lisbon, Portugal, 17–21 September 2015; pp. 1753–1762.

32. Lin, Y.; Shen, S.; Liu, Z. Neural relation extraction with selective attention over instances. In Proceedings of the 54th Annual Meeting of the Association for Computational Linguistics, Berlin, Germany, 7–12 August 2016; pp. 2124–2133.

33. Niu, X.; Sun, X.; Wang, H. Zhishi.me-weaving chinese linking open data. In Proceedings of the International Semantic Web Conference, Bonn, Germany, 23–27 October 2011; pp. 205–220.

34. Auer, S.; Bizer, C.; Kobilarov, G. Dbpedia: A nucleus for a web of open data. *The Semantic Web*; Springer: Berlin/Heidelberg, Germany, 2007; pp. 722–735.

35. Bollacker, K.; Evans, C.; Paritosh, P. Freebase: A collaboratively created graph database for structuring human knowledge. In Proceedings of the 2008 ACM SIGMOD International Conference on Management of Data, Vancouver, BC, Canada, 9–12 June 2008; pp. 1247–1250.

36. Wikidata: A Free Collaborative Knowledge Base. Available online: https://ai.google/research/pubs/pub42240 (accessed on 5 July 2019)

37. Chen, W.; Zhang, Y.; Isahara, H. Chinese named entity recognition with conditional random fields. In Proceedings of the Fifth SIGHAN Workshop on Chinese Language Processing, Sydney, Australia, 22–23 July 2006; pp. 118–121.

38. Rudas, I.J.; Pap, E.; Fodor, J. Information aggregation in intelligent systems: An application oriented approach. *Knowl.-Based Syst.* **2013**, *38*, 3–13. [CrossRef]

39. Felfernig, A.; Wotawa, F. Intelligent engineering techniques for knowledge bases. *AI Commun.* **2013**, *26*, 1–2.

40. Martinez-Gil, J. Automated knowledge base management: A survey. *Comput. Sci. Rev.* **2015**, *18*, 1–9. [CrossRef]

41. Zhou, P.; Shi, W.; Tian, J. Attention-based bidirectional long short-term memory networks for relation classification. In Proceedings of the 54th Annual Meeting of the Association for Computational Linguistics (Volume 2: Short Papers), Berlin, Germany, 7–12 August 2016; Volume 2, pp. 207–212.

42. Teaching Machines to Read and Comprehend. Available online: http://papers.nips.cc/paper/5945-teaching-machines-to-read-and-comprehend (accessed on 5 July 2019)

43. Bahdanau, D.; Cho, K.; Bengio, Y. Neural machine translation by jointly learning to align and translate. *arXiv* **2014**, arXiv:1409.0473.

44. Xu, K.; Ba, J.; Kiros, R. Show, attend and tell: Neural image caption generation with visual attention. In Proceedings of the International Conference on Machine Learning, Lille, France, 6–11 July 2015; pp. 2048–2057.

MDPI

St. Alban-Anlage 66

4052 Basel

Switzerland

Tel. +41 61 683 77 34

Fax +41 61 302 89 18

www.mdpi.com

Applied Sciences Editorial Office

E-mail: applsci@mdpi.com

www.mdpi.com/journal/applsci